PROCEEDINGS OF THE 2014 INTERNATIONAL CONFERENCE ON BIOMEDICAL ENGINEERING AND ENVIRONMENTAL ENGINEERING (ICBEEE 2014), WUHAN, CHINA, 24–25 DECEMBER 2014

Biomedical Engineering and Environmental Engineering

Editor

David Chan
ACM Macau Chapter, Macau

CRC Press
Taylor & Francis Group
Boca Raton   London   New York   Leiden

CRC Press is an imprint of the
Taylor & Francis Group, an **informa** business

A BALKEMA BOOK

CRC Press/Balkema is an imprint of the Taylor & Francis Group, an informa business

© 2015 Taylor & Francis Group, London, UK

Typeset by MPS Limited, Chennai, India

Published by: CRC Press/Balkema
P.O. Box 11320, 2301 EH Leiden, The Netherlands
e-mail: Pub.NL@taylorandfrancis.com
www.crcpress.com – www.taylorandfrancis.com

ISBN: 978-1-138-02805-0 (Hardback)
ISBN: 978-1-315-68548-9 (Ebook PDF)

Table of contents

Environmental science and technology

Environmental sustainability

Preface

This proceeding contains the accepted papers from the 2014 2nd International Conference on Biomedical Engineering and Environmental Engineering (ICBEEE 2014). ICBEEE 2014 has been held on December 24–25 2014, Wuhan, China. This conference series is a forum for enhancing mutual understanding between Biomedical Engineering and Environmental Engineering field.

This proceeding is enriched by contributions from many experts representing many industry and academic establishments worldwide. The researchers are from different countries and professional. The conference will bring together the researchers from all over the world to share their new findings, thus to promote academic exchanges.

This conference received significantly attention from all over the world. This volume represents accepted papers which are of good quality, but also related to the theme of the conference. ICBEEE 2014 features unique mixed topics of Bioinformatics and Computational Biology, Biomedical Engineering, Environmental Science and Technology, Environmental sustainability.

All of the accepted papers have been passed the strict peer-review process before publication by CRC Press/Balkema (Taylor & Francis Group). Reviews were conducted by expert referees to the professional and scientific standards. This book covers the subject areas of Bioinformatics and Computational Biology, Biomedical Engineering and Environmental Science and Technology.

We would like to express our deep gratitude to all the keynote speakers, authors, referees, exhibitors, technical co-sponsoring organizations, committee members, and many others on whom the success of this prestigious event depends.

Finally we wish all the authors and attendees of ICBEEE2014 have a unique, rewarding and enjoyable memory at ICBEEE 2014 in Wuhan, China.

With our warmest regards

Publication Chair
David Chen, *ACM Macau chapter, Macau*

Organizing Committee

General Chair

Yiyi Zhouzhou, *Azerbaijan State Oil Academy, Azerbaijan*
Mark Zhou, *Hong Kong Education Society, Hong Kong*

Program Chairs

Prawal Sinha, *Department of Mathematics & Statistics, Indian Institute of Technology Kanpur, India*
Harry Zhang, *SMSSI, Singapore*

Publication Chair

David Chen, *ACM Macau chapter, Macau*

International Committee

Dongce Ma, *University of London, England*
Kucbel Marek, *ENET – Energy Units for Utilization of non Traditional Energy Sources, Ostrava, Czech Republic*
Ni Sheng, *Macau University of Science and Technology, Macao*
Omar Altuwaijri, *King Saud University, Saudi*
Xuan Xiao, *Jingdezhen Ceramic Institute, China*
Haiyan Chen, *ChangChun University, China*
Jianyang Lin, *First Hospital of China Medical University, China*
Damu Wei, *Southwest Jiaotong University, China*
Kai Li, *Jilin University, China*
Hui Li, *Ocean University of China, China*
Tao Jiang, *National University of Defense Technology, China*
Zhaoxia Xu, *Shanghai University of Traditional Chinese Medcicine, China*
Yanpeng Mao, *Shandong University, China*
Jun-wu Fang, *Hubei University of Science and Technology, China*
Rui Zhang, *Tianjin Bohai Vocational Technology College, China*
Ping-fan Liao, *Hubei University of Science & Technology, China*

Bioinformatics and computational biology

Irspot-GCFPseNC: Identifying recombination spots with pseudo nucleotide composition

Xuan Xiao & Yuan-ting Wu
Jingdezhen Ceramic Institute, Jingdezhen, China

ABSTRACT: Meiosis is a life process necessary for sexual reproduction in eukaryotes, which led to the exchange of recombinant DNA, thereby increasing the genetic diversity of the offspring. The meiotic recombination, in accordance with the frequency of occurrence of high and low genome regions genome, is divided into hotspots and cold spots. For this, effective methods to identify hot spots and cold spots have become current needs. The related information can provide evolutionary studies with useful insights into recombination mechanisms and the genome evolution process. In this article, we developed the predictor through a GM Model and a complexity method to identify recombination hotspots and cold spots. At the same time, we used the feature selection method to improve its effects. It is an effective method for identifying the recombination spots and their complementary effect on existing methods.

1 INTRODUCTION

Genetic recombination during meiosis is in two stages: the first stage is meiosis prophase crossover; the second stage is late non homologous chromosomes free combination. Genetic recombination provides a rich source of biological variation, and it is one of the important reasons for the formation of biological diversity. Recombination pushes the life evolution process; it also greatly improves the efficiency of biological evolution. This phenomenon of recombination in the genome is not random, but occurs at different frequencies at different locations. So there are crucial recombination hotspots and cold spots, both of which reveal the mechanism of recombination forecasts. But it is difficult to predict these different recombination areas in genome. In order to strengthen the effectiveness that distinguishes these crucial and not so crucial areas, the increment of diversity combined with quadratic discriminate analysis (IDQD), based on sequence k-mer frequencies [1] and iRSpot-PseDNC [2], was proposed. Recently, predictions of protein have appeared in many articles using GM models [3, 4] and their complexity [5] has proved interesting.

The gray system theory was proposed by Professor Deng Julong in 1982. It is a deepening development of systems thinking. The theory has attracted much attention in the international community, and has been given very high appraisal. Many young scholars have joined the ranks of gray system theory study, making today's theories and methods widely used in different disciplines, and different areas of research. After a few years, its theoretical research and applied research has made great progress, it has been quickly applied to the economy, society, environment, agriculture, and other fields. Especially in the core part of the GM model, it has been deeply and widely used. The GM model is a predictive and established model of the grey system, which an important content of the grey system theory. For many practical applications, using GM model predictions, there are many successful examples.

In recent publications a new prediction algorithm, called the complexity measure factor, has been developed. It is a pseudo amino acid composition better reflecting the impact of the sequence order effect. There are several ways to evaluate the complexity of the method, such as the complexity measure by Lempel and Ziv, and stochastic complexity [6]. The complexity method has been successfully used to predict protein and DNA sequences.

Compare their methods, in order to strengthen predication the accuracy and effectiveness of recombination spots that we presented a new predictor called "iRSpot-GCFPseNC" It is combining GM model, the complexity measure factor, feature selection and the pseudo nucleotide composition of the DNA sample.

In a recent comprehensive review [7] and in a series of recent publications [8] a useful method or predictor in biological information system was developed, and we need to consider the following steps: (1) establish testing and training model on a benchmark dataset; (2) an effective intrinsic correlation analysis that truly reflects its goal to develop a mathematical expression in biological samples; (3) the development of a robust algorithm (or engine) for testing; (4) cross-validation tests, objective evaluation of the expected accuracy of the prediction; (5) create a user friendly web-server, open to the public for prediction methods.

In the following, let us elaborate some of these procedures.

2 DATASETS AND METHODS

2.1 Benchmark data set

Using the yeast sequence test performance of our proposed method, we selected 490 recombination hotspots and 591 recombination cold spots in this study. The benchmark data set S taken from Liu et al. [1], can be expressed by the following equation:

$$S=S^-\cup S^+ \tag{1}$$

where the subset S^+ is representative for hotspots and the subset S^- is representative for cold spots; $\cup$ represents the symbol of union of two subsets. Considering the convenience of the reader's reading, we set up S^+ have the 490 hotspots sequences and S^- have 591 cold spots sequences.

2.2 GM model

Using the digital encoding of each nucleotide as a digital signal. Set $A=10; C=20; G=30; T=40$. Digital coding used in this paper only considers distinguishing each nucleotide, without considering its physical and chemical properties. For each DNA sequence we use a GM (2, 1) model to calculate the gray coefficient. After the count we get the GM (2, 1) model of the model parameters a, b, c. The calculation process can be seen in the literature [9]. In this article the related calculations are implemented in Matlab 2007a.

2.3 Feature extraction

A DNA sequence we set to D, it has N nucleotides; i.e.

$$D_1=P_1P_2P_3P_4P_5\ldots P_N \tag{2}$$

where P represents the n ($n=1,2,3,\ldots N$) nucleotide location in the sequence. As a simple and effective method, the vector occurrence frequency was widely used for feature extraction. We can get a set of feature vectors, based on a mononucleotide composition in the DNA sequence, and we have

$$D_1=[f'(A)\ f'(T)\ f'(G)\ f'(C)]^T \tag{3}$$

where $f'(A)$, $f'(T)$, $f'(G)$, and $f'(C)$ are respectively as the number of A, C, G, T in the DNA sequence that appears; the symbol T is the transpose operator. We can see that the sequence information using equation (3) represents the DNA sequence of this sequence that will be lost. If the DNA sequence is with the dinucleotide composition to express, we have

$$D_2=[f^{''}(AA)\ f^{''}(AT)\ f^{''}(AG)\ f^{''}(AC)\ \ldots\ f^{''}(CC)]^T$$
$$=[f_1^{''}\ f_2^{''}\ f_3^{''}\ f_4^{''}\ldots f_{16}^{''}] \tag{4}$$

where $f''(AA)$, $f''(AT)$, $f''(AG)$, $f''(AC)\ldots f''(CC)$ are respectively for the normalized occurrence frequencies of AA, AC, AG, AT, and so forth in the DNA sequence. If using the trinucleotide composition to represent a DNA sequence, we have

$$D_3=[f^{'''}(AAA)\ f^{'''}(AAT)\ f^{'''}(AAG)\ f^{'''}(AAC)\ \ldots\ f^{'''}(CCC)]^T$$
$$=[f_1^{'''}\ f_2^{'''}\ f_3^{'''}\ f_4^{'''}\ldots f_{64}^{'''}] \tag{5}$$

where $f'''(AAA)$, $f'''(AAT)$, $f'''(AAG)$, $f'''(AAC)\ldots$ $f'''(CCC)$ are respectively for the normalized occurrence frequencies of AAA, AAC, AAG, AAT, and until CCC in the DNA sequence. As we know, trinucleotide composition is an amino acid. After testing, the effect of triples has the best results.

According to the theory of Zhou's pseudo-amino acid composition, we established the GM (2, 1) model. Through the relevant factors and trinucleotide combining the 64 components, based on GM (2, 1) model, The DNA sequence with pseudo nucleotide sequence is expressed as:

$$X=[x_1 x_2 x_3 x_4\ldots\ x_{64}\ x_{65} x_{66} x_{67}]^T \tag{6}$$

where:

$$X_i = \begin{cases} \dfrac{f_i}{\sum\limits_{k=1}^{64} f_k + w\sum\limits_{k=1}^{3} p_k}, & (1\le i\le 64) \\[2em] \dfrac{w(k-64)^* p(k-64)}{\sum\limits_{k=1}^{64} f_k + w\sum\limits_{k=1}^{3} p_k}, & (65\le i\le 67) \end{cases} \tag{7}$$

where $f_i(i=1,2,3,\ldots,64)$ are respectively as the number of trinucleotide in the DNA sequence that appears. P_k ($k=1, 2, 3$) are gray coefficient a, and b, and c. ω is the weight factor. In this article, ω was taken as 0.8 to achieve better results.

2.4 Complexity measure factor

In this study, we need to encode nucleotide. Considering the size of molecular weight and DNA base properties, we select the coding mode: $A=01$, $C=00$, $T=10$, $G=11$. So DNA sequence can be represented in a digital signal.

In this we use the Ziv-Lemple complexity method, which uses the synthesis minimum step to represent the sequence.

In the process of synthesis, only two steps are allowed: (1) Add a new symbol to ensure that each has a unique sequence of strings; (2) From the synthesized sequence copy the string of the longest. For example, the sequence S = 0001101001000101, it Ziv-Lemple complexity:

$$\begin{cases} H(S)=0\_001\_1\_0\_100\_1000\_101 \\ C_{LS}(S)=6 \end{cases} \tag{8}$$

For a given DNA sequence, we can generate 6 vector group $\varphi1$, $\varphi2$, $\varphi3$, $\varphi4$, $\varphi5$, $\varphi6$. One is the 01 sequence complexity, the other five numbers create pseudo amino acid composition as follows:

$$\overline{\varphi}_{\lambda+1} = \frac{1}{M-\lambda} \sum_{k=1}^{M-\lambda} p(k)p(k+\lambda), \lambda = 1,2,3,4,5 \quad (9)$$

Chou's [8] articles have detailed descriptions of the following method: an expression of a DNA sequence through 22D vector spaces.

$$P = [p_1 \, p_2 \, p_3 \, p_4 \ldots p_{22}]^T \quad (10)$$

where T is the transpose operator, and we have:

$$P_k = \begin{cases} \dfrac{f_k}{\sum\limits_{j=1}^{16} f_j + w \sum\limits_{j=1}^{6} \varphi_j}, & (1 \leq k \leq 16) \\[4mm] \dfrac{w(k\text{-}16)*\varphi(k\text{-}16)}{\sum\limits_{j=1}^{16} f_j + w \sum\limits_{j=1}^{6} \varphi_j}, & (17 \leq k \leq 22) \end{cases} \quad (11)$$

where $f_k (k = 1,2,3,\ldots,16)$ are respectively for the normalized occurrence frequencies of nucleotide in the DNA sequence according to the formula (4). ω is is the weight factor. In this article, ω was taken as 0.8 to achieve better results.

2.5 KNN algorithm

KNN is an effective method for the classification of information. In many articles it can be seen that a KNN algorithm was applied. This paper uses the K-nearest neighbouring algorithm for prediction. In this article, we use the statistical test method, jackknife.

2.6 Feature selection

Feature selection actually contains two aspects: feature extraction and feature selection. Feature extraction is a kind of data transformation to a low dimensional space from a high dimensional space, to reduce the dimension of the object. Feature selection means removing redundancy from a set of characteristics or features related to dimension reduction. Both are often used in combination. The role of feature selection is to improve their classification performance. Good feature sets typically have the following characteristics: be clearly distinguished, reliable, independent, and small in number.

The basic methods of feature selection are: generating first feature subset (algorithm), then the subset evaluation (evaluation criteria). The methods of forming a subset of features are: the exhaustive method, heuristic method, and stochastic method.

We know that not all of the extracted features help in classification, so the feature selection process always has been a classification problem. Its goal of feature

selection is to find the feature subset, and to maximize performance prediction. We usually rely on some heuristics to overcome the complexity of the exhaustive search. In this study, we used Sequential Forward Selection (SFS) [11]. It mainly includes the following steps: (1) Use one classifier (in this case, KNN), and prediction accuracy estimation. (2) Select the most accurate among all features as the first feature. (3) Select the feature, from among all the unselected features, together with the selected features that gives the highest accuracy. (4) Repeat the previous process until the accuracy is good enough.

3 RESULTS

The prediction of recombination spots is a meaningful and challenging task. Identifying recombination spots is still important owing to the fact that they may represent some special biological significance worth our attention. In this study, the predictor, based on the GM model, was proposed. One important step in developing a useful prediction method is to objective the evaluation of its performance or expected success rate. Now let us face this problem. To facilitate this, one method is compared with other methods, and we use the data sets in the literature [1].

The following equation (2) is generally used to test the quality of a predictor; in many articles you can see its index [1, 2].

To facilitate understanding, the equation is expressed as follows:

$$\begin{cases} Sn = 1 - \dfrac{N_-^+}{N^+} \\[3mm] Sp = 1 - \dfrac{N_+^-}{N^-} \\[3mm] Acc = 1 - \dfrac{N_-^+ + N_+^-}{N^+ + N^-} \\[3mm] Mcc = \dfrac{1 - \left(\dfrac{(N_-^+ + N_+^-)}{N^+ + N}\right)}{\sqrt{\left(1 + \dfrac{N_-^- - N_+^-}{N^+}\right)\left(1 + \dfrac{N_+^- - N_-^+}{N^-}\right)}} \end{cases} \quad (12)$$

where N^+ represents the number of samples of hotspot sequences; N_-^+ represents the hotspot errors, which are predicted to be cold spot numbers of the samples; N_- represents the number of samples of cold spot sequences; N_+^- represents the cold spot errors predicted to be the hotspots number of samples [8].

4 CONCLUSIONS

In a recent article you can see the following three cross-validation methods are used to test the prediction of outcome indicators: independent dataset test, sub-sampling test, and jackknife test. Also in recent publications and articles, of the above three methods, the jackknife test is considered to be the most effective and objective statistical method.

Table 1, lists the accuracy in the prediction of the Jackknife test.

Table 1. A comparison of between iRSpot-GCFPseNC with the other method.

Predictor	Test method	Sn (%)	Sp (%)	Acc (%)	MCC
iRSpot-GCFPseNC[a]	Jackknife	90.49	82.27	85.38	0.709
iRSpot-PseDNC[b]	Jackknife	73.06	89.49	82.04	0.638
IDQD[c]	5-fold cross	79.40	81.00	80.30	0.603

[a]The paper used: $\omega = 0.8$; [b]From Chou et al. (2); [c]From Liu et al. (1).

According to the formula (12), from the Table 1 we can see predicted results of our predictor than iRSpot-PseDNC and IDQD [1] better; such as Sn Acc and Mcc in the four metrics. From this we can understand that this prediction is able to become an effective tool to distinguish DNA recombination hotspots and cold spots or become an auxiliary tool. The proposed iRSpot-GCFPseNC has better low-dimensional expression of the sequence information advantage. Based on this development predictor, prediction accuracy is greatly improved. In addition, this study also has carried on the beneficial exploration of the application of Grey Theory in bioinformatics. We believe that the grey theory for the G Protein Coupled Receptor (GPCRs) and the prediction of protein structure prediction type also have good application prospects. As it can be expected that the pseudo amino acid method of Chou is applied to DNA or RNA, the pseudo nucleotide sequence may also be used to solve the related problems of other genomes. These will be the future research projects for the author.

ACKNOWLEDGMENT

The authors wish to thank the anonymous reviewers for their constructive comments, which were indeed very helpful for strengthening the presentation of this study.

REFERENCES

Chou K.C. Some remarks on protein attribute prediction and pseudo amino acid composition. Journal of Theoretical Biology (2011). Volume 273, pp. 236–247.

Chou K.C. Prediction of protein cellular attributes using pseudo-amino-acid. Protein, pp. 246–55.

Chou K.C., Zhang C.T. Prediction of protein structural classes. Crit Rev Biochem Mol Biol. (1995), 30 (4), pp. 275–349.

Guoqing Liu, Jia Liu, Xianjun Cui, Lu Cai. Sequence-dependent prediction of recombination hotspots in Saccharomyces cerevisiae. Journal of Theoretical Biology, 293 (2012), pp. 49–54.

Hui L., Jie Y., Kuo-Chen Chou. Prediction of protein signal sequences and their cleavage sites by statistical. Biochemical and Biophysical Communications 338 (2005), 1005–1011.

Pu Wang (2013) NRPred-FS: A Feature Selection based Two-level Predictor for Nuclear Receptors.

Wei C., Peng M.F., Hao L. and Chou K.C. iRSpot-PseDNC: identify recombination spots with pseudo dinucleotide composition. Nucleic Acids Research, 2013, Vol. 41, e68.

Wei Lin, Xuan, Xiao (2007) Based on the GM (1, 1) protein level two types of structure prediction. Computer Engineering and Applications, 43(34), pp. 41–45.

Xuan Xiao (2011) Predicting secretory proteins based on protein Hasse matrix image. Computer Engineering and Applications, 2011, 47(32), pp. 170–172.

Xuan Xiao, S. Shao. Using complexity measure factor to predict protein subcellular location. Amino Acids (2005), Volume 28, pp. 57–61.

Yury L.O., Vladimir P.F. Construction of stochastic context trees for genetic texts. In silico biology 2(2002), pp. 233–247.

Biomedical Engineering and Environmental Engineering – Chan (Ed.)
© *2015 Taylor & Francis Group, London, ISBN: 978-1-138-02805-0*

Transcriptome analysis of *sapium sebiferum* seed in three development stages using a Hiseq 2000 illumina sequencing platform

Haiyan Chen
ChangChun University of Science and Technology, Jilin, China

Yun Liu
Beijing Key Laboratory of Bioprocess, College of Life Science and Technology,
Beijing University of Chemical Technology, Beijing, China

ABSTRACT: Hiseq 2000 illumina sequencing technology was employed to investigate the profile of gene expression and its functional gene in *sapium sebiferum* seed with a substantial expressed sequence tags dataset. A total of 69215 unigenes and 32086 unigenes with protein function annotations were obtained. The unigenes with function annotations could be classified into 25 functional-categories and accounted for 46.36% of all unigenes. The all-unigenes were queried against the KEGG pathway database, and 6837 unigenes were given pathway annotations and related to 277 pathways, including metabolism, genetic information processing, and cell metabolism pathways, etc. GO function analysis showed that 14979 unigenes belonged to a molecular function, 24043 unigenes to a biological process, and 19093 unigenes to cellular components. The results from our work can contribute to analysing the quantity of gene expressions and regulation patterns, as well as to elucidating the diversity of gene expression in different developmental stages, which also provide the theoretical basis and technical guidance in the molecular genetic breeding of *sapium sebiferum*.

Keywords: *Sapium sebiferum*; transcriptome sequencing technology; unigene; pathway

1 INTRODUCTION

Sapium sebiferum, a typical oil-containing plant in China, has high oil content within the range of 20% to 50% and shows great economic value for its wide applications [1]. It has been demonstrated that *S. sebiferum* is versatile owing to its lipid as the replacement of cocoa butter [2], its flowers as a superior honey nectar source, leaves containing active components for insect protection and as an insecticide, wood for furniture, and oil from its kernel as edible oil and biofuel [1]. After being extracted, the oil from the seeds, and the oilseed cake can be used as an organic fertilizer. It is noteworthy that *S. sebiferum* also shows effective functions in soil fixation and water conservation since its roots are over 2.6 m deep and stretch horizontally over 10 m. Because of its high growth rate and strong adaptability, *S. sebiferum* is widely distributed as a natural habitat across China, Japan and India, and also grows well in the southern coastal United States [3]. Fast-growing *S. sebiferum* reaches maturity within approximately 3–4 years, and then it can generate economic yields in its productive lifespan of more than 40 years.

Biodiesel, an important alternative biofuel, has been attracting increasing attention due to the serious depletion of fossil-based fuels and the bad environmental pollution caused by their combustion [4, 5]. Nowadays, the bottleneck of biodiesel in large-scale industrialization lies in oil sources and price. Therefore, it is of significance to develop alternative woody oil-rich raw materials for biodiesel production. The properties of the components in biodiesel may guide the genetic modification of existing oilseed crops to optimize biodiesel fuel properties [6]. In our previous work, we have reported on the physicochemical properties, especially the Fatty Acid (FA) profile and the triglycerides (TGs) structure of stillingia oil from *S. sebiferum* seed kernel, and the biodiesel production and its properties from Chinese vegetable tallow oil [1]. However, no information on the genome of *S. sebiferum* has been available up to now, although the genomes of some other woody oil plants such as rape, *Jatropha curcas*, and oil-tea camellia have been available in literatures [7, 8].

Transcriptome conveys the identity of each expressed gene and its level of expression for a defined population of cells, including codong RNA and non-coding RNA and being modulated by external and internal factors. It differs from the exome since it covers only those RNA molecules found in a specified cell population, and usually includes the amount or concentration of each RNA molecule in addition to the molecular identities [8]. In this work, Hiseq 2000 illumina sequencing technology is employed to investigate the transcriptome of clonal growing *S. sebiferum* "Luo 22-1" seeds and analyse the quantity of gene expression and regulation pattern, aiming to establish a wide molecular genetic information platform. The findings in this work will provide the theoretical basis in genetic modification and molecular directed breeding of high oil content variety of *S. sebiferum*.

2 MATERIALS AND METHODS

2.1 *Materials*

S. sebiferum "Luo 22-1" seeds in three different growing periods were obtained from Jiufeng National Forest Park in Wuhan, Hubei province, China. The seeds were gathered in July (seeds starting to swell, and being labelled as Luo (7)), August (seeds in meiosis, and being labelled as Luo (8)) and October (weeds towards maturity, and being labelled as Luo (10)), respectively. All samples were immediately preserved in liquid nitrogen after collection.

2.2 *RNA isolation, cDNA library construction and transcriptome sequencing*

Total RNA was extracted from the samples using Trizol Reagent (Invitrogen, USA) according to the manufacturer's instructions [9]. RNA quanlity was assessed by spectrophotometer before proceeding RNA-seq, and the ratio of 28S:18S should be greater than 1.8. The Oligo (dT)-magnetic beads were used to enrich poly (A) mRNA. Following purification, the mRNA was fragmented into smaller pieces using divalent cation. Serving these short fragments as templates, first-strand cDNA was synthesized using SuperscriptTM III reverse transcriptase and random hexamer (N6) primers. Then, the second strand cDNA was synthesized using RNaseH and DNA polymerase I. The obtained short double cDNA fragments were purified. After ending reparation and A-tailing, the short cDNA fragments were connected to the illumina paired-end adaptors and purified with magnetic beads. To prepare the cDNA sequencing library, suitable ligation products were amplified using Illumina primers and Phusion DNA polymerase. Finally, Illumina Hiseq2000 system (Illumina) was applied to sequence from both the 5' and 3' ends. The fluorescent image outputs from the sequencing machine were transformed by the base calling into sequence data, which were called raw reads.

2.3 *Analysis of RNA-seq data and function annotation*

To obtain high-quality reads, the raw reads were cleaned by removing the sequences corresponding to the sequencing adapters and reads with low quality ($Q \leq 5$), and then assembled and rearranged by contigs and scaffolds to obtain full unigenes in certain

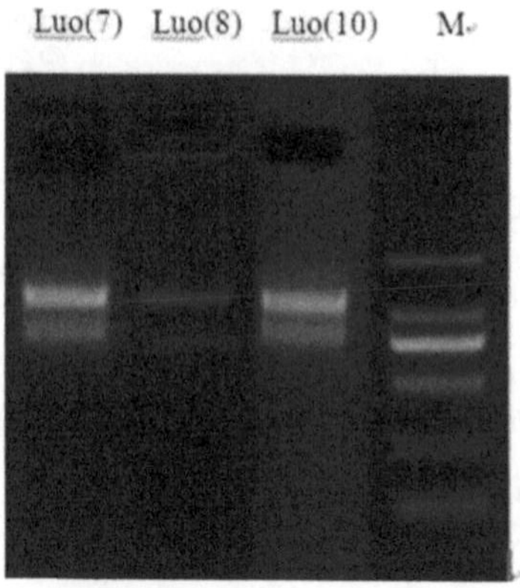

Figure 1. Electrophoresis results of RNA.

Table 1. RNA detection result of A260/A280 and concentration.

Samples	Luo(7)	Luo(8)	Luo(10)
A260/A280 (28S:18S)	2.07	1.92	2.18
Concentration (ng/μL)	249.3	384.4	55.83

length distribution [9]. All Unigenes, remapped by more than 5 reads, were aligned using BLASTX to a public protein databases NR, Swiss. Prot, KEGG and COG [9]. Next, GO functional classification of all unigenes was analysed to determine the distribution of the gene function at macro level.

3 RESULTS

3.1 *RNA quality detection*

Figure 1 showed that the extracted RNA bands by SDS-PAGE were bright, clear and sharp in three different stages, and that the calculated values of 28S:18S were ranging from 1.9 to 2.2. It suggested that the quality of RNA was suitable for sequencing. Furthermore, no residual DNA fragments in the lanes revealed that DNA was fully precipitated in fractional precipitation. The concentration of RNA extracted from *S. sebiferum* seed Luo (8), Luo (7) and Luo (10) was 384.4 ng/μL, 249.3 ng/μL and 56 ng/μL, respectively. The RNA concentration variances in different collection periods were probably due to the varying RNA transcriptions and expressions in meiosis and mitosis of seed formation and development.

3.2 *Unigenes sequencing and assembly*

Using Hiseq 2000 illumina technology, a total of 69215 unigenes were obtained in three different stages samples (Fig. 2).

Length of these unigenes mainly ranged between 200 bp and 1700 bp, and the number of gene above

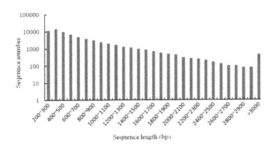

Figure 2. Length distribution of all-unigenes.

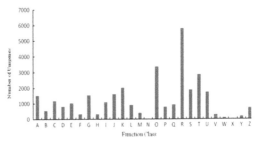

Figure 3. Function classification of unigenes.

3000 bp was only about 441 (Fig. 2). Most of these unigenes (94.7%) were distributed in the 200 bp–1700 bp region, and 4.26% for 1700–3000 bp. Recently, by sequencing clones from three non-normalized cDNA libraries, 32,521 unigenes sequences were obtained and most of them were used for floral transcription factors prediction from lycoris longituba [10]. Therefore, this transcriptome dataset provided a useful resource for future analyses of genes related to *S. sebiferum* seed development. To the best of our knowledge, this was the first comprehensive study of the transcriptome of *S. sebiferum* seed.

3.3 *COG annotation*

Assignments of COG were used to predict and classify the possible functions of the unique sequences [11]. Based on sequence homology, 32086 unigenes (46.36%) of all 69215 assembled unigenes had COG functional classification, and the COG-annotated putative proteins were classified into twenty-five functional categories (Fig. 3). The general function prediction was the only category represented in the largest group (5797, 18.07%), which was known to be involved in processing the precursors of storage proteins into mature proteins [12], followed by posttranslational modification, protein turnover and chaperones (3362, 10.48%) and signal transduction mechanisms (2847, 8.87%). The aforementioned categories affected the intense catabolic activity within germinating seeds related to the mobilization of protein reserves [13]. Only a few unigenes belonged to cell modility and a nuclear structure (16 and 2 unigenes, respectively). Notably, 946 and 413 unigenes were assigned to secondary metabolites biosynthesis, transport and catabolism, and to cell wall/membrane/envelope biogenesis, respectively.

Note: A: NA processing and modification; B: Chromosome structure and dynamics; C: Energy production and conversion; D: Cell cycle control, cell division and chromosome partitioning; E: Amino acid transport and metabolism; F: Nucleic acid transport and metabolism; G: Carbohydrate transport and metabolism; H: Coenzyme transport and metabolism; I: Lipid transport and metabolism; J: Translation, ribosomal structure and biogenesis; K: Transcription; L: Replication, recombination and repair; M: Cell wall/membrane/envelope biogenesis; N: Cell

modility; O: Posttranslational modification, protein turnover and chaperones; P: Inorganic ion transport and metabolism; Q: Secondary metabolite biosynthesis, transport and catabolism; R: General function prediction only; S: Function unknown; T: Signal transduction mechanisms; U: Intracellular trafficking, secretion and vesicular transport; V: Defence mechanism; W: Extracellular structures; Y: Nuclear structure; Z: Cytoskeleton.

3.4 *KEGG pathway mapping*

To understand the biological pathways that might be active in *S. sebiferum* seeds, the unigenes were compared against the KEGG database [14]. The results showed that of all 69215 unigenes, 6837 unigenes (9.88%) had significantly matched and were assigned to 277 KEGG pathways (Table 1). The most represented pathway annotations were metabolisms (2777 unigenes with 127 pathways), followed by genetic information processing (2133 unigenes with 27 pathways), cellular processes (610 unigenes with 18 pathways), and environmental information processing (292 unigenes with 24 pathways). Furthermore, the largest category metabolism included carbohydrate metabolism (473 unigenes), energy metabolism (438), amino acid metabolism (396), enzyme families (296), lipid metabolism (220), glycan biosynthesis and metabolism (189), nucleotide metabolism (168), metabolism of cofactors and vitamins (180), metabolism of other amino acids (141), metabolism of terpenoids and polyketides (122), biosynthesis of other secondary metabolites (85) and xenobiotics biodegradation and metabolism (69). It was considerable to highlight enzyme families and lipid metabolism were dedicated to oil accumulation, and the acylglycerols of lipid metabolism acted as an energy reserve in many organisms and were the major components of seed storage oils [8]. And genetic information processing, the second largest category, consisted of folding, sorting and degradation (645 unigenes), translation (529), replication and repair (512) and transcription (446). Of cellular processes, transport and catabolism (256 unigenes) were dominating, followed by cell growth and death (181), cell motility (108) and cell communication (65). In addition to these, the environmental information processing also had some subcategories: signal

9

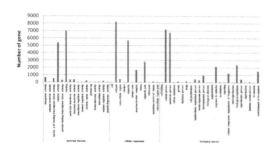

Figure 4. Go function annotation and classification of unigenes.

Table 2. Pathway annotation and classification of unigene.

Pathway annotation	Annotation and classification	Pathway quantity	Unigene quantity
Cellular Processes	Cell Communication	4	65
	Cell Growth and Death	7	181
	Cell Motility	2	108
	Transport and Catabolism	5	256
Environmental Information Processing	Membrane Transport	4	61
	Signal Transduction	14	171
	Signalling Molecules and Interaction	6	60
Genetic Information Processing	Folding, Sorting and Degradation	10	645
	Replication and Repair	9	512
	Transcription	4	446
	Translation	4	529
Metabolism	Amino Acid Metabolism	14	396
	Biosynthesis of Other Secondary Metabolites	12	85
	Carbohydrate Metabolism	15	473
	Energy Metabolism	9	438
	Enzyme Families	3	296
	Glycan Biosynthesis and Metabolism	13	189
	Lipid Metabolism	14	220
	Metabolism of Cofactors and Vitamins	12	180
	Metabolism of Other Amino Acids	7	141
	Metabolism of Terpenoids and Polyketides	12	122
	Nucleotide Metabolism	2	168
	Xenobiotics Biodegradation and Metabolism	14	69
Organismal Systems	Circulatory System	2	49
	Development	2	22
	Digestive System	5	31
	Endocrine System	7	138
	Environmental Adaptation	4	110
	Excretory System	4	36
	Immune System	12	93
	Nervous System	3	48
	Sensory System	4	9
Human Diseases	Cancers	15	61
	Cardiovascular Diseases	4	24
	Immune System Diseases	2	31
	Infectious Diseases	10	137
	Metabolic Diseases	2	34
	Neurodegenerative Diseases	5	203

transduction (171 unigenes), membrane transport (61) and signalling molecules and interaction (60).

KEGG pathway analysis and COG annotation contributed to predict potential genes and their functions at transcriptome level. These annotations, together with COG analysis, would be a valuable resource for further research on specific processes, structures, functions, and pathways in *S. sebiferum* seed.

3.5 *GO function classification*

To functionally categorize *S. sebiferum* seed unigenes based on NR annotation, Gene Ontology (GO) [15] analysis was performed. A total of 58115 unigenes (83.96%) of all 69215 assembled unigenes were available. Of these, assignment of contigs to the biological process made up the majority (24043, 41.37%), followed by cellular components (19093, 32.85%) and molecular functions (14979, 25.77%).

The distribution of the sub-categories in each main category was shown in Figure 4. In the biological process category, cellular process and metabolic process represented the majority of its unigenes, indicating that these unigenes were involved in some important metabolic activities in *S. sebiferum* seed [9]. Meanwhile, it was noteworthy to realize that both oxidoreductase activity and auxin biosynthesis were key processes in early plantlet establishment, related to energy uptake from seed reserves and growth [8]. Interestingly, only a few of the unigenes were assigned to carbon utilization, biological adhesion, and pigmentation. Within the cellular component category, cell factors and organelle were prominent. Several predicted cellular components were identified as membrane-associated proteins [8]. In the molecular function category, binding represented the most abundant term, followed by catalytic activity. The important signal transduction abilities were also classified, such as transporter activity and molecular transducer activity, and the transposable elements were highly active in developing seeds [8].

4 CONCLUSION

S. sebiferum, with a high oil content ranging between 20% and 50%, has become the significant oil source for biodiesel production in China. In this study,

Hiseq 2000 illumina sequencing technology has been employed to evaluate transcriptome gene expression profiling. Protein databases NR, Swiss. Prot, KEGG and COG were used to analyse the function, classification, and metabolic pathway of unigenes. It is concluded that the results of this work have laid the theoretical basis to explain the molecular regulation mechanism of fatty acid accumulation in *S. sebiferum* seed, as well as the technical director of its genetic breeding.

ACKNOWLEDGEMENT

Authors thank the financial funds from the National Science & Technology Pillar Program during the Twelfth Five-year Plan Period (2011BAD22B08).

REFERENCES

Conesa A, Gotz S, Garcia-Gomez JM, et al. Blast2GO: a universal tool for annotation, visualization and analysis in functional genomics research [J]. Bioinformatics, 2005, 21(18): 3674–3676.

Costa GGL, Cardoso KC, Bem DLEV, et al. Transcriptome analysis of the oil-rich seed of the bioenergy crop Jatropha curcas L. [J]. BMC Genomics, 2010, 11: 462–471.

Gressel J. Transgenics are imperative for biofuel crops [J]. Plant Science, 2008, 174(3): 246–263.

He QL, Cui SJ, Gu JL, et al. Analysis of floral transcription factors from Lycoris longituba. Genomics, 2010, 96: 119–127.

Hu F, Wei FX, Wang ZC, et al. Enzymatic preparation of cocoa butter equivalent (CBE) from Chinese tallow oil using response surface methodology (RSM) [J]. Food Research and Development, 2010, 31: 94–97.

Huang HH, Xu LL, Tong ZK, et al. De novo characterization of the Chinese fir (Cunninghamia lanceolata) transcriptome and analysis of candidate genes involved in cellulose and lignin biosynthesis [J]. BMC Genomics, 2012, 13: 648–662.

Kanehisa M, Goto S, Kawashima S, et al. The KEGG resource for deciphering the genome [J]. Nucleic Acids Res, 2004, 32: 277–280.

Knothe G. Improving biodiesel fuel properties by modifying fatty ester composition [J]. Energy Environment Science, 2009, 2: 759–766.

Lin P, Cao YQ, Yao XH, et al. Transcriptome analysis of Camellia Oleifra Abel Seed in four development stages [J]. Molecular Plant Breeding, 2011, 9: 498–505.

Liu Y, Shang SW, Xin HL, et al. Enzymatic production of biodiesel from Chinese tallow kernel oil [J]. Applied Chemical Industry, 2008, 37: 977–980.

Liu Y, Xin HL, Yan YJ. Physicochemical properties of stillingia oil: Feasibility for biodiesel production by enzyme transesterification [J]. Industrial Crops and Products, 2009, 30: 431–436.

Purkrtova Z, Jolivet P, Miquel M, Chardot T. Structure and function of seed lipid body associated proteins [J]. Comptes Rendus Biologies, 2008, 331(10): 746–754.

Wang R, Xu S, Jiang YM, et al. De novo sequence assembly and characterization of Lycoris aurea transcriptome using GS FLX Titanium platform of 454 pyrosequencing [J]. Plos One, 2013, 8(4): 60449–60459.

Yin Y, Huo GH, He XL. Antioxidant activity study on Chinese tallow leaves (Sapium sebiferum) extract in vitro [J]. Food Science, 2003, 24: 141–144.

Zheng J, Xu L, Liu Y, et al. Lipase-coated K_2SO_4 microcrystals: Preparation, characterization, and application in biodiesel production using various oil feedstocks [J]. Bioresource Technology, 2012, 110: 224–231.

Biomedical Engineering and Environmental Engineering – Chan (Ed.)
© 2015 Taylor & Francis Group, London, ISBN: 978-1-138-02805-0

The establishment of a 4-33 week male rats' lung weight database based on CSI

Jianyang Lin & Jinjie Xu
Department of Pharmacy, The First Hospital of China Medical University, Shenyang City, China

Yonggang Li
School of Automobile and Transportation, Shenyang Ligong University, Shenyang City, China

ABSTRACT: The mathematical method of CSI (cubic spline interpolation) is adopted to establish the database of 4-33 week male rats' lung weights. The anatomical data of male rat lung weight is from the National Natural Science Foundation of China, for experimentally establishing and analytically studying the SD of rat's organ characteristics for a foundation database for pharmacokinetics physical simulation under Grant #81302841. The data processing is accomplished by the software MATLAB R2010a. At first, the normality test is carried out for the data of each week. As a result, these data are consistent with normal distributions. Then, interval estimation of confidence level 0.95 is done to the arithmetic mean of each group of data thus getting their confidence of the upper and lower limits. At last, the confidence of upper and lower limit confidence is interpolated by cubic spline interpolation. Through this method, the range of the arithmetic mean of the lung weight is achieved.

1 INTRODUCTION

The researchers weighed 4-33 week male rats respectively and killed and dissected the male rats according to the related content of laboratory animal medicine for the purpose of getting the weights of their lungs. The data of every week contain 17 samples. But these data are discrete values. The values of those days between every two weeks cannot be obtained through the discrete values. It should be noted that, there is a complex function between the weight and the number of days. So in order to solve the above problem, the approximation expression of the function must be constructed from the known data. A common method is data interpolation, which stems from practice and is widely used in practice. With the development of the computer and the improvement of computer skills, this numerical method has played an increasingly important role in the field of national production and scientific research. The data interpolation is an important method of functional approximation. In most of the production and scientific experiments, the relational expression $y = f(x)$ between the independent variable x and the dependent variable y could not be written directly. What we can achieve is just the values at several points. When we want to know the functional values outside the observation points, we need to estimate the functional values at those points [1, 2]. Values in the unknown points can be determined on the basis of the original data by using interpolation methods [3].

As a very good engineering applications software, MATLAB has the advantage of a powerful numerical calculation function, strong engineering and drawing functions and ease of use [4]. Using MATLAB to carry on data processing can greatly reduce the computational workload and decrease the error caused by calculation and graphing, thus achieving relatively accurate results [5]. So in this paper MATLAB is adopted to carry out data processing.

1.1 *The definition of interpolation [6]*

The function $y = f(x)$ has a definition in the interval [a, b] and the value of y in the n + 1 nodes $a \leq x_0 < x_1 < \ldots < x_n \leq b$ is $y_0, y_1, \ldots, y_n$ respectively. If there is a simple function P(x) which can make $P(x_i) = y_i (i = 0, 1, \ldots, n)$, P(x) is called the interpolation function of f(x) in the node $x_0, x_1, \ldots, x_n$ where node $x_0, x_1, \ldots, x_n$ are called interpolation nodes, the interval [a, b] containing the interpolation nodes is called the interpolation interval, f(x) is called the interpolated function. The way of solving the interpolation function is called interpolation.

If P(x) is an algebraic polynomial whose number does not exceed n, namely $P(x) = a_0 + a_1 x + \ldots + a_n x^n$ where a_i is a real number, P(x) is called interpolation polynomial and the corresponding interpolation method is called polynomial interpolation. If P(x) is a piecewise polynomial, the corresponding interpolation method is called piecewise interpolation.

From the interpolation definition, the interpolation function is in fact a curve passing n + 1 nodes (x_i, y_i)(i = 0, 1, . . . ,n).

1.2 Cubic spline interpolation [7–10]

Cubic Spline Function Interpolation (CSI) is defined as follows. The function f(x) has a group of nodes $a = x_0 < x_1 < x_2 < \ldots < x_n = b$ in the interval [a, b] and the corresponding function value is $y_0, y_1, \ldots, y_n$ respectively. If s(x) has the following properties: (1) s(x) is a polynomial whose number does not exceed 3 in each subinterval; (2) s(x), s'(x) and s''(x) are continuous in the interval [a, b]. So s(x) is a cubic spline function; (3) Adding the condition $s(x_i) = y_i$ ($i = 0, 1, 2, \ldots, n$), s(x) is called cubic spline interpolation function.

In the cubic spline, cubic polynomial is needed to be found to approximate the curve between each pair of data points. In terms spline, these data points are called breakpoints. Because two points can only decide a straight line and the curves between two points can be approximated by an infinite number of cubic polynomials, the constraints of cubic polynomial are added in cubic spline to make the result unique. Making the first and second order derivative of each cubic polynomial equal at the breakpoints can better determine all the internal cubic polynomials. In addition, approximation polynomial slope and curvature are continuous through these breakpoints. However, the first and last cubic polynomials have no adjoint polynomials other than the first and last breakpoints. So other methods must be used to determine other constraints. Among these methods, the most commonly used method is employing the not-a-knot condition. This condition forces the third derivative of the first and second cubic polynomials as well as the last and penultimate cubic polynomials equal.

The commonly used interpolation methods include Lagrange interpolation, Newton interpolation, Hermite interpolation and cubic spline interpolation. Among these methods, the cubic spline interpolation is widely used in engineering applications because this method has the advantages of high accuracy, a smoother curve, strong convergence, and stability, thus eliminating morbidity better and this method, which needs only a small amount of calculation, is simple and practical. Therefore, the cubic spline interpolation is adopted in this paper to carry out the data processing.

2 DATA PROCESSING

2.1 Normality test of data

There are many ways of normality test such as the normal probability paper inspection act, the Lilliefors inspection act and the Jarque-Bera inspection act [11]. The normal probability paper inspection act which is realized through the normplot function in MATLAB is used in this paper to inspect the normality of 17 data every week. Because of the paper space limitation, this article lists only the test results of the first 4 weeks as shown in Figure 1. The normality test results of the other weeks have similar results.

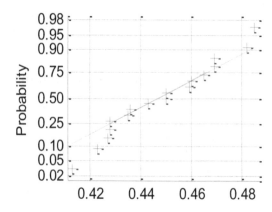

Figure 1. The normality test results of the fourth week.

As can be seen from Figure 1, the data of each week are substantially distributed in a straight line. So we can consider that the data of each week are consistent with a normal distribution.

2.2 Interval estimation of arithmetic mean

The arithmetic mean and standard deviation of data for each week are calculated by using the mean function and std function as shown in Table 1 where E stands for the arithmetic mean and S the standard deviation.

And the next work to be done is the interval estimation of the arithmetic mean. According to the knowledge of the interval estimation, the smaller confidence interval has the higher the estimation precision; the larger confidence level has the higher the estimation credibility, but obviously the two indicators are contradictory. The usual practice is to make the confidence intervals as small as possible at a certain confidence level [12]. Considered comprehensively, the interval estimation of the arithmetic mean for the data of every week is done by the normfit function [13] in MATLAB with the confidence level. So the upper and lower confidence limit of the arithmetic mean of the data from every week are gained as shown in Table 2 where θ_1 and θ_2 stands for the upper and lower confidence limit respectively.

The data of Table 2 are drawn into a two-dimensional map as shown in Figure 2 where the blue dots and the green dots stand for the upper and lower confidence limit of interval estimation respectively. According to Figure 2, the male rat lung weight is on the overall upward trend. But those data are discrete and there is no way to achieve the weight of a certain day between every two weeks. Therefore the data remain to be processed further.

2.3 Cubic spline interpolation

Cubic spline interpolation has high precision, minor average error and maximum error, good anastomosis with the original data, and a stability calculation process [14]. The curves generated by cubic spline interpolation can pass all the sampling points and the

Table 1. The arithmetic mean and standard deviation of data for 4-33 week.

Week	4	5	6	7	8	9	10	11	12	13
E	0.448	0.630	0.830	1.078	1.233	1.285	1.471	1.532	1.600	1.645
S	0.022	0.048	0.105	0.152	0.177	0.137	0.178	0.118	0.195	0.143

Week	14	15	16	17	18	19	20	21	22	23
E	1.802	1.755	1.934	1.982	1.935	2.007	2.031	2.056	2.261	2.188
S	0.187	0.202	0.180	0.187	0.185	0.206	0.207	0.225	0.215	0.223

Week	24	25	26	27	28	29	30	31	32	33
E	2.241	2.225	2.328	2.401	2.355	2.445	2.410	2.518	2.476	2.523
S	0.211	0.240	0.216	0.178	0.238	0.182	0.304	0.250	0.238	0.233

Table 2. The interval estimation of the arithmetic mean with the confidence level of 0.95.

Week	4	5	6	7	8	9	10	11	12	13
θ_1	0.445	0.622	0.812	1.052	1.203	1.263	1.441	1.512	1.568	1.621
θ_2	0.452	0.638	0.847	1.103	1.262	1.308	1.500	1.552	1.633	1.668

Week	14	15	16	17	18	19	20	21	22	23
θ_1	1.771	1.721	1.903	1.950	1.904	1.973	1.997	2.019	2.225	2.150
θ_2	1.834	1.789	1.964	2.013	1.966	2.042	2.066	2.094	2.297	2.225

Week	24	25	26	27	28	29	30	31	32	33
θ_1	2.205	2.185	2.292	2.371	2.315	2.415	2.359	2.476	2.437	2.485
θ_2	2.276	2.265	2.364	2.431	2.395	2.476	2.461	2.560	2.516	2.562

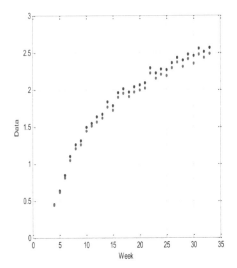

Figure 2. Interval estimation figure of arithmetic mean with the confidence level 0.95.

is making the entire curve smooth on these points, the cubic spline interpolation is selected in this paper. The upper and lower confidence limit of the arithmetic mean is interpolated by the cubic spline function named csapi. The interpolation results are shown in Fig. 3. In this figure six points, which are used to represent the number of days between every two weeks, are set between every two weeks and the interpolated data of these points are also calculated.

According to Figure 3, the weight data of every day in 4-33 week can be obtained. For example, if we want to know the data of six days between 21 week and 22 week, we can take the abscissa as 21.14, 21.29, 21.43, 21.57, 21.71, 21.86. The scope of the arithmetic mean can be read out from Figure 3. The scope are [2.046, 2.121] [2.079, 2.154], [2.115, 2.189], [2.151, 2.224], [2.184, 2.256], [2.209, 2.281].

method of cubic spline interpolation successfully overcomes the disadvantages of other methods which just reflect the overall shape of the curve and neglect the details' features [15]. For the purpose of assuring that each curve is continuous on the connection points and

3 CONCLUSION

In this article, the cubic spline interpolation is carried out on the data of the male rats' lung weights. The conclusions can be obtained as followings:

  I. The cubic spline interpolation is suitable for the data processing of this study since it produces smooth curves.

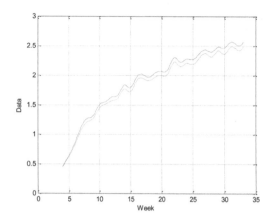

Figure 3. Cubic spline interpolation of the interval estimation of the arithmetic mean.

II. A range of the arithmetic mean of the 4-33 week male rats' lung weights has been accomplished. This scope can be used to verify the correctness of the repeatability experiment data thus providing a reference for future researches.

III. The curves in this paper can be used as a standard to test and verify whether the male rats of each group grow normally or not. Take a certain number of male rats of a random group, dissect the male rats and weigh the weight of the lungs to compare with the curves achieved in this paper. If the data is consistent with the curves, there is no need to dissect more male rats. This practice can reduce the cost of experiments to a certain extent.

ACKNOWLEDGEMENTS

This work is supported by National Natural Science Foundation of China under Grant #81302841.

REFERENCES

Bao Yanke, Li Na, 2008. Mathematical statistics and MAT-LAB data processing [M]. Liaoning: North-eastern University Press Ltd, 123–127.

Chen Shaoyun, 2014. MATLAB method for interpolation and fitting comparison [J]. Journal of Mudanjiang College of Education, (7):70.

Chen Hongying, Li Huifen, He Jing, Yang Lei, 2013. MAT-LAB application in the afterwards data processing of launch vehicle telemetry [J]. Journal of Telemetry, Tracking and Command, 34(1):71–75.

Chen Hong, Liu Hai, Qiao Shenghua, Wang Yafei, 2013. Vehicle travel data analysis based on cubic spline interpolation [J]. Automobile Technology, (8):54–57.

Dragon Technology. MATLAB engineering calculations and analysis [M]. Beijing: Tsinghua University Press, 2011 (p. 138, 157–158).

Gong Li, Lian Jinhua, Liu Zhongru, Li Aifeng, 2014. Study on excavator standard working curve based on cubic spline interpolation [J]. Mining Machinery, 42(6):37–39.

Hu Rui, 2012. Comparison of several interpolation algorithms in the calculation of aerodynamic data [J]. Science Mosaic, (4): 34–37.

Sun Quicken, Zhou Yahoo, Ming Chuang, Cui Jinmen, 2014. Subpixel edge detection method based on cubic spline interpolation [J]. Control Engineering of China, 21(2):290–293.

Xue Zechun, Liang Mingli, Li Lianzhi, Zhang Xianxi, Li Dacheng, 2014. Study on separation of overlapping peaks in application of cubic spline interpolation and continuous 'Mexh' wavelet transform [J]. Journal of Liaocheng University (Nat. Sci.), 27(1):50–54.

Yan Lin, 2005. Mathematical and experimental: Application of MATLAB and QBASIC [M]. Beijing: Science Press, 139.

Yu Haijun, Chen Jinyi, 2013. Research and comparison of three interpolation methods [J]. Journal of Henan Science and Technology, (10): 210–211.

Zhang Defeng, 2011. MATLAB numerical analysis and simulation case [M]. Beijing: Tsinghua University Press, 82.

Zhang Defeng, 2012. MATLAB numerical analysis [M]. Beijing: Machinery Industry Press, 250–254.

Zhaofeng, 2012. Experimental data processing based on MATLAB [J]. Development & Innovation of Machinery & Electrical Products, 25(2):95–96.

Zhou Bo, Xie Donglai, Zhang Xianhai, 2010. MATLAB scientific computing [M]. Beijing: Machinery Industry Press, 198.

Biomedical Engineering and Environmental Engineering – Chan (Ed.)
© 2015 Taylor & Francis Group, London, ISBN: 978-1-138-02805-0

To discuss the development of Sanda from safety

Runjie Qu
School of WuShu, Shenyang Sport University, Shenyang, Liaoning, China

ABSTRACT: Safety is the premise of all sports activities. Since Sanda is a combat against sport in Wushu, and the main means to win is hitting and knocking each other, how to guarantee the safety of athletes in the fierce confrontation case becomes more important. By means of literature review, comparative method and logical reasoning method, the thesis studies the development direction of Sanda in the security in China, and explores that security is the premise of Sanda sustainable development, Sanda rules should be modified based on security, and then puts forward the security measures of athletic Sanda. In addition, with the concept of "light sport", the study puts forward the constructive suggestions to promote Sanda.

Keywords: Sanda; Security; Light sport; Rules; Referees

1 INTRODUCTION

Sanda is a modern competitive sport, with two unarmed people kicking, hitting, and wrestling based on certain offensive and defensive techniques and rules. The main purpose is hitting and wrestling, which is its characteristic. The two kinds of competition form are sport competition and commercial competition. There are many differences in the rule, sites, clothing, gear and other aspects. To pursuit stimulation and ratings, especially in the commercial competition, greater harm tends to bring to athletes. With the vigorous development of Sanda, athletes competitive level constantly improves, and the competition often leads to intense. It is possible that athletes would be KO by one hitting or kicking. Certainly, the audience is easy to be attracted by the fierce competition, but the athletes' safety and health is worried about. Therefore, how to ensure the safety of athletes has plagued every professional. This thesis main studies Wushu Sanda sports development from security, analyzes safe trouble in concept and in practice, and find ways to avoid it in order to provide references for sustainable development, the popularization and the promotion of Sanda.

2 SAFETY HIDDEN TROUBLESAFETY IS THE PREMISE OF SUSTAINABLE DEVELOPMENT OF SANDA

In 1952, Chairman Mao, in the founding conference of the all-china sports, wrote "To develop sports, and improve the people's physical fitness". This inscription is the essence of the Chinese Sports, which established national sports and pointed out the final direction of sports development. Recently, in the national mass sports advanced commendation congress, the national leader Jinping Xi advanced that the development of sports is the fundamental principle and task of sports work. To improve the people's physical fitness is the ultimate purpose, therefore any sports should be on how to improve the health of the participants.

Sanda is fight against race and is the representative of the Chinese traditional sport, the teenagers are attracted by the essential characteristics of the sport. At present, our country actively strives to make Wushu (including Taolu and Sanda) – a representative of the Chinese national tradition sports, join in Olympic Games, and make people enjoy it. This aim gives higher requirements to the development of Sanda, especially the safety of athletes in the competition and training. Sanda is attacking each other, so it's hard to avoid sports injury such as abrasion and contusion and so on. At present, there are two kinds of competition form: competitive sports and commercial sports. Competitive sports wear protective devices of a game, so the game is not as intense as commercial sports, and the security is better than the commercial competition. While commercial competition has been promoted at home and abroad. But in order to improve the television rating, the organizers or sponsor tend to modify rules, and let go of some rules to disable methods (with the knee to attack), which greatly reduce the safety factor of the participants. In the long run, it's hard to avoid injury accident happening, which is against the purpose of the sports. Therefore, the athletes' personal safety should be paid more attention whatever competitive or commercial that is also a primary precondition for sustainable development of Sanda.

3 TO ANALYZE THE INTERNAL AND EXTERNAL CAUSES OF SANDA SAFETY HIDDEN TROUBLE

Some reasons lead to Sanda injury and accident harm. According to the nature, the reasons can be divided into internal problems and external problems, collectively known as the internal cause and external cause. Analyzing the causes Objectively has the vital significance in preventing damages and improving the security of Sanda.

3.1 The internal cause

3.1.1 Unreasonable preparation activities
Preparation activities are a series of warm-up exercises before training or competition. To make the body from static step by step into the working state, the main purpose is into the motion state as soon as possible. There are some mainly unreasonable aspects as following: no exercises, inadequate preparation, too much activities and untargeted preparation activities.

3.1.2 Partial overload
Partial overload means that the training load is beyond the physiological limits of athletes. The athletes' physical qualities are different. Then, the arrangement of the training load should vary from person to person. Some coaches constantly increase the exercise load in order to make athletes get good scores in matches. This makes some athletes "wear out". The abitual dislocation is the result of the typical partial overload.

3.1.3 Ignorance of the injure
Sanda coaches and athletes must learn the theory of sports injury in training. To take all kinds of auxiliary method in training, and improve the power of the vulnerable parts, is meaningful to reduce the damage of the vulnerable parts.

3.1.4 Nonstandard technical actions
Due to technical shortcomings and errors lead to the technique is not standard. These errors violate the law of human activity, and the mechanics principle caused tissue damage. To streng the training of the standard technology In the sanda sports training. In order to achieve action automation technology, thousand of repeat is necessary.

3.1.5 Inattention
Inattention is one of the important cause to get injury. Sanda is competitive sport with fierce against. Then requiring the athletes always alert, highly concentrated, and observing the opponent's every move in the game. Not taken, there will be a knockout or wounded. Sanda athletes must streng the psychological quality improving the attention, reducing the occurrence of accidents.

3.2 External cause

3.2.1 Gear
Accoding to sanda competition rules that athletes must wear helmet, chest protector and supporter below, shields, knuckles and strap. Removing shin pads and spats back. Guarantee the security of the game. However commercial game removes helmet, chest protector, which increased the risk of injury.

3.2.2 Ring
There is no ropes competition games, and 2m wide protection pad that ensure the safety when athletes fall off ring. Commercial games borrowed with boxing ring or narrow its original site area. Increasing and improving the intensity of the games. There's no doubt the risk of injury has increased.

3.2.3 The rules
Some organisers of commercial games consensus rules, in order to increase the atmosphere of the game, regardless the security of the athletes. Let go of the rules of the disable methods (such as knee) lead directly to the participants the personal safety. Some games put different weight level athletes play together, such as the "sanda" and "kung fu king". This violate the rules of competition seriously, also brings the game a higher risk.

3.3 Strengthen physical training and enhance physical reserves

Take 2 out of 3 sets of sanda competition, each game for 2 minutes, 1 minutes rest between innings; Some commercial games require played three or five innings, combined with the characteristics of sanda movement against fierce, the athlete's physical stamina is undoubtedly a great challenge. If lack of athlete's physical ability, even if the technique, tactical use of flexible and skilled, will also affect the results of the competition, and even deadly accidents. So athletes in the usual training to strengthen the training of the physical quality, strengthen energy reserves.

3.4 Increase the ability of on-the-spot referees

Sanda competition, the referee staff is mainly composed of the deputy director, the director, the stage the referee, edge of the referee. Including the director, in the case of security is mainly responsible for field athletes, winning, to step down, forced seconds, etc., on the smooth progress of the referee for the game plays an important role, he is not only to promote athletes play in a positive way, and to protect the athlete's personal safety. This requires that the director and the stage referees should keep a clear head and observant and watch to see if there is damage accidents, and dealt with in a timely manner.

3.5 Set the rules of the contest, we must give full consideration to the personal safety of athletes

Sanda commercial competition activities, on the one hand, to expand the social influence of sanda, on the other hand also makes the athletes have more opportunity to exercise at the same time, also certain degree to bring economic benefits. But make the rules of the game of the business game, must give top priority to the athletes' personal safety, to give full consideration to the athletes' physical load demand, resilient force withstand the competition, whether there is to enable the methods of offensive and defensive training in preparation for all of these is to ensure safety of basic conditions of the competition.

4 HOW TO AVOID THE HAPPENING OF SANDA INJURY ACCIDENT

"Light sport", also known as "happy sports" or "sports" entertainment, is along with the continuous development of social economy and people living standard rise, to the leisure, entertainment as the goal of happy, relaxed sports of the new situation. Sports is an important part of social culture, due to the diversity of social culture form, the existence of competitive sports also need a variety of forms. The introduction of "light sport", opened up a new path to the development of competitive sports. For example, futsal, soft tennis, beach volleyball, and so on, these are light reflected the idea of sport. The concept of "light sport" is the "soft" development of competitive sports, the simplified technique, reduce the exercise load, location equipment, the rules of the simplification, etc.

In the rapid development of modern society, compared with other competitive projects, sanda movement must be diversified road, draw lessons from the development mode of Korea taekwondo, absorb the light sports idea, promote the public participation, the increasing population of sanda, sanda can continuously promote and popularize, this is the need of the development of sanda sport itself, is also the needs of the development of modern civilization.

Throughout the development of sanda movement, due to the influence of traditional ideas and media publicity of sanda business events, causing a lot of people think sanda is fierce USES harsh, hindered the popularization and promotion of the sport. At present, our country the development of sanda project basically is given priority to with professional development, the public school sanda sanda and not ideal.

In public school sanda sanda or the activity, can be used to shorten the game time, prohibit struck head, wear more gear, nudges, etc., thus decrease the difficulty of the game at the same time improve the practitioners of the safety factor. Practitioners, to guarantee the personal safety of reduces the fear psychology, will no doubt attract more people to attend in sanda sports activities.

4.1 Stepping up efforts in supervising the medical supervision

What medical supervision is monitor the athletes' health in training and matches by applicating medical knowledge and technology, to prevent and treat sports injury, to ensure athletes keep healthy physical condition in training and games, all aim to improve the performance. Therefore, the state general administration of sports wushu sports management center was established in September 2012, "national sanda medical supervision committee", and develop the wushu sanda medical supervision and regulation. But for now, medical institutions and personnel is scattered. The implementation of the supervision and regulation is not enough in place, and there is no set up independent medical monitoring system. The author thinks that improving the security of sanda competition must increase the intensity of sanda medical supervision and regulation. Registering the athletes' health databases recently, recording players' injured in the game. Checking the qualification in accordance with the regulations, and improving the athletes' and coaches', safety consciousness. These are the most effective measures to avoid sports injury accident.

4.2 Raise the level of training and tactics

Not only including the correct specification sanda technique, to improve the level of sanda sports training, arranging reasonable exercise. Pay attention to recover training after injured. Arranging the offensive and defensive technology training of the proportion is more important. To improve the athletes' defensive ability and consciousness of attack and defense conversion, only in this way can make the athletes in the match do advances freely, and to reduce the probability of hit on the KO in the game.

4.3 Strengthen physical training and enhance physical reserves

Take 2 out of 3 sets of sanda competition, each game for 2 minutes, 1 minutes rest between innings; Some commercial games require played three or five innings, combined with the characteristics of sanda movement against fierce, the athlete's physical stamina is undoubtedly a great challenge. If lack of athlete's physical ability, even if the technique, tactical use of flexible and skilled, will also affect the results of the competition, and even deadly accidents. So athletes in the usual training to strengthen the training of the physical quality, strengthen energy reserves.

4.4 Increase the ability of on-the-spot referees

Sanda competition, the referee staff is mainly composed of the deputy director, the director, the stage the referee, edge of the referee. Including the director, in the case of security is mainly responsible for field athletes, winning, to step down, forced seconds, etc., on the smooth progress of the referee for the game plays an important role, he is not only to promote athletes play in a positive way, and to protect the athlete's personal safety. This requires that the director and the stage referees should keep a clear head and observant and watch to see if there is damage accidents, and dealt with in a timely manner.

4.5 *Set the rules of the contest, we must give full consideration to the personal safety of athletes*

Sanda commercial competition activities, on the one hand, to expand the social influence of sanda, on the other hand also makes the athletes have more opportunity to exercise at the same time, also certain degree to bring economic benefits. But make the rules of the game of the business game, must give top priority to the athletes' personal safety, to give full consideration to the athletes' physical load demand, resilient force withstand the competition, whether there is to disable the methods of offensive and defensive training in preparation for all of these is to ensure safety of basic conditions of the competition.

5 THE CONCEPT OF "LIGHT SPORT" REVELATION TO THE SANDA DEVELOPMENT

"Light sport", also known as "happy sports" or "sports" entertainment, is along with the continuous development of social economy and people living standard rise, to the leisure, entertainment as the goal of happy, relaxed sports of the new situation. Sports is an important part of social culture, due to the diversity of social culture form, the existence of competitive sports also need a variety of forms. The introduction of "light sport", opened up a new path to the development of competitive sports. For example, futsal, soft tennis, beach volleyball, and so on, these are light reflected the idea of sport. The concept of "light sport" is the "soft" development of competitive sports, the simplified technique, reduce the exercise load, location equipment, the rules of the simplification, etc.

In the rapid development of modern society, compared with other competitive projects, sanda movement must be diversified road, draw lessons from the development mode of Korea taekwondo, absorb the light sports idea, promote the public participation, the increasing population of sanda, sanda can continuously promote and popularize, this is the need of the development of sanda sport itself, is also the needs of the development of modern civilization.

Throughout the development of sanda movement, due to the influence of traditional ideas and media publicity of sanda business events, causing a lot of people think sanda is fierce USES harsh, hindered the popularization and promotion of the sport. At present, our country the development of sanda project basically is given priority to with professional development, the public school sanda sanda and not ideal.

In public school sanda sanda or the activity, can be used to shorten the game time, prohibit struck head, wear more gear, nudges, etc., thus decrease the difficulty of the game at the same time improve the practitioners of the safety factor. Practitioners, to guarantee the personal safety of reduces the fear psychology, will no doubt attract more people to attend in sanda sports activities.

6 CONCLUSION

Safety is the first condition to sustainable development of sanda sports and competitive sports games or commercial competition, the revision sanda competition rules should be around security, including gear wear, use methods, disable methods, such as game innings key provisions are particularly important. Strengthen health supervision, improve the level of referees of board of cutting, strengthen the safety of the athletes, coaches, education, is the guarantee of athletic sanda security; The development of sanda, not only need to competitive sports arena, you also need to national fitness, the concept of "light sport" enlightenment, sanda sports and pointed out the direction of the promotion.

REFERENCES

Pi-xiang Qiu, etc. The Chinese wushu tutorial (part ii) [M]. Beijing: people's sport publishing house, 2004.

The Chinese wushu association. Wushu sanda competition rules and the referee law [M]. Beijing: people's sport publishing house, 2013.

Rui-qi Zhu, highlighting. Athletic sanda, light sport sanshou teaching reform – school of rational thinking [J]. Journal of guangzhou sports institute, 2006 (6).

Mai-jiu Tian, sports training word solution [M]. Beijing: Beijing sport university press, 2002.

Http://Baike.Baidu.Com/

Biomedical engineering

Biomedical Engineering and Environmental Engineering – Chan (Ed.)
© 2015 Taylor & Francis Group, London, ISBN: 978-1-138-02805-0

A study into the mechanics of the ossicular network in a starfish

Dongce Ma, Xi Zhang & Yuanbo Li
School of Engineering and Materials Science, Queen Mary University of London, UK

ABSTRACT: The starfish body wall has attracted the focus of biologists and material researchers as it has a variability of mechanical properties. People find that this is mainly contributed by the Mutable Connective Tissue (MCT) in the starfish body wall. The anatomical structure of starfish has been already researched clearly in the field of biology. In this report, the project of measuring the mechanical properties of starfish ossicular networks has been summarized in detail. In this project nanoindentation as a method was used to test the micromechanics and was combined with qBEI and EDX to analyse the variations in the chemical compositions of different ossicle regions. The mechanisms of the research methods were summarized and the experimental results of elastic modulus and magnesium and calcium content in the mineralised tissue were analysed.

1 INTRODUCTION

This project investigated the structural/mechanical properties and control mechanisms of a remarkable smart biomaterial-mutable connective tissue in the body walls of starfish. A long-term goal mentioned in the application of the project will be to utilize the mechanistic understanding of this biological system to inspire the design of novel dynamic synthetic materials, with potential applications in medicine, which would contain many potentials. If the mechanism of how the peptides control elasticity is determined, then we might be able to apply this new research into introducing various other applications using similar mechanisms as from the peptides in sea cucumbers and urchins, to reduce wrinkles by increasing the stiffness and softness of the collagen in our skins in the hope of looking younger and healthier (Queen Mary University of London, 2012).

2 MATERIALS AND METHODS

2.1 *Sample preparation*

The common starfish (Asteriasrubens) were cultivated in artificial seawater in an aquarium at Queen Mary University of London. One of them about 6 cm in diameter was selected as an experimental object. Most of the biological samples are soft and contain large amount of moisture and the sample will be placed in the vacuum system of a Scanning Electron Microscope (SEM). To avoid any failure and deformation during the test, the sample needs to be dehydrated and embedded in a medium. The intact starfish was well embedded in Polymethyl Methacrylate (PMMA) resin,

which is a transparent polymer material with considerable toughness and will not be broken up or dissolved in the SEM. One of the arms was cut from the bulk of the starfish sample using a power-driven saw. A sample of the tip of starfish arm was cut in the same way. Then the cut part was polished using a polishing machine with a turntable. The materials for polishing were Silicon carbide (SiC) abrasive paper with decreased grain size(median particle diameter of 6.5 μm, 2.5 μm) and diamond solution (less than 1 μm). A cuboid sample was produced which has a smooth surface of the cross section of the starfish arm tip and has been carefully preserved in a clear plastic sample container to avoid scratches of the polished surface.

2.2 *Nanoindentation tests*

Nanoindentation technology provides a method of testing the mechanical properties of a material by using a computer to control and apply continuous changing load and monitor the depth of the penetration in the material. The depth of nano-scale can be obtained since the ultra-low load was applied and the monitoring sensor has a resolution in displacement of less than 1 nm. A prescribed load was applied to a circular conical type diamond indenter which contacted a cross section surface of the sample. In this project, for Area-1 and Area-2, the maximum load was prescribed as 3,000 mN and for Area-3, Area-4 and Area-5 was 2,000 mN. The depth of penetration into the sample was recorded as the load was applied to the indenter (Figure 1).

On the basis of Oliver and Pharr's theory (1992), introduced by Poon (2008), the mechanical properties of a material can be derived from the measure load-displacement and loading/unloading curves through appropriate dataanalysis. The load and displacement

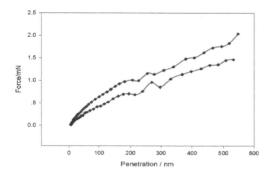

Figure 1. A typical nanoindentation curve graph of the 81th in the 100 points at Area-5. The corrected data were recorded and the loading curve(upper) and unloading curve(lower) were obtained.

for a conical indenter was derived by Sneddon (1948) as

$$p = \frac{2E_s \cdot \tan\alpha}{\pi(1-v_s^2)} h^2 \quad (1)$$

where p is the load measured by the indenter, E_s and V_s are the Young's modulus and Poisson's ratio of the indented material, α is the half angle of the indenter tip, and h is the displacement of the indenter. And further derived by Fischer-Cripps (2004) as

$$\frac{dP}{dh} = \frac{2\sqrt{A_c}E_s}{\sqrt{\pi}(1-v_s^2)} \quad (2)$$

A_c is the projected contact area of the indenter.

The contact depth h_c is defined as

$$h_c = h - \frac{2(\pi-2)}{\pi} \cdot \frac{P}{dP/dh} \quad (3)$$

By formula (1), (2) and (3), the projected contact area at this load can be obtained:

$$A_c = \pi \tan^2\alpha \cdot h_c^2 \quad (4)$$

A_c is a function related to h_c and $\tan 2\alpha$. h_c are recorded by computer. As defined, the hardness is written as follows

$$H = \frac{P_{max}}{A_c} \quad (5)$$

P_{max} is the maximum load applied during the indentation. The reduced modulus E_r can be gotten from

$$E_r = \frac{\sqrt{2\pi}}{2\beta} \cdot \frac{S}{\sqrt{A_c}} \quad (6)$$

β is a non-dimensional correction factor to account for deviations from the original stiffness equation (the value of β is 1 here), S is the experimentally measured stiffness of the unloading data.

E_r is a physical quantity combined with the sample indentation modulus E_s and the indenter tip modulus E_i, given by

$$\frac{1}{E_r} = \frac{1-v_s^2}{E_s} + \frac{1-v_i^2}{E_i} \quad (7)$$

v_s and v_i are the Poisson ratios of the sample and the indenter respectively. The value of the Poisson ratios $v_s = 0.20$ and $v_i = 0.07$, and the elastic modulus of the indenter $E_i = 1150\,GPa$ were used here. For each contacted target point, the processes of loading and unloading were carried out alternatively since the indenter contacting the sample surface. The time interval between each loading and the previous unloading was 0.2 s. The indenter was maintained at the maximum load for 2 min to eliminate the effect of residual creep in the tissue.

Area-1:

An area on the right ambulacral skeleton and under the third ossicle notch was selected to be tested firstly, which was named as Area-1. A prescribed maximum load of 3000 mN was applied to the indenter in this region. The micro measurement was performed under the pattern that the landed points were set to make up a 7×7 matrix, which means there would be 49 testing points landing on Area-1, and the interval was specified as 15 μm. The maximum load was set as 3000 mN.

Area-2:

In order to explore the difference of mechanical property (elastic modulus) In deferent regions of the starfish ossicles, other five regions were also be tested. The second region (Area-2) selected to be measured was the carinal ossicle, which is on the top position in the starfish ray. The measurement was performed under the same setting as Area-1 except that a 9×9 matrix was specified instead of 7×7. The maximum load was 3000 mN.

Area-3:

A relatively large ossicle of the right marginal ossicles was selected as the third region to be tested. The setting was adjusted as a 10×10 matrix and an interval of 10 μm. The maximum load was changed to 2000 mN.

Area-4:

The outermost ossicle of the left reticulum ossicles was the fourth region marked as Area-4. The 10×10 matrix and interval of 10 μm were still applied in this test. The maximum load was not changed.

Area-5:

The region at the lower right position, under the fourth ossicle notch of the right ambulacral skeleton was selected as the fifth and last area, marked Area-5. The setting and specifications were not changed from Area-4.

2.3 Chemical composition analysis: qBEI/EDX

A thin carbon layer was coated on the surface on the sample by Vacuum Carbon Evaporation(CED 030

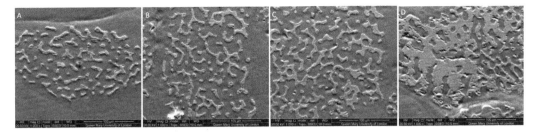

Figure 2. The 3D morphology of the surface of the starfish sample (A) SE Image of Area-2 (carinal ossicle). (B), (C) and (D) the same regions.

Balzers, Liechtenstein) for qBEI and EDX analysis. The coated sample was placed into the chamber of SEM, which was operated at an acceleration voltage of 20.00 KV. Then it was observed in the vacuum environment and imaged at 200× magnifications. In principle, the Quantitative Backscattered Electron Imaging(qBEI) is based on the amount of electrons backscattered from a thin surface layer (Roschger, et al. 2003) and is proportion alto the weight concentration of the minerals (calcium and magnesium) within the selected ossicle. The intension of backscattered electron signals, which is reflected as the grey level, was calibrated using the "atomic number (Z) contrast" (Roschger, et al. 1998). The average ZZ_{mean}, of a compound can be calculated using the formula given by Roschger, et al. (1998):

$$Z_{mean} = \frac{\Sigma(N_i \times A_i \times Z_i)}{\Sigma(N_i \times Z_i)} \qquad (8)$$

where N_i is the number of the ith atom, A_i is the relative atomic mass of the ith atom and Z_i is the atomic number of the ith atom (Lloyd, 1987).

In the common operation, carbon (C, $Z = 6$) and aluminium (A_l, $Z = 13$) with specific backscattered electron(BE) grey level are selected for calibration (Roschger, 1998 and Fratzl-Zelman, et al. 2009). The energy dispersive X-ray spectroscopy(EDX) is a chemical microanalysis technique used conjunction with Scanning Electron Microscope (SEM) (Shrivastava, 2003). It is based onthe fact that the characteristic x-ray emitted from the sample surface where the SEM electron beam with high energy bombardment can be detected by a $S_i(L_i)$ detector (Johansson, 1988) to be observedand confirmed. The characteristic X-rays are generated when the electron beam with high energy interact with the shell electrons of the sample atoms. The result is that an inner shell electron is ejected and the removal of this electron ionizes the atom temporarily until an outer shell electron drops into the vacancy to make the atom stable. Different elements will release unique series of X-rays, which are the characteristic X-ray (Bozzola and Russell, 1999). Based on the classes of the characteristic X-rays, the elemental composition of the selected volume is also identified. The phases as small as 1 μm or less can be analysed. The elemental energy X-ray analysis was operated at 10 kV and in a working distance

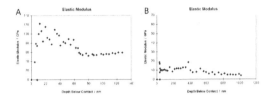

Figure 3. Micro-indentations on tissue. (A) The indenter landed on the hard tissue of Area-1. (B) Indenter contacted on a point of soft tissue or embedding medium.

of 10 mm. In this project, the chemical formula of the mineral in the ossicles can be expressed as the form of $Ca_xMg_yCO_3$ as the content of magnesium is much less than calcium, where $x + y = 1$. The value of x and y can be calculated by the ratio of calcium and magnesium from the data of EDX, and the mineral concentration of these 3 regions are also obtained from the EDX reports. The mass fraction of mineral ($Ca_xMg_yCO_3$) in the 3 regions were calculated through the mass fraction of calcium and magnesium measured by EDX in a digital environmental scanning electron microscope with a four-quadrant back-scattered electron detector. The fractures may occur in vivo and be in regeneration if the smooth trabeculae could be observed by a SEM (Moureaux, et al. 2011). The secondary electron (SE) image is usually used to characterize the morphology of the sample surface. So the SE images were also taken in this project which revealed the 3D topologies of the ossicles (Figure 2).

3 RESULTS AND DISCUSSION

3.1 Nanoindentation

For each indentation, a series data of h_c (abscissa axis) and elastic modulus (vertical axis) can be obtained. According to these data, a scatter diagram which like shown in Figure 3.

Combining each scatter diagrams of the same area (Area-1, e.g.) together in one diagram, we got the scatter diagram of one region we measured (Figure 4A). From the diagram in Figure 4A, the substances with lower elastic modulus of less than 20 GPa could be identified clearly, which are the organic tissue filled in the pores in ossicles and embedding medium

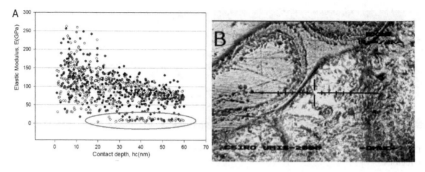

Figure 4. (A) The scatter diagram of Area-1. (B) A screenshot image of one point on which indenter landed (the crosshair in the centre) through a light microscope.

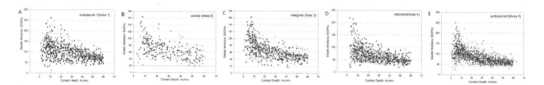

Figure 5. Scatter diagrams of the five regions.

(PMMA). Observed and compared with the corresponding scatter diagram through a monitor connected to a light microscope combined with the nanoindentation point by point, the points where the indentor landed with lower modulus less than 25 GPa were soft tissues or embedding medium; the points with higher modulus greater than 40 GPa were hard mineral tissue; the rest of the points with modulus between 25 and 40 GPa were generated as the indenter landed on the edges of hard tissue and soft tissue. The deepest depth of hc in soft tissue could reach 550–1200 nanometres (nm), and in hard tissue could be 80–180 nm. In this research, what we really focused on are the mechanical properties of the hard mineralized tissue. So the data series with low modulus were eliminated and we got the scatter diagrams with only hard tissue of all the five tested regions (Figure 5). The data are denser in Area-1(A) and Area-5(E) because the mineralized parts are expansive and massive in the ambulacral skeletons and the indenter has a larger probability landing on the hard tissue. On the contrary, the points in Area-2 show a sparser distribution since the carinal ossicle has more pores and the cross section is made up of bitty patches and a large area of organic parts. Only 12 in the 81 points were landed on the hard parts. According to these scatter diagrams, we find that the Young's modulus has a significant trend of decline in every areas we measured. In order to further verify this trend, the histograms with error bars of Standard Deviation (SD) were drawn for them (Figure 6).

When the contact depth h_c reached some value, about 40 nm, the modulus had a trend to be constant. The tops (mean values) of all the columns in each area were connected together into a smoothed curve in a graph (Figure 7), which showed the deference in elastic modulus among the deferent regions.

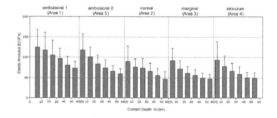

Figure 6. E-hc histogram with error bar of the five tested regions.

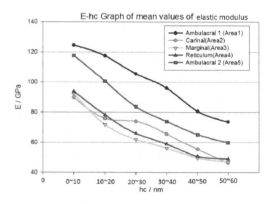

Figure 7. The mean value of elastic modulus (E,GPa) decreases with the contact depth(chin).

3.2 Chemical composition analysis

Four elements were measured above the detection limit in the ossicles: O, C, Ca, Mg (in increasing atomic abundance order) were shown in Table 1.

Table 1. The values of x, y and Z_{mean}.

	x	y	Chemical formular	Z_{mean}
Area-3	0.918	0.081	$Ca_{0.918}Mg_{0.081}CO_3$	12.3097
Area-4	0.908	0.092	$Ca_{0.908}Mg_{0.092}CO_3$	12.2744
Area-5	0.894	0.106	$Ca_{0.894}Mg_{0.106}CO_3$	12.2288

Table 2. The porosities of all ossicles.

	Min (%)	Max (%)	Means	Porosity
Carinal				79.9
Area-2				79.9
Reticulum ossicles	68.8	75.1	73.6	
Area-4				71.3
Marginal ossicles	72.5	76.6	74.6	
Area-3				76.6
Adambulacral ossicles	69.4	70.2	69.8	
Ambulacral ossicles	50.4	70.5	61.0	
Area-1				53.5
Area-5				59.5

Using Image J, the mean grey level of the BE images of five indented regions canbe obtained under the same grey level calibration. The result is Area-2 (117.97) > Area-4 (105.68) > Area-3 (104) > Area-5 (102.67 > Area-1 (101.08) (carinal > reticulum > marginal > ambulacral).

3.3 Porosity

According to the SEM images, it showed that the ossicles of the starfish contain many small pores (Moureaux, et al. 2011). The function of these pores may be similar to the bones of humans. They are necessary for hard tissue (ossicles) formation because 'they allow migration and proliferation of osteoblasts and mesenchymalcells' (Karageorgious, 2005).

The porosity, ϕ, is defined as the percentage of void space in a solidmaterial (Karageorgious, 2005), which can be written as

$$\varphi = \frac{V_p}{V_B} \times 100\% \qquad (9)$$

where V_p is the volume of pores and other void space in the material, V_B is the bulk volume of the material. Porosity can be calculate indirectly by the dense degree, D

$$\varphi = 100\% - D = \left(1 - \frac{V_s}{V_B}\right) \times 100 \qquad (10)$$

where V_s is the volume of solid (mineral part in this report) in the material. In some narrower sense, porosity is a morphological property of a porous material such as bones in humans and ossicles in starfish. The Young's modulus of bone is well known to be reduced by porosity (Currey, 1988). Higher porosity decreased mechanical properties of porous material. A measurement of estimating the porosity of the ossicles was made and the results are listedin the Table 2.

According to Table 2, we can obtain the order of the porosity of ossicles roughly, from large to small, which is carinal > marginal > reticulum > ambulacral. Corresponding the specific areas we measured previously, the order can be expressed as Area-2 > Area-3 > Area-4 >Area-5 > Area-1. In some local areas of the ambulacral ossicles, the porosity was even lower at 44.4%.

4 CONCLUSIONS

This research summarized the measurement of the mechanics of the ossicular network in star fish using nanoindentation and qBEI/EDX, containing both theoretical and experimental respects. The elastic modulus of mineralized hard tissue was discussed relatively deeply and the chemical analysis could be improved and many problems were exposed.

In conclusion, from Figures 3.5 and Table 3.1, the elastic modulus of ambulacral ossicle (Area-1 and Area-5) is significantly larger than the other three regions. There is no obvious difference in carinal, marginal and reticulum ossicles. The strongest parts in the ossicles, the ambulacralossicles, comprise almost all the weight of the starfish body. Their special properties still need be further studied.

Having obvious advantages in Mg/Ca ratio and porosity, ambulacral ossicles have the largest value in elastic modulus reflected in mechanical properties. However, effected by the Mg/Caratio, mineral concentration and porosity, the other three regions showed relatively complex results and behaviour.

The researches based on the foundation of previous studies, more experimental facts are being discovered. Researches into the mechanics of ossicular networks in starfish can provide opportunities for further exploration into the body walls of starfish. The organic tissue and inorganic tissue can perform different properties in the behaviour of the of body wall, which also make our understanding of the behaviour of the whole body wall, the mutable connective tissue, and the ossicular network much clearer.

REFERENCES

Bozzola, J.J., Russell, L.D. Electron Microscopy: Principles and Techniques for Biologists. Jones & Bartlett Learning. 1999; pp 374.

Curry, J.D. The effect of porosity and mineral content on the Young's modulus of elasticity of compact bone. Journal of Biomechanics. 1988; 21(2), 131–139.

Fratzl-Zelman, N., Roschger, P., Gourrier, A. Combination of nanoindentation and quantitative backscattered electron imaging revealed altered bone material properties associated with femoral neck fragility. Calcified Tissue International. 2009; 85(4), 335–343.

Johansson, S.A.E. PIXE: A novel technique for elemental analysis. John Wiley & Sons Ltd. 1988; pp 16.

Karageorgiou, V. Kaplan, D. Porosity of 3D biomaterial scaffolds and osteogenesis. *Biomaterials.* 2005; 26(2005), 5474–5491.

Lloyd, G.E. Atomic number and crystallographic contrast images with SEM:A review of backscattered electron techniques. *Mineralogical Magazine* 1987; 15(359), 3–19.

Moureaux, C., Simon, J., Mannaerts, G. Effects of field contamination by metals (Cd,Cu, Pb, Zn) on biometry and mechanics of echinoderm ossicles. *Aquatic Toxicology* 2011; 105(3–4), 698–707.

Oliver, W.C., Pharr, G. Man improved technique for determining hardness and elastic-modulus using load and displacement sensing indentation experiments. *Journal of Material Research*1992; 7(6), 1564–1583.

Poon,B., Rittel, D., Ravichandran, G. An analysis of nanoindentation in linearly elastic solids. *International Journal of Solids and Structures* 2008; 45(24), 6018–6033.

Queen Mary, University of London. 2012. Marine animals could hold key to looking young: Sea cucumbers, sea urchins can change elasticity of collagen. Science Daily. http://www.sciencedaily.com-/releases/2012/10/121001132150.htm

Roschger, P., Fratzl, P., Eschberger, J., Klaushofer, K. Validation of quantitative back scatted electron imaging for the measurement of mineral density distribution in human bonebiopsies. *Bone* 1998; 23(4), 319–326.

Roschger, P., Gupta, H.S., Berzlanovich, A., et al. Constant mineralization density distribution in cancellous human bone. *Bone* 2003; 32(3), 316–323.

Shrivastava, S. Medical device materials: Proceedings from the materials and processes for medical devices conference. *ASM International* 2003; pp 89.

Sneddon, I.N. Boussinesq's problem for a rigidcone. Mathematical Proceedings of the Cambridge Philosophical Society, 1948; 44(4), 492–507.

Biomedical Engineering and Environmental Engineering – Chan (Ed.)
© 2015 Taylor & Francis Group, London, ISBN: 978-1-138-02805-0

Research progress of the algorithms for segmentation of retinal images using optical coherence tomography

Shuaijun Ding, Jinle Shi, Pan Yang, Lei Xiao & Jianming Hu
College of Physics and Electronic, Chong Qing Normal University, China

ABSTRACT: Optical coherence tomography (OCT) is a novel, non-destructive, non-invasive and high-resolution medical imaging technology, providing the micro-structural information of bio-tissue and blood flow. Being a powerful ocular imaging technique, OCT can obtain retinal cross-sectional images of retina. Segmentation of the layered structure of retina is contributive to the diagnosis and treatment of ophthalmic diseases. The basic retinal imaging modalities and image segmentation algorithms are presented in this paper. Also the advantages and disadvantages of segmentation algorithms of retinal images using Optical Coherence Tomography are analyzed. The development tendency of the algorithms for segmentation of retinal images using Optical Coherence Tomography is predicted in the end.

1 INTRODUCTION

Retina can be seen as the image plane of ophthalmic imaging system and receiver of the optical information. Anatomically, retina can be divided into 10 layers: retinal pigment epithelium (RPE), photoreceptor layer (PRL), external limiting membrane (ELM), outer nuclear layer (ONL), outer plexiform layer(OPL), inner nuclear layer (INL), inner plexiform layer (IPL), ganglion cell layer (GCL), nerve fiber layer (NFL) and inner limiting membrane (ILM). Henry Gray proposed the retinal anatomical structure as shown in figure 1. Obtaining the accurate structural and functional information of retina is important to the research of visual function and the diagnosis of ophthalmic diseases.

Moreover, the indispensable information of diagnosing the ophthalmic diseases was acquired, such as age-related macular degeneration (AMD), glaucoma, diabetic retinopathy (DR) and other ophthalmic diseases. More and more researchers devote themselves to processing the retinal images.

The ophthalmoscope invented by Hermann von Helmholtz in 1851 brought revolutionary breakthrough in the field of ophthalmology. Later, fluorescein angiography and the indocyanine green dye angiography technique made it possible to observe the patterns of blood flow clearly within the eye, getting a better understanding of the pathology of ophthalmic diseases. Jackman et al acquired the first in vivo imagining of human retina in 1886. Current retinal imaging modalities mainly contain adaptive optics fundus camera, ophthalmoscope, optical coherence tomography (OCT), fundus fluorescein angiography, etc. The traditional retinal imaging modalities can only obtain two-dimensional cross-sectional imaging of the retina. Then in 1991,OCT was proposed by Huang et al. OCT is a novel, noninvasive and nondestructive optical medical imaging modality, which can obtain three-dimensional images of the retina by using a reconstruction algorithm. Due to the advantages of three-dimensional imaging, OCT has been widely applied to clinical ophthalmology. The imaging modality is able to detect the fine structure of the retina within the organization, with the axial resolution unaffected by human parallax error, improving the early diagnosis of retinal diseases significantly, promoting better understanding of pathogenesis of ophthalmic diseases, and thus facilitating the monitoring and treatment of diseases.

Image segmentation is a common image preprocessing operation, which is a critical and challenging work. The purpose of image segmentation is to identify the physical objects, such as anatomical organs or

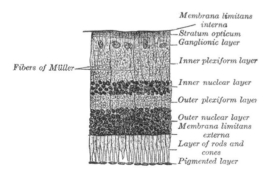

Figure 1. Retinal anatomical layer structure.

pathological entities, describe the objects and locate the position of target areas. Image segmentation is a technique by which images are partitioned into some specific and non-overlapping regions, regarding some unique characteristics, such as intensity or texture. There are many traditional image segmentation techniques in the literature, such as threshold processing, segmentation based on region and morphological watershed segmentation, etc. The layered structure information can be acquired when those algorithms are applied to segment the retinal images using OCT, providing an early prediction of retinal diseases and monitoring the development of retinal diseases. This paper summarizes the main algorithms for segmentation of retinal images using OCT.

2 ALGORITHMS FOR SEGMENTATION OF RETINAL IMAGE USING OCT

Researchers have proposed many algorithms for segmentation of retinal images using OCT in ophthalmology, which helps improve the application effect of OCT technology in clinical ophthalmology. The progress of OCT from the time domain to the frequency domain propels the research and treatment of retinal disease. However, when OCT is employed to get retinal images, there are many shortcomings as follows: (1) speckle noise of OCT reduces the image contrast, limits the applications of OCT in the diagnosis of ophthalmic diseases, decreases the quality of the retinal image, increases the complexity of the image analysis, and hinders the precise identification of the boundary layers and specific characteristics of retina; (2)retinal absorption and scattering effects of light decreases the imaging depth; (3) Hemoglobin in the retinal blood vessels has strong absorption of light, resulting in vascular shadow and low optical contrast. In order to segment the retinal image using OCT more accurately, and get more precise thickness, these shortcomings should be taken into account for segmentation algorithms.

The purpose of segmentation of retinal images is to determine the boundary layers of the retina and measure the thickness of the main cell layer, the inner retina and the pathological retina. Compared with the layered structure of normal retina, we can know better the pathogenesis of retinal diseases, and Zeimer et al proposed the macular thickness as the diagnostic criterion for glaucoma. It's the key process from image processing to image analysis, segmenting retinal image quickly and accurately and detecting more retinal cell layers. There are thousands of segmentation methods, no uniform one available. So it's necessary to design the image segmentation method according to priori knowledge. Segmentation approaches of retinal image depend on the number of features of the retina to be segmented, the most of which are based on gradient or intensity information. Segmentation approaches are classified into the following categories: image

segmentation based on boundary, image segmentation based on region, and image segmentation based on combined particular theories. This chapter summarizes the common segmentation algorithms for the retinal images using OCT.

2.1 Segmentation based on boundary

Detecting each pixel and its neighbors and determining whether the pixel is on the boundary, we can determine the boundaries of the image, and those pixels with desirable characteristics are the edge points. Image segmentation based on boundary is to utilize the discontinuity of pixel gray in different regions to find the edge points, then connect these edge points to form contour as thus to constitute divided region for image segmentation. Segmentation based on edge detection can be divided into serial edge detection and parallel edge detection in terms of processing sequence. The edge points of serial edge detection are determined by the validation results of the previous pixel, while the edge points of parallel edge detection depend on the pixels under detection and those adjacent pixels. Differential operator method and boundary searching method are traditional edge detection techniques. The new edge detection technologies include fitting detection method, iterative method, model method based on morphology, method based on fuzzy mathematics and so on.

In 1995, Hee et al for the first time utilized a 1-D edge detection approach for segmenting retinal images using OCT, which is the earliest edge detection algorithm based on the changes of intensity, getting precise thickness of total retina and RNFL. With a similar approach, Huang et al segmented the retinal images to measure the thickness of retina and outer retina choroid complex (ORCC), with animal and human retina as experimental subjects. Bagci et al designed a segmentation algorithm based on a 2-D edge detection approach, getting six retinal layer structures and their thickness. The approach not only inhibited the shot noise but also increased imaging depth, getting more accurate segmentation results. Given the spatial characteristics of the image, Gregori et al proposed a iterative edge detection algorithm, which locates automatically or interactively the complex geometry and topology of macular pathologies in the retinal image using TDOCT for the first time, able to precisely locate the boundaries of main anatomical layers in the internal retina. The previous quantitative analysis focused on the total thickness of retina and inner and outer retina. Ruggeri et al utilized a 3-D segmentation algorithm to detect the IPL and RPE boundaries in 2007 by means of an iterative process with initial assumption repeatedly evaluated, getting the quantitative layered structure. In 2007, Dai Qing, Sun Yan et al used an edge detection algorithm to segment the retinal images, making the gradient information of edge as the kernel, determining the layer boundaries by the shortest path searching and getting six retinal layer structures successfully.

The difficulty of segmentation approach based on edge lies in the conflict between noise immunity and detection accuracy. Pseudo-edge produced by noises leads to unreasonable contours while detection precision gets improved, and the improvement of noise immunity will result in contour missing inspection and position deviatio. On the other hand, the edge detection can't cope with the edge blur and subsistent noise of images. Given the noise of images, Gaussian function has been used in image smoothing filtering technique, which results in edge blur effect and gradual edges. With the expansion of the region and noise suppression, edge blur effect is enhanced and there are more noises. While edges are being detected, edge points use only local information of the image, so the connectivity of image edge and closure of target region get worse. Meanwhile there are some broken edge regions and more noises. Edge detection has the advantages of accurate positioning and fast computing speed, capable to get edge points of the image easily. However, edge detection is weak for high-precision retinal image segmentation. So edge detection must be used together with other algorithms.

2.2 Image segmentation based on region

Image segmentation based on region is an approach to search for target region directly, connecting the pixels with similar characteristics together to obtain the final segmentation region. There are two basic forms of image segmentation based on region, such as region growing and region splitting and merging. Region growing is a procedure to merge the pixel clusters or regions starting from a single pixel, in accordance with regional characteristics. Region splitting and merging is a procedure to split the image into various regions and then merge the regions with similar characteristics starting from the global, in accordance with the consistency of regional characteristics. Region growing, a semi-automatic segmentation algorithm, needs growing points and criterion. It is mainly used for the segmentations of retinal blood vessels. For example M. Elena et al used second-order directional derivative and region growing algorithm to segment retinal blood vessels. Cabrera et al utilized an algorithm based on gradient vector flow snake mode to segment retinal TDOCT images, reducing speckle noises through a nonlinear anisotropic diffusion filter. This approach segmented retinal images semi-automatically and determined the cystoids and sub-retinal fluid regions. Later, they took advantage of a peak search algorithm based on coherent information on the retinal structures to segment images automatically or semi-automatically, getting seven retinal layers.

Threshold segmentation is one of the earliest segmentation techniques based on region, as well as the most intuitive image segmentation technique. Images processed by threshold segmentation techniques use gray level to represent the characteristics of each pixel point. We set different thresholds depending on the differences of gray level between the target and the background of image, so that pixels are divided into several classes. The key of threshold segmentation is to determine an appropriate threshold, to determine each pixel of the image belonging to the target region or the background, thus resulting in a corresponding binary image, and extracting the target region from the background based on 1-D or 2-D gray histograms. When the threshold value is too high, too many target pixels are classified as the background region. When the threshold value is too low, pixels belonging to the background are too few. The threshold generally has three forms: global threshold, local threshold and dynamic threshold. Generally, there are many methods to determine global threshold value, such as peak analytics of grey histogram, minimum error method etc. Specifically, the algorithm based on threshold segmentation creates image partition with typical features, such as intensity or gradient. Threshold algorithms can also be divided into algorithms based on edge (Canny edge detection and Laplacian edge detection), algorithms based on region (region growing algorithm) and mixed algorithms (watershed algorithm).

In 2000, George et al got retinal images using TDOCT (Humphrey 2000 OCT) ,then segmented the retina and choroid capillary structure by using the dual-threshold, and performed median filtering and image homogenization using Nagao filter. In 2004, Herzog et al proposed an adaptive threshold algorithm based on edge maximization and smoothness constraints, and segmented the retinal images using TDOCT (Humphrey 3000 OCT), extracting the boundary between ONH and the retina automatically. Ishikawa et al used the histogram of reflectivity by axial scan to segment the retinal images by means of median filter and adaptive threshold method, getting four retinal layer boundaries. Similarly, Ahlers et al used an adaptive threshold technique and an intensity peak detection technique to segment the retinal SDOCT images of the patient with RPE detachments, eliminating the thinner glass membrane by filtering and getting the boundary between inner and outer retina precisely. Fabritius et al used a threshold value based on the signal strength to segment the retinal SDOCT image and determined the boundary between the inner and outer retina.

Threshold segmentation of retinal images has the advantages of simple calculation, high operational efficiency, and fast speed. Simple threshold values are sensitive to noises and intensity inhomogeneity of retinal OCT images, because it does not consider the spatial characteristics of the image. Segmentation algorithms based on region are different from threshold segmentation algorithms. Both the similarity of pixels and the adjacency of regions need to be taken into account in the former segmentation algorithms. So image segmentation based on region can eliminate noises effectively, having strong robustness and high segmentation accuracy. On the other hand, the image segmentation algorithms based on region overcomes the discontinuity of image segmentation space effectively by using the local spatial information, but the image may be over-segmented, damaging the boundaries of the region.

2.3 *Image segmentation based on particular theories*

There has not yet been a universal segmentation algorithm to perform image processing. Along with new theories and methods of every discipline, there have been some segmentation techniques based on combined particular theories and methods. Image segmentations based on combined particular theories include image segmentations based on mathematical statistics, image segmentations based on graph theory and support vector machine (SVM), dynamic programming, genetic algorithms and so on.

Markov Random Field (MRF) is a common statistical method and a conditional probability model, in which the probability of each pixel is only related to adjacent points. The segmentation technology of using MRF to segment retinal images has made a great progress in the past two decades. In 2001, Koozekanani et al introduced the MRF model, which has stronger robustness than the standard threshold algorithm, and then used a 1-D edge detection method to segment the retinal images using the OCT of $10\mu m$ resolution, getting the boundary between the inner and outer retina. In 2006, Boyer et al proposed a new methodology based on Markov model and parabolic model of the cup geometry, determining the boundary between choroid and ONH and getting the thickness of RNFL and ONH. In 2008, Shrinivasan et al used the modified algorithm based on Markov model to segment retinal images using UHR-OCT, making quantitative measurements of the outer retinal morphology and getting six retinal layer structures.

SVMs are typical supervised learning models, able to achieve the optimization of small samples, high dimension and nonlinear. In 2007, Fuller et al used a multi-resolution hierarchical SVM to segment 3-D retinal OCT images semi-automatically, measuring the thickness of PRL and the whole retina. Fuzzy C-means (FCM) is one of the most widely used fuzzy clustering algorithms. In 2008, Mayer et al used FCM to segment automatically retinal OCT images from five normal adults and seven patients with glaucoma, reducing shot noises with a 2-D mean filter and finally getting precise thickness of NFL.

Dynamic programming decomposes the problem into several sub-problems. we can get the solutions of the original problems, according to the solutions of sub-problems. In 2008, Tan et al proved the effects of glaucoma on the thickness of the inner retinal layers of the macular with a 2-D gradient approach using dynamic programming, segmenting automatically the retinal images into many layered structures. In 2009, Mishra et al proposed a modified active contour algorithm, which is based on sparse dynamic programming and two-step kernel optimization. Combined with a variety of particular theories, this approach

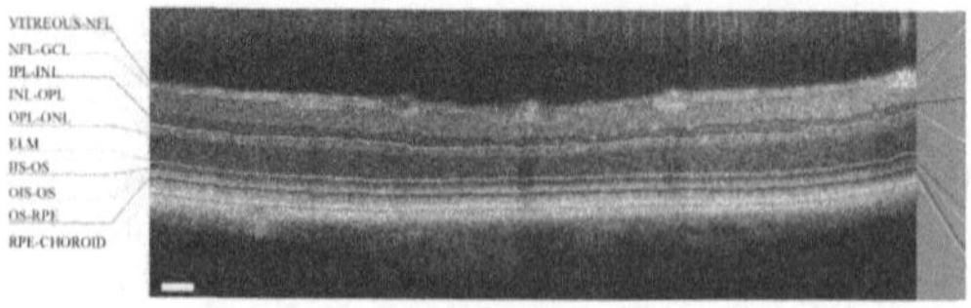

Figure 2. Ten retinal layer boundaries in a WT mouse SD-OCT B-scan.

can segment the images with low image contrast and irregular shaped structures. In 2010, Hu et al used a graph-theoretic methodology to segment RPE/Bruch membrane complex. In the same year, Chiu et al use an algorithm based on graph theory and dynamic programming to segment automatically the retinal data from 10 healthy subjects, getting seven retinal layers and shortening the time of image processing.

With the development of various particular theories, more and more image segmentation methods have been proposed. The combined utilization of traditional image segmentation methods and particular theories contributes to accurate retinal OCT image segmentations. In 2011, Mwanza et al obtained retinal OCT images from patients with glaucoma, and segmented the macular layers automatically with the ganglion cell analysis algorithm (GCA) built-in CIRRUS HD-OCT based on graph search. The methodology quantified the retinal thickness, measured the thickness of GCL and IPL complex, and assessed the reproducibility, promoting the diagnosis and monitoring of glaucoma. In 2014, Srinivasan et al proposed an automatic segmentation approach using sparse based denoising, SVM, graph theory and dynamic programming(S-GTDP) to segment retinal OCT images of mice. They used simultaneous denoising and interpolation (SBSDI) and block-matching and 3D filtering (BM3D) algorithm to denoise the retinal SDOCT images, and segmented the images according to segmentation theories, finally getting ten retinal layer boundaries as shown in figure 2.

With the particular theories applied in image segmentation technology, image segmentation technology is getting more perfect, which has the advantages of high accuracy, automation and better robustness and adaptability. Image segmentation technology combined with particular theories makes retinal OCT image segmentation automatic, accurate, rapid, and better adaptability and robustness. The technology further improves the performance, efficiency and versatility of segmentation algorithm, getting precise retinal layer structure. Along with the breakthroughs and innovations of new theories and new methods, retinal image segmentation technology has also made great success.

3 CONCLUDING REMARKS

OCT technology has made great progress during the past 30 years. Segmentation technology of retinal

OCT image has also made certain achievements. However, image segmentation based on edge detection may not be able to detect the best edges of the target layer in retinal OCT images. Image segmentation based on region may result in over-segmentation. Also the image segmentation technology based on particular theories also has its own limitations. Multi-mode and multi-method image segmentation algorithms are being developed by combining image segmentation algorithms with particular theories. These algorithms are used for the segmentation of retinal OCT images, achieving more accurate segmentation results, facilitating the further research of the clinical relationships between the changes of retinal structure and ophthalmology and non-ophthalmic diseases, and laying a solid foundation for the diagnosis and treatment of retinal diseases.

REFERENCES

Ahlers, C., Simader, C., Geitzenauer, W., Stock, G., Stetson, P., Dastmalchi, S & Schmidt-Erfurth, U.2008. Automatic segmentation in three-dimensional analysis of fibrovascular pigmentepithelial detachment using high-definition optical coherence tomography. *Br. J. Ophthalmol.* 92, 197–203.

Bagci, AM., Shahidi, M., Ansari, R., Blair, M., Blair, NP & Zelkha, R. 2008. Thickness profiles of retinal layers by optical coherence tomography image segmentation. *American journal of ophthalmology* 146(5): 679–687.

Bischoff, P.M. & Flower, R.W. 1985. Ten years experience with choroidal angiography using indocyanine green dye: a new routine examination or an epilogue. Documenta ophthalmologica 60(3): 235–291.

Boyer, K.L., Herzog, A. & Roberts, C. 2006. Automatic recovery of the optic nervehead geometry in optical coherence tomography. Medical Imaging, IEEE Transactions on 25(5): 553–570.

Cahan, D., Hermann von Helmholtz. 1993. the foundations of nineteenth-century science. Univ of California Press.

Chiu, SJ., Li, CT., Nicholas, P., Toth, CA., Izatt, JA, & Sina Farsiu. 2010. Automatic segmentation of seven retinal layers in SDOCT images congruent with expert manual segmentation. Optic Express. 18(18): 19413–19428.

Dai, Q & Sun, Y. 2011. Automated layer segmentation of optical coherence tomography images. in Biomedical Engineering and Informatics (BMEI), 2011 4th International Conference on. IEEE.

Drexler, W. & Fujimoto, J.G. 2008. Optical coherence tomography: technology and applications. Springer.

Faber, D.J., et al. 2003. Light absorption of (oxy-) hemoglobin assessed by spectroscopic optical coherence tomography. Optics letters 28(16): 1436–1438.

Fabritius, T., Makita, S., Miura, M., Myllyla, R. & Yasuno, Y. 2009. Automated segmentation of the macula by optical coherence tomography. Optics express 17(18): 15659–15669.

Fernandez, D.C. 2005. Delineating fluid-filled region boundaries in optical coherence tomography images of the retina. Medical Imaging, IEEE Transactions on 24(8): 929–945.

Fuller, AR., Zawadzki, RJ., Choi, S., Wiley, DF., Werner, JS. & Hamann, B. 2007. Segmentation of Threedimensional Retinal Image Data. IEEE Transactions on Visualization and Computer Graphics 13(6), 1719–1726.

Geeraets, W.J., et al. 1960. The loss of light energy in retina and choroid. Archives of ophthalmology 64(4): 606–615.

George, A., Dillenseger, JA., Weber, A. & Pechereau, A. 2000. Optical coherence tomography image processing.Investigative ophthalmology and visual science 41(pp): S173–S173.

Gonzalez, R.C. & Woods, R.E. 1992. Digital imaging processing. Massachusetts: Addison-Wesley.

Gonzalez, R.C. & Woods, R.E. 2002. Digital image processing second edition. Beijing: Publishing House of Electronics Industry.

Gray, H., C.M. Goss & D.M. Alvarado. 1973. Anatomy of the human body. Lea & Febiger Philadelphia.

Gregori, G. & Knighton, R. 2004. A robust algorithm for retinal thickness measurements using optical coherence tomography (Stratus OCT). Investigative Ophtalmology and Visual Science 45(5): 3007.

Group, D.R.S.R. 1981. Diabetic retinopathy study. Association for Research in Vision and Ophthalmology.

Hee, M.R. 1997. Optical Coherence Tomographyof the eye. Thesis (PhD). Massachusetts Institute of Technology, Source DAI-B 58/04. 1952.

Herzog, A., Boyer K.L. & Roberts, C. 2004. Robust extraction of the optic nerve head in optical coherence tomography, in Computer Vision and Mathematical Methods in Medical and Biomedical Image Analysis. Springer 395–407.

Hu, Z., Abramoff, MD., Kwon, YH., Lee, K., Garvin, M. 2010. Automated Segmentation of Neural Canal Opening and Optic Cup in 3-D Spectral Optical Coherence Tomography Volumes of the Optic Nerve Head. Invest Ophthalmol Vis Sci. Jun 16. [Epub ahead of print] PubMed PMID: 2055–4616.

Huang, D., et al. 1991. Optical coherence tomography. Science 254(5035): 1178–1181.

Huang, Y., et al. 1998. Relation of optical coherence tomography to microanatomy in normal and rd chickens. Investigative Ophthalmology & Visual Science 39(12): 2405–2416.

Ishikawa, I., Stein, DM., Wollstein, G., Beaton, S., Fujimoto, JG. & Schuman, JS. 2005. Macular Segmentation with Optical Coherence Tomography. Investigative Ophthalmology and Visual Science (IOVS) 46(6), 2012–2017.

Kittler, J., Illingworth, J. & Föglein, J. 1985. Threshold selection based on a simple image statistic. Computer vision, graphics, and image processing 30(2): 125–147.

Koozekanani, D., Boyer, K. & Roberts, C. 2001. Retinal thickness measurements from optical coherence tomography using a Markov boundary model. Medical Imaging, IEEE Transactions on 20(9): 900–916.

Liang, J., Williams, D.R. & Miller, D.T. 1997. Supernormal vision and high-resolution retinal imaging through adaptive optics. JOSA A 14(11): 2884–2892.

Lim, L.S., et al. 2012. Age-related macular degeneration. The Lancet. 379(9827): 1728–1738.

Martinez-Perez, M.E., et al. 1999.Segmentation of retinal blood vessels based on the second directional derivative and region growing. in Image Processing, 1999. ICIP 99. Proceedings. 1999 International Conference on. IEEE.

Mayer, MA., Tornow, RP., Bock, R., Hornegger, J. & Kruse, FE. 2008. Automatic Nerve Fiber Layer Segmentation and Geometry Correction on Spectral Domain OCT Images Using Fuzzy C-Means Clustering. Invest. Ophthalmol. Vis. Sci. 49: E-Abstract 1880.

Mishra, A., Wong, A., Bizheva, K. & Clausi, DA. 2009. Intraretinal layer segmentation in optical coherence tomography images. Opt. Express 17(26), 23719–23728.

Mwanza, J.-C., et al. 2011. Macular ganglion cell–inner plexiform layer: automated detection and thickness reproducibility with spectral domain–optical coherence tomography in glaucoma. Investigative ophthalmology & visual science 52(11): 8323–8329.

Novotny, H.R. & Alvis, D.L. 1961. A method of photographing fluorescence in circulating blood in the human retina. Circulation 24(1): 82–86.

Peiming Zhang, et al. 2009. Advances in high-resolution retinal imaging technology. *Laser & Optoelectronics Progress* 46(1): 30–36.

Roorda, A. 2000. Adaptive optics ophthalmoscopy. Journal of Refractive Surgery 16(5): S602–S607.

Ruggeri, M., et al. 2007. In vivo three-dimensional high-resolution imaging of rodent retina with spectral-domain optical coherence tomography. Investigative ophthalmology & visual science 48(4): 1808–1814.

Sahoo, P.K., Soltani, S. & Wong, A. 1988. A survey of thresholding techniques. Computer vision, graphics, and image processing 41(2):2 33–260.

Schmitt, J.M., Xiang, S. & Yung, K.M. 1999. Speckle in optical coherence tomography. Journal of biomedical optics 4(1): 95–105.

Shrinivasan, Y.B. & van Wijk, J.J. 2008. Supporting the analytical reasoning process in information visualization. in Proceedings of the SIGCHI Conference on Human Factors in Computing Systems. ACM.

Srinivasan, P.P., et al. 2014. Automatic segmentation of up to ten layer boundaries in SD-OCT images of the mouse retina with and without missing layers due to pathology. Biomedical optics express 5(2): 348–365.

Standring, S. 2008. Gray's anatomy. The anatomical basis of clinical practice, 2008. 39.

Tan, O., Li,. G., Lu, AT., Varma, R., Huang, D. 2008. Advanced Imaging for Glaucoma Study Group. Mapping of macular substructures with optical coherence tomography for glaucoma diagnosis. Ophthalmology. Jun 115(6): 949–56.

Tanuj, D. and I. Parul. 2011.Pearls in Glaucoma Therapy. JP Medical Ltd.

Wade, N.J. 2007. Image, eye, and retina (invited review). JOSA A 24(5): 1229–1249.

Xiang, S., L. Zhou & J.M. Schmitt. 1998. Speckle noise reduction for optical coherence tomography. in BiOS Europe'97. International Society for Optics and Photonics.

Xinzheng Xu, et al. 2010. New theories and new methods of image segmentation. *Acta Electronica Sinica* 38(2): 76–82.

Zeimer, R., et al. 1996. A new method for rapid mapping of the retinal thickness at the posterior pole. Investigative Ophthalmology & Visual Science 37(10): 1994–2001.

Biomedical Engineering and Environmental Engineering – Chan (Ed.)
© *2015 Taylor & Francis Group, London, ISBN: 978-1-138-02805-0*

Application of subcostal Transversus Abdominis Plane block in postoperative analgesia in patients undergoing gastrectomy

Kai Li, Li Yang, Longyun Li & Guoqing Zhao
Department of Anaesthesiology, China-Japan Union Hospital, Jilin University, Jilin, China

ABSTRACT: To evaluate the effect of subcostal Transversus Abdominis Plane (TAP) block on postoperative analgesia and short-term prognosis. Forty patients were divided into TM and M groups. Ultrasound-guided bilateral subcostal TAP block was performed by injection of ropivacaine or PBS using a patient-controlled intravenous analgesia pump. Postoperative hypoalgesia plane was T6–T10 in TM group. The postoperative time to press the PCIA pump was (175.85 ± 8.99) min and (64.10 ± 27.37) min in M group. Sufentanil consumption at postoperative 0–12 h and 12–24 h and total sufentanil consumption were significantly lower than M group. Patients had a significantly higher degree of satisfaction than the M group. Out of bed time was (30.28 ± 2.01) h and (39.55 ± 2.56) h and the first flatus time was (48.95 ± 3.83) h and (64.40 ± 4.58) h in M group. It can offer 6 h analgesia time, enhance analgesic effect within postoperative 24 h, decrease the dosage of opiod drugs and promote intestinal tract function and mobility of recovery.

Keywords: Ultrasonography transversus abdominis plane, nerve block, gastric cancer, postoperative analgesia

1 INTRODUCTION

For patients undergoing open radical gastrectomy, the analgesic effect of single intravenous injection was poor. Acute upper abdomen pain after surgery severely reduced the comfortableness and affected respiration, off-bed time and the first flatus time and even progressed into neuropathic pain. In 2001, Rafi *et al* proposed the concept of Transversus Abdominis Plane (TAP) block based upon dissection knowledge, which has been widely applied in the analgesia of lower abdominal operations. Hebbard *et al* suggested that subcostal TAP was efficacious for upper abdominal operations. Our preliminary experiment validated the good compound analgesic effect of subcostal TAP during surgery. However, whether subcostal TAP could effectively supplement the postoperative intravenous analgesia and affect the quality of postoperative recovery remains unstudied, which is the starting point of this study.

2 MATERIALS AND METHODS

2.1 General data

The experimental procedures were approved by Ethical Research Committee in our hospital. Forty patients with ASA(American Society of Anestheiologists) Grade I to II undergoing open radical gastrectomy between July 2013 and December 2013 were enrolled in this study, aged between 42 and 75 years, weighing between 48 and 75 kg. All participants were randomly assigned into a transversus abdominis plane group (TM group; n = 20) and general anaesthesia group (M group; n = 20). Patients with an allergy to topical anaesthesia drugs, abdominal trauma, mental abnormality, abnormal coagulation, communication disorders and difficulty in puncture were excluded from this clinical trial. All surgical procedures were completed by one single surgeon. Prior to the operation, all patients had normal abdominal walls. They were informed about the analgesia strategy and the use of an analgesia pump. Informed consents were obtained from all participants and their relatives.

2.2 Methods

2.2.1 Induction and maintenance of general anaesthesia

After entry into the operation room, patients' peripheral veins were cut open, ECG, blood pressure, carbon dioxide of the expiration tip, and pulse oxygen saturation were measured. Conventional anaesthesia induction was conducted by 0.02 mg/kg midazolam, 0.6 mg/kg rocuronium, 0.3 μg/kg sufentanil and 2.0 mg/kg propofol.

Mechanical ventilation was performed after tracheal intubation at a tidal volume ranging from 6 to 8 ml/kg. Intraoperatively, MAC was maintained between 1.2 and 1.5 by inhalation of sevoflurane, BIS value 45–60 and PetCO2 35–45 mmHg. Intraoperatively, analgesia and vascular active drugs were supplemented based on actual situations, and rocuronium was intermittently added. The addition of analgesia agents was discontinued at 1 h before the conclusion of surgery. Postoperatively, patients were transferred into the post-anaesthesia care unit.

2.2.2 Emergence from anaesthesia and implementation of postoperative analgesia

After transferring to the post-anaesthesia care unit, patients received mechanical ventilation to maintain the status of anaesthesia. The number was randomly assigned automatically by computers. The drugs, including 60 ml of 0.333% ropivacaine (TM group) or equivalent amount of PBS (M group), were prepared by the assistants who were not involved with the operation Ultrasound-guided subcostal transversus abdominis plane block was performed (Sonosite M-Turbo, probe 5–10 MHz), making the medication a fully formed continuous low intensity plane between the internal abdominal oblique muscle and the transversus abdominis. After the block, the process of emergence from anaesthesia was initiated.

After leaving the post-anaesthesia care unit, patients were connected with a disposable patient-controlled intravenous analgesia pump containing 3 μg/kg fentanyl and the maximal use time was 48 h.

2.3 Data record

Postoperative evaluation was conducted by one doctor who was blind to this study. The data recorded included: (1) The pain sensation of the abdominal wall at postoperative 1, 2, 4, 6, 12 and 24 h was measured by blunt needle methods. (2) Postoperative time to press patient-controlled intravenous analgesia pump and consumption of fentanyl at different time intervals. (3) The level of sedation was assessed by Ramsay grading: 1 point denotes irritation and anxiety; 2 points means good directional force and quiet cooperation; 3 points denotes drowsiness but can deliver correct instructions; 4 points represents sleep status but awakenability; 5 points denotes poor response to awakening; 6 points means deep sleep status and no reaction to awakening. The range of 2 to 4 points was defined as satisfactory sedation and 5–6 points as excessive sedation. (4) The first flatus time postoperatively was utilized to calculate the time of intestinal recovery. (5) Off-bed time. (6) Patients' overall degree of satisfaction to the analgesia plan was recorded after the analgesia treatment (expressed as a scale of 0–100). (7) The incidence of nausea, vomiting, respiratory depression and alternative adverse events were observed and recorded and effective treatment was delivered in a timely manner.

2.4 Statistical analysis

SPSS 19.0 statistical software package was utilized for data analysis. Normally distributed measurement data were expressed as means ± standard deviation. Non-normally distributed measurement data were denoted

Table 1. Comparison of general data in patients between two groups.

Study group	M group	TM group
Age (year)	58.55 ± 5.14	59.10 ± 6.93^a
Weight (kg)	66.05 ± 4.06	65.05 ± 3.37^a
Operation time (min)	163.80 ± 9.05	162.85 ± 9.33^a

$^aP > 0.05$ denotes compared with M group

Table 2. The plane of pain sensation loss within postoperative 6 h in the TM group.

Time	1 h	2 h	4 h	6 h
Plane	6.05 ± 0.58	7.10 ± 0.72	8.10 ± 0.64	9.75 ± 0.70

as the median (quartile). Measurement data were statistically analysed by the t-test. Enumeration data were assessed by the chi-square test. $P < 0.05$ was considered to be of statistical significance.

3 RESULTS

3.1 General situation

All 40 patients successfully underwent puncture. Postoperative observation and the management of pain sensation was conducted smoothly. The age, weight and operation time did not significantly differ between the two groups, as illustrated in Table 1.

3.2 The results revealed that patients in the M group had no abnormality in pain sensation of abdominal wall. The pain sensation plane in the TM group: the mean plane representing alleviation or loss of pain sensation within postoperative 1 h reached T6.05, T7.10 at 2 h, T8.10 at 4 h and T9.75 at 6 h. In the TM group, merely 7 patients had slight numbness in abdominal wall at postoperative 12 h, and showed no alleviation or loss of pain sensation compared with alternative sites and organs. The pain sensation in the abdominal walls of all patients had fully abated at postoperative 24 h, as shown in Table 2.

3.3 In the M group, the postoperative time to press a patient-controlled intravenous analgesia pump was (64.10 ± 27.37) min, significantly earlier than (175.85 ± 8.99) min in the TM group ($P < 0.05$). In the M group, the consumption of fentanyl at postoperative 0–12 h and 12–24 h was significantly larger compared with their counterparts in the TM group ($P < 0.05$), whereas no statistical significance was noted at postoperative 24–36 h and 36–48 h between two groups in terms of the consumption of fentanyl. The total consumption of fentanyl in the TM group was significantly less than that in the M group, as illustrated in Table 2.

3.4 Compared with the M group, the postoperative off-bed time in the TM group was (30.28 ± 2.01) h, significantly earlier than (39.55 ± 2.56) h ($P < 0.05$). The

Table 3. Comparison of postoperative fentanyl consumption at different time intervals between two groups (n = 20).

Group	M group	TM group
0–12 h	53.60 ± 6.68	27.30 ± 2.89[b]
12–24 h	37.25 ± 5.42	20.85 ± 3.54[b]
24–36 h	19.45 ± 2.01	16.30 ± 1.87[c]
36–48 h	12.85 ± 1.60	9.70 ± 1.30[c]
Total	123.15 ± 9.35	84.15 ± 4.44

[b]$P < 0.05$ and [c]$P > 0.05$ denote compared with M group

Table 4. The incidence of adverse events, status of postoperative recovery, and patients' degree of satisfaction.

Group	M group	TM group
First off-bed time	39.55 ± 2.56	30.28 ± 2.01[d]
First flatus time	64.40 ± 4.58	48.95 ± 3.83[d]
Degree of satisfaction to analgesia	77.40 ± 7.68	88.65 ± 4.08[d]
Sedation score	2.25 ± 0.97	2.35 ± 0.49[d]
Nausea n (%)	6 (30)	3 (15)
Vomiting n (%)	2 (10)	1 (5)

[d]$P < 0.05$ denotes compared with M group

first flatus time in the TM group was (48.95 ± 3.83) h after surgery, significantly earlier than (64.40 ± 4.58) h in the M group ($P < 0.05$). In addition, patients in the TM group had significantly higher degree of satisfaction towards postoperative analgesia (88.65 ± 4.08) compared with their counterparts in the M group (77.40 ± 7.68) with a statistical significance. The sedation scores between the two groups were within the range of satisfactory sedation (2.35 ± 0.49 for the TM group and 2.4 ± 0.8 for the M group) with no statistical significance. No statistical significance was observed between the two groups regarding the incidence of postoperative nausea, vomiting, or other clinical complications, as shown in Table 4.

4 DISCUSSION

4.1 Ultrasound-guided subcostal transversus abdominis plane block has been widely applied in China. However, the adoption of block plane remains debated. Barrington *et al* demonstrated that single and multiple injections of drugs generated different block planes. Multiple sites of injection could enlarge the spreading scale of medication and block more ganglion segments. Anja Ulrike Mitchel *et al* adopted ultrasound-guided subcostal transversus abdominis plane block in 20 conscious volunteers and the maximal area of pain sensation block reached T4-L4. In this study, ultrasound-guided subcostal transversus abdominis plane block with injection of 60 ml of 0.333% ropivacaine at multiple sites. The results revealed that

the mean plane of alleviation or loss of pain sensation within postoperative 1 h was T6.05, T7.10 within 2 h, T8.10 within 4 h, T9.75 within 6 h and the loss of pain sensation was observed until the 12 h whereas partial patients felt numbness in abdominal wall. At postoperative 24 h, all patients had a normal recovery of the abdominal wall. In this experiment, the middle incision was made during open radical gastrectomy from the xiphoid to 2 cm below umbilicus. Consequently, ultrasound-guided subcostal transversus abdominis plane block can offer 6 h analgesia time for such incision. McDonnell *et al* found similar findings that the pain sensation in the abdominal wall was gradually alleviated within 90 min after block and fully recovered at postoperative 24 h. In recent years, multi-mode analgesia has been recommended to achieve good analgesia and reduce the adverse events as possible. Subcostal transversus abdominis plane block yields good efficacy with stable block plane and target blocks the sensation nerves on the surface between the internal abdominal oblique muscle and an internal abdominal oblique muscle, thereby exerting an analgesia effect on abdominal incision. Compared with intravenous and epidural analgesia, it may reduce the incidence of adverse events, such as respiratory depression, urine retention, nausea, vomiting and postoperative intestine blockage, etc. Moreover, it could achieve better analgesia effect. It is supposed that this technique might exert good analgesia effect during upper abdomen surgery, but this hypothesis remains to be further investigated.

4.2 The results obtained revealed that postoperative time to press a patient-controlled intravenous analgesia pump was (64.10 ± 27.37) min in the M group, significantly earlier than (175.85 ± 8.99) min in the TM group, indicating that topical block could achieve relatively good analgesia effect without adjuvant treatment. Moreover, the consumption of sufentanil at postoperative 0–12 h and 12–24 h in the M group was significantly lower compared with those in the TM group (both $P < 0.05$), whereas no statistical significance was noted at postoperative 24–36 h and 36–48 h between two groups (both > 0.05). These findings suggested that the injection of 60 ml of 0.333% ropivacaine during subcostal transversus abdominis plane block could provide relatively analgesia effect enduring as long as postoperative 24 h, which greatly reduced the required amount of opiod drugs for analgesia purpose postoperatively. Nevertheless, the block effect may sharply vary according to drug, concentration and volume. Previous studies revealed that transversus abdominis plane block could offer at least 48 h analgesia effect, indicating that albeit it is a safe and reliable block approach, various effects produced by different methods remain to be investigated.

4.3 The off-bed time in the TM group was (30.28 ± 2.01) h, significantly earlier than (39.55 ± 2.56) h in the M group ($P < 0.05$). In addition to full analgesia effect, subcostal transversus abdominis plane block also blocks the nerve T6-L1 which dominates the sensation of the frontal abdominal wall.

Compared with epidural analgesia, it averts the sensation disorders or movement block of the lower extremity. Pain sensation block of the abdominal wall positively affects postoperative off-bed time and significantly reduces the incidence of postoperative complications, thereby bringing benefits to postoperative recovery and the patients' quality of life. Besides, the first flatus time in the TM group was (48.95 ± 3.83) h and (64.40 ± 4.58) h in the M group $(P < 0.05)$. On one hand, it is a topical block surgery, which causes traumatic stress and inhibits the slow intestinal movement. On the other hand, it can yield relatively good analgesia effect. Thus, the required dosage of intravenous analgesia medication is greatly reduced. In particular, opiod drugs have a systemic effect. A large dose of opiod drugs is likely to lead to enteroparalysis, urine retention, and other complications. Consequently, it can accelerate the first flatus time to promote recovery as early as possible.

4.4 The sedation scores in two groups are within the range of satisfactory sedation grade, suggesting that the postoperative analgesia methods adopted between both groups are successful. But, patients in the TM group have a significantly higher degree of satisfaction to postoperative analgesia than their counterparts in the M group $(P < 0.05)$, indicating that it is comfortable and highly accepted by patients. Moreover, no statistical significance was seen between two groups regarding the incidence of adverse events, probably due to simultaneous delivery of a certain dose of ramosetron combined with postoperative analgesia.

Ultrasound technique has been widely applied in nerve block. In this study, ultrasound-guided bilateral subcostal transversus abdominis plane block is employed to greatly reduce the incidence of potential surgical complications possibly caused by blind puncture. Instead, it can precisely reach the dissection site and explicitly display the continuous plane between the internal abdominal oblique muscle and the internal abdominal oblique muscle, which ensures the efficacy and accuracy of puncture. At present, novel progress has been achieved in China regarding the nerve block on abdominal wall. The traditional Petie triangle block technique has been merely applied in lower abdomen open surgery. After persistent modification, subcostal nerve block has been utilized in the postoperative analgesia of laparoscope surgery of the upper abdomen and gastric cancer surgery. Sound efficacy has been obtained. However, the block effect of subcostal transversus abdominis plane block has not been investigated in a systemic manner. This study is designed to adopt subcostal transversus abdominis plane block in combination with general anaesthesia in gastric cancer patients undergoing open surgery, and to evaluate the quality of postoperative analgesia and postoperative recovery, and to measure the area of postoperative block plane and reduction time. All these findings indicate that this technique can enhance the quality of postoperative recovery, maintain relatively good analgesia plane within postoperative 6 h, significantly reduce the required dosage of opiod drugs, accelerate the first flatus time, yield no anaesthesia-related complications, and achieve satisfactory analgesia effect. Consequently, ultrasound-guided bilateral subcostal transversus abdominis plane block is a safe and efficacious method for topical block.

REFERENCES

Rafi AN. Abdominal field block: a new approach via the lumbar triangle [J]. Anaesthesia, 2001, 56: 1024–1026.

Hebbard P. Subcostal transversus abdominis plane block under ultrasound guidance [J]. Anesthesia and Analgesia, 2008, 106: 674–675.

L Yang, K Li, YL Li et al. Application of ultrasound-guide multiple subcostal transversus abdominis plane block in open gastrectomy [J]. Chinese journal of laboratory diagnosis, 2013, 17(8): 1477–1479.

Barrington MJ, Ivanusic JJ, Rozen WM et al. Spread of injectate after ultrasound-guided subcostal transversus abdominis plane block: a cadaveric study. [J]. Anaesthesia, 2009, 64(7): 745–750.

Anja UM, Henrik Torup, Egon G Hansen, et al. Effective dermatomal blockade after subcostal transversus abdominis plane block.[J]. Dan Med J, 2012, 59(3): A4404–4405.

McDonnell JG, O'Donnell BD, Farrell T, et al. Transversus abdominis plane block: a cadaveric and radiological evaluation. Reg Anesth Pain Med, 2007, 32: 399–404.

O'Donnell BD. The transversus abdominis plane (TAP) block in open retropubic prostatectomy. Reg Anesth Pain Med, 2006, 31:91.

Farooq M, Carey M. A case of liver trauma with a blunt regional anesthesia needle while performing transversus abdominis plane block. Reg Anesth Pain Med, 2008, 33: 274–275.

Carney, J, McDonnell, J G, Ochana, A et al. The transversus abdominis plane block provides effective postoperative analgesia in patients undergoing total abdominal hysterectomy [J]. Anesthesia & Analgesia, 2008, 107(6): 2056–2060.

McDonnell JG; Curley G, Carney J. The analgesic efficacy of transverses abdominis plane block after caesarean delivery: a randomized controlled trial [J].Anaesthesia and Analgesia, 2008, 106(01): 186–191.

El Dawlatly AA, Turkistani A, Kettner SC et al. Ultrasound-guided transversus abdominis plane block: description of a new technique and comparison with conventional systemic analgesia during laparoscopic cholecystectomy. [J]. British journal of anaesthesia, 2009, 102(6): 763–767.

YQ Wu, ZS Jin, QM Liu, et al. Comparison of degree of pain in patients after radical gastrectomy under different anaesthetic regimens [J]. Chinese Journal of Anaesthesiology, 2012, 32(1): 74–77.

Biomedical Engineering and Environmental Engineering – Chan (Ed.)
© 2015 Taylor & Francis Group, London, ISBN: 978-1-138-02805-0

Mifepristone plays a leading role in progesterone receptor expression of canine endometrial stromal cells

P.Y. Zhou, T. Jin, M. Du, Y.W. Yue & W.Z. Ma
College of Animal Science and Veterinary Medicine, Tianjin Agricultural University, Tianjin, China

ABSTRACT: Many diseases of the canine uterus are related to the level of a progesterone receptor expression in endometria. The aim of this research was to investigate the effects of mifepristone on a progesterone receptor expression in isolated endometrial stromal cells. In the present study, 24 wells of endometrial stromal cells were isolated from 7 healthy anestrus bitches and randomly divided into eight groups of three, which were added into eight kinds of culture mediums containing different steroid hormones. After cultured 48 h, the cells were tested for a progesterone receptor expression by a quantitative polymerase chain reaction. As a result, mifepristone significantly increased the progesterone receptor expression ($P < 0.01$). After analysis of P values of some coupled groups related to mifepristone, we found mifepristone played a leading role in the joint effects of the three steroid hormones on keeping the progesterone receptor expression at a normal level.

Keywords: Canine; progesterone receptor; endometrial stromal cells

1 INTRODUCTION

Canine cystic Hyperplasia-Endometritis Complex (CEH) has been clinically recognized as a common disease, which is responsible for infertility problems (Miller-Liebl et al. 1994). In spite of the high incidence of this disease and many efforts to cure it, the pathogenesis is still not fully understood. Many authors consider bacterial and hormonal interaction to be the initial phase in the development of CEH (Hardy & Osborne 1974). In dogs, an increase in plasma estrogen concentration has been proved relevant to an increase in the number of uterine Progesterone Receptor (PR), while elevated plasma progesterone concentration reduces PRs (Johnston et al. 1985). PR in isolated canine endometrial cells can be also regulated by estrogen and progesterone in this way (Galabova-Kovacs et al. 2004). What is different in humans, however, is that the progestin exerts an up-regulation of PR by increasing the steady-state level of PR mRNA specifically in human endometrial stromal cells (Tseng & Zhu 1997). These results suggest that the mechanisms of progesterone regulation in canine endometria may differ from those in human's, especially in the Endometrial Stromal Cells (ESCs).

Mifepristone, one of the progesterone receptor blockers, possesses an exact effect on abortion in dogs (Linde-Forsberg et al. 1992), but its mechanism of action on hormonal receptors has been rarely mentioned and reported. For these reasons, we took canine ESCs as cell models to research the regulatory effects of steroid hormones on the PR expression.

In the present study, we investigated the expressions of PR in the canine ESCs under treatments of progesterone (R5020, Sigma), 17β-estradiol (Sigma), and mifepristone (Sigma) in order to understand their regulatory mechanisms.

2 MATERIALS AND METHODS

2.1 *Isolation and culture of canine ESCs*

In all experiments, we strictly followed the Directive 2010/63/EU on the protection of animals used for scientific purposes. And the research was approved by Committee on Animal Experiments of Tianjin Agricultural University. Fresh uteri were obtained from 7 health anoestrus bitches (aged 4 ± 2) presented for routine ovariohysterectomy at the Clinic of the Affiliated Animal hospital of Tianjin Agricultural University, Tianjin, China. The isolation and purification of canine ESCs were carried out according to the previous method (Zhang et al. 1995). After resuspended in modified DMEM/F12 medium (Hyclone, Beijing, China) with 20% Fetal Bovine Serum (FBS, Gibco BRL) supplemented, single ESCs were placed into a 24-well plate (Corning Costar Corporation, Beijing, China) and kept at 37° with an atmosphere of 5% CO_2 in air and 95% humidity.

2.2 *Treatment with steroid hormones on canine ESCs*

As primary ESCs overgrew culture clusters, they were passaged into three 24-well plates at 1:3 split ratio and cultured in modified DMEM/F12 medium. When passaged ESCs covered 80% of each cluster, the modified

Table 1. Treatments with steroid hormones on eight canine groups of ESCs.

Groups	17β-estradiol (mol/L)	Progesterone (mol/L)	Mifepristone (mol/L)
C	0	0	0
E	10^{-7}	0	0
P	0	10^{-6}	0
M	0	0	10^{-6}
EP	10^{-7}	10^{-6}	0
EM	10^{-7}	0	10^{-6}
PM	0	10^{-6}	10^{-6}
EPM	10^{-7}	10^{-6}	10^{-6}

Table 2. Primers used in relative quantification of mRNA.

Primer name	Prime sequence	Product length
Canine β-actin	F 5'AGGCTGTCCTGTCCCTGTATG-3' R 5'-CCAGGTCCAGACGCAAGAT-3'	132
Canine PR	F 5'-ATTCCAAATGAAAGCCAAGC-3' R5'-TAAGACTCGTCAGCAAAGAACT-3'	172

DMEM/F12 medium needed to be replaced by seven kinds of hormone-treated mediums, which consist of DMEM/F12 without phenol red (Hyclone, Beijing, China) and different steroid hormones. In the study, 24 wells of ESCs were randomly divided into eight groups of three. To be distinguished from each other easily, the groups were named with different capital letters, including C (control group), E (17β-estradiol), P (progesterone, R5020) and M (mifepristone). For example, group EPM indicated ESCs in this group were cultured in mediums containing 17β-estradiol, R5020, and Mifepristone. Others were read in the same way and the eight groups' treatments were show in Table 1. After being cultured for 48 h, the treated ESCs were tested for the expression levels of PR.

2.3 Relative quantification of PR mRNA

The total RNA of ESCs was extracted with TRIZOL Reagent (Invitrogen). Then, 1 μg RNA was used to synthesize cDNA with First Strand cDNA Synthesis Kit (GeneCopoeia lnc, USA) according to manufacturers recommended methods. Quantitative polymerase chain reaction (qPCR) was performed using qPCR Mix (GeneCopoeia lnc, USA) in a final volume of 20 ul. The sequences of the primers used for amplifications of PR and β-actin were newly designed with the primer 5.0 software (Table 2) and synthesized by Beijing Genomics Institute. The PCR was performed by the Bio-Rad CFX Connect Real-Time PCR Detection System and analysed by the Bio-Rad CFX-96 system. The optimal annealing temperature range, amount of cDNA and the number of PCR cycles for qPCR were determined in preliminary experiments. The following PCR protocol for the amplification of PR and β-actin was used: denaturation program (94° for 10 min), amplification and quantification program repeated 50 times (94° for 15 s, 58° for 10 s, 72° for 30 s with a single fluorescence measurement), melting curve program (72° to 94.5° with a heating rate of 0.5° per 10 s and a continuous fluorescence measurement) and finally a cooling step to 25°.

Pfaffl's model of the comparative Cq value (the quantification cycles needed for reaching the threshold fluorescence) method taking the relative expression ratio (R) as the target gene expression was applied

for the analysis of the fluorescent data obtained from real-time PCR (Pfaffl 2001).

2.4 Statistical analysis

The data obtained from qPCR were analysed with Fisher's protected Least Significant Difference (LSD) for the intergroup comparisons ($P < 0.05$). Prior to analysis, the data of R were analysed by the Levene test for homogeneity ($P > 0.05$). The Spearman correlation coefficient (r) was used to assess the correlation between 2 quantitative variants. All statistical analyses were performed using the SPSS software package (version 11.5, SPSS, Inc., Chicago, IL).

3 RESULTS

3.1 Isolation and culture of canine ESCs

The primary ESCs obtained from 7 bitches showed comparable behaviour during their culture. They showed that the fibroblast characteristic elongated form were attached to the culture dishes after 24 h (Fig. 1A). During the first 3 days they formed small islets, after 4-6 days in culture a confluent monolayer was established (Fig. 1B).

3.2 Relative standard curves and determination of amplification efficiency

Five gradient concentrations of cDNA consisted of $1/20^4$ dilution, $1/20^3$ dilution, $1/20^2$ dilution, $1/20$ dilution and original. Then real-time PCR efficiencies (E) were calculated according to the equation: $E = 10^{[-1/slope]}$ (Bustin, 2000), while the slopes had been given in Bio-Rad CFX-96 system (Fig. 2). Investigated transcripts showed the high real-time PCR efficiency rates; for β-actin, 2.01 and PR, 2.04 in the investigated five concentration gradients of cDNA with high linearity (Pearson correlation coefficient $r > 0.95$).

3.3 Normalized PR mRNA expressions in different hormone-treated ESCs

Figure 3 showed normalized mRNA expressions of PR with ratio (R) of a target gene, which was calculated as the following equation:

$$Radio = \frac{(E_{target})^{\Delta Cq_{target}(control-sample)}}{(E_{ref})^{\Delta Cq_{ref}(control-sample)}}$$

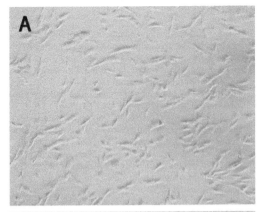

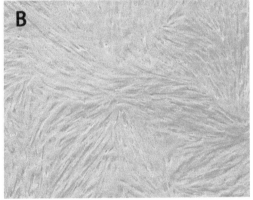

Figure 1. Inverted phase-contrast micrograph of adherent primary ESCs after 24 h × 200 (A) and confluent monolayer formed by canine ESCs after 4 d × 200 (B).

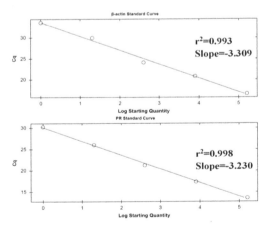

Figure 2. The standard curves were generated by the Bio-Rad CFX-96 system.

(Pfaffl, 2001). In the equation, R is calculated based on amplification efficiency (E) and Cq.

Deviation of unknown sample was versus a control, and was expressed in comparison to a reference gene. PR expressions of Group E and Group EP were significantly different from those in Group C (P < 0.01)

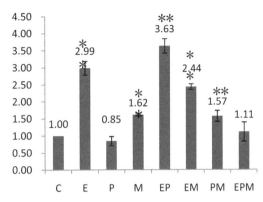

Figure 3. PR mRNA expressions were normalized with ratio (R). Expressions of each treated group were compared with Group Control's through LSD analysis. '*' indicated 0.01 < P < 0.05; '**' indicated P < 0.01.

Table 3. P values of coupled groups related to mifepristone.

Coupled groups	P values
E-EM	7.4×10^{-4}
P-PM	5.8×10^{-5}
EP-EPM	2.2×10^{-12}

and they were also the highest two groups among eight groups. Group M's PR was expressed much more than Group C's (P < 0.01).

In addition, there was a distinct relationship between the effects of mifepristone and its combination with 17β-estradiol or progesterone on PR receptor levels. Table 3 illustrates the relationship with P values of coupled groups related to mifepristone obtained from LSD analysis. As for PR mRNA expression, P value of the coupled groups were much less than 0.01 indicating that the changes in corresponding to steroid hormones were evident between groups treated with mifepristone-included and mifepristone-excluded combination.

4 DISCUSSION

To reach high accuracy and reproducibility, we analysed target genes expressions with Pfaffl's model of the comparative Cq value method, in which control levels were included to standardize each reaction run with respect to RNA integrity, sample loading and inter-PCR variations. Furthermore, 58°, the common optimal annealing temperature for the three pair of primers, was determined in preliminary experiments. Therefore, qPCR runs of all the samples proceeded in a 96-well plate to avoid the error from different batches.

Galabova-Kovacs et al. have demonstrated that β-estradiol induced up-regulation of PR and progesterone induced its reduction in ESCs. They also found ESCs were responsive to all doses of β-estradiol and progesterone in three days. Thus we chose a large dose of steroid hormones to stimulate ESCs for 48 h. In Fig. 3 we found 17β-estradiol stimulation of ESCs (Group E) increased PR levels up to 3-fold as Group C, while progesterone stimulation (Group P) slightly declines its expression. It is consistent with the results of Galabova-Kovacs et al. As for mifepristone, it highly significantly increased the PR level ($P < 0.01$). In addition, our study also showed that mifepristone greatly influenced PR expressions of Group EPM, whose expressions were similar to those uncontrol group's. Then the PR expressions remained at normal levels when the three steroid hormones were applied in combination to the ESCs.

In conclusion, by comparing different hormone-treated ESCs, we reported that mifepristone, especially combined with progesterone and 17β-estradiol, was able to balance the PR expressions in ESCs. Further studies will be needed to confirm the effects of mifepristone on other cells in canine endometrium. In the near future hormone combination method may be provided as an effective means of precaution for canine CEH.

ACKNOWLEDGEMENTS

We would like to thank Tianjin Agricultural University for all the experiment instruments and reagents used in the research. And we were also grateful for the help of the Affiliated Animal Hospital of Tianjin Agricultural University. This study was supported by Tianjin Agricultural University Experiments and Teaching Centre for Agricultural Analysis.

REFERENCES

Bustin, S.A., 2000. Absolute quantification of mRNA using real-time reverse transcription polymerase chain reaction assays. J. Mol. Endocrinol. 25: 169–193.

Galabova-Kovacs, G., Walter, I., Aurich, C., Aurich, J.E., 2004. Steroid receptors in canine endometrial cells can be regulated by estrogen and progesterone under in vitro conditions. Theriogenology 61 (5): 963–976.

Hardy, R. M., Osborne, C.A., 1974. Canine pyometra: pathophysiology, diagnosis and treatment of uterine and extra-uterine lesions. J. Am. Anim. Hosp. Assn. 10: 245–268.

Johnston, S.D., Kiang, D.T., Seguin, B.E., Hegstad, R.L., 1985. Cytopasmic estrogen and progesterone receptors in canine endometrium during the estrous cycle. Am. J. Vet. Res. 46: 1653–1658.

Linde-Forsberg, C., Kindahl, H., Madej, A., 1992. Termination of mid-term pregnancy in the dog with oral RU486. Journal of Small Animal Practice 33 (7): 331–336.

Miller-Liebl, D., Fayrer-Hosken, R., Caudle, A., Downs, M., 1994. Reproductive tract diseases that cause infertility in the bitch. Vet. Med. 89: 1047–1054.

Pfaffl, M.W., 2001. A new mathematical model for relative quantification in real-time RT-PCR. Nucleic acids research 29 (9): e45.

Tseng, L., Zhu, H.H., 1997. Regulation of progesterone receptor messenger ribonucleic acid by progestin in human endometrial stromal cells. Biology of reproduction 57 (6): 1360–1366.

Zhang, L., Rees, M.C., Bicknell, R., 1995. The isolation and long-term culture of normal human endometrial epithelium and stroma. Journal of Cell Science 108 (1): 323–331.

Biomedical Engineering and Environmental Engineering – Chan (Ed.)
© *2015 Taylor & Francis Group, London, ISBN: 978-1-138-02805-0*

The fabrication and character of a plant hard capsule made from corn starch treated with isoamylase

Hui Li & Chun-Hu Li
Chemical Engineering College, Ocean University of China, Qingdao, Shandong Province, China

Nan-Nan Wang
College of Medicine, Guizhou University, Guiyang, Guizhou Province, China

Lin Wu
Qingdao Technical College, Qingdao, Shandong Province, China

Yi-Jun Jiang & Xin-Dong Mu
Qingdao Institute of Biomass Energy and Bioprocess Technology, Chinese Academy of Sciences, China

ABSTRACT: A novel plant capsule was made from corn starch modified by isoamylase with PVA, alginate and carrageenan as auxiliary materials. Then the capsule film was tested for its property of sensitivity to humidity disintegration. It was indicated that, the process of starch expansion and gelation can be controlled very well, when the corn starch is treated by 0.4 mg isoamylase for 5 min. The viscosity of the starch solution was reduced effectively as well. The film-forming characters and strength of the starch film were enhanced, by added 3% carrageenan, 0.2% PVA and 0.5% alginate, also, the disintegration time of the capsule will be controlled effectively and reasonably.

1 INTRODUCTION

Two-piece hard capsules have been used for drug delivery for nearly 200 years in the area of medicine, health care products and functional foods (Jones, 2005). At present, gelatin in hard hollow capsules is the main raw material of hard hollow capsules in the market (Hai-yang, 2010). Gelatine is a collagen protein, which originates from the bone and skin of cattle and pigs, as the characters of it, such as film-forming ability and thermo-reversibility, make it suitable for capsule manufacturing (Jones, 2004). However, the molecular structure of gelatine is not perfect, such is the crossing-linking reaction of gelatine (Digenis, 1994). These drawbacks affect the collapse of the capsule and the stability of the drug, which would be breakable, fragile and metamorphic when in long-term storage in a low humidity environment (Tao, 2005). Furthermore, gelatine capsules are strongly rejected by vegetarians, Muslims and some other people in the world because of their faith and religion (Cun-yan, 2005). In recent years, with frequent global animal endemic diseases outbreak, and the poison capsule scandal has been exposed in some countries, so the study of plant capsules which would replace gelatine capsules, is developing rapidly, and must be an inevitable developing trend for medicine and food industry.

As one of the most important purely natural raw materials for food, starch has so many advantages such as being safe, nontoxic (Tharanathan, 2005), cheap, biodegradable, etc. So it has been widely used as a ingredient for food and medicine, and is considered to be the most promising raw material for capsules as a substitute for gelatine (Vinod, 2000 & Stepto, 1997). It is well known that starch is a heterogeneous material contains two micro structures: linear (amylose) and branched (amylopectin) (Liu, 2009; Whistler, 1984 & Liu, 2009). Based on the study of starch based materials and related food science the researches on starch capsules has had a dramatic breakthrough (Eith, 1987; Gohil, 2004 & Vilivalam, 2000). While, because the gelation properties of starch can't meet the demand of the production process of capsules, gels and other additive are needed.

Bae (Bae, 2008) prepared plant capsules with starch from green beans, sweet potatoes and water chestnuts by impregnation and moulding methods. Traditional dip moulding methods have several disadvantages, such as the starch solution viscosity is so high that it's hard to control the thickness of capsules and realize the automatic production (Liang, 2013). When Liu Y and Zhou C (Yu, 2009) prepared cassava starch capsules, the viscosity of the starch solution was reduced by adjusting the pH with sodium hypochlorite. However, the method regulating the viscosity of starch solution by controlling pH indirectly was relatively difficult, so large scale production was limited. Furthermore, using an enzyme reaction is also a method of reduceing the viscosity of the gel of starch (Li-tian, 1980). When the starch was modified by isoamylase, amylopectin and α-1, a 6-glycosidic bond of the glycogen were hydrolysed to amylose with different lengths (Xiang-yun, 2010). Previous studies (Biliaderis, 1985; Jovanovich, 1999; Liu, 2009; Raphaelides, 1988 & Modliszewski, 2011) have found that amylose and carrageenan could form an amylase-carrageenan complex, which is stable up to 90°C in excess water solution.

This research was to prepare a new kind of medical hard capsules with corn starch which was modified by amylase instead of traditional gelatine capsules. And the mechanical properties of the capsules was enhanced by adding an additive such as sodium alga acid and PVA overcoming, the defects of traditional gelatine capsule preparation and laying a solid foundation for the research and development of new plant capsules.

2 EXPERIMENT

2.1 Materials

Carrageenan (K-type) was produced by Qingdao Teck Whye Marine Biotechnology (China), Poly(vinyl alcohol) (PVA) was purchased from KURARY (Japan), sodium alginate corn starch, glycerol, sorbitol KCl were all pharmaceutical grade from Qingdao Huanghai Pharmaceutical Co. (China) Ltd. Isoamylase (1 U/mg), glucoamylase (100 U/mg), α-amylase (BR) were all ibuprofen from Aladdin (China).

2.2 Technological process

2.2.1 Plasticizing and sol preparation

Glycerol, sorbitol, PVA and sodium alginate were added into a flask with 100ml deionized water and kept stirred and heated to 70°C. Then, followed by the addition of carrageenan under stirring for 30 min, corn starch was added under stirring at 400 r/min, and the temperature was then increased and kept at 90°C for 40 min. The mixture was gradually well-distributed and transparent.

2.2.2 Enzyme treatment and PH adjustment

The mixture was cooled to 55°C and kept at that temperature for about 20 min. Then a certain amount of isoamylase glucoamylase, and α-amylase respectively was added into the mixture prepared before, stirring for some time. According to the amount of enzyme added, append a certain amount of 1 M NaOH solution under stirring for about 3 min to make the enzyme activate, then add the same amount of HCl to regulate pH to neutral.

2.2.3 Dipping plastic moulding

Pour the solution into 100 ml beaker and kept it at 60°C, then dipped the gel with moulds to prepare the capsules with uniform thickness. Meanwhile, the rest of the mixture was cast into a film. Then the moulds and films were dried an air-circulating oven at 50°C.

2.2.4 Modification and storing

After drying the product for 10 h, the capsules were pulled out from the moulds and trimmed into standard. Then the capsules and films were preconditioned in a climate chamber at 25°C and 50% RH for at least 24 h prior to testing and packing.

2.3 Melding speed and shell thickness measurement

The time from dipping the gel solution to solidify into pieces was recorded by a timer, and named it as gelation time representing moulding speed. The unit of moulding speed is second. The thickness of the capsules shell was measured by vernier caliper. The date was averaged over 6–8 specimens.

2.4 Mechanical testing

Tensile tests of the starch films were performed on an electronics universal tensile testing machine (Testometic, UK) at a cross-head speed of 50 mm min^{-1}. Rectangular specimens of 15 mm × 50 mm were used for tensile tests and their thicknesses were measured using a micrometre before they were placed in the fixture. For each sample, a minimum of three measurements was taken, and the average result was taken.

2.5 Viscosity measurements

The viscosity of the starch solutions was measured at 60°C using a Brookfield viscometer (NDJ-5s, Shanghai Fang Rui instrument Co. Ltd.) operating at 30 rpm for all samples. The viscometer spindle was immersed into the solution for about 2 min to achieve thermal equilibrium between the solution and spindle with continued shearing. Five measurements were recorded for each solution, and the average values were calculated.

2.6 Moisture sorption of the capsules (He, 2013)

Moisture absorbability of the starch films and the capsules shells were measured in an environment that contained a saturated sale solution to give the desired relative humidity (RH) (NaNO$_3$: RH = 75%) The films were cut into pieces, and the initial weights of them were tested after they had been dried to a constant weight in a vacuum oven at 50°C for 24 h. The weighted samples were then placed in incubators at 25°C at 75% RH and they were weighted at intervals until the weight of them became constant. The tests were performed in quadruplicate and the value of moisture absorbability (MA) was calculated as follows:

$$MA\% = \frac{W - W_0}{W_0} \times 100\%$$

Table 1. The effects of mixed ratio to the viscosity of starch gel and the tensile strength of the film.

Blend ratio (m_1: m_2)	Melding time/s	Viscosity/ MPa·s	Thickness/ mm	Strength/ MPa
11:2	11 ± 0.04	4361 ± 0.30	0.06 ± 0.002	16.7 ± 0.022
10:3	7 ± 0.09	4537 ± 0.16	0.08 ± 0.001	17.0 ± 0.016
9:4	5 ± 0.11	4952 ± 0.24	0.09 ± 0.001	169 ± 0.013
8:5	3 ± 0.03	5084 ± 0.33	0.1 ± 0.003	174 ± 0.017

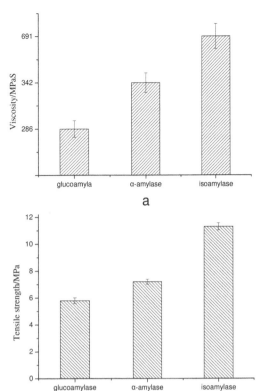

Figure 1. Sol viscosity (a) and membrane strength (b) after different amylase processing.

where W_0 and W are the dry weight of the starch pieces and its weight at a certain time in the specific saturation humidity respectively.

2.7 Brittle broken performance

30 hollow capsules were placed into petri dishes, then placed the dishes into incubator at 25°C, and 50% RH for 24 h. Stand a glass tube on a board whose thickness was 2.5 cm, then put these capsules into the tube, and make a 20 g cylindrical Jordan farmar fall from the top of the tube. At the same time, observe whether the capsules were broken and get the number of the fractured ones to calculate the breakage ratio. The breakage ratio must be not more than 30% (Xiao-ju, 2008).

2.8 Measurement of disintegrating performance

On the basis of Chinese Pharmacopoeia (National pharamacopoeia committee, 2005), peristalsis of the stomach was simulated to test the disintegrating performance of the capsules. According to the prescribed inspection method, the disintegration time limit (DTL) (YY/T 0188) refers to the time that solid preparation disintegrate to less than $2.0\,mm^2$ in liquid medium takes. Put a certain number of capsules cut as the standard requirements into the pool of the disintegration tester full of $37 \pm 0.5°C$, pH6.8 liquid phosphate buffer, which was used as slaking fluid. For each sample, a minimum of five measurements were taken, and the average result was calculated.

3 RESULTS AND DISCUSSION

3.1 Research of the mixed ratio

Corn starch and carrageenan were the main raw materials, and their total concentration was 13%. The two materials, whose qualities were respectively marked as m_1 and m_2, were mixed in a different proportion (m_1: m_2), after that, keep stirring and heating at the gelatinization temperature for 40 min. The viscosity of the blending ratio and the tensile strength of the samples are shown in the Table 1.

Table 1 shows that both of the mixtures could be uniform and the tensile strength of each wasn't much different than others. When the proportion was

10:3 or 9:4, the thickness was relatively smooth and conformed to the requirements of capsules, and the viscosity of them were both about 4700 MPa·s. But considering factors such as cost, 10:3 was selected as the best blend ratio as the basic material for the capsule.

3.2 Comparing performance of the starch modified with different enzymes

In order to satisfy the process of making hard capsules, it is important to reduce the viscosity of the gel solution to a proper value. In this paper, the enzyme was used to control the viscosity of the solution due to the advantage of low cost and safety for the enzyme method. As we all know, there are mainly three different type of amylase (Ming-li, 2008), which are the isoamylsae, α-amylase and glucoamylase respectively.

In our experiment, these three enzymes were employed to modify the corn starch respectively. Based on the above results, the optimal starch gel was selected, the gel was treated by these three enzymes for 5 min severally, and then detected the viscosity of the gel and the tensile strength of the film. The final results are shown in Figure 1.

Figure 1 presents that all the films were not strong like before (see Table 1), while the viscosity of the gel modified with isoamylase was more than two times higher than the one modified with the other two enzymes, and the tensile strength of the film modified with isoamylase is also relatively high. The reason might be that glucoamylase can starch macromolecules into glucose molecules. So the viscosity of the sol was the lowest and the strength of the film is also low and fragile. When α-amylase was used, glucose, fructose, dextrin and some other monosaccharide or polysaccharide with short chain length were produced to drop the sol viscosity rapidly (Xi-jun, 2009). As a result of the existence of polysaccharide, the tensile strength of the film was larger than the one treated by glucoamylase. While, in the process of isoamylase working, amylopectin and α-1, 6-glycosidic bond of the glycogen were hydrolyzed to amylose with different lengths which can maintain the character of the corn starch, so the film formed is stronger than the one treated by other enzymes (Xiang-yun, 2010).

By comparing these three kinds of enzyme, isoamylase was selected to modify corn starch. As a result, the viscosity of the sol was well reduced, also, the relatively higher tensile strength of the film was ensured.

3.3 The effect of the time and amount of the enzyme working

In our experiments, the amount of the isoamylase in the starch gel was controlled by changing the added-volume of the constant enzyme solution (2mg/ml), and then dealt with the starch solution for a different time. After that, the sol viscosity and the maximum tensile strength were measured, so the relationship between the amount of isoamylase and the viscosity or strength was got and the results are shown in Figure 2.

It was known from the Figure 2 that the more the isoamylase was added into the solution, the lower the viscosity was, while, the lower the tensile strength was. The reason was more isoamylase increased the velocity for starch decomposition, and then reduce the viscosity and strength of the film. The starch chain gradually shortened to generate dextrin as the reaction progress, resulting in the decrease of the viscosity of the sol and the tensile strength of the films. According to the optimal viscosity for capsule making, the appropriate viscosity ranged 3500 from 2500 MPa·s. So the optimum dosage of the enzyme solution was determined at 0.2 ml, which could ensure the viscosity of sol well controlled.

The reaction time was also evaluated in our experiments. As seen from the Figure 3, the viscosity of the sol kept decreasing within 10min, and the strength of the obtained film decreased rapidly to 60% of its original. After 10min, the viscosity and strength were gradually constant, which suggested the reaction was probably finished, and at this time the viscosity was about 2000 MPa·s and the strength was nearly 8 MPa. According to the technical requirement of dipping glue process, the ideal working time was set as 5 min.

In conclusion, the optimum amount of isoamylase is 0.4 mg, which could reduce the viscosity to an ideal

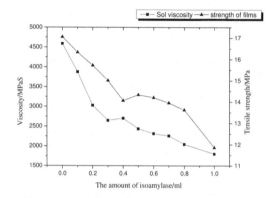

Figure 2. Effect of isoamylase quality to sol viscosity and film strength.

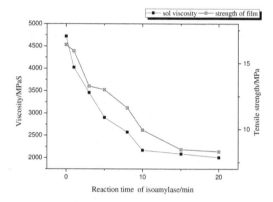

Figure 3. Sol viscosity and film strength at different reaction time of isoamylase.

range and the strength could also be ensured. The viscosity of the modified sol was decreased about 42.5% than the sol without isoamylase, which is significant for the production of hard capsule shell.

3.4 Determination of the concentration of the auxiliary materials

It was observed that the membrane strength declined due to the isoamylase addition, so it is necessary to increase the film strength by introduction of some other additives in the experimental process. In our experiments, polyvinyl alcohol (PVA), glycerine, sodium alginate (Wei-jiang, 2012) were chosen to improve the film's strength. Therefore, an orthogonal test was designed based on these three factors, and the test level values are shown in Table 2. What is more, Table 3 shows the design of the experiments and the results of it.

According to the result of orthogonal experiment, it's known that A was the most important factor for the quality of the capsules, then C was followed and B was the last. When PVA and sodium alginate were added, the tensile strength of the films increased significantly, but on the other hand, they also increased the viscosity

Table 2. Factor and levels of orthogonal test.

Levels	A/%	B/%	C/%
	PVA	Glycerinum	Sodium alginate
1		0.5	
2	0.2	1	0.2
3	0.4	1.5	0.5

Table 3. Results of orthogonal experiments.

Factor A	B	C	Strength/ MPa	Viscosity/ MPa	Brittle ratio/%
1	1	1	16.45	3561	62.5
2	1	2	16.79	3667	68.1
3	1	3	17.18	3929	73.9
4	2	1	18.34	3828	73.7
5	2	2	19.18	3794	76.5
6	2	3	17.96	3645	74.2
7	3	1	20.77	3803	72.8
8	3	2	18.52	4105	73.4
9	3	3	19.24	4338	70.6
k1	68.167	69.333	69.033		
k2	74.800	71.667	71.133		
k3	71.267	73.233	74.067		
R	6.633	3.9	5.034		

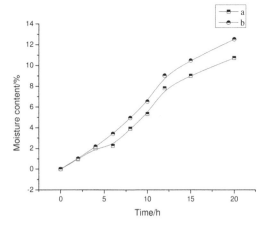

Figure 4. The comparison of MA between modified film (a) and pure film (b).

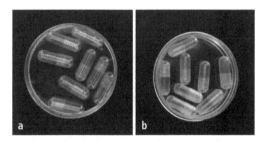

Figure 5. Comparing starch capsules (b) with gelatine capsules (a).

of the starch sol. Glycerol has excellent water-retaining ability, so the moisture absorbability of the capsules increased as the content of glycerol increased. Because official PVA is a security water-soluble polymer, in the hydroxyl group, hydrogen bonding was formed between the starch and PVA, so the tensile strength of the complex film increased obviously.

In conclusion, the optimum process condition was A2B2C3, and in other words, when 0.2%(w/w) PVA, 1%(w/w) glycerol, and 0.5%(w/w) sodium alginate were added, the integrated performance of the capsules worked relatively well in controlling the viscosity and the tensile strength in a reasonable range.

3.5 Moisture sorption of the capsules

The starch sol was prepared as the optimal proportion in duplicate, and was marked as the one with added PVA and other additive and modified by isoamylase for 5 min as Sol(a), while the one without any additive was marked as Sol(b). Then films with the thickness of 0.08–0.10 mm respectively were made and prepared to be tested.

Corn starch is a polysaccharide that readily absorbs moisture, because its macromolecules contain many hudroxyl groups. It is preferred for reducing the ability of the capsules for water adsorption. Then the moisture absorbability of the starch films at RH of 75% was tested and the results are given in Figure 4. When modified by isoamylase, corn starch was hydrolysed to amylase, the structure of the film produced would be more compact than the one unmodified, so the moisture permeability was poor. When PVA and other additives were mixed, the structure would be the most compact and relatively waterproof. In addition, the moisture sorption of the modified film was lower than that of the starch films.

3.6 Measurement of the disintegrating performance

The contrastive photographs between the capsules prepared and the gelatine capsules are shown in Figure 5, which shows that the starch capsules are not as hyaline as the gelatine capsules. The reason may be that the dextrinization process of the starch was incomplete so that the solution was uneven. Starch granules stayed in the sol, so the light was hindered. Due to the low light transmittance, the sunscreen was needless. Not only the cost was reduced, but the original character of the drugs in capsules was protected.

It is important for capsules to release the drug in it within 10 min. So the disintegrating performance of the capsule was evaluated. The disintegrating performance can be charactered with BJ-4A Disintegration Tester (Tianjin chong hing electronic equipment Co. Ltd.). Chose pure corn starch capsules without any additive, modified starch capsules without additive, modified starch capsules with additive and gelatine capsules to compare their disintegration phenomenon, and the results are shown in Table 4.

Table 4. The comparison of several capsules' disintegrating performance.

Time (min)	Disintegrating performance			
	Capsule A*	Capsule B*	Capsule C*	Capsule D*
1	Lose shape	Crack	——	——
2	Crack	Collapse	Lose shape	——
3	——*	Dispersed	Crack	Lose shape
4	Collapse	Dissolved	——	——
5	——	——	Collapse	——
6	Dispersed	——	——	Crack
8	Dissolved	——	Dispersed	Collapse
10	——	——	Dissolved	——

*A: Pure corn starch capsules without any additive.
*B: Modified starch capsules without additive.
*C: Modified starch capsules with additive.
*D: Gelatin capsules.
*——No obvious phenomenon.

As seen from Table 4, when there was no additive, the DTL of the pure corn starch capsules was much longer than those modified by isoamylase. When the capsules without modification were set into slaking fluid, they began to curl, shrink and then deform in turn in the first 2 min, then a crack appeared. At last they turned into pieces at 4 min, and then dispersed and dissolved in the buffer completely. While the modified capsules collapsed into pieces in 2 minutes, then gradually dispersed and quickly dissolved in the buffer. The reason might be that some polysaccharide with a short chain was contained in the modified capsules, so the disintegrating speed could have been reduced effactually. When the additives, such as PVA and sodium alginate, were added, the DTL of capsules was prolonged, due to the improvement of the capsules structure. Even so, they could still disintegrate and dissolve completely within 10 minutes to satisfy the application (Xiao-ju, 2008), that is to say the DTL can be well controlled by adjudging the composition of additives. Compared with gelatine capsules, the starch capsules can adjudge the disintegrating time according to the release time of the drug.

4 CONCLUSION

(1) When modified by isoamylase, the viscosity of the corn starch sol dropped to about 3000 MPa·S, which could meet the requirements of the dipping glue process and be advantageous to control the thickness of the capsule shell.
(2) The addition of 3% carrageenan, 0.2% PVA, 0.5% sodium alginate, etc. ensured the gel property of the starch sol, and increased the tensile strength and toughness of the film when the water absorption performance was stable.
(3) When the concentration of the sol was 13%, the surface of the capsules could be smooth, and the thickness could be uniform, the brittle broken performance was also very well to control.

In conclusion, corn starch modified by isoamylase can be used as a raw material for the preparation of medicinal plant capsules. Also, a new kind of plant capsule, which is superior and conforming to the standard of medicinal material, can be prepared by the method of using isoamylase to tailor corn starch.

REFERENCES

Bae, H., Cha, D., Whiteside, W. & Park, H. 2008. Food Chem, 106: 96–105.
Biliaderis, C.G., Page, C.M., Slade, L. & Sirett, R.R. 1985. Thermal behavior of amylose–lipid complexes. Carbohydrate Polymers, 5, 367–389.
Cunyan Zhang, Wang, C.P. & Chun-long Wang, 2005. New capsule materials – hydroxypropyl methyl cellulose (J). Journal of Chinese pharmacy, (12): 891–892.
Digenis, G.A, Gold, T.B. & Shah, V.P. 1994. Cross-linking of Gelatin capsules and its relevance to their in-vivo in vivo performance. J. Pharm. Sci., 83(7): 915–921.
Eith, L., Stepto, R.T.F., Wittwer, F. & Tomka, I. 1987. Injection-moulded drugdelivery systems. Manufacture Chemical, 58, 21–25.
Guo Weijiang & Tang Hesheng, 2012. The development of new Marine plants hard hollow capsule (J), Food and Machinery, 1(28): 235–237.
Gohil, U.C., Podczeck, F. & Turnbull, N. 2004. Investigations into the use of pregelatinised starch to develop powder-filled hard capsules. International Journal of Pharmaceutics, 285, 51–63.
Hai-yang Zhang, Lv Yi & Ni Yue. 2010. The study of the preparation of microcapsule by conforming gelatin to CMC (J). Journal of food and machinery 26(5): 44–47.
Jones, B.E. 2005. The Manufacture and Properties of Two-piece Hard Capsules. In: Podczeck F. & Jones, B.E. eds. Pharmaceutical capsules. London: Pharmaceutical Press, 79–100.
Jovanovich, G. & Añón, M.C. 1999. Amylose–lipid complex dissociation. A study of the kinetic parameters. Biopolymers, 49, 81–89.
Jones, B.E. 2004. The manufacture and properties of two-piece hard capsules. In F. Podczeck, & B.E. Jones (Eds.), Pharmaceutical capsules (pp. 79–100). London: Pharmaceutical Press.
Liu Yu & Zhou Chuan-xin, 2009. Preparation process of cassava starch glue and its plant gum, vegetable capsule: China, 101338045(P), 01–07.
Liu, H.S., Yu, L., Chen, L., & Li, L. 2007. Retrogradation of corn starch afterthermal treatment at different temperatures. Carbohydrate Polymers, 69, 756–762.
Liu, P., Yu, L., Liu, H.S., Chen, L., & Li, L. 2009. Glass transition temperature of starch studied by a high-speed DSC. Carbohydrate Polymers, 77, 250–253.
Liu, H.S., Yu, L., Simon, G., Dean, K. & Chen, L. 2009. Effects of annealing on gelatinization and microstructures of corn starches with different amylose/amylopectin ratios. Carbohydrate Polymers, 77, 662–669.
Modliszewski, J.J., Kopesky, R. & Sewall, C.J. US patent, A1,0203124.2101-08-12.
Ming-li kang, 2008, Amylase and the function mode (J). Food Engineering, 3(9): 11–15.
National pharmacopoeia committee. 2005. The pharmacopoeia of the People's Republic of China (M). Beijing: Chemical Industry Press: the appendix 72.
Raphaelides, S. & Karkalas, J. 1988. Thermal dissociation of amylose–fatty acid complexes. Carbohydrate Polymers, 172, 65–82.

Stepto, R.F.T. 1997. Thermoplastic starch and drug delivery capsules. Polymer International, 43, 155–158.

Tharanathan, R.N. 2005. Starch-value addition by modification. Critical Reviews in Food Science and Nutrition, 45, 371–384.

Vilivalam, V.D., Illum, L. & Iqbal, K. 2000. Starch capsules: An alternative system for oral drug delivery. Pharmaceutical Science & Technology Today, 3, 64–69.

Vinod, D. Vilivalam, Lisbeth lllum and Khurshid lqbal. 2000. Starch capsule: an alternative system for oral drug delivery. Research focus (J) 2(2), 64–69.

Whistler, R., Bemiller, N.J. & Paschall, F.E. 1984. Starch: Chemistry and technology. Orlando: Academic Press.

Xijun Lia & Xu Liu, 2009. The preparation of resistant starch by using potato starch using α-amylase (J). Grain and Oil, 2: 11–15.

Xiao-ju Zhang, Fa-tang Jiang. 2008. Performance detection of plant hard hollow capsules (J). Journal of chemical and biological engineering, 25(12): 77–78.

Yongqiang He, Xingrui Wang, Di Wu. 2013. Biodegradable amylase films reinforced by grapheme oxide and Polyvinyl alcohol. Materials Chemistry and Physics, (142):1–11.

YY/T 0188. 8-1995. The drug inspection procedures section 8. The general principles of the preparation (S).

Zhang liang, Wang Yanfei & Hong-sheng Liu, 2013. the research and development on natural macromolecular medicinal capsule (J). Journal of polymer. (1): 1–10.

Zhang Litian, 1980. The separation of amylase and amylopectin. Strach and its Suger, (01).

Zhou Tao, 2005, non gelatine capsules (J), Science and technology of gelatin, 25(3): 45.

Zhu Xiangyun, 2010. The research on the properties of corn starch hydrolyzed by isoamylase enzyme and its film-forming (D), Tianjin, China, and Tianjin University.

Biomedical Engineering and Environmental Engineering – Chan (Ed.)
© *2015 Taylor & Francis Group, London, ISBN: 978-1-138-02805-0*

MRI investigations into internal fixations of the lumbar vertebrae with titanium alloy

Z.-W. Huang, B. Xiao, L.-S. Zhong & J.-Z. Wan
Department of Biomedical Engineering, Luzhou Medical College, Luzhou, Sichuan, P.R. China

ABSTRACT: Objectives: MRI scanning of patients after internal fixation surgery of the lumbar vertebrae was performed to explore the efficacy of different MRI sequence combinations. Methods: Twelve patients were examined by MRI using the following scan sequences: (1) T2-weighted Fast Spin Echo (FSE) sagittal fat suppression (fs) sequence, (2) T2-weighted Fast Spin Echo sagittal sequence, (3) T1-weighted Fast Spin Echo sagittal sequence and (4) T2-weighted Fast Spin Echo transverse sequence. The quality of the images was subsequently evaluated. Results: The fixation objects exhibited low-intensity signals in both T1-weighted and T2-weighted images with well-delineated edges of high signal intensity. The spinal cord and canal were also clearly visible. Artefact interference was minimised by a combination of sequences (1) and (2). The quality of all of the images met diagnostic requirements. Conclusions: Use of the combination of sequences described in this study may improve clinical MRI examination after internal fixation surgery.

1 INTRODUCTION

Magnetic Resonance Imaging (MRI) plays an important role in the evaluation and follow-up treatment for spine surgery patients. In recent years, there has been an increase in the prevalence of internal fixation implant surgery, requiring postoperative MRI examinations. Metal implants alter MRI signals, causing changes in the resulting images that affect any diagnosis. Therefore, there is little value in performing postoperative MRI examinations on patients with these implants. In the current literature, methods to minimise the affect artefacts mainly involve manipulating images acquired, using a single imaging sequence. There is no precise and feasible method to manage artefacts when scanning sequence combinations are used. In addition, the constraints of clinical work limit experimentation with sequence combinations. Therefore, it is urgent that appropriate, reasonable and feasible scanning sequence combinations be developed for patients, who receive internal fixation surgery of the lumbar vertebrae with titanium alloys, to minimise artefact interference. In this study, MRI examination of twelve patients was performed to investigate a scanning scheme that addresses this clinical need.

2 MATERIALS AND METHODS

2.1 Objects of study

From December 2012 to March 2013, 12 patients (8 male and 4 female) ranging in age from 22–63 years

Table 1. The specific information of the patients in our study.

Fixation sits	Male	Female	Total number
L5~S1	2	1	3
L4~S1	1	0	1
L1~L3	5	3	8

old were randomly selected from West China Hospital of Sichuan University. The patients had received internal fixation surgery of the lumbar vertebrae with a titanium alloy, and they were conscious during MRI examination. The specific patient information is shown in Table 1.

2.1.1 Selection criteria
1) The patients were Chinese and 18–75 years old.
2) All of the patients had internal fixation surgery of the lumbar vertebrae with a titanium alloy.
3) The patients were conscious during scanning.

2.1.2 Exclusion criteria
1) Pregnant woman.
2) Patients with contraindications for MRI examination.
3) Patients who had internal fixation surgery of the spine using non-Titanium alloy.

2.2 Examination instruments and materials

2.2.1 MR scanner

Siemens Trio 1.5 T I-class MR scanner (Siemens AG, Erlangen, Germany).

2.2.2 Picture Archiving and Communication System (PACS)

Siemens Syngo PACS (Version 35A, Siemens AG, Erlangen, Germany).

2.3 Scans

2.3.1 Scanning steps

(1) The patient was placed in supine position on the scanner bed and entered the magnet bore headfirst with hands positioned naturally to the side of the body. The scout scan (cross-section, sagittal section and coronal section) of the lumbar spine was performed using a GRE sequence. (2) A T2-weighted Fast Spin Echo sagittal fat suppression (T2-FSE-sag-fs) sequence was used in the scout scan of coronal sections. A scan of the median sagittal plane was performed in which the centre was positioned at the spinal cord of the lumbar vertebrae. The scout scan of the sagittal plane was centred at position 3 of the waist. (3) Rescans were performed according to the procedure in step (2) without fat suppression. The other parameters remained unchanged. (4) A T1-weighted Fast Spin Echo sagittal (T1-FSE-sag) sequence scan was performed with 3-mm slice thickness, 3-mm layer spacing, 10 layers, and the positioning line of T1 was copied. (5) A T2-weighted Fast Spin Echo transverse (T2-FSE-tra) sequence scan was performed. When a lesion was evident, the lesion site was scanned. In addition, scanning of at least two screw position levels was performed to investigate the relationship between the screw and the surrounding tissue.

2.3.2 The optimised scan sequences and parameters

The following is the optimised scanning scheme that was adopted for this study.

(1) Turbo spin echo (TSE) sagittal T2-weighted imaging (T2WI) fs sequence: $TR/TE = 3100$ ms/ 96 ms, $FOV = 380$, slice thickness $= 3.0$ mm, bandwidth $= 260$ Hz/Px, and echo train length $(ETL) = 19$.

(2) TSE (turbo spin echo) sagittal T2WI sequence: $TR/TE = 3100$ ms/96 ms, $FOV = 370$, slice thickness $= 3.0$ mm, bandwidth $= 260$ Hz/Px, and $ETL = 19$.

(3) TSE sagittal T1-weighted imaging (T1WI) sequence: $TR/TE = 600$ ms/93 ms, $FOV = 380$, slice thickness $= 3$ mm, bandwidth $= 252$ Hz/Px, and $ETL = 16$.

(4) TSE transverse T2WI sequence: $TR/TE = 3500$ ms/ 94 ms, $FOV = 210$, slice thickness $= 3.0$ mm, bandwidth $= 255$ Hz/Px, and $ETL = 19$.

2.3.3 Image analysis

(1) Investigating the imaged artefacts of patients subjected to internal fixation with a titanium alloy.

First, the images were searched for moving artefacts, chemical shift artefacts, etc. Next, the artefact size was approximated according to the following procedure: 1 If the edges of the titanium alloy objects were unclear, these edges were classified as having low signal shadows, high signal shadows, or mixed signal shadows. 2 If the signal of the internally fixed objects could not be identified, they were replaced by semi-circular or irregular shapes. 3 The effect of the artefact on adjacent vertebrae, tissue, and structures was evaluated.

(2) Identifying relevant structures and lesions for clinical diagnosis.

In the T1-FSE-sag sequence images, the anatomical structure of the tissue could be clearly identified. These images were used to determine the location of the internally fixed objects after surgery. In a previous study of 10 patients with nerve root damage after pedicle screw fixation surgery, who were examined by MRI, T1-weighted imaging was effective in identifying the position of the screw [3]. In the T2-FSE-sag/tra sequence images, the lesions had high signal intensity. In the T2-weighted sagittal plane images with fat suppression, the lesions were easily identified. However, they had more metal artefact interference than T2-weighted sagittal plane images without fat suppression. Therefore, the sequences with and without fat suppression were mutually complementary and could be used to prevent errors in identifying lesions caused by interference. In T2-FSE-sag/tra sequence images, the internal fixation object was surrounded by a high signal shadow with a clear demarcation line. The correct placement of the screw was seen as well as the relationship between the screw, the spinal cord, nerves and surrounding tissue. All of the images met clinical diagnostic requirements and were of good quality.

3 RESULTS

The scan sequences were performed according to the following scheme: (1) T2-FSE-sag-fs, (2) T2-FSE-sag, (3) T1-FSE-sag and (4) T2-FSE-tra. This combination was advantageous because it is highly repeatable, rational, and feasible for MRI examination of patients after internal fixation surgery of the lumbar vertebrae with a titanium alloy. MRI examinations of the patients did not cause any adverse reactions. In MRI images, artefacts from the internal fixation objects were visible. In the T2-weighted images without fat suppression, chemical shift artefacts between the cerebrospinal fluid in the dural sac and the fat behind the dural sac appeared as a black line (Figure 1). An obvious motion artefact was present in the image of one patient. The MRI images used sequence identification of the shapes of the fixed objects. Moreover, T1WI and T2WI of these objects revealed low-intensity signal areas with clearly delineated edges of high signal intensity. In the T2-weighted sagittal plane images

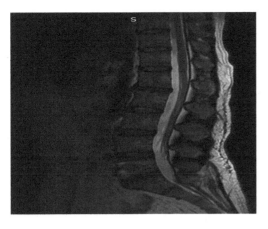

Figure 1. A sagittal T2-weighted image (without fat suppression) of the lumbar vertebrae.

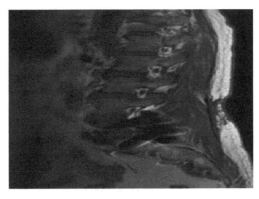

Figure 3. A sagittal T1-weighted image of the lumbar vertebrae of Patient A.

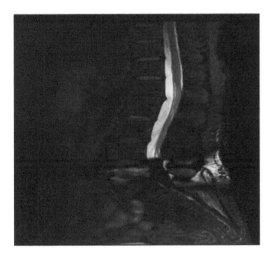

Figure 2. A sagittal T2-weighted image (with fat suppression) of the lumbar vertebrae.

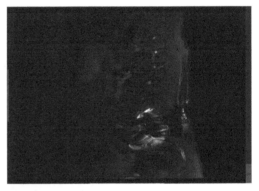

Figure 4. A sagittal T2-weighted image (with fat suppression) of the lumbar vertebrae of Patient A.

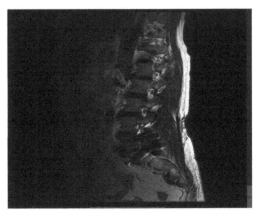

Figure 5. A sagittal T2-weighted image (without fat suppression) of the lumbar vertebrae of Patient A.

without fat suppression, the spinal cord and the structures in the spinal canal were visible, and artefact interference caused by internal fixation objects was small. Although the lesion detection rate was improved in the T2-weighted sagittal plane images with fs, the artefacts from the pedicle screw tails were large. This caused the doctors to focus on adjacent vertebral laminas and facet joints (Figure 2). Artefact interference was minimised through analysis of scans obtained using different sequences. It was convenient to observe the structures adjacent to the fixation objects, such as the lateral face of the spinal canal, vertebral artery, intervertebral foramen, spinal nerve, and prominent posterolateral intervertebral disc. The image quality met diagnostic requirements.

Figures 3–6 show the MRI images of Patient A, while Figure 7 shows his X-ray image.

Figure 1: In the sagittal T2-weighted image (without fat suppression) of the lumbar vertebrae, a black line was evident between the cerebrospinal fluid in the dural sac and the fat behind the dural sac that

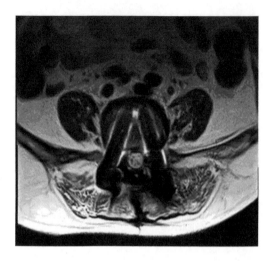

Figure 6. A transverse T2-weighted image (without fat suppression) of the lumbar vertebrae of Patient A.

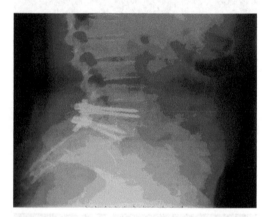

Figure 7. Plain film X-ray of the lumbar vertebrae of Patient A.

corresponded to a chemical shift artefact. After fat suppression, the black line disappeared (Figure 2). However, the artefact interfered with the display of the intervertebral discs of L_{4-5} and L_5-S_1, the L_5 vertebral body and the corresponding plane of the vertebral canal.

Figure 3: In the sagittal T1-weighted image of the lumbar vertebrae of Patient A, the screw appears as wide, belt-like low-intensity signal surrounded by a narrow, belt-like low-intensity signal shadow. The interference from the artefacts was minimal. The other anatomical structures of the lumbar region were well-defined.

Figure 4: In the sagittal T2-weighted image (with fat suppression) of the lumbar vertebrae of Patient A, reducing the high-intensity signal from fat in the tissue improved the detection rate of the lesions. However, the artefacts generated by the screw significantly interfered with visualisation of the sacral vertebrae. In addition, the muscle tissue and the epidermis fat

behind the screws had fan-shaped artefacts, which reduced the lesion detection rate in these areas.

Figure 5: In the sagittal T2-weighted image (without fat suppression) of the lumbar vertebrae, which is shown in the same plane as Figure 4, the screw artefact was significantly reduced. The influence of the artefact on the sacral vertebrae and the muscle behind the screws was eliminated. Therefore, the lesions were more easily identified, which could reduce misdiagnosis.

Figure 6: In the transverse T2-weighted image (without fat suppression) of the lumbar vertebrae of Patient A, the shape of the screw artefact was identical to that in the sagittal plane. The screw was in the correct location, and the relationship between the screw and the surrounding structures was evident. This image supplemented the cross-sectional view, which was visible by X-ray (Figure 7).

4 DISCUSSION

4.1 Artefacts

The multifaceted and parameterised nature of MRI, the complexities of the imaging procedure and long imaging times lead to various types of artefacts. These include artefacts arising from the magnetic field, gradient field, Radio Frequencies, image processing and human body [5].

Magnetic field homogeneity is fundamental to the quality of MRI images. Several factors may cause the main magnetic field to vary, which causes magnetic field-related artefacts. It is important to understand the source of artefacts and take appropriate measures to prevent them to improve the accuracy of MRI-based clinical diagnosis.

4.2 Safety of MRI examination after internal fixation surgery with a titanium alloy

Titanium alloys are non-ferromagnetic and consequently do not generate deflection stress in static magnetic fields. In addition, the heating effect of the current on the implant and the surrounding tissue is negligible [1]. In this study, none of the patients had an adverse reaction to MRI examination. Furthermore, MRI did not cause displacement of the internal fixation object during the study or afterwards. Therefore, patients subjected to internal fixation surgery with a titanium alloy may safely be examined by MRI [6].

4.3 Selection of the 1.5 T MRI

The Signal to Noise Ratio (SNR) in MRI images increases as the main magnetic field strength increases. However, scanners with high magnetic field strengths are subject to more artefacts caused by human physiological movement and metallic foreign bodies than scanners with low magnetic field strengths. Previous studies of body implant safety have focused mainly on magnetic fields of 1.5 T or less. A recent study reported that while some metal implants were only weakly magnetic in 1.5-T fields, they were strongly magnetic in 3.0-T fields [7]. Therefore, a Siemens 1.5 T MR scanner was used in this experiment to ensure high SNR of the images, to minimize image artefacts and to prevent any potential danger to the patient.

4.4 *T2-weighted FSE*

The titanium alloy implants caused artefacts in images generated using GRE, SE and FSE sequences. Among these sequences, T2-weighted GRE generated the largest artefacts, indicating that doctors should try to avoid using this sequence in the clinic. In an examination of the cervical vertebrae, Xinhau et al. reported that, in sagittal and cross-sectional T2-weighted FSE images, the spinal cord was clearly visible and the cerebrospinal fluid was characterised by a high-intensity signal. The structure and contours of the vertebral bodies and attachments could also be clearly identified. Furthermore, the nerve root was more visible in T2-weighted FSE images than in T2-weighted GRE images [4]. Therefore, T2-weighted FSE was used in the scanning solution.

4.5 *The application of the fat suppression technology of STIR*

In this study, STIR technology was investigated because it has a low dependency on magnetic field strength and does not require magnetic field homogeneity. Also, fat suppression by STIR is good when the FOV value is large. The characteristics of the artefacts caused by internal fixation objects acquired by T2WI with and without fat suppression were different. STIR technology with and without fat suppression was used in the same patients (all other parameters were unchanged). In the images obtained without fat suppression, the artefacts significantly decreased. The location and shape of the internal fixation objects were the same as those observed in plain film X-rays, indicating that the quality of the image did not adversely affect the clinical diagnosis (Figure 5). In contrast, when fat suppression was used, the artefacts caused by the internal fixation objects greatly increased. The adjacent tissue was partially concealed, and the relationship between the screw and the cortex of the bone was unclear (Figure 4). However, lesions were more evident using this imaging sequence. Therefore, combination of the two sequences may reveal postoperative complications better than either sequence alone and enable effective follow-up [8].

4.6 *Parameter optimization in MRI examination of patients with titanium alloy internal fixation objects*

The turbo spin echo sequence is the most commonly used sequence in magnetic resonance imaging [2].

Occasionally, additional parameters, such as fat suppression, increases in bandwidth, and a reduction of the FOV value must be applied to obtain a satisfactory image. Frequency-encoding and slice-selection gradients are more prone to magnetic susceptibility of metallic materials than phase-encoding gradients. The direction of the frequency-encoding gradient can control the direction of the artefacts. Shifting the artefacts away from the region of interest (ROI) as much as possible is important when selecting the direction of the frequency-encoding gradient. In addition, the imaging plane should be chosen to avoid or minimise the inclusion of metallic materials. Increases in echo train length and shorter echo train spacing cause reduction in artefacts. However, the SNR of the image decreases when the echo train is too long. Therefore, a medium echo train length (ETL= 10-20) was used for the sequences in this study. In addition, artefacts are reduced as the slice thickness increases. Therefore, parameters should be optimally adjusted in order to control any artefacts, taking into account the critical factors that must be balanced to improve image quality.

5 CONCLUSIONS

Patients who had internal fixation surgery of the lumbar vertebrae with a titanium alloy were evaluated by MRI. A combination of imaging sequences was investigated, including T2-FSE-sag-fs, T2-FSE-sag, T1-FSE-sag and T2-FSE-tra. Use of this MRI scanning scheme could improve the clinical evaluation of patients. Compared with conventional methods for scanning the lumbar vertebrae, the images acquired in this study were more effective in revealing changes after internal fixations of the spine. When doctors analyse postoperative MRI images, they should be fully aware of the artefacts induced by internal fixation, objects containing titanium, and how they affect the accuracy of acquired images. Doctors can combine MRI results with clinical manifestations of the patient to improve diagnosis

REFERENCES

Dai, Liyang. 2002 The imaging manifestation of common complications of spinal surgery. The Journal of Cervicodynia and Lumbodynia. 3: 3–5.

He, Shisheng & Yang, Yonghua. 2007. The influence of the internal fixation system of the posterior cervical spine with titanium alloy on MRI. Orthopedic Journal of China. 11: 23–25.

Huang, Qun & Yang, Haitao. 2007. To study the application after the surgery of the spinal internal fixation. Radiologic Practice. 7: 22–25.

Huang, Xinhua & Peng, Guangming. 2001 The comparison between the FSE image and GRE T2-weighted image in the cervical spine examination. Journal of Practical Radiology.12: 2–5.

Liu, Juan. 2009. To study the artifacts of different dental metallic materials and the artifacts' control in MRI examinations, Master's thesis, Tianjing Medical University.

Luo, Wenbing. 2010. The diagnosis of degenerative joint disease with X-ray plain film and MRI. Modern Diagnosis & Treatment. 5: 21–24.

Yang, C.W & He, C.S. 2008. The technology progress of the spinal metallic implants' artifacts in MRI. Journal of Spinal Surgery. 6: 16–18.

Zheng, C. & Diao, Q. 2010. The technology and clinical application of fat suppression in MRI. Chinese Medical Equipment Journal. 12: 1–5.

Biomedical Engineering and Environmental Engineering – Chan (Ed.)
© 2015 Taylor & Francis Group, London, ISBN: 978-1-138-02805-0

A digital design of a humanoid model of an auricle based on 3D printing

Tao Jiang, Jian-zhong Shang, Li Tang & Zhuo Wang
Laboratory of Mechanism Design, School of Mechatronics Engineering and Automation,
National University of Defense Technology, Changsha, P.R. China

ABSTRACT: Auricular deformity is a common clinical disease. The currently most used method in curing it is through operation, in which patient's costal cartilage is harvested and sculptured to an ear's shape based on a 2D paper model which mimics the normal ear on the opposite side. This method requires relative high costs and long period, and it is difficult to ensure the accuracy in the sculpturing process because of the non-intuition of 2D model. This paper explores a novel approach based on 3D printing, by which auricular shape data is collected through CT scan, the digital ear model is built by CAD software and finally the model is 3D printed. Through the experiments, the model was successfully built and printed by 3D printer of FDM principles, with high similarity of the real ear. This study provides solid reference to the clinical sculpture, and laid foundation for further research of transplanting bio-material 3D printed ear into the patient.

1   INTRODUCTION

Auricular deformity is a commonly seen clinical illness (Jiao et al. 2001). Based on clinical statistics, paediatric microtia occurs in every 1500 to 2000 births, which counts for 1/2090 of all auricular diseases (Yan 2000).

Autologous tissue transplanting surgery is the most frequently used treatment for this disease, in which patient's costal cartilage is harvested, a 2D paper auricular model is trimmed, being based on the shape of normal ear, after which the cartilage is sculptured to an auricular scaffold according to the paper model (see Figure 1) and then implanted under the periauricular skin, and finally the cavity is vacuumized to adhere the skin to the scaffold and form a normal auricle (Jiang et al. 2006). This surgery is very technique demanding, and costs high operation fees, and most importantly, surgeons have to sculpture a 3D model based on a 2D paper model, making the process operate rather slowly, and the accuracy cannot be preferably guaranteed.

With the development of 3D printing technology, it becomes feasible to rapidly manufacture such complex structures like auriculars. 3D printing is a follow-up development of Rapid Prototyping technology raised in 1980s. Compared with conventional manufacturing methods, 3D printing is a kind of additive manufacturing, in which digital models are sliced along the modelling orientation by CAD software, and thermally melting sticky materials are utilized to build the model slice by slice. This method is capable of manufacturing models with complex inner structures that are unable to be manufactured by traditional ways, simultaneously with a fast building speed and convenience of modification (Campbell et al. 2011). All the above advantages enable 3D printing to be spread widely in fields like

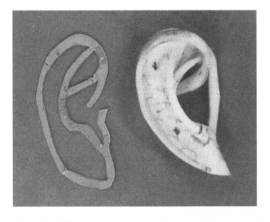

Figure 1.   2D paper model and cartilage model sculptured based on the paper in current operation.

astronautics, architecture, and medical equipment, etc (Wohlers Associates 2013).

In this paper, the original auricular digital data were acquired from CT scans. An auricular model was built and smoothed by reversed modelling technology, and a solid auricular model was 3D printed with high accuracy and at low cost, which could provide clinical surgeons with a 3D model reference from the sculptured process.

2   MODEL DESIGN AND OPTIMIZATION

2.1   CT data collection

A spiral CT scanner (Toshiba Aquilion/64) was used to scan the head of a 49-year-old male, from which

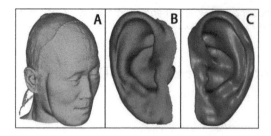

Figure 2. 3D model was rebuilt from CT images through Mimics, and the post-process was executed in Studio 2012. A: original CT data of the head; B: model of the normal side of ear; C: mirrored and optimized auricular model.

original 2D CT image sequences were acquired with DICOM format (.dcm). The scanning parameters were:

Matrix size: 512pxl × 512pxl;
FOV = 23.96 cm;
Pixel Size: 0.468 mm;
Algorithm: FC23;
Total number of slices: 332.

Images were imported into Mimics (Materializes Corp, Belgium) to filter necessary data and finally converted to a stereolithography (STL) format. The original CT data contain all scanned information (including bone, soft tissues, skin, fat, etc, see Figure 2A), while only skin information is required for the case of auricular modelling. To achieve best quality of the skin and simultaneously with minimal noise information, experiments were conducted and the results showed that the best thresholding was between −776 and −202 in this case. Moreover, the slices were cropped along the normal side of the auricle to maximally reduce useless information and the model was then built (see Figure 2B) and converted to a STL format for post-process.

2.2 3D solid model of auricle

The surface of the directly converted model was rough and spiky, the transition between the triangles were drastic, which was beyond the nature of a real auricle, hence a post-process was required.

The model was then imported into Studio 2012 (Geomagic Inc. US). First, the model was mirrored to obtain the opposite ear of the normal one. Then the triangle mesh was optimized to balance their distribution on the auricular surfaces. Model noise reduction and smoothness was implemented after the optimization, and small holes were filled, spiky objects were removed and regional mesh deformities were corrected afterwards. Finally, the auricular model, with relatively high surface quality, was obtained (see Figure 2C), the total triangle number of which is 89136, and average length of each edge is 0.37 mm.

At this point the model was already ready to be printed, as STL files are widely supported by common 3D printers nowadays. However a model with

this format cannot be recognized as a solid part in CAD software because of the lack of solid elements information (like surface info and curve info), which make it difficult for further analysis, and for this reason, the STL model needed to be solidified. Studio 2012 is capable of converting triangle models into Non-Uniform Rational β-Splines (NURBS) solid models. NURBS are commonly used to describe different curves and surfaces, which is an extension of β-splines and Bezier curve/surfaces (Wu & Hu 1994). The definition of NURBS is:

$$r(t) = \frac{\sum_{i=0}^{n} N_{ik}(t)\omega_i P_i}{\sum_{i=0}^{n} N_{ik}(t)\omega_i} \tag{1}$$

where the vector of node: $T = [t_0, t_1, t_2, \cdots, t_{n+k}]$, and

$$N_{i1}(t) = \begin{cases} 1, t_i = t < t_{i+1} \\ 0, others \end{cases}$$

$$N_{ik}(t) = \frac{t - t_i}{t_{i+k-1} - t_i} N_{ik-1}(t) + \frac{t_{i+k} - t}{t_{i+k} - t_{i+1}} N_{i+1-k}(t)$$

Similarly, a NURBS surface can be defined as:

$$r(u, v) = \frac{\sum_{i=0}^{m}\sum_{j=0}^{n} N_{ik}(u) N_{jl}(v)\omega_{ij} V_{ij}}{\sum_{i=0}^{m}\sum_{j=0}^{n} N_{ik}(u) N_{jl}(v)\omega_{ij}} \tag{2}$$

$$\begin{cases} U = [u_0, u_1, \cdots, u_{m+k}] \\ V = [v_0, v_1, \cdots, v_{n+l}] \end{cases}$$

The detailed definition of the parameters in the above equations can be found in references (Leslie & Wayne 1987), (Wayne 1983) and (Leslie 1991).

By defining and editing the node configuration of splines and surfaces, such as node position and the curvature at node, NURBS are capable of precisely describing complex object surfaces, which meet the requirements of auricular modelling.

A solidifying process was taken to the STL auricular model with the above mentioned methods, and some regional bad meshes, with relatively high long-width ratios, were manually adjusted to improve the quality of the surface. The solidified model was compared with the original triangle model to examine the error during the process. The results are shown below:

Maximum positive/negative error:

$\delta_{max}^+ = 0.071$ mm, $\delta_{max}^- = -0.053$ mm;

Average positive/negative error:

$\overline{\delta}^+ = 0.005$ mm, $\overline{\delta}^- = -0.004$ mm;

Overall average error: $\overline{\delta} = 0.001$ mm;

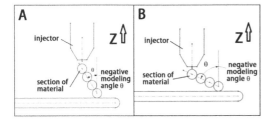

Figure 3. Printing process with different negative s modelling angle θ, where Z represents the modelling orientation. A: a small θ will not result in collapse; B: the θ is too large and the printed material has a high risk of collapse in the printing process.

Standard deviation: $\sigma = 0.007\,mm$.

No obvious visual error could be seen, and the existing minor error mainly distributed along the cutting edge of the auricle, which is the results of the elimination of sharp edges during the solidifying process, while this operation did not affect the exterior shape of the auricle.

The solidified model was then converted to IGES format by a general CAD port, and imported into SolidWorks for printing analysis and optimization.

2.3 Model optimization

As the complex structure of an auricle is rather different from general mechanical parts, printing orientation needs to be optimized in case of deformation or failure in the printing process. Hence, the position and angle parameters of model $[x, y, z, \theta, \phi, \gamma]$ require adjustments according to the feature of 3D printing. This paper proposes to optimize the parameters by the Draft Analysis function in SolidWorks (Dassault Corp, US).

As 3D printing with Fused Distribution Model (FDM) principle is to be used, the following three factors should be considered during the optimizing process:

(1) As FDM principle features with additive manufacturing slice by slice, if hanging structures exist, a support will need to be manufactured to hold them. Whereas certain damage to the model will be inevitable during the elimination of the support after the printing process, and to ensure the smoothness of the front surface of the auricle, support holding these surfaces should be avoided.
(2) The negative modelling angle needs to be reduced to a minimal value as a large negative modelling angle will induce the collapse of the material in the printing process because of gravity (see Figure 3).
(3) The inner side of the model should be filled with a robust but simple structure to minimize thermal deformation during the printing process as well as keeping the manufacturing time short.

The auricle model with .IGES format was imported to SolidWorks and draft analysis and it was then implemented. The drafting angle was set to be 0, and the model was then rotated along each axis according to

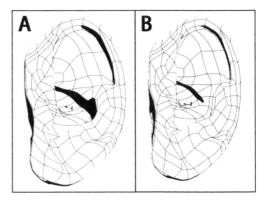

Figure 4. Drafting analysis results before and after optimization, where the white area is the positive drafting area, while the black is the negative drafting area.

Table 1. Printing parameters.

Printer	CubeX	Printing speed	40 mm/s
Material	PLA	Injector temperature	210°C
Slice thickness	0.1 mm	Platform temperature	55°C
Number of shell	5	Diameter of Material	1.75 mm
Filling density	25%	Import format	STL
Filling shape	hexagon		

the analysis results for the purpose of maximize the positive drafting area on the front side of the auricle (see Figure 4). Notably, since the surface structure is rather complex, it is impossible to make the entire front side of the positive drafting area.

3 EXPERIMENT AND DISCUSSION

3.1 Printing configuration

The printer was a CubeX from 3D Systems Corp. The maximum supported printing volume of the printer is $275\,mm \times 265\,mm \times 240\,mm$, and minimal slice thickness is 0.1 mm, which is very suitable for printing the model with high accuracy. The back side of the model was held by a support to avoid the material collapsing, while the front side was free of any support. The inner side of the model was filled with a hexagonal alveolate structure to enhance its intensity and reduce bending deformation when printing. 5 slices of shells were printed to ensure the surface quality. The printing room was closed to eliminate unexpected thermal flow and the platform temperature was maintained at 55 cent degree. More detailed printing parameters can be seen in Table 1.

The printing process took about 2 hours. The printed model had high surface quality, and no material collapse or bending deformation occurred (see Figure 5A). The support was then removed from the auricular model, and some minor surface spots were eliminated by a polishing process. The final auricular model and

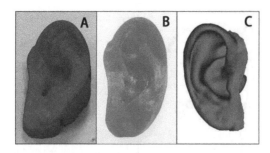

Figure 5. Completed printed auricular model. A: auricular model before the support was removed; B: finished auricular model after removing the support and eliminating surface spots; C: original digital auricular model in Mimics.

its comparison to the CT model can be seen in Figure 5B, C).

3.2 Discussion

The auricular model with PLA material was successfully manufactured by the procedures above, which was mirrored from the original CT auricular data without any changes in size or shape. Since the hexagonal alveolate structure was used as an inner filling structure and support structure, no obvious deformation occurred during the printing process and the printed model maintained a high accuracy. The printed model preferably described the complex structure of a human auricle, and compared with the current 2D paper model, this 3D model can provide a more vivid reference for surgeons, and subsequently improve the accuracy of the sculptured cartilage scaffold.

4 CONCLUSION

Through theoretical analysis and experimental practice, reversed modelling of an auricular model was completed, surface quality improvements and model orientation optimization, based on 3D printing principles were implemented, and finally the 3D auricular model was printed. The duration of the entire process can be completed in under 3 hours and the printed model has achieved a high enough quality to describe the complex structure of a real human auricle, while the total cost of the process is quite low. To summarize the analysis above, the research meets the requirements of rapid manufacture, high accuracy, and low cost.

3D printing of human auricles is an intersection with such fields as bio-engineering, material engineering, and mechanical engineering, and it is a hot issue in the world. It can be predicted that with further research, the 3D printed bio-material auricle can take the place of the cartilage one and be directly transplanted into patients' bodies. It will also be start a trend for curing auricular diseases like paediatric microtia.

REFERENCES

Campbell, T.A., Williams, C.B., Ivanova, O.S., Garrett, B. Strategic Foresight, Report No. 1, Strategic Foresight Initiative, Atlantic Council, 2011.

Chen-xian Yan. Pediatric Otorhinolarynogology. Tianjin: Tianjin Science and Technology Press, 2000: 127–131.

Hai-yue Jiang, Bo Pan, et al. Ear reconstruction using soft tissue expander in the treatment of congenital microtia. Chinese Journal of Plastic Surgery, 2006; 22(4): 286–289.

Piegl Leslie, Tiller Wayne. Curve and surface constructions using rational B-spline. CAD, 1987; 19(9).

Piegl Leslie. On NURBS: A survey. IEEE CG&A, 1991; 7.

Ting Jiao, Fu-qiang Zhang. The present situation of auricular prostheses. Chinese Journal of Dental Materials and Devices, 2001; 10(4): 213–215.

Tiller Wayne. Rational B-spline for curve and surface representation. IEEE CG&A, 1983; 9.

Wohlers Associates, Wohlers Reports 2013 - Additive manufacturing and 3D printing state of the Industry Annual Worldwide Progress Report, 2013, USA.

Zhong-ke Wu, De-rong Hu. Modeling Methods on NURBS. Journal of Beijing University of Aeronautics and Astronautics, 1994; 20(2): 198–206.

Biomedical Engineering and Environmental Engineering – Chan (Ed.)
© 2015 Taylor & Francis Group, London, ISBN: 978-1-138-02805-0

The effect of Compound-Maqin-Decoction on airway inflammation in asthmatic rats

Yan-Hong Xie, Zhao-Xia Xu, Xue-Liang Li, Peng Qian, Xue-Ping Li & Yi-Qin Wang
Shanghai University of Traditional Chinese Medicine, Shanghai, China

Na Li
Hubei University of Traditional Chinese Medicine, Wuhan, China

ABSTRACT: To research the intervention effect of Compound-Maqin-Decoction (CMD) on airways in asthmatic rats, we reproduced an asthma model of rats using ovalbumin and observed the effects of CMD, such as the general, pathologic changes in lung tissues, blood, lung coefficient, eosinophil (EOS) count in bronchoalveolar lavage fluid (BALF), and content of TNF-α and IL-6 in the serum of rats with asthma. We found that CMD could improve the appetite, activity, and pathologic features of asthmatic rats. After treatment with CMD, the lung coefficient in the CMD group decreased compared with the model and the dexamethasone groups; the difference was statistically significant. Compared with the model group, the WBC count decreased, the RBC count increased, the percentage of lymphocytes increased, and the percentage of monocytes decreased in the CMD group; the difference was statistically significant. Compared with the model group, the eosinophil count in BALF of rats in the dexamethasone and CMDgroups was decreased; the difference was statistically significant. Compared with the model group, the expression of TNF-p and IL-6 in the sera of rats in dexamethasone and CMD groups was decreased; the difference between the model and CMD groups was statistically significant. These results indicate that CMD can improve airway inflammation in asthmatic rats.

1 INTRODUCTION

At present, a wide variety of animal models exist for asthma, but no animal model can precisely replicate the pathophysiologic changes with human bronchial asthma; only some features which cause asthma are similar in humans. Airway inflammation is thought to be an important pathological basis for airway hyper-responsiveness [Wang JY. 2005], therefore an important therapeutic approach for asthma is to effectively control airway inflammation and reduce airway hyper-responsiveness. Compound-Maqin-Decoction (CMD) can regulate immune function, inhibit inflammation, relax bronchial tubes and relax muscles, control airway inflammation, and improve lung function in asthmatic guinea pigs. These findings have been confirmed by clinical and experimental studies over the past 20 years. In the current study, we collected the following data to determine the effect of CMD on airway inflammation in asthmatic rats: lung histopathologic changes; routine blood testing; lung coefficient; eosinophil (EOS) count in the Bronchoalveolar Lavage Fluid (BALF); and the serum Tumour Necrosis Factor (TNF)-α and Interleukin (IL)-6 expression.

2 MATERIAL AND METHODS

2.1 Animal grouping

Forty healthy SD male rats (Shanghai SLAC Laboratory Animal Co., Ltd, Shanghai, China), weighing 200 ± 20 g, were randomly divided into four groups, as follows: a normal control group; a model group; a dexamethasone group; and a CMD group.

2.2 Modelling method

The asthma rat model was prepared by injecting Ovalbumin (OVA) into the abdominal cavity and inhalation of aerosol based on the modelling method previously described [Wasemmn S et al, 1992; Lv GP et al, 1995]. On the first day after animal grouping, each rat in groups B, C, and D was injected intraperitoneally with 1 mL of solution containing 100 mg of OVA and 100 mg of aluminium hydroxide. After 14 days, an aerosol inhalation of 1% OVA was administered for 20 min at a flow rate of 2mL·min^{-1}, and continued for 28 days. The rats in group A were injected with 1 mL of 0.9% sodium chloride solution and an aerosol inhalation of 0.9% sodium chloride solution was administered with a flow rate of $2\,mL\cdot min^{-1}$, and this continued for 28 days.

2.3 Treatment intervention

Dexamethasone tablets (No. H3120793-01; Shanghai Xinyi Pharmaceutical Factory Co., Ltd., Shanghai, China) were prepared as a suspension ($0.32\,mg\cdot mL^{-1}$) with 0.9% sodium chloride solution. CMD consists of Zhimahuang (Ephedra; 4 g), Huangqin (Scutelleria; 9 g), Cangerzi (Xanthium; 9 g), Tianzhuzi (Nandina domestica; 9 g), Lameihua (Chimonanthus praecox flowers; 9 g), and Hutuiziye (Elaeagnus leaf; 9 g). These prescribed herbs were soaked in water for 1 h. Ten volumes of water were added during the first cooking time. Four volumes of water were added during the second cooking time. Then, the prepared decoctions from both cooking times were collected, evaporated, and concentrated into a $3\,g\cdot mL^{-1}$ crude drug solution.

The first day after asthma was induced, the rats in groups A and B were given normal saline by gavage, the rats in group C were given dexamethasone by gavage, and the rats in group D were given CMD decoction by gavage ($10\,mL\cdot kg^{-1}$ body weight once daily). After 4 weeks, the materials were collected for the detection of relevant indicators.

2.4 Apparatus and reagents

The following equipment was used: atomizer (Shanghai Fulin Medical Equipment Co. Ltd., Shanghai, China); low magnification ELX800 light absorption microplate reader (Bio-Tek, high-precision pipettes and pipette tips (10, 20, 200, and $1000\,\mu l$); and water bath and ADVIA 120 hemocytometer (Bayer Corporation).

The following reagents were used: OVA (No. F20110323; Shanghai Sinopharm Chemical Reagent Co. Ltd., Shanghai, China); aluminium hydroxide powder (No. 101001; Shanghai Sinopharm Chemical Reagent Co. Ltd.); 0.9% sodium chloride solution (No. 10123050; Shanghai Sinopharm Chemical Reagent Co. Ltd.); ELISA kit (Shanghai Dior Biotechnology Co. Ltd., Shanghai, China); rat TNF-α (No. CK-E30436); rat IL-6 (No. CK-E30437); and EOS direct counting solution (No. 20110608; Nanjing Jiancheng Bioengineering Institute, Nanjing, China).

2.5 Detection method and indicators

The respiratory rate, the presence or absence of dyspnea, cyanosis, cough, and other respiratory tract symptoms were studied in each group of rats. The lung coefficients [lung coefficient = lung weight (g)/body weight (kg)] were determined at the end of treatment, then the rats were killed and the lungs were removed and weighed.

Routine blood test results were compared in each group of rats. Abdominal aortic blood (1 mL) was collected into an anticoagulant tube, shaken, and tested by the Technology Experiment Center at Shanghai University of Traditional Chinese Medicine.

The preparation of BALF was as follows: a contralateral lung was douched, and the left lung was ligated; a puncture needle was inserted into the upper end of the trachea, and 5 mL of saline was used for irrigation three times; and BALF was recovered and centrifuged for 10 min at 4°C and 3000 r/pm. The precipitate was used for the EOS count. The EOS count was performed as follows: the BALF precipitate was collected and dissolved with $100\,\mu L$ of PBS; $20\,\mu L$ of suspension was obtained and added to 0.38 mL of the EOS counting solution; the mixture was allowed to stand for 5–10 min; $20\,\mu L$ of solution was put on the counting plate; cells were counted at low magnification; and the EOS were counted in 10 large squares of two counting chambers. The method used to calculate the EOS count was as follows: EOS/L = EOS count in 10 large squares $\times 20 \times 10^6$.

Serum was prepared for ELISA testing. The blood was collected and centrifuged for 10 min at 3000 r/pm. Then, the supernatant was collected and stored at $-20°C$. The serum expression of TNF-α and IL-6 was tested by ELISA. Steps were carried out strictly in accordance with instructions included in the ELISA kit. To increase the accuracy of data, within 15 min of adding the stop solution, the wavelength (OD) value was measured in a microplate reader at 450 nm. The OD value was recorded every 3 min for a total of three times. The average value of 3 OD readings was achieved. The standard curve (R2 values ≥ 0.99) was obtained using SPSS software. Then, the TNF-α and IL-6 concentrations in serum of each rat were calculated according to the curve.

2.6 Statistical analysis

All of the experimental data were analysed statistically by SPSS 18.0 statistical software. ANOVA was used for the analysis of differences between groups. The inspection level was $\alpha = 0.05$. $P < 0.05$ was considered to be a statistically significant difference.

3 RESULTS

3.1 General performance

The rats in the normal group exhibited good performance, were lively with greater mobility, had supple fur, gained weight, and had no evidence of respiratory impairment. The rats in the model group were more sluggish, moved more slowly, had less glossy fur, were restless after asthma was induced, had shortness of breath, sneezed and coughed, had mild cyanosis, had muscle twitching, and were unresponsive compared with the rats in the normal group. After treatment with dexamethasone and CMD, the asthma symptoms improved; the CMD group improved more than the dexamethasone group.

Table 1. Comparison of rat lung coefficients.

Group	n	Lung coefficient Range	Mean±SD	
Group A	10	6.52–8.14	7.33 ± 1.14	
Group B	10	7.72–10.58	9.15 ± 1.85	▲P = 0.008
Group C	10	9.25–11.67	10.46 ± 1.57	▲P = 0.000; ●P = 0.348
Group D	10	8.03–9.86	8.38 ± 1.73	▲P = 0.006, ●P = 0.759, ★P = 0.026

Note: ▲Compared with group A; ●compared with group B; ★compared with group C.

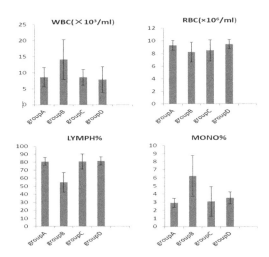

Figure 1. Comparison of routine blood tests results in rats.

3.2 Morphologic research involving bronchopulmonary tissues in asthmatic rats

The bronchopulmonary tissues of rats group A had no significant damage, the structure was normal; bronchial epithelial integrity was intact at all levels; there was no exudate within the lumen; and there was no significant inflammatory cell infiltration. The bronchial epithelial cell of the rats in group B had sporadic injuries, and a shedding of epithelial cells was noted in the bronchial cavities, the smooth muscle fibres had mild hyperplasia, and the pulmonary artery wall was thickened. The bronchial epithelial of rats in group C had mild edema, a large amount of exudate in the bronchial cavity, and a small amount of inflammatory cell infiltration in the bronchial walls. There was a small amount of inflammatory exudate in the bronchial cavities of the rats in group D.

3.3 Comparison of lung coefficients between rats

Compared with group A, the lung coefficients in groups B, C, and D increased (P < 0.05). Compared with group C, the lung coefficients in group D decreased (P < 0.05). The results are shown in Table 1.

3.4 Comparison of routine blood test results in rats

Compared with group A, the WBC count was increased, the RBC count was decreased, the percentage of lymphocytes was decreased, and the percentage of monocytes was increased in group B; the difference was statistically significant. Compared with group B, the WBC count was decreased, the RBC count was increased, the percentage of lymphocytes was increased, and the percentage of monocytes was decreased in group D; the difference was statistically significant. The WBC count was decreased, the percentage of lymphocytes was increased, and the percentage of monocytes was decreased in group C. With the exception of the RBC count, the difference was statistically significant. The results are shown in Figure 1.

Table 2. Comparison of the EOS count in BALF of rats.

Groups	EOS counting in BALF (×106/L) Range	Mean ± SD	P-value
Group (n = 10)	0.14–0.22	0.18 ± 0.06	
Group B(n = 10)	0.35–0.45	0.40 ± 0.06	▲P=0.000
Group C(n = 410)	0.29–0.35	0.32 ± 0.04	▲P=0.000; ●P = 0.009
Group D(n=9)	0.22–0.32	0.27 ± 0.07	▲P=0.001, ●P = 0.000, ★P = 0.076

Note: ▲Compared with group A; ●compared with group B; ★compared with group C.

3.5 Comparison of the EOS count in the BALF of rats

Compared with group A, the EOS count in BALF of rats in the other groups was increased; the difference was statistically significant. Compared with group B, the EOS count in the BALF of rats in groups C and D was decreased; the difference was statistically significant. Comparing group C with group D, there were no statistically significant differences. The results are shown in Table 2.

3.6 Comparison of the serum expression of TNF-α and IL-6 in rats

Compared with group A, the serum expression of IL-6 of rats in the other groups was increased; the difference between group A and groups B and C was statistically significant. Compared with group B, the serum expression of IL-6 in rats in groups C and D was decreased; the difference between groups B and D was statistically significant. The serum expression of IL-6 in group D

63

Table 3. Comparison of serum expression of IL-6 in rats.			
	Expression of IL-6 in resuming/ml)		
Groups	Range	Mean ± SD	P-value
Group A (n = 10)	0.34–0.47	0.41 ± 0.09	
Group B (n = 10)	0.46–0.57	0.52 ± 0.02	▲P = 0.003
Group C (n = 10)	0.46–0.56	0.51 ± 0.07	▲P = 0.006; •P = 0.8288
Group D (n = 9)	0.36–0.46	0.41 ± 0.07	▲P = 0.902, •P = 0.005, ★P = 0.008

Note: ▲Compared with group A; •compared with group B; ★Compared with group C.

Table 4. Comparison of serum expression of TNF-s in rats.			
	Expression of TNF-s in serum (ng/ml)		
Groups	Range	Mean ± SD	P-value
Group A (n = 10)	0.08-0.09	0.08 ± 0.01	
Group B (n = 10)	0.09-0.11	0.10 ± 0.01	▲P = 0.001
Group C (n = 10)	0.09-0.10	0.09 ± 0.01	▲P = 0.019; •P = 0.245
Group D (n = 9)	0.08-0.09	0.08 ± 0.01	▲P = 0.790, •P = 0.002, ★P = 0.036

Note: ▲Compared with group A; •compared with group B; ★compared with group C.

rats was decreased, compared to group C; the difference was statistically significant. The results are shown in Table 3.

Compared with group A, the serum expression of TNF-s in rats in the other groups was increased; the difference between group A and groups B and C was statistically significant. Compared with group B, the serum expression of TNF-s in rats in groups C and D was decreased; the difference between groups B and group D was statistically significant. The serum expression of TNF-sin rats in group D was decreased compared to group C; the difference was statistically significant. The results are shown in Table 4.

4   DISCUSSION

Organ weight and coefficient (organ weight/body weight × 100%) of experimental animals are major biological characteristics and an important basis for the identification of animal genetic quality. In biomedical research, the organ weight and coefficient can also be used to measure and reflect on the functional status of animals [Sun JX, et al. 2009]. The variation in organ coefficient often better reflects the organ toxicity of chemical poisons and circumstantial evidence of histopathologic changes, but it is also an important clue to determine the role of the target organ toxicant [Huang YQ, et al, 2003]. After modelling, the exudate of inflammatory lung tissues, cell edema, and capillary congestion can be a direct cause of the increased lung coefficient.

There is a close relationship between a variety of cytokines and asthma, and TNF-α and IL-6 are more closely related to airway inflammation and pathologic changes. TNF-α is secreted from activated macrophages and lymphocytes. TNF-α is one of the important factors in the pathogenesis of asthma, and an important mediator of the immune defence system and inflammatory responses [Rubira Garcia N. 2000]. TNF-α can promote a number of cytokines to cause a chain reaction within the body, thus participating in the physiologic process of injury to the body's

immune system [Li YQ. 2007]. Studies have shown that TNF-α has a biological role to enhance local vascular permeability, but also promotes the release of active substances [Brightling C, et al. 2008; Berry M, et al, 2007; Brightling C, et al, 2008], then it leads to the appearance of clinical symptoms of bronchial asthma. TNF-α not only promotes adhesion between EOS and endothelial cells, but also stimulates the secretion of airway smooth muscle cells of endothelial (ET)-1, which can exacerbate the contraction of smooth muscles and cause airway hyperresponsiveness. Busse et al. [Busse PJ, et al, 2009] showed that an anti-TNF-α antibody can significantly reduce airway inflammation and pathological changes in asthmatic rats. IL-6 has an important influence on the development of asthma and the inflammatory processes involved in damage, and produces a wide range of biological effects by promoting cell differentiation and proliferation of Thl7. Studies have shown that the IL-6 levels of patients with asthma are closely related to their symptoms and lung functions. The serum IL-6 levels during acute exacerbation of asthma were significantly increased, and IL-6 levels during exacerbation were significantly higher than the IL-6 levels in remission. Because the serum levels of IL-6 were decreased, IL-6 can be used as an indicator of acute exacerbation of bronchial asthma. In severe viral infections, IL-6 may become a target in the development of asthma. These studies showed that IL-6 and IL-8 can play a synergistic role to prompt proliferation of bronchial smooth muscle tissues [Kuo PL, et al, 2001; Shi YH, et al, 2010; Liu YM, et al, 2010].

Based on TCM pathogenesis, asthma is related to disorders in water metabolism, involving multiple organs (lungs, spleen, and kidneys). Phlegm is a pathologic product that causes asthma. Because phlegm stays in the lungs, an exogenous diet, an emotional state, and weariness causes asthma because phlegm blocks the airway and increases lung-qi. Clinical treatment focuses on staging and diffident syndromes. CMD consists of Zhimahuang (Ephedra; 4 g), Huangqin (Scutelleria; 9 g), Cangerzi (Xanthium; 9 g), Tianzhuzi (Nandina domestica; 9 g), Lameihua

(Chimonanthus praecox flowers; 9 g), and Hutuiziye (Elaeagnus leaf; 9 g).

Preliminary studies [Xu Zhaoxia, et al, 2013; Wang YQ, et al, 2006] confirmed that the prescription can significantly alleviate asthma symptoms and improve lung function. These results showed that CMD can ameliorate symptoms in asthmatic rats, such as their appetite, fur condition, excrement, and urine. CMD can effectively alleviate inflammatory infiltration of bronchopulmonary tissues, reduce the total number of leukocytes in the blood, decrease the pulmonary coefficient, reduce the number of EOS in BALF, and reduce the expression of TNF-α and IL-6 in serum. CMD relieves asthma symptoms by improving airway inflammation. In further research, we will explore the mechanisms underlying the effects of CMD on asthma based on the changes in cytokine expression and signal transduction pathways from airway epithelial cells.

ACKNOWLEDGMENTS

This study was supported by the Shanghai Natural Science Foundation (No. 14ZR1441500) and the Shanghai TCM Special Funds Project of Health and Family Planning (No. 2014JP029A).

REFERENCES

Berry M, Brightling C, Pavord I, et al. TNF-alpha in asthma [J]. Curr Opin Pharmacol, 2007 (3): 279–282.

Brightling C, Berry M, Amrani Y. Targeting TNF-alpha: a novel therapeutic approach for asthma [J]. J Allergy Clin Immunol, 2008, 121(1): 5–10.

Busse PJ, Zhang TF, Schofield B, et al. Decrease in airway mucous gene expression caused by treatment with anti-tumor necrosis factor [alpha] in a murine model of allergic asthma [J]. Ann Allergy Asthma Immunol, 2009, 103(4): 295–303.

Huang YQ, Zhang WC, Li HY. Effect of cadmium on body weight and organ coefficient of ovaries in female rat. Occupation and Health, 2003, 19(7): 6–9.

Kuo PL, Hsu YL, Tsai MJ, et al. Nonylphenol Induces Bronchial EPithelial Apoptosis via Fas- mediated Pathway and Stimulates Bronchial Epthelium to Secrete IL-6 and IL-8, causing Bronchial Smooth Muscle Proliferation and Migration [J]. Basic Clin Pharmacol Toxicol, 2011, 10: 1742–1784.

Li YQ. Clinical significance of Measurement of Changes of Serum TNF-α and plasma VP levels after treatment in elderly patients with bronchial asthma. J of Radioimmunology, 2006, 19(5): 379–380.

Liu YM, Nong GM, Li SQ. Effects of inhaled budesonide in early phase on airway inflammation and interleukin-6/signal transducers and activators of transcription 3 signaling pathway in asthmatic mice. J Appl Clin Pediatr, 2010, 25(16): 12220–1224.

Lv GP, Cui DJ, Guo YJ, et al. Introducing an experimental method of copying rat model of asthma. Chinese Journal of Tuberculosis and Respiratory Diseases, 1995, 18(6): 377–378.

Rubira Garcia N. Tumor necrosis factor [J]. Allergol Immunopathol (Madr), 2000, 28(3): 115–124.

Sun JX, An J, Lian J. Analysis of factors about the affect the experimental animal organs and organ weight coefficient [J]. Laboratory animal science. 2009, 26(1): 49–51.

Shi YH, Shi GC, Wang HY, et al. The prevalence of blood Th17 and CD4+CD25+ Treg cells in patient with bronchial asthma. Chinese Journal of Immunology, 2010, 26(2): 740–745.

Wang JY. 2005, Medicine. Beijing: People's Medicine Publishing House. 8: 49, 54.

Wang YQ, Li FF, Yan HX, et al. The regulatory role of the immune function of children with asthma about compound of Maqin prescription. Shanghai Journal of Traditional Chinese Medicine, 2006, 40(4): 37–38.

Waseman S, Olivenstein R, Renzi P M. The Relationship between Late Asthma tic Responses and Antigen-Specific Immunoglobulin. J Allergy Clin Immunol, 1992, 90: 661–669.

Xu Zhaoxia, Li Xueliang, Li Na, Qian Peng, Xu Jin, Wang Yiqin. Influence of Xuanfei-Guben Prescription on Mechanical Characteristics of Airway. Remodeling in Asthmatic Rats. Journal of Mechanics in Medicine and Biology. Vol. 14, No. 2 (2013) 1340014:1–8.

Xu Zhaoxia, Li Xueliang, Li Na, Qian Peng, et al. Studying the effect of different Traditional Chinese Medicine treatment which to the elasticity modulus of asthma rats' lung. Frontier and Future Development of Information Technology in Medicine and Education (ITME2013), July 19–21, 2013, P321–327.

Biomedical Engineering and Environmental Engineering – Chan (Ed.)
© 2015 Taylor & Francis Group, London, ISBN: 978-1-138-02805-0

A novel microfluidic chip with CCD-based fluorescence imaging for digital PCR

Maokai Yuan, Wei Jin, Zhiyu Wang, Yanan Xu & Ying Mu
Research Center for Analytical Instrumentation, Institute of Cyber-Systems and Control Department of Control Science and Engineering, The State Key Laboratory of Industrial Control Technology, Zhejiang University, China

ABSTRACT: Previously, our group developed a novel Self-Priming Compartmentalization (SPC) microfluidic chip. It has good performance at the sample injection speed and the uniform of the sample discretization, and it is very cost-effective. Therefore it becomes a good reaction platform for digital Polymerase Chain Reaction (PCR). The unique sample injection method promotes the development of microfluidic chip technology. Meanwhile, it puts forward more requirements on the fluorescence detection system. In this work, we develop a fluorescence detection system based on CCD and Labview. It realizes the function of image capture, processing and the resulting calculation for digital PCR on SPC chip. On the other hand, we apply the Sobel algorithm to the image processing, which improves the quality of edge extraction, and improves the anti-noise property of the system. The control experiment shows that for the same chip, our system has the equivalent result as the commercial imaging device (Maestro Ex IN-VIVO IMAGING SYSTEM), which needs manual calculation for the targeted DNA copies after taking photos. This proves the effectiveness of our self-developed microfluidic chip fluorescence detection system.

1 INTRODUCTION

Digital PCR technology has been developed rapidly and widely used in the medical, chemical, environment and food field in 21 century. Microfludic chip as the good digital PCR reactions platform, its character affects the PCR reaction and the accuracy and validity of the detection results. In recent years, scientists at home and abroad optimized the microfluidic chip in terms of manufactured materials, manufacture processes, structural designs, fluid drives and injection modes etc. This greatly promoted the development of the microfluidic chip at low cost and high efficiency. Now, digital PCR detection technology has found a mature theoretical basis. Optical detection, electrochemical detection, mass spectrometric detection are widely used by scientists to detect all types of microfluidic chips. Fluorescence detection, as the earliest detection mode, has become the most sensitive and the most widely used detection mode after the hardware optimization of light paths and sensors and the software application of image processing techniques.

In 2012, our laboratory research group designed a novel Self-Priming Compartmentalization (SPC) microfluidic chip. Its innovative injecttion mode got the approval of scientists at home and abroad, and it opened up the thinking of the design of microfluidic chips. Now, in order to build a complete experimental environment based selfpriming chip, the development of signal acquisition control and detection device and analysis system for this chip's characteristics is the most important task.

2 THE SPC MICROFLUIDIC CHIP

The Self-Priming Compartmentalization (SPC) microfluidic chip consists of seven layers. They are the auxiliary self-priming layer, glass bottom, the reaction layer, mediated layer, anti-evaporation layer, a thick PDMS reinforcement layer and glass upper layer[1]. The design principle is the porous permeability of PDMS material. After the chip is evacuated sufficiently, we allow part of the gas into the reaction chamber through the main channel, and then drop the sample into the inlet. The gas in the reaction chamber will access the interior of PDMS as before. This makes the air pressure in the reaction chamber lower, and the sample enters the reaction autonomously under the influence of the air pressure difference. Our group optimized the chip design and added a cover slip layer as a vapour proof measures. This greatly reduced the water loss of the sample during the PCR reaction. This kind of chip does not need the additional drive device during the injection process compared with the traditional injection mode. It greatly improves the portability of the chip. Meanwhile, we tested the sample injection speed, injection uniformity, the evaporation and the heating transfer characteristic, and all the parameters achieved good standards. Its design principle is shown in Figure 1.

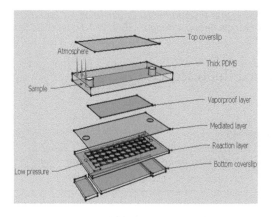

Figure 1. The structure and the principle of the self-priming compartmentalization (SPC) microfluidic chip.

Currently, we can detect four samples using the self-priming compartmentalization (SPC) microfluidic chip simultaneously, and the sample will be evenly distributed to 1280 reaction chambers which volume is 1.5 nL. We used the cDNA of 18sRNA extracted from A549 lung cancer cell as sample to prove the feasibility of the chip successfully[2].

3 FLUORESCENCE DETECTION SYSTEM

The digital PCR detection system is built according to the fluorescence detection principle. The Taqman oligonucleotide probe as the labelled fluorescent group, its 5' end tag the fluorescent group and its 3' end tag the quencher group. No fluorescence occurs when the probe is intact. And as the PCR reaction goes on, the probe is broken under the action of Taq polymerase 5'–3' exonuclease. Accepting the action of excitation light, the fluorophore will produce fluorescence. This experimental device uses FAM fluorophore. If the green fluorescence of which wavelength is 520 nm occurs under the excitation of blue light the wavelength of which is 480 nm. This proves that the PCR amplification acts in this micro reaction chamber, and the reaction result is positive. The CCD is used as an acquisition device to achieve the fluorescence image. After counting the number of positive chambers, we can calculate the sample starting copy number using Poisson distribution model. The comprehensive digital PCR detection platform includes three parts: adark room, an external power supply and a motor controlled module, information processing computer.

3.1 *Hardware design*

We built a hardware structure to achieve fluorescence excitation and acquisition. Its main components are LED, blue filters, lenses, green filters, CCD cameras, dark room, frame and computer. The excitation light source is two CREE brand blue LEDs each power is 3 W. The 470 nm and 520 nm spike filter used at

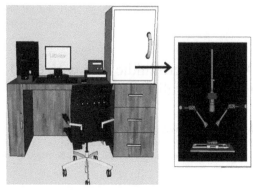

Figure 2. Schematic diagram of experimental environment.

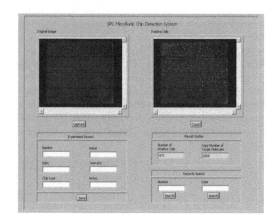

Figure 3. Software interface diagram.

excitation source and CCD respectively are custom-made at Shanghai Mega-9 company. The model of CCD is the Bigeye G-132B cool industrial camera produced from the German AVT company. This CCD can supply high resolution (1.3 million pixels) image with excellent signal to noise ratio depending on its cooling module which can make the working temperature lower to zero degree[3]. The digital image is then transferred to a computer by the Gige interface. The model of the lens is M3514-MP, and it is a kind of megapixels fixed focus industrial lens with a C-type interface. Its focal length is 35 mm. Its working distance is 142 mm. The computer is a PC with windows OS.

3.2 *Software design*

The software platform developed by the graphical programming language Labview and database technology achieve image acquisition and processing, data calculation and analysis, record storage of the whole experiment process and other functions. Using the image processing software package of the Labview, choose Acquire Image (IEEE1394 or Gige) mode to obtain the fluorescence image. Adopt filtering algorithm to process the original image and then display the processed image. Data calculation function includes

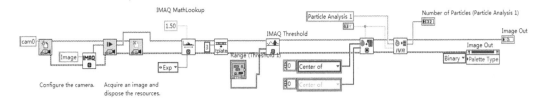

Figure 4. VI source of counting function.

highlights counting and reverse calculation results output. Record storage function save the experimental images, data, text, and information to a dedicated database. We query the experiment information by the experiment record number which is used as a primary key. The software interface is shown in Figure 3. The counting function VI source is shown in Figure 4.

The development of software makes effective image processing, and it filters out the noise signal. The appropriate feature extraction algorithm effectively guarantees the accuracy of the results. Database technology used to save all the Self-Priming Chips' experiments records, not only achieved office automation for the scientists, but also accumulated a rich experience data for biological information mining.

4 IMAGE PROCESSING

Under the same digital PCR reaction conditions, the image quality of the detection system depends on the optical path and image processing methods. In order to get a better detection results, we collimate the LED to avoid its non-uniformity.

Up to now, many domestic and foreign scientists have designed some complicated confocal optical paths, which can improve the image quality, but in the meantime, increase the system's complexity. In this situation, we focus on the image processing algorithms instead of optical design.

4.1 Median filtering

After image preprocessing, we applied the median filtering to format the new image. The median filtering algorithm is a non-linear signal processing technology based on the order statistical theory. Its basic principle is to replace a certain pixel's gray scale value from digital images or sequences with the mean value of all pixels from a neighbouring window. This method can eliminate typical speckle noises to achieve better visual results[4]. In this work, we choose a 3*3 median filtering window after repeated experiments the image being processed showed a high definition.

4.2 Edge extraction

A Sobel operator has been used widely to extract the edge of the image. The Sobel operator takes the gray scale weighted difference of each pixel's neighbourhood into account. Usually a 3*3 neighbourhood

template is chosen as a variance template. The following two parameters (as shown in formula (1)) represent horizontal and vertical matrices respectively, which are used to convolve the 2-d image for acquiring corresponding edge detection images, They then, calculate the two-dimensional gradient vectors of each pixel.

$$A = \begin{bmatrix} -1 & -2 & -1 \\ 0 & 0 & 0 \\ 1 & 2 & 1 \end{bmatrix} \quad B = \begin{bmatrix} 1 & 0 & -1 \\ 2 & 0 & -2 \\ 1 & 0 & -1 \end{bmatrix} \quad (1)$$

On the basis of the Sobel operator, a new edge extraction method named four-direction template Sobel operator has been proposed. It adds two more 3*3 matrices (as shown in formula (2)) than the basic Sobel operator. This algorithm significantly improves the filtration ability of background noises on the microfluidic chip.

$$C = \begin{bmatrix} 0 & 1 & 2 \\ -1 & 0 & 1 \\ -2 & -1 & 0 \end{bmatrix} \quad D = \begin{bmatrix} -2 & -1 & 0 \\ -1 & 0 & 1 \\ 0 & 1 & 2 \end{bmatrix} \quad (2)$$

Currently, improved image processing algorithms are common in the field of information technology and applied to biomedical engineering area extensively, but very few have been used to analyse biological reactions on microfluidic chips[5]. In this paper, we introduce the digital image processing algorithm into the fluorescent detection system and the results have proved the high accuracy of the system.

5 EXPERIMENTAL RESULT

On this chip, we use the AxyPrep Multisource Total RNA Miniprep Kit (Axygen Biosciences America Company); EASY Dilution (TaKaRa Japan Company); A549 lung cancer cell (Cell Resource Center of Shanghai Institute of Life Science Chinese Academy of Sciences) to make digital PCR reactions to detect the EGFR gene. According to the endpoint detection method, the fluorescence detection system counted the highlights and obtained that the number of positive chambers is 473. And by the poisson statistical formula[6], we can calculate that the concentration of target molecular copies are 6500 copies/uL. The result calculated by Labview software is shown in Figure 3. At the same time, for the same reaction chip, using the

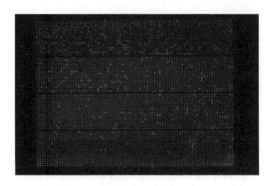

Figure 5. Image of the control experiment.

fluorescence imager Maestro Ex IN-VIVO IMAGING SYSTEM (CRI Maestro USA) as a control, we get the following resulting image, shown in Figure 5. The result calculated manually is the same as the output from the Labview software.

6 CONCLUSIONS

The effectiveness and accuracy of our self-developed system are confirmed by the experimental results. Our future work will focus on the novel image processing algorithms and apply them to our digital PCR chips. What's more, we will try our best to develop the artificial intelligence algorithms, which could meet more complicated image processing requirements and improve the performance of SPC microfluidic chip fluorescent detection system further.

ACKNOWLEDGEMENTS

This work was supported by the National Natural Science Foundation of China (31270907, 21275129), National key foundation for exploring scientific instrument (2013YQ470781) and the State Key Laboratory of Industrial Control Technology, Zhejiang University, China.

REFERENCES

2013. Bigeye-G_TechMan_V2.1.0_en, Technical manual.
Elizabeth A. Ottesen et al. 2006. Microfluidic Digital PCR Enables Multigene Analysis of Individual, Science, 314, 1464.
Mengchao Yang, Xingyu Jin, Maokai Yuan. 2014. Tumor Cell Detection Device Based on Surface Plasmon Resonance Imaging and Image Processing, Chemical Research in Chinese Universities 30 (2): 211–215.
Qiangyuan Zhu, Wenxiu Yang. 2013. Microfluidic Digital Chip for Absolute Quantification of Nucleic Acid Amplification, Chemical Journal of Chinese Universities, 34 (3): 545–550.
Yibo Gao. 2013. Study on the Self-priming and Discretization Microfluidic Chip for the Digital Nuclear Acid Amplification, Zhejiang University degree thesis.
Yajing Xiao. 2010. Research on the Detection Methods and Technology of Microfluidic Chip Based on Image Processing, Hebei University degree thesis.

Biomedical Engineering and Environmental Engineering – Chan (Ed.)
© 2015 Taylor & Francis Group, London, ISBN: 978-1-138-02805-0

An arterial compliance classification of the upper arm based on the cuff pressure signal

Yuqi Wang
Beihang University, Beijing, China
University of Missouri, Columbia, MO, USA

Yubo Fan
Beihang University, Beijing, China

Xi Chen, Daniel Credeur, Paul J. Fadel, Dong Xu, P. Frank Pai & Ye Duan
University of Missouri, Columbia, MO, USA

ABSTRACT: The arterial compliance which represents vessel stiffness is an important indicator of cardiovascular health. The purpose of this research is to develop a simple method to estimate arterial compliance with only an inflatable cuff. The cuff pressure was recorded during blood pressure measurement. Using only the cuff pressure signal, the change of volume, and the inner pressure and outer pressure for all beat cycles were computed. The volume-pressure curve was derived and fitted to an arctangent function. Taking the fitting coefficients as the input features, compliance classification was performed using a supported vector machine. To test this method, cuff blood pressure was measured on 5 subjects. The carotid compliance by ultrasound method was calculated and taken as the ground truth. A cross validation was performed on the data and the results showed that the cross validation accuracy was 81%. Therefore, the proposed classification method achieved a relative high degree of classification accuracy.

Keywords: cuff, compliance, stiffness, oscillometric pulse, supported vector machine, classification

1 INTRODUCTION

Arterial compliance is an important parameter that can reflect the health status of human cardiovascular systems. It represents the softness or stiffness of a vessel and is related to many diseases (Amar, J. 2001). Blood pressure is another index indicating cardiovascular health status. There are many blood pressure measuring devices using a cuff available on the market (O'Brien, E. 2001). If the arterial compliance can be also obtained during blood pressure measurement, it will be meaningful and useful. One measure can provide people with more information about their health, not only two pressure values.

Most non-invasive blood pressure measuring methods, like the Korotkoff sound method and the oscillometric method, employ an inflatable cuff to measure pressure (O'Brien, E. 1994). Cuff pressure waveforms may contain much information rather than only two pressure extremes. Therefore, it is used to estimate the arterial compliance of the upper arm in this paper.

Arterial compliance is defined as the change of volume divided by the change of transmural pressure (inner pressure minus outer pressure). In this paper, the compliance was estimated from the volume-pressure curve generated from the cuff pressure signal which is widely employed in common blood pressure measuring devices. Then, the volume-pressure curve was fitted to an arctangent function. Taking the fitting coefficients as the input features, an arterial compliance classification was performed using the supported vector machine classifier. The carotid compliance was also simultaneously measured by the gold standard method, the ultrasound method, and these compliance values were taken as the ground truth. Finally, a cross-validation was conducted to validate the classification accuracy.

2 METHODS

An important step in this method is to get the volume-pressure curve. The compliance at a certain pressure is essentially the local slope of the volume-pressure curve. In this research, we assume that the volume change is related to the amplitude of the oscillometric pulse of the cuff pressure, that the outer pressure is the cuff pressure, and that the inner pressure fluctuates between the diastolic and systolic blood pressure envelope which both decrease linearly during the cuff

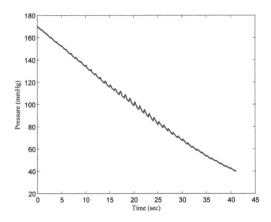

Figure 1. Original cuff pressure.

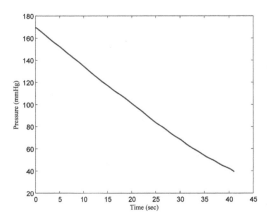

Figure 2. Decline trend of cuff pressure.

deflation. In one heart beat cycle, the change of the volume, outer pressure and inner pressure were separately estimated. Then, the first order derivative of the volume-pressure curve at each beat pulse could be calculated. Eventually, the volume-pressure curve can be obtained from the integral of the derivative.

2.1 Get the decline trend curve and oscillometric pulse from the cuff pressure

The original cuff pressure acquired consisted of the decline trend of cuff pressure, the oscillometric pulse and noise, as shown in Figure 1. The traditional method of separating the oscillometric pulse from the trend was the use of two digital filters: one low pass filter to get the decline trend, and one high filter to get the oscillometric pulse.

A novel method, the modified empirical mode decomposition method, was employed here to denoise the original cuff pressure signal (Huang, N. E. 1998, Pai, P. F. 2008) and decompose the cuff pressure into a decline trend and oscillometric pulse signal. The new method is suitable for non-stationary or non-linear signals, whereas the filter method can only be applied to only deterministic signal. Figure 2 shows the decline

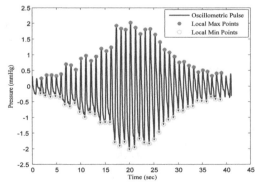

Figure 3. Oscillometric pulse.

trend signal of the cuff pressure. Figure 3 shows the oscillometric pulse signal of the cuff pressure during deflation. Both the decline trend and the oscillometric pulse signal look smooth.

2.2 To decompose the oscillometric pulse into beat cycles

The compliance estimation is based on each pulse signal. Therefore, the oscillometric pulse needs to be decomposed into beat cycles. The oscillometric pulse signal is periodic. As shown in Figure 3, the oscillometric pulse may vary in its maximum amplitude from beat to beat and the local dicrotic peaks may exist within one beat cycle. Thus, the traditional method of using the sign change of first directive to find peaks may result in incorrect local peaks to be found within one beat cycle.

To avoid the above problem, the global maximum point was first identified and the period of this oscillometric pulse was estimated from frequency spectrum of the signal. Then, the start point of this pulse which was this local minimum point just before this global point and the end point of this pulse which was the local minimum point just after this global point was found. According to the signal period estimated by the spectrum, adjacent unknown extrema were found one by one. One beat cycle was considered the oscillometric pulse between every two local minimum points. As shown in Figure 3, the upper stars are the local maximum points and the lower circles are the local minimum points. The oscillometric pulse between two adjacent circles is one beat cycle.

2.3 To estimate the systolic and diastolic blood pressure

In this research, the Systolic Blood Pressure (SBP), Diastolic Blood Pressure (DBP) and the Mean Arterial Pressure (MAP) were determined by the oscillometric method that is the most commonly used method of blood pressure measuring devices (Ramsey 1991). First, the oscillometric pulse amplitude of each beat cycle was calculated by subtracting the minimum pressure from the maximum pressure of this beat pulse. As

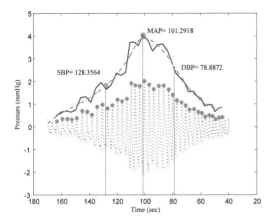

Figure 4. To determine systolic and diastolic blood pressure by oscillometric method.

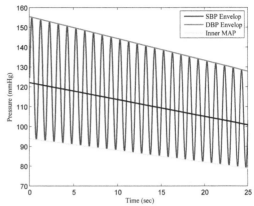

Figure 5. Inner pressure.

shown in Figure 4, the upper solid bell-shaped curve is the oscillometric pulse amplitude signal.

To calculate the SBP and the DBP, the global maximum oscillation amplitude was first acquired and its corresponding cuff pressure, obtained from the cuff pressure trend signal, was considered the MAP. The oscillation amplitude signal before the MAP point, which is the rising part of this signal, and the declining part of the oscillation amplitude signal after the MAP point, were separately fitted into four orders of polynomials. The dashed lines in Figure 4 show the fitted rising and declining curves. The SBP was determined by the cuff pressure when the oscillometric pulse has risen to 45% of its maximum oscillation and the DBP was determined by the cuff pressure when the oscillometric pulse fell down to 65% of its maximum oscillation. Figure 4 shows how to determine the systolic and diastolic pressures by the oscillometric method.

2.4 Calculate the local compliance of volume-pressure curve

To estimate the volume-pressure curve of the upper arm artery, the volume, outer pressure and the inner pressure should be known. For each beat cycle, the change of volume and the outer and inner pressure during cuff deflation were first estimated.

2.4.1 Volume change in one beat
It was assumed that the volume change in one beat, during the cuff deflation, was linearly related to the change of the oscillometric pulse in that beat (Forster, P. D. 1986, Ursino, M. 1996, Baker, P. D. 1997). The shape of the volume change curve is the same as the oscillometric pulse amplitude signal, as the upper solid curve shown in Figure 4.

2.4.2 Outer pressure
Obviously, the outer pressure exerted on the upper arm artery, during the cuff deflation, was the decline trend of cuff pressure, as is shown in Figure 2. In one beat cycle, the change of the outer pressure was only 2–4 mmHg, because the cuff deflation rate was 2–4 mmHg/second and the heart rate of a subject is typically 1 second.

2.4.3 Inner pressure
Some researchers believe that the inner blood pressure during cuff deflation varies between the diastolic and systolic pressure and for each beat, its maximum pressure is systolic pressure and its minimum pressure is diastolic pressure (Komine, H. 2012). The inner pressure change for one beat is defined as the difference between its maximum and minimum pressure in that beat cycle. Therefore, they supposed that the inner pressure change curve should be a constant line.

However, during blood pressure measurement, the blood vessel is completely closed when the cuff deflation just starts and then opens gradually when the cuff pressure drops to between the diastolic and systolic pressure. Finally, it will open completely. Therefore, we assume that both the local maximum and minimum pressure of the inner pressure in one beat is high when the blood vessel just opens at the high cuff pressure and is low when the blood vessel opens largely at the low cuff pressure, because the smaller open area of the blood vessel may carry a greater blood impulse. For simplicity, we assume that the SBP and DBP envelope of the inner pressure decreases linearly and will reach the overall systolic and diastolic pressure of this measurement at the end of the measurement.

For illustration only, based on the simple assumption that the inner pressure waveform in one beat cycle is the sine wave, the inner pressure is plotted in Figure 5. The upper decline line is the SBP envelope of the inner pressure and the lower decline line is the DBP envelope of the inner pressure. The middle decline line demonstrates the MAP.

The inner pressure change in one beat cycle, which equals the SBP minus the DBP of that beat, namely the pulse pressure of that beat, also decreases linearly during the cuff deflation and will drop to the pulse pressure at the end of the measurement.

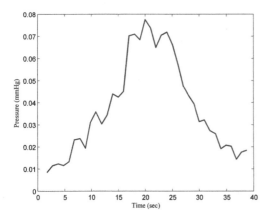

Figure 6.   Local compliance of all beat cycles.

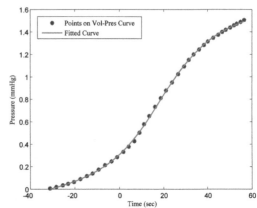

Figure 7.   The volume-pressure curve and the fitted curve.

2.4.4   *Transmural pressure*

The transmural pressure is the outer pressure minus the inner pressure. The change of transmural pressure in one beat equals the change of the outer pressure minus the change of inner pressure in one beat cycle. The change of outer pressure in one beat cycle which is just about 2–4 mmHg is very small compared with the change of the inner pressure in one beat which is approximately the pulse pressure.

2.4.5   *Local compliance of volume-pressure curve*

According to the above analysis, in one beat cycle, the change of volume, inner pressure, outer pressure, and the transmural pressure were obtained. After that, the local compliance in one beat cycle was defined as the change of volume divided by the change of transmural pressure in that beat (equation (1)). It is essentially the local slope or the first order derivative of the volume-pressure curve at that beat.

Local Compliance

$$= volume\,change/(inner\,BP\,change - outer\,BP\,change) \quad (1)$$

The local compliance of all beat cycles are shown in Figure 6.

2.5   *Get the volume-pressure curve*

The local compliance is the first derivative of a volume-pressure curve. Therefore, the volume-pressure curve can be obtained by integrating the local compliance curve. For discrete beats, the summation was used instead of the integral. The average inner pressure is the MAP signal, as the middle decline line shown in Figure 5. For one beat cycle, the transmural pressure that is the x-axis of the volume pressure is the difference between the MAP and the decline trend of cuff pressure. As shown in Figure 7, the star points are the points on the volume-pressure curve in all beat cycles.

2.6   *Fit the volume-pressure curve to arctangent function*

There were only about 40 points on the above volume-pressure curve which could reflect the arterial compliance at different transmural pressures. We expected to obtain the compliance value at more transmural pressure points and find some parameters to represent the overall compliance of the artery. Therefore, the volume pressure curve was fitted to an arctangent function shown in the following equation,

$$Vol(TP) = A * arc\tan(B * TP + C) + D \quad (2)$$

where Vol = the volume and TP = the transmural pressure.

The initial values of fitting coefficients A, B, C and D were set to 1, 0.04, 0.1 and 1. A least squares curve fitting method was employed. The fitted curve is the solid curve shown in Figure 7.

2.7   *Experiments*

The gold standard to measure the arterial compliance is the ultrasound method. The B mode ultrasound images of a carotid arterial were acquired simultaneously and meanwhile the pressure of the carotid artery was measured by a tonometer when the blood pressure of the upper arm was measured by an inflatable cuff. The diameter change of the carotid arterial could be calculated from the ultrasound images and hence the volume change. The pressure change could be calculated by the tonometer pressure. Therefore, the carotid compliance by the ultrasound method was obtained and taken as the ground truth.

The data of 5 subjects were acquired, including the cuff pressure in the upper arms, the ultrasound images of the carotid arteries and the carotid pressures from the tonometer. For each subject, 6 or 7 blood pressure measures were taken by a cuff. There was a total of 33 measures.

2.8 Classification of arterial compliance

Some researchers used only the fitting coefficient B, as the indicator of compliance (Komine, H. 2012). However, we believe that all fitting coefficients A, B, C and D are related to the compliance. At least, four parameters can provide more information than one.

To reveal the relationship of the four parameters with arterial compliance, classification of arterial compliance was performed on the data of the 33 measures. According to the compliance calculated by ultrasound, the data was divided into two categories, a group of high compliance and another group of low compliance. All four fitting coefficients were taken as input features and the supported vector machine classifier (Vapnik, V. N. 1998) with a linear kernel was employed for the classification.

To test the accuracy of the classification, an 11-fold cross validation was performed on these 33 labelled data.

3 RESULTS

The cross-validation accuracy of the above classification was 81.8%.

4 CONCLUSIONS

In conclusion, the proposed method for compliance classification of an upper arm artery based on cuff pressure and SVM can achieve a relatively high degree of accuracy. This proposed method offered an easily implemented way of estimating the arterial compliance of the upper arm without the need of using ultrasound.

ACKNOWLEDGEMENT

The authors would like to thank sincerely Jingqing Jiang for her suggestions and instruction.

REFERENCES

Amar, J., Ruidavets, J. B., Chamontin, B., Drouet, L., & Ferrières, J. 2001. Arterial stiffness and cardiovascular risk factors in a population-based study. *Journal of hypertension*, 19(3), 381–387.

Baker, P. D., Westenskow, D. R., & Kück, K. 1997. Theoretical analysis of non-invasive oscillometric maximum amplitude algorithm for estimating mean blood pressure. *Medical and biological engineering and computing*, 35(3), 271–278.

Forster, F. K., & Turney, D. 1986. Oscillometric determination of diastolic, mean and systolic blood pressure—a numerical model. *Journal of biomechanical engineering*, 108(4), 359–364.

Huang, N. E., Shen, Z., Long, S. R., Wu, M. C., Shih, H. H., Zheng, Q. & Liu, H. H. 1998. The empirical mode decomposition and the Hilbert spectrum for nonlinear and non-stationary time series analysis. *Proceedings of the Royal Society of London. Series A: Mathematical, Physical and Engineering Sciences*, 454(1971), 903–995.

Komine, H., Asai, Y., Yokoi, T., & Yoshizawa, M. 2012. Non-invasive assessment of arterial stiffness using oscillometric blood pressure measurement. *Biomedical engineering online*, 11, 6.

O'Brien, E., Waeber, B., Parati, G., Staessen, J., & Myers, M. G. 2001. Blood pressure measuring devices: recommendations of the European Society of Hypertension. *BMJ*, 322 (7285), 531–536.

O'Brien, E., & Fitzgerald, D. 1994. The history of blood pressure measurement. *Journal of human hypertension*, 8(2), 73.

Pai, P. F., Huang, L., Hu, J., & Langewisch, D. R. 2008. Time-frequency method for nonlinear system identification and damage detection. *Structural Health Monitoring*, 7 (2), 103–127.

Ramsey III, M. 1991. Blood pressure monitoring: automated oscillometric devices. *Journal of clinical monitoring*, 7(1), 56–67.

Ursino, M., & Cristalli, C. 1996. A mathematical study of some biomechanical factors affecting the oscillometric blood pressure measurement. *IEEE Transactions on Biomedical Engineering*, 43 (8), 761–778.

Vapnik, V. N., & Vapnik, V. 1998. *Statistical learning theory* (Vol. 2). New York: Wiley.

Biomedical Engineering and Environmental Engineering – Chan (Ed.)
© *2015 Taylor & Francis Group, London, ISBN: 978-1-138-02805-0*

A method of denoising and decomposing cuff pressure based on improved Empirical Mode Decomposition

Yuqi Wang
Beihang University, Beijing, China
University of Missouri, Columbia, MO, USA

Yubo Fan
Beihang University, Beijing, China

Xi Chen, P. Frank Pai, Dong Xu, Daniel Credeur, Paul J. Fadel & Ye Duan
University of Missouri, Columbia, MO, USA

ABSTRACT: Most noninvasive blood pressure measuring methods employ an inflatable cuff to measure pressure. An improved processing of cuff pressure signals can improve the accuracy of blood pressure measurement. In this paper, a novel method of denoising and decomposing the cuff pressure, based on modified Empirical Mode Decomposition (EMD), was proposed, which performed EMD on the cuff pressure superimposed onto a high frequency oscillatory wave. This method was tested on a simulated signal. Taking the simulated ideal signal as the true, the root mean squared errors (RMSEs) of the trend and oscillatory wave processed by both the EMD and the traditional filter method, were calculated. The result showed that the RMSEs of the trend and oscillatory wave implemented by the EMD method were both smaller than these by the filter method. Therefore, the proposed method produced a better performance than the filter method. They are efficient and effective signal denoising and decomposing methods.

Keywords: cuff pressure, trend, oscillometric pulse, decomposition, denoise

1   INTRODUCTION

1.1   *Importance of cuff pressure processing*

Blood pressure is one of the most important parameters that can reflect the health status of human cardiovascular systems. There are many noninvasive blood pressure measuring methods, most of which employ a cuff to measure pressure (Perloff, D. 1993, O'Brien, E. 1995, O'Brien, E. 2001). The Korotkoff sound method which is the gold standard of noninvasive blood pressure measurement employs an inflatable cuff wrapped around the upper arm to measure the pressure (Shevchenko, Y. L. 1996). The oscillometric method which is adopted by most of the commercial blood pressure measuring devices also measure pressure via a cuff (Forster, F. K. 1986, Baker, P. D. 1997). Therefore, a cuff pressure signal is an important signal for blood pressure measurement. It is meaningful to develop a good method of processing the cuff pressure.

1.2   *Empirical mode decomposition*

The Empirical Mode of Decomposition (EMD) method was proposed by Huang to analyse the nonlinear and non-stationary time series (Huang, N. E.

1998, Huang, N. E. 2005). Huang defines the intrinsic mode function as the function that has almost the same number of extrema and zero crossings and whose upper and lower envelopes are symmetric with respect to zero. EMD can decompose a complicated signal into a series of intrinsic mode functions. The intrinsic mode functions are essentially the basic oscillatory modes of the signal. The cuff pressure signal is the superimposition of the decline trend and oscillometric pulse signal. Thus, the EMD method can be employed to extract the oscillometric pulse and the decline trend of cuff pressure from the original cuff pressure signal.

1.3   *Use EMD to denoise and decompose cuff pressure*

The current method to process a cuff pressure signal is the use of some filters. In this paper, a novel method, based on modified EMD, was proposed to denoise the original cuff pressure signal and then decompose it into the decline trend and oscillometric pulse signal. There are some differences between the proposed method and the original EMD. First, a high amplitude and frequency signal was superimposed onto the original cuff pressure signal and EMD was performed on the compound signal instead of the original cuff pressure. Besides, a stricter criterion was applied to the extreme point of detection. To compare the proposed decomposition method with the traditional filter method, a simulated cuff pressure was constructed and decomposed by both the modified EMD and the traditional filter methods.

2 METHODS

2.1 Introduction of the original EMD

Huang introduced a method of how to perform EMD to decompose a signal into IMFs (Huang, N. E. 1998, Huang, N. E. 2005). It is described as the following:

- First, the local maximum and minimum points are identified.
- Then, the local maximum and minimum points are separately fitted to a cubic spline line to get the upper and lower envelope of the signal. The mean envelope is calculated by averaging the upper and lower envelope.
- The mean envelope is subtracted from the signal to get the oscillatory component. Ideally, this component should be an IMF. If it is not, it will be treated as the input signal and the above process will be repeated until some criteria are satisfied. Now, this oscillatory component is considered an IMF.
- Next, the IMF is subtracted from the signal to get the residue.
- Take the residue as the input signal and repeat the above procedure to obtain the next IMF until the residue is a monotone function.

Finally, the signal will be decomposed into a series of oscillatory IMFs and the last residue which is usually the trend of the signal.

2.2 The problem of performing EMD on the original cuff pressure

This original EMD is an effective method to decompose signals. However, if it is directly performed on the original cuff pressure, some incorrect local extrema will be found and hence the wrong upper and lower envelope will be generated, because the original cuff pressure contains much high frequency noise. The original cuff pressure acquired from a human subject is shown in Figure 1. It is the superimposition of the decline trend, oscillometric pulse, and noise.

Figures 2 and 3 show the result of performing the EMD directly on the original cuff pressure. Figure 2 shows the 'zoomed in' view of the peak area of one beat pulse of the original cuff pressure. The solid line is the original cuff pressure signal. High frequency noise can be obviously seen on the original cuff pressure. The stars are the local maximum points and the dashed line is the upper envelope which is obtained by fitting the stars to a cubic spline. The circles are

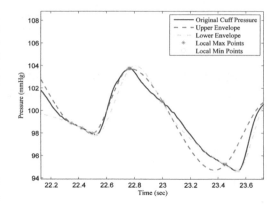

Figure 1. The original cuff pressure signal.

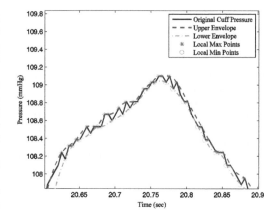

Figure 2. One peak area of the original cuff pressure.

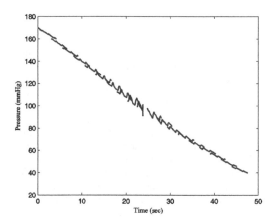

Figure 3. One beat of the original cuff pressure.

the local minimum points of the signal and the dash-dot line is the lower envelope. Since it shows only the area around a pulse peak, there should be only one local maximum point here. However, due to the high

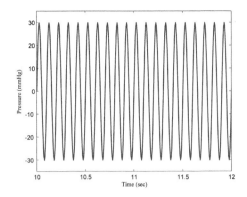

Figure 4. The added sine wave.

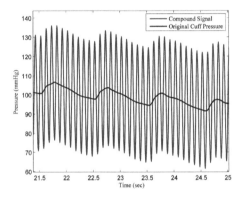

Figure 5. The compound signal.

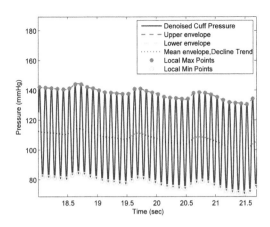

Figure 6. Performing EMD on the compound signal.

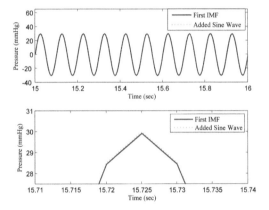

Figure 7. The first IMF and added sine wave.

frequency noise, many local maximum and minimum points are mistakenly found here.

Figure 3 shows the 'zoomed in' view of one pulse of the original cuff pressure. The signal is supposed to be located between the lower (dash-dot line) and upper envelope (dashed line). However, because of the existence of noise, many wrong local maximum and minimum points were found, resulting in the incorrect lower and upper envelope generated. Some parts of the lower envelope even go up above the upper envelope ridiculously.

2.3 Our improved EMD

To overcome the above problem, an improved EMD method was proposed. First, a high amplitude and frequency signal was superimposed onto the original cuff pressure and then EMD was performed on the compound signal to denoise the cuff pressure. Thereafter, EMD was performed again on the denoised signal to extract the decline trend and oscillometric pulse.

2.3.1 To add a high amplitude and frequency signal

To get rid of the noise's influence on the local extreme point identification, a high amplitude and frequency signal was superimposed onto the original cuff pressure. Here, a sine wave with the amplitude of 30 mmHg and the frequency of 10 Hz was superimposed onto

the original cuff pressure signal. The frequency of 10 Hz was selected because it was much higher than the frequency of the pulse and was a divisor of the sampling frequency of 200 Hz. By doing this, the dominant oscillatory component of the compound signal became the added sine wave rather than the high frequency noise. Figure 4 shows the added sine wave. As shown in Figure 5, the oscillatory solid curve is the compound signal and the middle curve is the original cuff pressure signal with noise.

2.3.2 To conduct EMD on the compound signal

The traditional EMD was performed on the compound signal to denoise the cuff pressure. As shown in Figure 6, the oscillatory solid curve is the compound signal. The upper dashed curve is the upper envelope and the lower dash-dot curve is the lower envelope. The middle doted curve is the mean envelope. It can be seen that all local extrema (stars and circles) were all correctly identified.

The first IMF was obtained by subtracting the mean envelope from the compound signal. It is shown in Figure 7. In Figure 7, the solid curve is the first IMF and the dotted curve is the added sine wave. The first

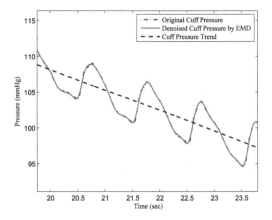

Figure 8.   Denoised cuff pressure by EMD.

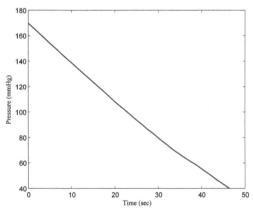

Figure 10.   Decline trend of cuff pressure.

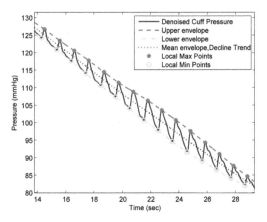

Figure 9.   Performing EMD on the denoised cuff pressure.

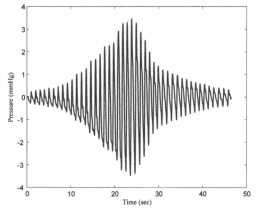

Figure 11.   Oscillometric pulse.

IMF is very similar to the added wave. The lower sub-figure shows the 'zoomed in' view of one peak area of the first IMF and the added sine wave. Little difference can be seen here, because the first IMF is essentially the sum of high frequency noise and the added sine wave.

The mean envelope is the denoised signal of the original cuff pressure, because noise was removed when the IMF containing the noise was subtracted from the compound signal. As shown in Figure 8, the solid curve is the mean envelope and the dash-dot curve is the original cuff pressure. It can be seen that the denoised signal is much smoother than the original cuff pressure.

2.3.3   To extract the trend and oscillometric pulse from denoised cuff pressure

In this step, the denoised cuff pressure was further decomposed into the decline trend and oscillometric pulse by EMD. The first IMF was subtracted from the original cuff pressure. The residue, namely the mean envelope signal in Figure 6, was treated as the input signal and another EMD was performed on it.

After the cuff pressure was denoised, the high frequency noise was removed and it was easier to find the

correct extreme points within one beat pulse. In spite of this, a stricter rule was employed for the peak identification to make sure that the correct local extrema could be found. A point was identified as a local maximum (minimum) point, only if the first derivatives of its two previous points were both positive (negative) and the first derivatives of its two following points were both negative (positive). In addition, the period of the pulse which was about one second was estimated by the frequency spectrum method and considered during the extrema detection.

Figure 9 shows the procedure of the use of the EMD to decompose the denoised cuff pressure. In Figure 9, the solid oscillatory curve is the denoised cuff pressure. The stars are the local maximum points and the circles are the local minimum points. It can be obviously seen that only one local maximum and two minimum points were found within each beat cycle. The dashed curve is the upper envelope and the dash-dot curve is the lower envelope. The denoised cuff pressure signal is completely located between the lower and upper envelope. The middle dotted curve is the mean envelope, which is also the decline trend of cuff pressure. The dashed declining line in Figure 8

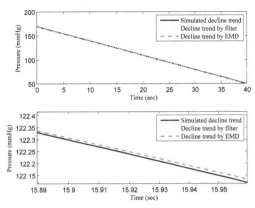

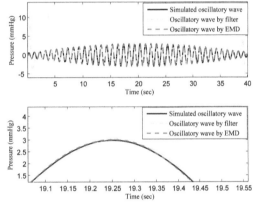

Figure 13. Simulated decline trend.

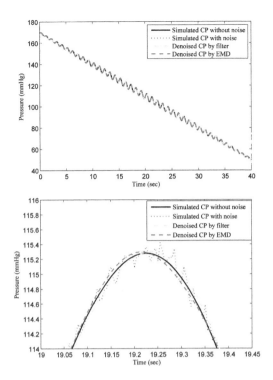

Figure 12. Simulated cuff pressure.

Figure 14. Simulated oscillatory wave.

is also this decline trend. It almost passes through the middle of every pulse and hence can reflect well the decline trend of the cuff pressure. The complete declining curve is shown in Figure 10. It is almost a linear decline line.

The IMF of this step obtained by subtracting the decline trend from the denoised cuff pressure is shown in Figure 11. It is in fact the oscillometric pulse that is expected to get. It looks smooth and is consistent with the general shape of the oscillometric pulse whose amplitude is typically a bell shaped curve. Finally, the original cuff pressure was successfully denoised and decomposed into the declining trend (Figure 10) and the oscillometric pulse (Figure 11).

3   SIMULATION EXPERIMENT

3.1   Simulated cuff pressure signal

To test this method, a simulated cuff pressure signal was constructed, which was the superimposition of a decline trend, an oscillatory wave and noise. The simulated trend was a linear decline line with the slope of −3 mmHg/sec and the intercept of 170 mmHg. The simulated oscillatory wave was a sine wave with the frequency of 1 Hz and the amplitude of 3 mmHg which was modulated by a slow frequency sine wave. The noise was the gauss white noise with the SNR of 20 dB. The doted curve in Figure 12 shows the simulated cuff pressure signal.

3.2   Traditional digital filter method

The traditional method to decompose the cuff pressure is the digital filter method (Jazbinsek, V. 2005, Komine, H., and Asai 2012). To denoise the cuff pressure signal, a 1024 order digital FIR low-pass filter, whose cuff-off frequency was 5 Hz, was designed and applied to the simulated cuff pressure. A 1024 order digital FIR low-pass filter with the cut-off frequency of 0.3 Hz was applied to the simulated cuff pressure to obtain the decline trend. A 1024 order digital FIR band-pass filter with the pass-band from 0.5 to 10 Hz was applied to the simulated cuff pressure to obtain the oscillometric pulse.

3.3   To compare EMD with the filter method

The simulated cuff pressure signal was denoised and decomposed by both the EMD and the filter method. Figures 12, 13 and 14 show separately the cuff pressure, the decline trend, and the oscillometric pulse signal. The solid, dashed and dash-dot curve represent separately the simulated ideal signal, the signal processed by EMD, and the signal processed by the filter method. The lower subfigures are the 'zoomed in' view of the signal in the upper subfigure.

Table 1. RMSEs of 3 signals processed by both the filter and EMD method.

	Cuff Pressure	Decline Trend	Oscillometric Pulse
By EMD (mmHg)	0.027	0.094	0.095
By Filter (mmHg)	0.1	0.2	0.17

To qualitatively compare the two methods, the simulated cuff pressure, trend and oscillatory wave were taken separately as the true and the Root Mean Squared Errors (RMSEs) of the three signals processed by both the filter and EMD method, with respect to the corresponding simulated true signal, were calculated. As shown in Figures 12, 13 and 14, the errors were the difference between the dashed and the solid curves, and the difference between the dash-dot and the solid curves. The gauss white noise was generated 100 times and the RMSEs of the three signals by both methods for the 100 times were calculated and showed in Table 1 in which the unit is mmHg.

4 RESULTS

As shown in Figures 12, 13 and 14, the denoised signal, the trend, and the oscillometric pulse obtained by filter and EMD methods are all similar to the original simulated signal, implying that good decomposition performance was achieved by both methods.

Strong fluctuations can be seen at the start and end parts of the filtered signal, which is called edge effect. This edge effect is unavoidable for the filter method. Although it also exists for the EMD method, it is much smaller and can be greatly reduced if the local extreme points are correctly found and fitted at the edges of the signal.

As shown in Table 1, the RMSEs of the cuff pressure, the decline trend, and oscillometric pulse signal, processed by EMD method, are all smaller than those of the corresponding signals by filter method, indicating that the EMD method produced a better performance.

5 CONCLUSIONS

The proposed method can efficiently decompose the cuff pressure into the trend and oscillometric pulse

by performing the EMD on a compound signal only two times. It has better performance than the traditional filter method. Especially, it may reduce greatly the edge effect if the local extrema at the edges of the signal are properly found and fitted. Therefore, it is a very efficient and effective method for cuff pressure denoising and decomposition. Furthermore, it can also be applied to the denoising and decomposition of any signal, which comprises a low frequency trend, a high frequency oscillatory component, and noise.

REFERENCES

Baker, P. D., Westenskow, D. R., & Kück, K. 1997. Theoretical analysis of non-invasive oscillometric maximum amplitude algorithm for estimating mean blood pressure. *Medical and biological engineering and computing*, 35(3), 271–278.

Forster, F. K., & Turney, D. 1986. Oscillometric determination of diastolic, mean and systolic blood pressure—a numerical model. *Journal of biomechanical engineering*, 108(4), 359–364.

Huang, N. E., Shen, Z., Long, S. R., Wu, M. C., Shih, H. H., Zheng, Q. & Liu, H. H. 1998. The empirical mode decomposition and the Hilbert spectrum for nonlinear and non-stationary time series analysis. *Proceedings of the Royal Society of London. Series A: Mathematical, Physical and Engineering Sciences*, 454 (1971), 903–995.

Huang, N. E., & Attoh-Okine, N. O. (Eds.). 2005. *The Hilbert-Huang transform in engineering*. CRC Press.

Jazbinsek, V., Luznik, J., & Trontelj, Z. 2005. Non-invasive blood pressure measurements: separation of the arterial pressure oscillometric waveform from the deflation using digital filtering. *IFBME proceedings of EMBEC'05*.

Komine, H., Asai, Y., Yokoi, T., & Yoshizawa, M. 2012. Non-invasive assessment of arterial stiffness using oscillometric blood pressure measurement. *Biomedical engineering online*, 11, 6.

O'Brien, E., Atkins, N., & Staessen, J. 1995. State of the Market A Review of Ambulatory Blood Pressure Monitoring Devices. *Hypertension*, 26(5), 835–842.

O'Brien, E., Waeber, B., Parati, G., Staessen, J., & Myers, M. G. 2001. Blood pressure measuring devices: recommendations of the European Society of Hypertension. *BMJ*, 322(7285), 531–536.

Perloff, D., Grim, C., Flack, J., Frohlich, E. D., Hill, M., McDonald, M., & Morgenstern, B. Z. 1993. Human blood pressure determination by sphygmomanometry. *Circulation*, 88(5), 2460–2470.

Shevchenko, Y. L., & Tsitlik, J. E. 1996. 90th Anniversary of the development by Nikolai S. Korotkoff of the auscultatory method of measuring blood pressure. *Circulation*, 94(2), 116–118.

Biomedical Engineering and Environmental Engineering – Chan (Ed.)
© 2015 Taylor & Francis Group, London, ISBN: 978-1-138-02805-0

The detection and identification of typical ECG wave feature points based on improved sparse separation

Xin-Xing Wang & Wei Zhang
Tongji University, Shanghai, China

ABSTRACT: The detection and recognition of typical Electronic-Cardiology (ECG) signal waves can lay a solid foundation for patients' heart monitoring and evaluation. This paper proposed and developed a new algorithm for ECG wave detection and recognition by extending sparse decomposition. The recognition is integrated into the detection process and can be deduced reliably. The atomic set used is generated by an expanding, shrinking, and shifting operation on the typical discrete waveform segment. The final combination is optimized through the classical Match-Pursuit (MP) algorithm. Results based on PhysioNet database show that the average detection rates of P wave, QRS complex and T wave can achieve 88.3%, 99.2% and 96.7%. One superiority of this algorithm is that it can detect the feature points of P and T waves directly without deduction or further classification like other algorithms. Furthermore, it can be adapted to arbitrary input ECG data lengths which make it more practical in real conditions.

1 INTRODUCTION

ECG reflects the electrical activity of the heart, contains useful information about the functional status of the heart, and provides important data reference for heart disease diagnosis, monitoring and evaluation [1], especially because of its non-invasive, low cost, easy to carry and high reliability. It has been appreciated extensively by fields like telemedicine and mobile health advances in recent years. Also, it is one of the key technologies to realize fully automated ECG data analysis and reliably extraction of typical wave groups as well as its key feature points from measured data in supporting, and monitoring automated distress, health assessment, and other applications [1].

In traditional methods, accurate determination of the QRS complex is essential for automatic ECG analysis. Once the positions of the QRS complexes are found, the locations of other components of ECG like P, T-waves and ST segment etc. are found relative to the position of QRS. In this sense, QRS detection provides the fundamentals for almost all automated ECG analysis algorithms.

Various methods have been applied to detect the QRS complex, such as derivative based algorithms, digital filtering techniques, template matching, nonlinear transform, and wavelet transform etc. Derivative based algorithms use the derivative of the ECG as a high-pass filter to assess the steep slope of the R wave, this method is easy to operate with low computational complexity, but is sensitive to noise interference and has poor robustness[2]. Digital filtering techniques enhance the QRS complexes by using a high-pass filter

or a band-pass filter[3,4]. Template matching methods take advantages of the great similarity between the respective signals of the cardiac periods, approximate the QRS complex to a fixed template.This method has good robustness, but the computation is complex, and needs to divide the ECG signal periodically by hand. Nonlinear transformation uses Hilbert transform to enhance the QRS complex[5,6]. Wavelet based QRS detection algorithm is based on searching the wavelet coefficients modulus maxima greater than an updated threshold[7,8]. There are also methods, based on artificial neural networks, genetic algorithms, syntactic methods, SVN etc.[9−12]. Due to the significant QRS complex features, the above algorithms can generally achieve high QRS complex detection accuracy. However, P, T-waves are difficult to detect because of the small amplitude and slope, and are easily affected by noise, and have large individual differences. Moreover, the current P, T-waves detect algorithms always rely on the accurate location of R-wave, which can easily accumulate errors, so the related algorithms still need to be improved. The essence of the foregoing algorithms such as wavelet transform is to look for the zero-crossing point or point mutations, due to noise ratio and other reasons, the zero-crossing point or point mutations is not fully consistent with the actual ECG feature point and need further recognition, i.e., the procession of ECG feature point detection and identification should be separated.

Signal sparse decomposition theory provides a new way for ECG waveform detection and identification[13]. In 2007, a paper[14] applied this theory to detect and identify the feature points of the

ECG waveform, claimed that the algorithm achieved 99% of detection rate. However, this paper has many restrictions on the input signal, firstly, before sparse decomposition the original signal should be segmented by cycle, which is often not feasible in practice automated analysis; secondly, the original signal is required to be a complete cycle rather than any band; finally, the iteration number is decided by the cycles of the input data, all the above constraints reduce the level of automatic ECG analysis. Although the detection rate of QRS complex is high, but the P-wave, T-wave detection rate was only 85%.

This paper improved the atomic set design according to the different characteristics of each ECG wave based on the sparse decomposition theory, and changed the drawback that the position of P, T-wave was inferred by the position of R-wave, improved the adaptive ability of the algorithm for any input ECG fragments and the recognition rate of the P, T-wave.

2 SPARSE DECOMPOSITION THEORY

The sparse decomposition theory describes that if the signal has sparse or compressive characteristics, the sparse signal can be reconstructed precisely or approximately through solving a nonlinear optimization problem. Thus, acquiring a best sparse signal is the basis and premise condition for applying CS theory. An orthogonal basis set dictionary is generally believed to produce the smallest set of sparse representations, but the pursuit of orthogonality can easily resulting in the lack of its physical meaning[15]. In natural language, a rich dictionary is beneficial to construct accurate sentences. Similarly, sparse representation of the complex signals can achieve better effects if we use an over-complete dictionary. In recent years, one of the research focuses of sparse representation is signal sparse decomposition with redundant dictionary. The key question is how to build an over-complete redundancy dictionary, and how to find the best linear combination of atomic items from the redundant dictionary to best represent the target signal[16].

Currently, in the Compressive Sensing area, the Matching Pursuit (MP) algorithm is the most commonly used signal sparse decomposition method. An MP algorithm applies a greedy strategy, that is, at each step of the matching pursuit, the time-frequency atoms in the redundant dictionary are selected one by one to find the optimal atom that matches the signal structure best. This process can be formalized as follows:

We define a dictionary, such that. The vector can be decomposed into.

Where is the residual vector after approximation in the direction of. Obviously and are orthogonal, so that the following relationship is satisfied.

To minimize, we must choose such that is maximum.

This yields the next step in the decomposition scheme.

By iteration we end up with the atomic decom position.

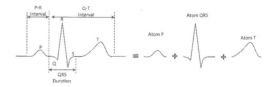

Figure 1.   A typical ECG signal.

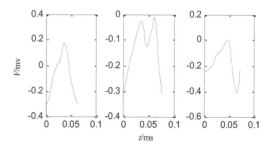

Figure 2.   All kinds of typical P waves.

As to the decomposition progresses, the error energy is gradually fading.

This paper applied this theory and the MP algorithm decomposes the target ECG into some typical wave components, and enables more accurate feature points detection, and recognition.

3 REDUNDANT ATOMIC SETS

ECG represents the electrical activity of the heart, which is characterized by a number of waves P, QRS, and T related to the heart's activity. Another wave, called U wave is also present but its importance is not yet identified, so it is generally not considered. A simple model is shown in Fig. 1.

The atomic set should contain components that express the characteristics of P wave, QRS wave group, and T-wave components. We extract discrete sequence from the preprocessed ECG signal according to the characteristics of each waveform segment, then expanding, shrinking and shifting the discrete sequence to construct the P wave, the QRS complex, and T-wave sub over-complete dictionary, where we execute the sparse decompose and reconstruct algorithms. The ECG signals are preprocessed by wavelet decomposition with db 4 wavelet.

3.1 *P atomic sets*

The P wave is the first wave of the ECG. It represents the spread of electrical impulses through the atrial musculature (activation or depolarization) and the typical duration is 80 ms–100 ms. According to the characteristics of P-wave and the observation of P waveform in PhysioNet database, we extracted three typical P-wave discrete sequences shown in Figure 2, and then normalize the sequences with the formula.

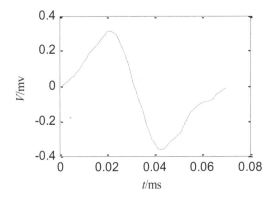

Figure 3. Typical QRS waves.

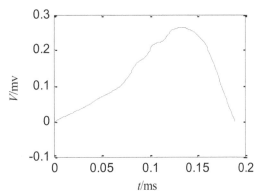

Figure 4. Typical T wave.

The P wave over-complete dictionary is constructed by expanding, shrinking and shifting the normalized sequences. Given as size factor and as shift factor, which means defines P-wave width and defines the horizontal position of P-wave, we construct the P wave over-complete dictionary by adapting these two parameters. When doing stretching transformation we can obtain the length of the P-wave after transformation, then the transformed discrete sequence of P-waves could be acquired by linear fitting methods. By experience, should range between 0.8 and 1.2, and the scope of is related to the length of the signal to be processed. The amplitude of wave segments is calculated based on the solution vector as the atoms used are all normalized.

3.2 QRS atomic sets

The depolarization of the ventricles is represented on the ECG signal by the QRS complex. Since the ventricles contain more muscle mass than the atria, there is an increased conduction velocity and the QRS complex is more pointed in shape than rounded. The QRS duration is 60 ms to 100 ms for a normal functioning heart. In case of any cardiac abnormalities, the QRS is wider and may be up to 200 ms. In general, QRS complex has three waves; a Q wave, an R wave, and an S wave.

Typical QRS complex discrete sequence is given in Figure 3. We construct QRS complex sub over-complete dictionary by expanding, shrinking and shifting the normalized discrete sequence by size factor and shift factor.

3.3 T atomic sets

The T wave in ECG Signal represents the repolarization (or recovery) of the ventricles. It can range from 50 ms to 250 ms depending upon the heart rate. The part before the highest point of the waveform is relatively flat, while the rear part is relatively short. Normally, its wave direction should be consistent with the R-wave direction, and the ratio of the amplitude

between T-wave and R-wave should be greater than 1/10.

Typical T-wave discrete sequence is given in Fig. 4. Similarly, we construct T-wave sub over-complete dictionary by expanding, shrinking and shifting the normalized discrete sequence by size factor and shift factor.

The method in this paper which uses discrete sequences of ECG signals as atoms is simple and easy to operate, moreover, this method avoids the drawbacks that generating function cannot correctly expresses the characteristics of an ECG signal.

4 FEATURE POINTS DETECTION AND IDENTIFICATION BASED ON IMPROVED MP

4.1 Proposed algorithm

Algorithm: ECG Feature Points Detection and Identification

    Input: The original signal
    Output: Position of the feature points
    Begin
// process of denoise based on wavelet
    wavelet_filter();
//Generate atomic sets, based on the characteristics of ECG waves
    For each ()
      generate()
    End
//execute sparse decomposition and reconstruction in the generated redundant dictionary, G is the solution vector, A is the sparse coefficient matrix of the matching atom, S is the vector of size factor, U is the vector of shift factor
    For each ()
      MP()
    End
//Process the situation where is not a complete cycle at the beginning and end, assume is the residual vector of sparse decomposition, and are the wave segments at the beginning and end of, and is the length of.

// Calculate the position of feature points of, assume the position of P, T-wave peak are respectively, the position of the starting point, peak point and end point of the QRS complex are respectively, the location of the five feature points in the atomic sets are respectively,

For each ()

End
End

Wherein the type of feature points can be decided corresponding to the type of atoms.

4.2 Key technology: feature points detection and identification

Among the ECG over-complete dictionary, the iterative selection method is used to select the atom mostly matches the signal or residual signal, and the sparse coefficient, size factor and the shift factor of the selected atom is noted. If the sparse coefficient of the selected atom is greater than 0.8 times of the sum of all other sparse coefficients, then the selected atom is effective and represents a characteristic wave.

So that all effective atomic scale factors composed of vector S, all effective atomic shift factors composed of vector U, assume that the position of a feature point in the atomic set is, then for each.

Where is the position of the feature point in the original ECG.

4.3 Key technology: adaptive algorithm for the incomplete ECG period

Find the maximum and minimum values of among the vector S, and then we select the wave segment of the residual signal from the starting point to the location of the minimum as the beginning part of the reconstructed signal, and set the wave segment from the maximum to the end of it, as the end part of the reconstructed signal. This method provides a good solution to the problem of an incomplete ECG period processing.

5 EXPERIMENTAL RESULTS EQUATIONS

5.1 Data source description

We have implemented the proposed algorithm described in Chapters 2 and 3, using the PhysioNet free access data files of physiologic signals. The NIF ECG database contains a series of 55 multichannel ECG recordings, each recording contains 2 thoracic signals, 3 or 4 abdominal signals, the recordings were collected at 2007 with 1 kHz sampling rate. Another relatively common used ECG database is MIT-BIH Arrhythmia Database, which contains 48 half-hour excerpts of two-channel ambulatory ECG recordings, and was obtained from 47 subjects studied by the BIH Arrhythmia Laboratory between 1975 and 1979. Due to the different acquisition time, quality of the ECG signal of these two databases are different, obviously, NIF ECG

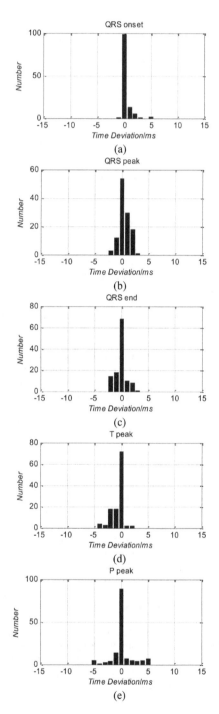

Figure 5. Histogram of time deviation between the detection results of the proposed approach and manually annotations results of twenty NIF ECG data sets: (a) QRS onset, (b) QRS peak, and (c) QRS end, (d) T peak, (e) P peak.

database can better reflect the essential characteristics of ECG. Therefore, our experiment is based on NIF ECG database, we intercept wave segments of ECG signal from any starting point to the continuous 2000

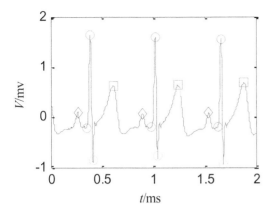

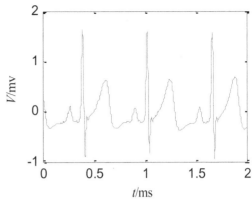

Figure 6. Results of processing 'ecgca595' from NIF ECG data sets (○, represent QRS feature points, □ represent T peak, ◇ represent P peak).

Figure 7. Reconstructed signal of 'ecgca595' from NIF ECG data sets.

sampling points (sampling time is 2s) as the original signal.

5.2 Results

Detection rate of feature points can be calculated by comparing them with the manual calibrated information of the original ECG. Assume the manual calibrated position of a feature point is, and the position obtained from the proposed algorithm is, and is the detect error of the feature point, then Make if, we believe that the feature point is correctly identified. Make, also that is the detection rate of feature points, is the number of feature points. The variance of the detection time error is defined as follow. The smaller the variance, the greater the stability of the algorithm.

We tested 20 recordings, about 130 heart beats which were all selected randomly from the NIF ECG database. Statistics found that the detection rate of QRS complex, T, P-wave is relatively 99.2%, 96.7% and 88.3%. The variance of the detection time error of PQRST is relatively 2.381, 0.7077, 0.3732, 0.6595 and 1.0439, which shows the good stability of the algorithm. Fig. 5 shows the detection time error distribution of each characteristic point.

Results of "ecgca595" in NIF ECG database is shown in Figure 6.

The reconstructed ECG of "ecgca595" in NIF ECG database is shown in Fig. 7. Fig. 8 shows the process of signal decomposition and reconstruction executed by the proposed algorithm in detail. Each component represent the reconstructed QRS complex, P, T-wave and the possible incomplete wave segments in the original ECG.

The above figures show that our algorithm can always be carried out correctly and accurately even if the input ECG fragments include an incomplete heart period. Compared to the original signal, the reconstructed signal little little difference, but the difference is within acceptable limits, and does not interfere with the detection and identification of the ECG feature points.

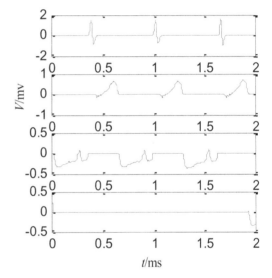

Figure 8. Detection steps of MP approach.

6 CONCLUSIONS

Experiments have verified that the algorithm we proposed reached a high detection rate of the ECG feature points. In the process of building atomic sets, considering that the generation function cannot represent the characteristics of the ECG signal wave suitably, while the use of training methods will increase the complexity of the algorithm, so this paper uses the discrete ECG signal sequence as atomic sets and also obtained good results. The process of signal decomposition and reconstruction based on the generated atomic sets has no special requirements from the reconstruct order of the P wave, QRS complex and T-wave. Namely, the identification of P and T-wave is entirely based on their own characteristics rather than on the identification of the QRS complex, which greatly improves the identification rate of the P, and T-waves compared to the traditional methods. Moreover, the algorithm will have a wide range of applications as it has no limits for the original ECG signal, and is highly automated as there is no need to input any parameters during operation.

ACKNOWLEDGEMENT

The author gracefully thanks the funding from the National Science Foundation of China (Grant No. 61004100) and the Returned Overseas Fund in 2012.

REFERENCES

[1] Wen-li Chen. Study on ECG Signal Detection and Analysis Algorithms Using Wavelet Transform [D]. Sichuan Normal University, 2008.

[2] Gupta R, Mitra M, Mondal K, et al. A Derivative-Based Approach for QT-Segment Feature Extraction in Digitized ECG Record [C]. Kolkata: 2011. 63–66.

[3] Chee-Ming T, Salleh S H. ECG based personal identification using extended Kalman filter [C]. Kuala Lumpur: 2010. 774–777.

[4] Al-Kindi S G, Ali F, Farghaly A, et al. Towards real-time detection of myocardial infarction by digital analysis of electrocardiograms [C]. Sharjah: 2011. 454–457.

[5] Mukhopadhyay S K, Mitra M, Mitra S. ECG signal processing: Lossless compression, transmission via GSM network and feature extraction using Hilbert transform [C]. Bangalore: 2013. 85–88.

[6] Mukhopadhyay S K, Mitra M, Mitra S. Time plane ECG feature extraction using Hilbert transform, variable threshold and slope reversal approach [C]. Kolkata, West Bengal: 2011. 1–4.

[7] Doudou W, Zhengyao B. An improved method for ECG signal feature point detection based on wavelet transform [C]. Singapore: 2012.1836–1841.

[8] Mazomenos E B, Biswas D, Acharyya A, et al. A Low-Complexity ECG Feature Extraction Algorithm for Mobile Healthcare Applications [J]. Biomedical and Health Informatics, IEEE Journal of, 2013, 17(2): 459–469.

[9] Sadhukhan D, Mitra M. Detection of ECG characteristic features using slope thresholding and relative magnitude comparison [C]. Kolkata: 2012. 122–126.

[10] Salih S K, Aljunid S A, Aljunid S M, et al. High speed approach for detecting QRS complex characteristics in single lead electrocardiogram signal [C]. Mindeb: 2013. 391–396.

[11] Tae-Hun K, Se-Yun K, Jeong-Hong K, et al. Curvature based ECG signal compression for effective communication on WPAN [J]. Communications and Networks, Journal of, 2012, 14(1): 21–26.

[12] Mehdi B, Khan T, Ali Z A. Artificial neural network based electrocardiography analyzer [C]. Karachi: 2013. 1–7.

[13] Yan Z, Heming Z, Tao L. Sparse decomposition algorithm using immune matching pursuit [C]. Beijing: 2012. 489–492.

[14] Chun-guang Wang, He-tao Huang. The Detection and Recognition of Electrocardiogram's Waveform Based on Sparse Decomposition [J]. Chinese Journal of Biomedical Engineering, 2008(02): 224–228.

[15] Ahani S, Ghaemmaghami S. Image steganography based on sparse decomposition in wavelet space [C]. Beijing: 2010. 632–637.

[16] Ghaffari A, Babaie-Zadeh M, Jutten C. Sparse decomposition of two dimensional signals [C]. Taipei: 2009. 3157–3160.

Biomedical Engineering and Environmental Engineering – Chan (Ed.)
© 2015 Taylor & Francis Group, London, ISBN: 978-1-138-02805-0

A low-cost, compact, LED-based animal fluorescence imaging system

Yamin Song, Chao Zhang & Fuhong Cai
Centre for Optical and Electromagnetic Research, Zhejiang Key Laboratory of Sampling Technologies, Zhejiang University, Hangzhou, China

ABSTRACT: A fluorescence imaging system using Light Emitting Diodes' (LEDs) arrays of several wavelengths as excitation light sources, is presented. By properly choosing the LEDs and the corresponding spectral filters to get appropriate excitation wavelengths, the system can be utilized for fluorescence imaging applications, including small animal fluorescence imaging. Because of the utilization of LEDs as excitation sources, this system can be compact and low-cost. Also, the LEDs are installed on plug-in light sockets. The LEDs can be easily replaced for different applications. Finally, the liver, kidney, and heart of a mouse processed by a fluorescent agent are imaged by the system.

1 INTRODUCTION

A small animal imaging system is commonly used for biomedical imaging systems. Users can observe biological processes of these animals, such as tumour growth and metastasis, the development of infectious diseases, expression of specific genes etc. This optical imaging system requires appropriate light sources. At present, common light sources for fluorescence imaging mainly include high-pressure mercury lamps and xenon lamps, but these light sources generally have the disadvantages of being expensive, having a short lifetime, and requiring complex operation steps. For example, a mercury lamp emits strong ultraviolet and blue-violet light, enough to excite various kinds of fluorescent material, and therefore, is widely used in fluorescence imaging. However, a mercury lamp takes a long time to start and after turning off, it needs to completely cool before restarting. Its lifetime is generally 200 hours; 400 hours is the longest. In addition, with a mercury filling inside the lamp, when the lamp is damaged or loses efficacy, the lamp would pollute the environment.

Compared with the mercury lamp and xenon lamp, an LED produces more concentrated spectral energy and is the most energy-efficient device to produce monochromatic. In addition, an LED is a safe, reliable, and environmentally friendly source. An LED only excites spectra in a narrow band, and multiple LED bands enable the LED light array source to be applied multi-colour fluorescence applications. For fluorescence imaging, excitation light centred at the peak absorption wavelength of the fluorescent agent is most desirable. When a broadband light source, eg, a xenon lamp, is used as a fluorescence imaging light source, most of the optical energy is removed by the filter, causing a waste of energy. For fluorescence imaging applications of a non-microscopic type, there is a need

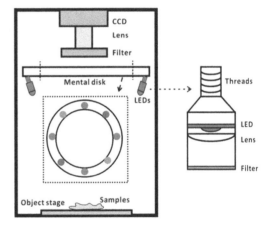

Figure 1. Schematic diagram of the LED-based fluorescence system and detailed structure of the LED header.

for improved light sources that combine flexibility and low cost with high performance.

In this paper, we designed a multi-spectral LED array structure. In front of each LED, there was a collimating lens and a filter, making the light illuminate uniformly, and narrowing the spectral width. With the LEDs' array structure, we constructed a compact small animal fluorescence imaging system, achieving the fluorescence imaging of an animal's internal organs.

The paper is organized as follows. We begin by introducing our designed fluorescence imaging system the LEDs' array structure. We then demonstrate the processing of the internal organs of a mouse and present the fluorescence image taken by our system. We conclude the paper with a discussion on the advantages and potential of our system.

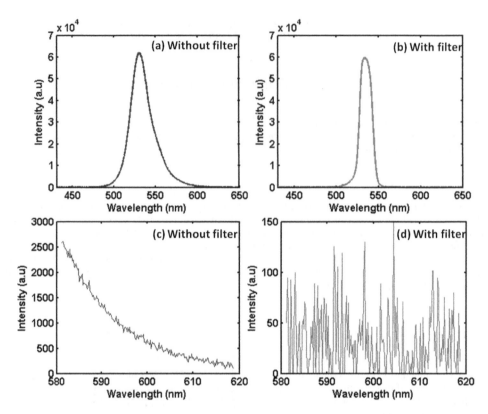

Figure 2. Spectra of 530 nm LED without and with filter. (a)–(b) LED spectrum with and without a filter between 430 nm and 650 nm. (c)–(d) LED spectrum with and without a filter between 580 nm and 620 nm.

2 MATERIALS AND METHOD

2.1 *Imaging systems*

The LED-based fluorescence system is given in Figure 1. The light structure mainly consists of a hollow metal disk, installing eight LEDs and an LED driver circuit. These eight light sockets are distributed uniformly in the edge of the metal disk, each having a designed angle of 19 degrees with the vertical direction depending on the size of the system. Each light socket can hold a light header through the designed helical structure. The filter before the CCD camera was fixed on a filter holder, which can be inserted straight into the filter slot, making the filter convenient to exchange, depending on different applications.

The samples were placed on a black object stage. The induced fluorescence of the samples was first filtered by the high-pass filter in front of the CCD camera to eliminate the excitation light, and then captured by a CCD camera. The LEDs array can be controlled by a LABVIEW-based program during the measurement, depending on the specific fluorescence agents used in the experiment. The emission intensities of the LEDs were controlled by a data acquisition card through a driving circuit. With the driving circuit, the current of every LED can be accurately controlled, ensuring stabilization of the light intensity.

2.2 *LEDs array design*

To be specific, each LED is installed in a light header with a lens and a filter in its front, as shown in the right part of Figure 1. A lens ensures the light is illuminating uniformly and the energy is more concentrated.

A known characteristic of an LED, relative to laser, is its boarder spectral bandwidth. For example, the excitation sources chosen to induce fluorescence excitation in our experiment were two identical 530 nm LEDs depending on the fluorescent agent we used; the LED's spectra are shown in Figures 2(a) and (c). Without using a proper filter in front of the LED, its spectrum near 600 nm is relatively strong, which will add a certain amount of noise to the emission fluorescence spectrum and eventually influence the experiment results. In order to optimize the above light source spectrum, in front of each 530 nm LED a filter was placed, ranging from 510 nm to 550 nm. The Optical Density (OD) of the used filter is 6 between 600 nm and 800 nm. The spectra for a 530 nm LED with a filter are shown in Figures 2(b) and (d). As shown in Figure 2(b), the spectrum near 600 nm was reduced significantly.

At present, four kinds of LEDs with peak wavelengths of 530 nm, 650 nm, 750 nm, and 850 nm are chosen for the system to induce the red or infrared fluorescence, which is commonly used in biological fluorescent imaging. The reason for choosing these four wavelengths is that fluorescence wavelengths excited by these four wavelengths are usually greater than 600 nm. Within this spectral band, imaging depth is deeper and imaging will not easily be disturbed by auto-fluorescence of food. Also, for certain situations where there is a need for using more than one wavelength to induce the fluorescent agent, our designed system is quite applicable.

3 RESULTS AND DISCUSSION

3.1 *Preparation of mouse fluorescence samples*

The fluorescent agent we used was Tetraphenylethene Triphenylamine Fumaronitrile (TTF), as described in detail in our previously reported work. The AIE dots of TTF were injected intravenously (train vein) in male BALB/c mice and they diffused to different organs due to the blood circulation. Major organs including the heart, liver, spleen, lungs, and kidney were resected after a proper time, and preserved in formalin.

3.2 *Fluorescence images*

The TTF solution showed an absorption spectrum peak at 500 nm and a weak red fluorescence with an emission peak at 624 nm. The CCD camera used here was of 2/3 inches, and for this camera, the chosen camera lens was a prime collimating lens with a focal length of 16 mm and an aperture (Max) of f/1.4 Along pass filter with a cut-off wavelength at 600 nm in front of the CCD camera was adopted to extract the fluorescence signals from the AIE dots.

The fluorescence image of the above samples of organs taken by the system is shown in Figure 3. The exposure time of the CCD camera is 100 ms. As can be seen from the result, these AIE dots assembled mainly in the liver, because the liver is closely related to the metabolic function of animals, so we can see that TTF fluorescent agents can be metabolized and can be used for future study in biomedicine.

4 CONCLUSION

In this paper, we have presented an LED-based animal fluorescence imaging system, which is low-cost, compact, and convenient to use. The system can be applied to various applications related to fluorescent imaging, e.g., fluorescent imaging of small animals. To improve the uniformity and the Signal to Noise Ratio (SNR) of imaging, the collimating lens and filters were installed in the front of the LED lights. By using this imaging system, we obtained the fluorescence images

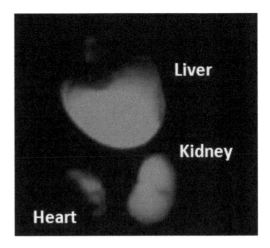

Figure 3. Fluorescence image of the organs samples taken by the system.

of the samples of mouse organs. In this experiment, the exposure time of the CCD is 100 ms, indicating the real time imaging application of the system. In addition, the LED component is designed as a plug-in structure, which allows users to conveniently extend the excitation wavelength range to violet and infrared, according to their needs in the future.

ACKNOWLEDGEMENT

This work was supported by the National Basic Research Program (973) of China (2011CB503700).

REFERENCES

Dasgupta, P. K., Eom, I. Y., Morris, K. J., & Li, J. 2003. Light emitting diode-based detectors: Absorbance, fluorescence and spectroelectrochemical measurements in a planar flow-through cell. *Analytica Chimica Acta* 500(1): 337–364.

Dong, Y., Liu, X., Mei, L., Feng, C., Yan, C., & He, S. 2014. LED-induced fluorescence system for tea classification and quality assessment. *Journal of Food Engineering* 137: 95–100.

Gioux, S., Kianzad, V., Ciocan, R., Gupta, S., Oketokoun, R., & Frangioni, J. V. 2009. High power, computer-controlled, LED-based light sources for fluorescence imaging and image-guided surgery. *Molecular imaging* 8(3): 156.

Wang, D., Qian, J., Qin, W., Qin, A., Tang, B. Z., & He, S. 2014. Biocompatible and Photostable AIE Dots with Red Emission for In Vivo Two-Photon Bioimaging. *Scientific reports* 4.

Biomedical Engineering and Environmental Engineering – Chan (Ed.)
© *2015 Taylor & Francis Group, London, ISBN: 978-1-138-02805-0*

Automatic recognition of mitral leaflet in echocardiography based on Auto Context Model

Feng Zhu, Xin Yang & Meng Ma
School of Electronic Information and Electrical Engineering, Shanghai Jiao Tong University, Shanghai, China

Lipin Yao
Shanghai Xin Hua Hospital, Shanghai, China

ABSTRACT: Due to the serious noise and poor image quality of ultrasound images, it is a difficult task to identify the mitral annulus (MA) and mitral leaflet (ML) in echocardiography. This paper presents an automatic algorithm based on Auto Context Model (ACM) for ML recognition. The algorithm is also applicable for MA hinge point identification. Different from other iterative algorithm frameworks, ACM takes advantage of the classification map as context information. It extracts features from both the local image and the classification map. In this paper, we extract LDB (Local Difference Binary) feature from the local image and local context feature from the context image. SVM (Support Vector Machine) is adopted as classifier. The algorithm firstly recognizes the leaflet and MA hinge points and then locates the leaflet more accurately according to the relative position of the two parts. The algorithm has been tested on RT3DE (Real Time Three-dimensional Echocardiography) datasets of ten patients. The experiment shows an accurate result within a 1.6 ± 2.3 pixels error of MA hinge point recognition compared with the expert manual segmentation result. The algorithm also achieves a precise ML identification result during both the diastole and systole periods.

1 INTRODUCTION

MA is an important part of the heart, playing a role in controlling the blood flow between the left ventricle (LV) and the left atrium [4]. MV structure and motion morphology has great significance for the diagnosis of valvular heart disease [2]. Clinically, ultrasound is a common technique used to observe the shape and trajectory of MA. Compared with MRI and CT, ultrasound is cheaper, faster and harmless to humans.

Many MV segmentation algorithms haves been proposed so far. Zhou et al. [17] put forward an ML tracking method by mapping the echocardiography sequence into a low-rank matrix. Martin et al. [7] use two connected active contours to track ML based on the prior knowledge that the MV movement can be divided into leaflet movement and muscle movement. Schneider et al. [12–14] present an MA segmentation algorithm using graph cuts in 3D ultrasound. Song et al. [15] utilize local context information and the position relationship between LV and MA to segment MA. The majority of these methods are semi-automatic or time consuming, which makes them limited for real time application.

In this paper, we present an automatic algorithm based on ACM for ML recognition. The algorithm adopts Histogram Intersection Kernel Support Vector Machine (IKSVM) as the classifier. An obvious speed improvement can be obtained via applying the method suggested in Reference [6].

The algorithm was evaluated on RT3DE datasets of ten patients provided by Shanghai Xin Hua Hospital. The dataset was captured by Philips Sonos 7500 ultrasonic system with a spatial resolution of $208 \times 160 \times 144$. Each patient's dataset includes RT3DE during a cardiac cycle, which contains 13 to 22 frames. For each frame of the RT3DE dataset, we select 20 slices near the standard four-chamber slice as samples. 15 slices are used as training samples and the other 5 slices are testing samples.

The innovation behind our work is as follows: Firstly, we present an automatic ML recognition algorithm from the perspective of pattern recognition. Secondly, our algorithm is also applicable in identifying MA hinge points, which are often needed as initial conditions in the methods aforementioned.

2 METHOD

2.1 *Auto Context Model*

Auto Context Model (ACM) is an algorithm based on iterative learning. In recent years, ACM has been successfully applied to the segmentation of brain [16], prostate [5] and liver [3]. Here we introduce it into the recognition of ML.

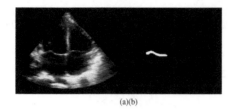

(a)(b)

Figure 1. The training sample: (a) is the sample and (b) is the manual labelled classification map of (a).

2.1.1 The training of ACM

Given the echocardiography sequences $\{X_i\}$ and the corresponding labeled classification maps $\{Y_i\}$, firstly construct the sample dataset:

$$S = \left\{ \left(y_{ij}, X_i(j) \right), i = 1,2,\ldots,m, j = 1,2,\ldots,n \right\} \qquad (1)$$

where y_{ij} means the class of the jth pixel in X_i and $X_i(j)$ means the image block centered at pixel j in X_i. m is the number of images while n is the pixel number of each image. The first classifier C_1 is learned from the local image. Then we get the new classification map P_i^0 and dataset:

$$S_1 = \left\{ \left(y_{ij}^0, \left(X_i(j), P_i^0(j) \right) \right), i = 1,\ldots,m, j = 1,\ldots,n \right\} \qquad (2)$$

After that, ACM extracts features not only from the local image, but also from the classification map created by the previous classifier which is utilized as context information to train the next classifier. The algorithm iterates and finally a sequence of classifiers is obtained:

$$C_t = \begin{cases} C(y_i|X_i(j)) & t = 1 \\ C(y_i|X_i(j), P_i^{t-1}(j)) & t = 2,\ldots,T \end{cases} \qquad (3)$$

T is the number of iterations. In this paper, we set $T = 4$ and adopt SVM as the classifier.

The points in the white region of Figure 1(b) are positive samples and the others are negative samples. However, it is not necessary to train all the pixels in $\{X_i\}$ in the actual training process. Otherwise it will be time-consuming and the negative sample points near the ML edge may bring bad influence on the accuracy of the classifier. Here, we adopt a strategy of randomly selecting double the amount of positive points as negative samples.

2.1.2 The testing of ACM

The testing process is similar to the training process. Firstly, classify the input image with C_1 to obtain the classification map P_i^1. Then extract features both from the sample and the classification map. Sequentially employ C_i to classify the sample. The last classification map P_i^T is the recognition result.

2.2 Feature design

The principle of feature selection should be directed to the image characteristics. Texture-based features such

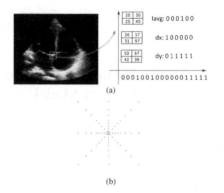

(a)

(b)

Figure 2. LDB and local context feature: (a) is LDB feature and (b) is the local context feature.

as LBP (Local Binary Patterns) [9] and HOG (Histogram of Oriented Gradient) [2] are not applicable for MV description due to the serious speckle noise and inconspicuous texture information of echocardiography. In addition, the pixel intensity of MV is similar to the ventricular wall and myocardial tissue. Directly extracting intensity as feature also may not achieve a good result.

As previously mentioned ACM needs to extract two kinds of features. Here we extract the LDB (Local Difference Binary) [10] feature from local image and the local context feature from the classification map. LDB is a kind of binary string feature which can be computed in two steps. Firstly, divide the image patch centred at the pixel into n × n blocks. Secondly, compare the average intensity and gradient in the x, y direction of each block with other blocks. As the computation of intensity and gradient can be accelerated by the integral image, LDB computes very quickly.

The Local context feature [7] adopts a sparse sampling strategy. Eight rays lead out from the point with a 45 degree interval. Then sample the average intensity in each ray at points with a distance sequence {1, 3, 5, 7, 9, 13, 17, 23} far from the interested point. Mean filter is used to remove high frequency noise.

2.3 Speed up of the histogram intersection kernel SVM

In this paper, we employ histogram intersection kernel SVM (IKSVM) as the classifier. The Histogram intersection kernel is a kind of additive kernel. The following part will briefly introduce how to accelerate its classification process using the method in [6].

Given a dataset, $\{(x_i, y_i), x_i \in \mathrm{IR}^n, y_i \in \{1, -1\}, i = 1,\ldots,n\}$, the goal of SVM is to find an optimal hyperplane to separate the multi-class data. The decision function is:

$$W(x) = \mathrm{sgn}\left(\sum_{l=1}^{m} \alpha_l y_l K(x_l, x) + b \right) \qquad (4)$$

where $x_l \in IR^n$ is the support vector and m denotes its total quantity. The definition of the histogram intersection kernel is $K_{\min(X,Z)} = \sum_{i=1}^{n} \min(x_i, z_i)$. We can transfer W(x):

$$W(x) = \mathrm{sgn}\left(\sum_{l=1}^{m} \alpha_l y_l K(x_l, x) + b\right)$$

$$= \mathrm{sgn}\left(\sum_{l=1}^{m} \alpha_l y_l \left(\sum_{i=1}^{n} \min(z_i, x_{l,i})\right) + b\right)$$

$$= \mathrm{sgn}\left(\sum_{i=1}^{n} \left(\sum_{l=1}^{m} \alpha_l y_l \min(z_i, x_{l,i})\right) + b\right) \quad (5)$$

$$= \mathrm{sgn}\left(\sum_{i=1}^{n} h_i(z_i) + b\right)$$

Let $\bar{x}_{li}$ represent the ith element of x_l in ascending order, while $\bar{\alpha}_l$ and $\bar{y}_l$ represent the corresponding parameters. For $h_i(s)$, let r represent the max index of $\bar{x}_l$ that is smaller than s. Then transfer $h_i(s)$ and we have:

$$h_i(s) = \sum_{l=1}^{m} \alpha_l y_l \min(s, x_{l,i})$$

$$= \sum_{1 \le l \le r} \bar{\alpha}_l \bar{y}_l \bar{x}_{l,i} + s \sum_{r < l \le m} \bar{\alpha}_l \bar{y}_l = A_i(r) + sB_i(r) \quad (6)$$

As can be seen from the formula above, $h_i(s)$ is independent of A_i and B_i. If we compute A_i and B_i in advance and presort the support vectors, the algorithm complexity of W(x) will fall from O(nm) to O($n\log m$).

3 EXPERIMENT

The algorithm was evaluated on RT3DE datasets of ten patients with a spatial resolution of $208 \times 160 \times 144$. Each patient's dataset includes 13 to 22 frames of RT3DE during a cardiac cycle. For each frame RT3DE dataset, we select 20 slices near the standard four-chamber slice as samples.

This section sequentially describes the identification of the mitral annulus hinge points and the leaflet. MA hinge points are useful for multi-modality registration with CT and MRI. Traditional MA segmentation algorithms often can split the overall MA part, but fail to distinguish the MA hinge points from the leaflet. Our work is helpful for improving this situation.

3.1 MA hinge points recognition

We extract LDB, the coordinates of the sample point, pixel value as local image features, and extract local context features as context image features. During the experiment, we set the iteration number T = 4. The blue regions in Figure 3 show the identification result of the MA hinge points using different methods and

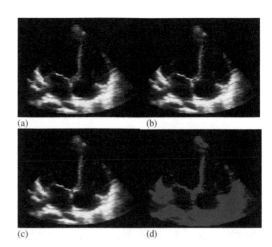

(a)　　　　　　　　(b)

(c)　　　　　　　　(d)

Figure 3. MA hinge point identification result contrast: (a) ACM+ local context + LBP features, (b) method in [15] SVM + local context features, (c) ACM + local context + histogram features and (d) ACM + LBP features.

Table 1. MA hinge point identification error.

Left MA Hinge Point		Right MA Hinge Point	
x	y	x	y
2.0	1.8	2.9	1.2

features. In order to obtain an accurate position, K-Means algorithm is applied to the blue point sets. The red cluster centroid points can be seen as the hinge points.

As seen in Figure 3, the recognition accuracy of our algorithm is higher than the other three. The LDB features provide better description ability by simultaneously using pixel information and gradient information. Besides, ACM combines the local image features and context features. On the other hand, the LBP feature is not applicable for MA hinge points recognition due to the heavy noise of ultrasound image.

Table 1 shows the identification error compared with the expert manual segmentation result.

3.2 ML recognition

Clinically, doctors usually diagnose valvular heart disease by observing the structure and movement morphology of the ML [8]. Figure 4 shows the ML identification result created by each classifier. The identification result is accurate and the wrong candidate points are gradually removed with the increase of iteration number T.

As seen in Figure 4(e), further processing is still required. Observe the echocardiography. The ML lies between the MA hinge points and its intensity is higher than other parts in this region. Based on this priori knowledge, we select the top 30% brightest points as

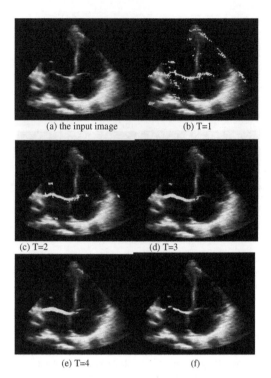

(a) the input image   (b) T=1

(c) T=2   (d) T=3

(e) T=4   (f)

Figure 4. ML identification result: (b) – (e) are the identification results after each iteration, and (f) is the final result
obtained by (e) after the filtering and thinning process.

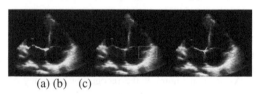

(a) (b)   (c)

Figure 5. ML identification result contrasts: (a) our method,
(b) method in [17], and (c) method in [7].

the final leaflet points. A thinning process is necessary
to remove the effects of ultrasound artefacts.

Figure 5 shows the results of our method and methods in [17] and [7]. Our method achieves an accurate
result. The method in [17] can not identify the ML
precisely. The authors proposed an opinion that the
ML region in an echocardiography sequence can be
mapped into a low-rank matrix. Maybe the correctness of this opinion is related to specific samples. The
method in [7] works well in the segmentation of the
long leaflet, but the overall effect of our method is
better.
Figure 6 shows the identification results of four
different patients.

4   CONCLUSION

In this paper, we propose an automatic algorithm
based on ACM for MA hinge points and ML recognition. According to the characteristics of MA, ACM

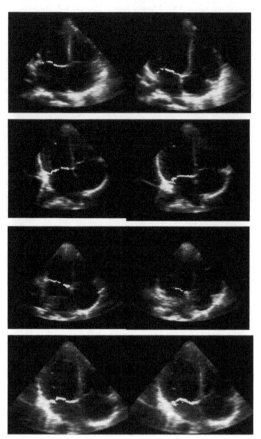

Figure 6.   The MA hinge point and ML identification results
of four patients.

takes advantage of the contextual information by
extracting LDB features from the local image, and
local context features from the classification map.
The experiment shows an accurate result. While there
exists plenty of parallel computation, our algorithm
can still be applied in real time applications via CUDA
acceleration [10, 11].

ACKNOWLEDGEMENTS

This paper has been partially supported by National
Basic Research Program of China (2010CB732506)
and Shanghai Basic Research Program (12JC1406600).

REFERENCES

[1] Dalal N. & Triggs B. 2005. Histograms of oriented
    gradients for human detection. 2005IEEE Computer
    Society Conference on Computer Vision and Pattern
    Recognition 1: 886–893.
[2] Tahta S.A., Oury J.H. & Maxwell J.M. 2002. Outcome
    after mitral valve repair for functional ischemic mitral
    regurgitation. *The Journal of heart valve disease* 11(1):
    11–8.

[3] Ji H., He J. & Yang X. 2013. ACM-based automatic liver segmentation from 3-D CT images by combining multiple atlases and improved mean-shift techniques. *IEEE journal of biomedical and health informatics* 17(3): 690–698.

[4] Kunzelman K.S. & Cochran R.P.& Chuong C. 1993. Finite element analysis of the mitral valve. *The Journal of heart valve disease*2(3): 326–340.

[5] Li W., Liao S. & Feng Q. 2012. Learning image context for segmentation of the prostate in CT-guided radiotherapy. P*hysics in medicine and biology* 57(5): 1283.

[6] Maji S., Berg A.C. & Malik J. 2013. Efficient classification for additive kernel SVMs. *IEEE Transactions on Pattern Analysis and Machine Intelligence* 35(1): 66–77.

[7] Martin S., Daanen V. & Troccaz J. 2006. Tracking of the mitral valve leaflet in echocardiography images. Biomedical Imaging: Nano to Macro 181–184.

[8] Nkomo V.T., Gardin J.M.& Skelton T.N. 2006. Burden of valvular heart diseases: a population-based study. *The Lancet* 368(9540): 1005–1011.

[9] Ojala T., Pietikainen M. & Maenpaa T. 2002. Multiresolution gray-scale and rotation invariant texture classification with local binary patterns. *IEEE Transactions on Pattern Analysis and Machine Intelligence* 24(7): 971–987.

[10] Pharr M., Fernando R. 2005. *Gpu gems 2: programming techniques for high-performance graphics and general-purpose computation.* Boston:Addison-Wesley

[11] Ryoo S., Rodrigues C.I. & Baghsorkhi S.S. 2008. Optimization principles and application performance evaluation of a multithreaded GPU using CUDA. 13th ACM SIGPLAN Symposium on Principles and practice of parallel programming. ACM: 73–82.

[12] Schneider R.J., Perrin D.P. & Vasilyev N.V. 2009. Mitral annulus segmentation from three-dimensional ultrasound. IEEE International Symposium on Biomedical Imaging: From Nano to Macro: 779–782.

[13] Schneider R.J., Perrin D.P. & Vasilyev N.V. 2010. Mitral annulus segmentation from 3D ultrasound using graph cuts. *IEEE Transactions on Medical Imaging* 29(9): 1676–1687.

[14] Schneider R.J., Tenenholtz N.A. & Perrin D.P. 2011. Patient-specific mitral leaflet segmentation from 4D ultrasound. Medical Image Computing and Computer-Assisted Intervention–MICCAI 2011 : 520–527.

[15] Song W., Xu W. & Yang X. 2013. Automatic detection of mitral annulus in echocardiography based on prior knowledge and local context. Computing in Cardiology Conference (CinC):1139–1142.

[16] Tu Z. & Bai X. 2010. Auto-context and its application to high-level vision tasks and 3d brain image segmentation. *IEEE Transactions on Pattern Analysis and Machine Intelligence* 32(10): 1744–1757.

[17] Zhou X., Yang C. & Yu W. 2012. Automatic mitral leaflet tracking in echocardiography by outlier detection in the low-rank representation. 2012 IEEE Conference on Computer Vision and Pattern Recognition : 972–979.

Biomedical Engineering and Environmental Engineering – Chan (Ed.)
© 2015 Taylor & Francis Group, London, ISBN: 978-1-138-02805-0

High-throughput monitoring of genetically modified soybean by microarray-in-a-tube system

Jingjing Yu, Chuanrong Hou, Rongliang Wang & Quanjun Liu
State Key Laboratory of Bioelectronics, Southeast University, Nanjing, China

ABSTRACT: An effective, easy and high-throughput method of monitoring GMOs is urgently needed, because of the new approvals of approved GM events rapidly increasing. Here, a microarray-in-a-tube system was developed for the GM soybean monitoring, which was sensitive, accurate, simple and high-throughput. In this system, multiplex amplification was integrated with hybridization, without washing and amplification product leakage, which could avoid the false negatives and positives. The 9 GM soybean events were successfully amplified simultaneously, with the detection sensitivity of this system reached 10^2 copies per reaction. Results showed that the microarray-in-a-tube system can be applied for effectively monitoring GM soybeans.

Keywords: Microarray-in-a-tube, genetically modified soybean, monitor

1 INTRODUCTION

The genetically modified (GM) crops have increased more than 100-fold from 1.7 million hectares in 1996 to over 175 million hectares in 2013, among which about 79% is GM soybean (James 2013). Although the contribution of GM crops is self-evident, the safety of genetically modified organisms (GMOs) is still controversial (Kamle & Ali 2013, Kearns et al. 2014), GM soybeans were used for food and feed, especially. It has proved that an effective and accuracy monitor method can ensure the safety use of GM soybean (He et al. 2014, Ladics et al. 2014, Leguizamón 2014).

Variety of methodologies have been developed for the monitoring of GM contents, which focus on the presence of foreign DNA or protein[3]. Although PCR is the popular and generally applied technique owing to its high sensitivity, low cost, and easy handling (Holst-Jensen et al. 2003), concern has intensified regarding cross-contamination, particularly where amplified DNA product is concerned.

Here, we developed a microarray-in-a-tube system taking advantage of highly efficient multiplex PCR and sealed chip strategy to monitor the GM soybeans, with advantages of high efficiency, simplified amplification process and preventing pollution. This system is composed of a PCR for multiplex amplification of 12 target DNA fragments and integrated microarray. In this system, all the commercialized import and export GM soybeans can be detected. The results obtained from the microarray-in-a-tube system were consistent with expectations. This system could be applied to effectively monitor the majority of the GM soybean in a single test.

2 MATERIAL AND EXPERIMENTAL PROCEDURE

2.1 Plant materials and DNA isolation

All the plant materials and plasmids of the GM soybean were genially offered by Jiangsu Entry-Exit Inspection and Quarantine Bureau (JSCIQ), and non-transgenic soybean materials were purchased from local markets. Genomic DNA of plant materials were extracted and purified using a DNA extraction kit (Qiagen 51304, Germany). The concentration of the DNA samples was evaluated by a NanoDrop 1000 spectrophotometer (NanoDrop Technologies, LLC, Wilmington, DE).

2.2 Chemicals and reagents

All the chemicals were purchased from Sigma-Aldrich (Missouri, America). Taq DNA polymerase, dNTPs and buffer were purchased from TaKaRa (Dalian, China), sample liquid was purchased from Capitalbio Corporation (Beijing, China), Hybridization solution contains $6 \times$ SSC, 50% Denhardt, and 0.2% (w/v) SDS.

2.3 Primers and probes

All the specific probes and primers used in this study were according to the database of European Union Reference Laboratory for GM Food and Feed (http://gmo-crl.jrc.ec.europa.eu/gmomethods). These genes were chosen according to the imported GM soybeans of Ministry of Agriculture. Each forward primer was modified by adding a 5′ universal sequence

(5′-CCACTACGCCTCCGCTTTCCTCTCTATGA-3′),
and each reverse primer was modified by adding
another 5′ universal sequence (5′-CTGCCCCGGGTT
CCTCATTCTCT-3′). The sequence of positive position probe was 5′-NH$_2$-poly (T)$_{20}$-cy3-3′, and positive
quality probe was 5′-NH$_2$-poly(T)$_{12}$-GAAAAATAAA
CTGTAAATCATATTC-3′. The universal primer
sequences were using BLAST checked to avoid
cross-reaction. All the primers and probes were
synthesized from Invitrogen Co., Ltd. (Shanghai,
China).

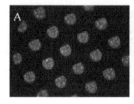

Figure 1. The pattern of the array. There were 9 kinds
of DNA microarray probes represented MON89788 (1),
A5547-127 (2), A2704-12 (3), GTS-40-3-2 (4), DP305423-1
(5), DP356043-5 (6), Lectin (7), CV127 (8), MON87701
(9), P-positive position probe, N-negative control probe,
Q-positive quality control probe.

2.4 Multiple PCR reaction

Microarray-in-a-tube system-PCR was performed on
an in a thermal cycler (Eastwin Biotech, China) with
25 μL of reaction mixture, including the following
reagents: 2 μL DNA (5 μg/ml), 7.5 μL TaKaRa PCR-
Taq master mix (Dalian, China), 5 μL primer mix and
10.5 μL of dd H$_2$O. The PCR program was set as following: 94°C for 5 min; 5 cycles of 94°C for 30 s, 60°C
for 30 s; 30 cycles of 94°C for 30 s, 68°C for 30 s, 5
cycles of 94°C for 30 s, 55°C for 30 s; and 72°C for
5 min; 4°C for forever (Han et al. 2006).

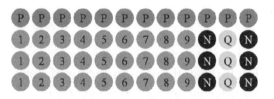

Figure 2. The coupling effect of probe. A. The effect of
direct spotting; B. The effect of spotting after coupling and
washing. The concentration of positive position probe is
50 μM.

2.5 Microarray-in-a-tube chip

In our research, the microarray chip was produced
in the cap of PCR tube. After washing, modified
by organic chemical reagent, probes and microarray
buffer (Capitalbio, Beijing, China) mixed up with a
final concentration of 50 μM, and printed on the cap
of PCR tube by a Smart Array 48 Microarray Spotter.
Fig. 1 showed Microarray chip.

Microarray Chip printed probes then incubated
an hour treatment at 37°C. For the hybridization
experiment, PCR products were hybridization solution containing the complementary sequence of hybrid
quality 10 nM, 0.1 μL, it was performed in half an
hour treatment at 40°C. Following the hybridization,
the microarray images were recorded by a fluorescence
microscope (Nikon ECLIPSE E 200, Japan). The signal to noise ratio (SNR) was used to predicate the
hybridization results, and SNR of the positive spot was
set as >1.5.

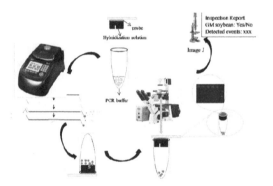

Figure 3. The protocol of microarray-in-a-tube system.

3 RESULTS

3.1 Microarray-in-a-tube chip quality verification

In the microarray-in-a-tube chip system, the process
for preparing the aldehyde modification was optimized from Liu's method (Liu et al. 2007). The
aminated oligonucleotide probe could be set well on
the cap (Fig. 2), and the SNR was more than 2. The
space and size of spot was set as 200 μm, 100 μm,
respectively. The capacity of the array which immobilized on the inner surface of the cap was about 300
spots, and the detection throughput was more than 200
probes in a single test.

3.2 The simplification of the microarray-in-a-tube system

The present microarray-in-a-tube system was modified from the previous designed (Liu et al. 2007).
The distinction of mold design was the place of the
hybridization solution pool, and its structure was compact and easy to transport and operate. There are five
components: the preparation of microarray, multiple
PCR amplification, upside-down hybridization, fluorescence imaging, data analysis. Briefly, the whole
operating process was fairly simple. The experiment
oparetors put the DNA sample into the eppendorf
tube, after amplification, take the tube upside down
for hybridization, acquired image by fluorescence
microscope and analyze the data by Image J software
(Fig. 3). In this system, all operations were operated
in the sealed eppendorf tube to prevent DNA product
pollution, and spent only three hours in single test.

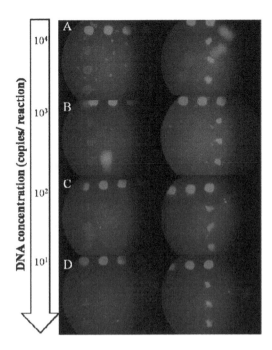

Figure 5. The detection accuracy of GM soybean. A, MON89788 plasmid DNA; B, A2704-12 genomic DNA; C, lectin plasmid DNA; D, GTS-40-3-2 plasmid DNA. All the fluorescence signal image was under the fluorescent microscope (4×).

Figure 4. The sensitivity detection of MON8977 GM soybean. A, 10^4 copies; B, 10^3 copies; C, 10^2 copies; D, 10 copies. The red spot on the horizontal line was position control, the red spot on the first left vertical rows was target gene hybridization spot, the last was positive quality control. All the fluorescence signal images were under the fluorescent microscope (10×).

3.3 The detection sensitivity of GM MON89788 soybean

To measure the detection sensitivity of the microarray-in-a-tube system, the plasmid DNAs of GM event MON89788 were tested, and the concentration was 10^4, 10^3, 10^2, 10 copies per reaction, respectively (Fig. 4). The results were performed that all of the expected MON89788 GM contents were successfully detected, and all the SNR was more than 1.5.

3.4 The detection accuracy of GM soybean events

To test the performance of the microarray-in-a-tube system for GM soybean monitoring, four types samples were prepared and analyzed (genomic and plasmid DNA). The results were consistent with that of real-time PCR (the results were supported by JSCIQ). It was showed that the system was stable and the date gained was accurate (Fig. 5), in spite of the positive spot was not intact and lined in alignment.

4 DISCUSSION

In the past two decades, the development of GM crops has been increasing (Liu et al. 2007). Following successful and rapid commercialization, regulation of GM crops has been one of the most important food issues (Kearns et al. 2014). To respond to these concerns, guidelines and regulations have been issued by many countries throughout the world, labeling of the GMOs seems to be a promising proposition towards the acceptance of GM crops (D'Silva & Bowman 2010). This requirement makes it necessary to develop effective methods for detection of GMO. With the development of newer transgenic crops, detection methods are also likely to be improved.

In this study, we established a microarray-in-a-tube system of high-throughput multiplex PCR integrated with hybridization. This detection system will cover the commercialized GM crops with the certificate issued by China Import and Export Commodity Inspection Bureau. According to the requirement of customer and research, the detection sites in the system can be increased. This system could be the capable of covering of all known GMO events (James 2013), and flexibly updated.

Compared to other methodology (Kamle & Ali 2013, Shao et al. 2014), in addition to the sensitivity, simplification, high-throughput, and accuracy, the microarray-in-a-tube system has other advantages. The most significant character is integration to avoid a false positive caused by PCR amplicon pollution. In the meanwhile, the low cost is attractive, and the price can be controlled within 1 dollar in mass commercialize production. So far, this system has also been applied to detection of human papillomavirus (HPV) genotypes and foodborne pathogenic bacteria. In a word, the microarray-in-a-tube system is a very promising research and industrialization.

ACKNOWLEDGMENT

This study was financial supported from National Basic Research Program of China (2011CB707600), the National Natural Science Foundation of China

(61071050, 61372031), and Tsinghua National Laboratory for Information Science and Technology (TNList) Cross-discipline Foundation.

REFERENCES

D'Silva, J., Bowman, D. M. 2010. To Label or Not to Label? – It's More than a Nano-sized Question. *European Journal of Risk Regulation* 1:420–427.

James, C. 2013. Global Status of Commercialized Biotech/GM Crops: 2013. *ISAAA Brief* No. 46. ISAAA: Ithaca, NY.

Han, J., Swan, D.C., Smith, S.J., Lum, Sh.H., Sefers, S.E., Unger, E.R., Tang, Y.W. 2006. Simultaneous amplification identification of 25 human papillomavirus types with templex technology. *Journal of clinical microbiology* 44(11): 4157–4162.

He, Z., Xia, X., Peng, S., Lumpkin, T.A. 2014. Meeting demands for increased cereal production in China. *Journal of Cereal Science* 59: 235–244.

Holst-Jensen A., Ronning S., Lovseth A., Berdal K. 2003. PCR technology for screening and quantification of genetically modified organisms. *Anal Bioanal Chem* 375: 985–993.

Kamle, S., Ali, Sher. 2013. Genetically modified crops: detection strategies and biosafety issues. Gene 522: 123–132.

Kearns, P., Suwabe, K., Dagallier, B., Hermans, W., Oladini-James, C. 2014. Genetically modified organisms, environmental risk assessment and regulations. *J. Verbr. Lebensm.* 9 (Suppl 1): S25–S29.

Ladics, G.S., Budziszewski, G.J., Herman, R.A., Herouet-Guicheney, C., Joshi, S., Lipscomb, E.A., McClain, S., Ward, J.M. 2014. Measurement of endogenous allergens in genetically modified soybeans. *Regulatory Toxicology and Pharmacology* 70: 75–79.

Leguizamón, A. 2014. Modifying Argentina: GM soy and socio-environmental change. *Geoforum* 53: 149–160.

Liu, Q., Bai, Y., Ge, Q., Zhou, S., Wen, T., Lu, Z. 2007. Microarray-in-a-tube for detection of multiple viruses. *Clinical Chemistry* 53(2): 188–194.

Shao, N., Jiang, S.M., Zhang, M., Wang, J., Guo, S.J., Li, Y., Jiang, H.W., Liu, C.X., Zhang, D.B., Yang, L.T., Tao, S.C. 2014. MACRO: A combined microchip-PCR and microarray system for high-throughput monitoring of genetically modified organisms. *Anal. Chem.* 86: 1269–1276.

Biomedical Engineering and Environmental Engineering – Chan (Ed.)
© 2015 Taylor & Francis Group, London, ISBN: 978-1-138-02805-0

Tissue characterization: The influence of ultrasound depth on texture features

M.A. Alqahtani
Department of Biomedical Technology, King Saud University, Riyadh, Kingdom of Saudi Arabia

D.P. Coleman & N.D. Pugh
Department of Medical Physics and Clinical Engineering, University Hospital of Wales, Cardiff, UK

L.D.M. Nokes
Institute of Medical Engineering and Medical Physics, School of Engineering, Cardiff University, Cardiff, UK

ABSTRACT: In diagnostic ultrasound, the echographic B-scan texture is an important area of investigation since it can be analysed to characterize the histological state of internal tissues. An important factor requiring consideration when evaluating ultrasonic tissue texture is the ultrasound depth. The aim of this study is to investigate the influence of depth on texture features. The left leg medial head of the gastrocnemius muscle of five healthy subjects were scanned using a 3D linear array transducer. The depth effect was tested by manually marking two regions, A and B, with varying depths within the muscle boundary. Texture features, including first order statistics, second order statistics, the Autoregressive (AR) model, and the wavelet transform parameters, were extracted from the images. One-way ANOVA was used to test the significant difference between the various settings. The results show that the gray level, the variance, and the run length matrix were significantly lowered when the depth increased. However, all the texture parameters showed no significant difference between depths A and B ($p > 0.05$), except for the gray level, the variance and the run length matrix ($p < 0.05$). This indicates that the gray level, variance, and run length matrix are highly depth-dependant, whereas the parameters co-occurrence matrix, the gradient, the AR model, and the wavelet are depth independent.

1 INTRODUCTION

The main purpose of carrying out an imaging procedure is for the definition of tissue features and characteristics [3]. The term 'texture analysis' simply means procedures for analysing grayzones in images. Texture analysis of ultrasonic images can also be used for the characterization and the differentiation of the muscle structure between athletes and untrained individuals [7]. [1] depicted that, with regards to ultrasonic images, the tiredness and fatigue of muscles may be revealed through the application of texture analysis. The effect of attenuation with the size of the region of interest, gain, and dynamic range are important variables to consider as they can influence the analysis of texture features. These sources of variability have to be considered carefully when evaluating image texture, as different settings might influence the resultant image.

Ultrasound depth is an important parameter to consider; the strength of the ultrasound echo signal will weaken with depth due to the attenuation of the ultrasound signal by tissue. In clinical B-scan ultrasound, the texture features of an organ or region are highly dependent on the distance between the tissue and the transducer. The texture features might not be the same

from the top to the bottom of an image for the same tissue. This is due to the intervening tissue layers that are present between the transducer and the tissue layer of interest [5]. Moreover, the strength of the attenuated ultrasound signal decreases with depth, and the returning echo from farther structures will be weaker than those returning from closer layers [4]. The ultrasound machine attempts to compensate for this effect by using Time Gain Compensation (TGC) to boost the level of the signal from deeper depths.

The creation of an image from echoes is based on sophisticated computer analysis of the returning echoes. However, due to the attenuation of the sound pulse with depth of penetration, the amplitude of returning echoes is affected. The more delayed (more distant) echoes need to be amplified to compensate for the energy attenuated by the extra distances they travel. This adjustment is termed 'timegain compensation' and is an integral part of image generation [2], [6]. There is a necessity for these returning echoes to be amplified, compressed, and subtracted in order to formulate images. The resolution of the ultrasound is directly proportional to the frequency of the incident sound, but the depth of penetration is inversely proportional to this frequency [2], [6]. Therefore, with higher frequency probes, structures that are more superficial may be studied in greater detail than deeper structures [8]. This study investigates the influence of depth on textural features.

2 MATERIALS AND METHODS

2.1 Participants

In this study, 10 healthy volunteers (8 male/2 female, mean age: 25 {range 18–38}) were chosen. The left medial head of the gastrocnemius muscle for all subjects was scanned using the Toshiba ultrasound system with a 3D transducer. The left leg was chosen because no significant features were noted between the right and the left legs. Informed consent was obtained from all participants after explaining the aim, protocol, and procedures in the study. All volunteers declared having normal health, with no history of any musculoskeletal disorders. The study was approved by the local ethics committee. The 3D ultrasound sweep was performed in the central (middle) region of the muscle. Following this, four 2D images were extracted from the 3D data set, with a distance of 10mm between them. All the scans were performed using the standard protocol.

2.2 Protocol

Ultrasound scans were performed according to a standard protocol for all the subjects. During the ultrasound scanning procedure, the participants were made to lie in the prone position on the scanning couch, adjusting the ankle position to bring them level with the end of the setting of the couch; the knees were fully extended. To locate the area of interest, a skin marker was used to draw lines in order to determine the medial and lateral portions of the gastrocnemius muscles. The underlying muscle was divided into three regions (proximal, central, and distal) as shown in Figure 1.

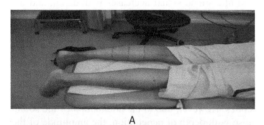

A

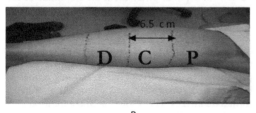

B

Figure 1. (A) Participant in prone position; (B) Sites demarcated for scanning; D: distal region, C: central region, P: proximal region.

The distal region lies at the furthermost end of the muscle where the muscle attaches to the Achilles tendon. The central region covers the area between the proximal region, where the muscle attaches to the knee, and the distal region. The central regions were chosen to be the RegionsOf Interest (ROI). The proximal region occupies the proximal end of the gastrocnemius muscle. The average length of the gastrocnemius muscle was 20.5 cm therefore; the distal and central regions were measured to be 6.5 cm. The point where the muscle attached to the Achilles tendon was chosen as a reference point for measuring the length of the distal and central regions. These reference points were chosen because the proximal end of the gastrocnemius muscle is difficult to visualize due to the complexity of overlapping muscle and tendons that cover the posterior aspect of the knee joint.

A three-dimensional ultrasound sweep was performed on the central region (C), and four slices were collected with a 10 mm distance between them. Only one slice was chosen for the investigation, as a previous study on 10 subjects had shown that the middle part of the gastrocnemius muscle is homogenous. All the scanning was performed at the same anatomically-defined locations with the subject in the same position. The subjects were asked to keep their muscle fully relaxed during the scanning as contraction of the muscle can result in an increased muscle diameter and decreased echo intensity. The scanning was performed at different settings of gain and dynamic range. Imaging was repeated once with the dynamic range being constant and the gain range varying between 70 and 90 dB (70-75-80-85-90), and again with a constant gain setting while varying the dynamic range through the same values. Five images were captured at each sitting, then the average was calculated to minimize the variations. When evaluating the effects of a given setting, the other settings were kept constant. All scanning procedures were performed by the same investigator.

2.3 Image analysis

The depth effect was tested by manually marking two regions, A and B, with varying depths using the computer mouse for each image. The regions A and B were defined within the muscle boundary. The size of both ROI was 280 * 20 pixels, and the distance between region A and B was kept constant at 5 mm. The distance between the skin and region A was kept constant at 10 mm.

2.4 Statistical analysis

The Shapiro-Wilk test was used to test the normality of the data due to the small sample size of the study (P > 0.05 was considered normal distributed). The paired t-test was used to test the depth effect for the normally distributed data, and the Wilcoxon-Mann-Whitney test was used for the non-normally distributed data. (P value of less than

0.05 was considered statistically significant). The SPSS software for Windows Version 16 (SPSS Inc., Chicago, Illinois, USA) was used for the statistical analysis.

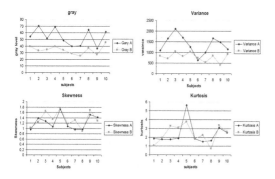

Figure 2. Line graph showing the dispersion of gray level values, variance, skewness and kurtosis at depth A and depth B, for the ten participants.

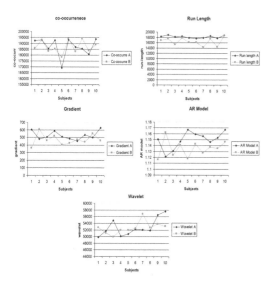

Figure 3. Line graph showing the dispersion of the co-occurrence matrix, the run length matrix, the gradient, the auto-regressive model, and the wavelet transform values at depths A and depth B, for the ten participants.

3 RESULT

Figure 2 shows the values of the first order statistical parameters: the gray level, variance, skewness, and kurtosis.

Figure 3 illustrates the second order statistical parameters: the co-occurrence matrix, the run length matrix and gradient, the autoregressive model, and the wavelet transforms at two different depths, A and B, of the gastrocnemius muscle (central region).

The results show that the gray level, the variance, and the run length matrix were significantly lowered when the depth increased. For instance, the gray level mean value decreased from 53.54 to 34.38 and the variance from 1374.19 to 789.46. The run length matrix decreased from 18175.71 to 16419. They decreased with depth as the strength of the ultrasound echo signal weakens with depth, due to the attenuation of the ultrasound signal by tissue. The other texture parameters showed similar values at different depths.

The mean value for each texture parameter at both depths is displayed in Table 1 (the upper and lower Confidence Intervals (CI) = 95% of the mean). All the texture parameters showed no significant difference between depths A and B ($p > 0.05$) except for the gray level, the variance, and the run length matrix ($p < 0.05$). This indicates that the gray level, the variance, and the run length matrix are depth-dependant.

4 DISCUSSION

This study shows the importance of considering the influence of ultrasound depth prior to making quantitative judgments about clinical significance of texture analysis results. The first order statistic parameters: the gray level, the variance, and the run length matrix should be extracted for analysis using standardised depth levels; in other words, for defining ROI on images, in order to extract these parameters the depth must be kept constant. These texture parameters are depth-dependant; this is due to the phenomenon of attenuation of the ultrasound signal. The ultrasound signal decreases with depth due to attenuation of the ultrasound beam by different tissue interfaces and, therefore, the echo returning from structures closer

Table 1. Mean values (upper and lower CI = 95% of the mean) and p-values for texture parameters for depths A and B (*p < 0.05 significant difference).

Texture Parameters	Depth A	Depth B	p-value
Gray level	**53.54** (44.64–62.45)	**34.38** (29.93–38.83)	**0.000***
Variance	**1374.19** (1069.76–1678.63)	**789.46** (645.81–933.10)	**0.002***
Skewness	**1.23** (1.04–1.42)	**1.34** (1.13–1.54)	**0.118**
Kurtosis	**2.35** (1.46–3.24)	**2.34** (1.65–3.11)	**0.926**
Co-occurrence matrix	**187286.81** (181753.56–192820.06)	**186999.52** (183126.54–190872.49)	**0.880**
Run length matrix	**18175.71** (17840.64–18510.78)	**16419.37** (15386.25–17452.50)	**0.001***
Gradient	**528.33** (486.64–570.01)	**468.97** (416.85–521.08)	**0.113**
AR Model	**1.135** (1.125–1.145)	**1.497** (1.139–1.160)	**0.082**
Wavelet	**52745.77** (50826.04–54665.51)	**52639.72** (51335.23–53944.21)	**0.918**

to the transducer head will be stronger than echoes returning from deeper structures. However, the other texture features showed no significant differences with depth A and depth B.

5 CONCLUSION

The results and analysis of this study indicate that the parameters co-occurrence matrix, the gradient, the AR model, and the wavelet are depth independent, whereas the gray level, the variance, and the run length matrix are highly influenced by depth. Therefore, when analysing these parameters it is important to keep the ROI at the same level of depth.

REFERENCES

[1] Basset O, Ramiaramanana V, Chirossel P, Hernandez A and Gimenez G (1994) Characterisation of muscle tissue during an effort by texture analysis of ultrasonic images. In *IEEE Ultrasonics Symposium Proceedings*. 3, 1455–1458.

[2] Kremkau F W (2002) *Diagnostic Ultrasound: Principles and Instruments*. 6th edition. Philadelphia: W.B. Saunders Company.

[3] Lerski R A, Straughan K, Schad L R, Boyce D, Bluml S and Zuna I (1993) MR image texture analysis–an approach to tissue characterization. *Magn Reson Imaging*. 11 (6) 873–887.

[4] Muzzolini R E (1996) *A Volumetric Approach to Segmentation and Texture Characterisation of Ultrasound Images. PhD thesis.*

[5] Powis R and Powis W (1984) *A Thinker's Guide to Ultrasonic Imaging*. Urban and Schwartzenberg inc., Baltiomore, Maryland.

[6] Riley W A (1996) *Ultrasonic B-mode imaging systems*. New York: Neurosonology 1st edition.

[7] Sipila S and Suominen H (1991) Ultrasound imaging of the quadriceps muscle in elderly athletes and untrained men. *Muscle Nerve*. 14 (6) 527–533.

[8] Walker F O, Cartwright M S, Wiesler E R and Caress J (2004) Ultrasound of nerve and muscle. *Clin Neurophysiol*. 115 (3) 495–5.

Biomedical Engineering and Environmental Engineering – Chan (Ed.)
© 2015 Taylor & Francis Group, London, ISBN: 978-1-138-02805-0

Alzheimer's disease rats behaviour research and the influences of NF-κBp65 and caspase-3 expression with the treatment of MCI-186

Xiaofan Yang, Ning An & Weina Zhao
Hongqi Hospital of Heilongjiang Province, Mudanjiang Medical College, Department of Neurology, Mudanjiang, Heilongjiang, China

Zhaoping Zang
Department of Neurology, Qiqihar Medical College, First Affiliated Hospital, Qiqihar, Heilongjiang Province, China

Pei Chen
Pathophysiology Department, Mudanjiang Medical College, Mudanjiang, Heilongjiang, China

ABSTRACT: [Objective] Study the influence of MCI-186 on the behaviour of AD Rats and hippocampal tissue NF-κBp65 and caspase-3. [Method] Use the bilateral hippocampal to inject Aβ25-35 to prepare the AD model, including the large-dose MCI-186 treatment group, the low-dose MCI-186 treatment group, the model group, the sham operation group, and the normal control group. Test the Morris water maze to explore the changes of learning and memory function of rats. Use immunohistochemical technology to detect the expression of NF-κBp65 and caspase-3 protein. [Results] MCI-186 could improve the learning and memory function of AD rats and inhibit neuronal apoptosis of the neuron and the expression of NF-κBp65 and caspase-3. [Conclusion] The test indicates that MCI-186 treatment can significantly improve the learning and memory functions of AD Rats and inhibit the apoptosis of the hippocampal neuron of rats.

Keywords: MCI-186, Alzheimer's disease, NF-κBp65, Caspase-3, [CLC] R743.32 [Document code] A

1 INTRODUCTION

AD (Alzheimer's Disease) has become the fourth killer following cardiovascular disease, cancer, and cerebrovascular disease, which poses a serious threat to human health[1]. According to the statistics, there were 36.5 million people with dementia in 2010 and new cases each year reach about 7.7 million, mainly in low and middle-income countries[2–3]. There are complex pathogenic factors for AD, including the behaviour of patients, dietary patterns, blood pressure, dyslipidemia, and metabolic diseases, etc.[4].

Edaravone (MCI-186) is a novel free radical scavenger[5–7] which could inhibit lipid peroxidation and cell apoptosis by clearing·OH, thereby protecting brain cells (vascular endothelial cells and nerve cells).

This experiment intends to inject Aβ25-35 into the bilateral hippocampus to make an AD rat model[8]. Then, the Aβ model rats will be treated using MCI-186 at different doses. By using the Morris water maze method to conduct a place navigation test and a spatial probe test, the aim is to detect the learning and memory ability of rats. The statistical analysis indicates the impact of MCI-186 treatment on the behaviour of the AD rat model; the immunohistochemical method is used to detect the expression of nuclear transcription factor κB p65 (NF-κB p65 and caspase-3 protein) around the CA1 region of rats. Then statistical analysis for the data is conducted to determine the treatment of MCI-186 for the neuron apoptosis of the AD rat hippocampus.

2 MATERIALS AND METHODS

2.1 *Reagents, animals, and grouping*

Aβ25-35 (Beijing Bioss Biotechnology Co. Ltd.), Edaravone (MCI-186) (Nanjing Simcere Company), Rabbit Anti-Active caspase-3, Rabbit Anti-NFκBp65 (Beijing Bioss Biotechnology Co. Ltd.), PV-6001 Goat anti-rabbit IgG antibody HRP multimers, and ZLI-9031 concentrated DAB kit (Zhongshan Golden Bridge biological Engineering Co. Ltd.).

Use the Morris water maze to select 60 active Wistar male rats aged between 3–4 months, and weighing of 200–250 g. Take random allocation: Group A (control group); Group B (sham operation group, inject artificial cerebrospinal fluid to the bilateral hippocampal with a stereotaxic technique); Group C (model group, inject the aged Aβ25–35 to the bilateral hippocampal with a stereotaxic technique); Group D (low-dose MCI-186 group, model + low-dose MCI-186); and

Group E (high-dose MCI-186 group, model + high-dose MCI-186).

2.2 Methods

1.2.1 Preparation and evaluation of animal models: conduct water deprivation for rats 4 h before the operation, and food deprivation 12 h before the operation. Based on the stereotaxic atlas of the rat brain, anesthetized rats are fixed on the stereotaxic instrument. Select the dorsal blade of the bilateral hippocampal dentate gyrus as the injection region. Use a micro syringe to inject 1ul Aβ25–355 mmol/l solution (injection speed of 0.2 ul/min), and retain the needle for 5 min to ensure the adequate dispersion of the solution. Then inject the other side in the same way. Then conduct straticulate saturation for the skin. Inject ACSF of an equal volume to the hippocampus in Group B. Carry out a place navigation test and a spatial probe test on the 4th day after modelling, once a day and 3 days in a row. Then identify the model and conduct statistical analysis.

1.2.2 Experimental drugs and treatment methods: on the 7th day after the model preparation, conduct a tail vein injection with MCI-186 for the Groups E and D once a day with a dose of 6 mg/kg and 3 mg/kg, the concentration of 1.5 mg/ml, total 14d; for Group C and Group B, after the successful model preparation, conduct a tail vein injection with normal saline 3 mg/kg. Group A is the control group.

1.2.3 Determination of learning and memory function: use the Morris water maze to determine the learning and memory function of the rats. Carry out a place navigation test and a spatial probe test to detect the number of times that it comes across the original platform within the escape latency and 2 min. Then conduct experiment respectively on the 10th d, 17th d, 24th d, twice a day, and 4 days in a row, and calculate the average score.

1.2.4 Materials, HE staining, and immunohisto-chemistry: conduct immunohistochemical detection of the expression of NF-κBp65, and the caspase-3 protein in the hippocampal CA1 region with an SP two-step method.

1.2.5 Statistical methods: adopt the positive cell count method to observe rats. Select 3 sections on the same part of each rat and 5 representative and non-overlapping visions in the hippocampal CA1 region under a 400 time light microscope, to count the number of positive cells in each vision; their average value will be used as the statistics of each animal. All measurement data is expressed by the mean ± standard deviation ($\bar{x} \pm S$). Then utilize SPSS17.0 software to conduct a paired t-test, repeat measurements, and one-way variance, to analyse data.

3 RESULTS

3.1 Determination of learning and memory function

For the selected Wistar rats, after seven days of model preparation with the Aβ25–35 hippocampal injection,

Table 1. The comparison of the escape latency and the number platform crossings before and after the modelling ($\bar{x} \pm S$, n = 30).

	The Number of Rats	Escape Latency	The Number of Platform Crossing
Before the Modelling	30	14.20 ± 3.43	8.21 ± 0.51
After the Modelling	30	59.02 ± 9.27*	2.22 ± 0.50*

Compared with the period before the modelling, *P < 0.01

their escape latency is prolonged and the number of platform crossings is significantly reduced. The paired t-test indicates the significant differences (P < 0.01). This means that the AD rat model with Aβ25–35 hippocampal injection damages the central cholinergic system of rats, coupled with an apparent decrease of learning and memory function, as shown in Table 1.

Comparison of the escape latency and the number of platform crossings of each group on the 14th day, 21st day, and 28th day after the modelling, using repeated measurements and variance analysis: ① Compared with Group A, there is no significant difference in the escape latency and the number of platform crossings of Group B (P > 0.05). ② Compared with Group A, the escape latency and the number of platform crossings of Group C show an obvious decrease with significant differences (P < 0.01). ③ Compared with Group C, the escape latency of Group D and Group E shows a significant decrease (P < 0.01) and the number of platform crossings shows a great increase (P < 0.01). After MCI-186 treatment for AD model rats, their learning and memory function has been greatly improved; ④ Compared with Group A, the escape latency of Group D and Group E has significant increase and the number of platform crossing shows great decrease, both with significant differences (P < 0.01). This indicates that the learning and memory function of rats with dementia has not yet returned to normal state after MCI-186 therapy; ⑤ The escape latency and number of platform crossing of Groups D and E have no significant difference (P > 0.05). This indicates that the increase in the MCI-186 dose fails to improve learning and memory function of AD rats, as shown in Table 2 and Table 3.

3.2 Comparison of immunohistochemistry

2.2.1 NF-κB p65 subunit immunocytochemical positive granules are located in the cytoplasm and nucleus. ① In the hippocampal CA1 region of Groups A and B, the p65 subunit immunohistochemical staining neurons are small in number and have a light colour, with no significant difference (P > 0.05). ② Compared with Group A, the number of positive neurons in Group C shows an apparent increase, coupled with having a dark colour, especially in the nucleus

Table 2. Morris water maze test results of escape latency of each group ($\bar{x} \pm S$, n = 6, second).

Group/time point	14d	21d	28d
Group A	22.61 ± 3.54	20.47 ± 3.45	19.34 ± 2.54
Group B	23.43 ± 3.48*	21.45 ± 2.50*	21.93 ± 4.11*
Group C	54.56 ± 6.74#	53.58 ± 5.03#	56.53 ± 7.98#
Group D	35.95 ± 5.36△	31.62 ± 4.65△	29.46 ± 5.00△
Group E	37.52 ± 4.94△⋆	32.56 ± 5.52△⋆	29.43 ± 5.14△⋆

Compared with Group A, * p > 0.05, # p < 0.01; compared with Group C, △p < 0.01; compared with Group D, ⋆ p > 0.05

Table 3. Morris water maze test results of the number of platform crossing of each group ($\bar{x} \pm S$, n = 6, number of times).

Group/time point	14d	21d	28d
Group A	7.86 ± 0.68	8.22 ± 0.90	8.85 ± 0.77
Group B	7.69 ± 0.51*	7.91 ± 0.55*	8.21 ± 0.61*
Group C	2.08 ± 0.25#	2.06 ± 0.13#	2.33 ± 0.12#
Group D	5.47 ± 0.52△	6.13 ± 0.30△	6.00 ± 0.54△
Group E	5.35 ± 0.50△⋆	5.73 ± 0.72△⋆	5.95 ± 0.59△⋆

Compared with Group A, * p > 0.05, # △p < 0.01; compared with Group C, P < 0.01; compared with Group D, ⋆ p > 0.05

Table 4. Immunohistochemical indicators of rats in each group ($\bar{x} \pm S$, n = 6).

Group/time point	NF-κB	Caspase-3
Group A	12.43 ± 3.69	37.88 ± 10.49
Group B	14.0 ± 2.97*	40.30 ± 7.77*
Group C	33.9 ± 7.18#	148.33 ± 23.21#
Group D	20.75 ± 5.42△	79.66 ± 13.02△
Group E	17.41 ± 2.94△⋆	68.02 ± 13.71△⋆

Compared with Group A, * p > 0.05, # p < 0.01; compared with Group C, △p < 0.01; compared with Group D, ⋆ p > 0.05

(P < 0.01). ③ Compared with Group C, the number of positive neurons in Groups D and E decreases significantly, and have a light colour, especially in the nucleus (P < 0.01). ④ The number of positive neurons in Groups D and E shows no significant difference (P > 0.05), as shown in Table 4.

2.2.2 The expression of casepase-3 in neuronal cytoplasm. ① Casepase-3-immunocytochemical positive neurons in the hippocampal CA1 region of rats in Groups A and B are smaller in number and have a light colour, with no significant difference (P > 0.05). ② Compared with Group A, the number of positive neurons in Group C increase greatly, and have a dark colour (P < 0.01). ③ Compared with Group C, the

number of positive neurons in Groups D and E significantly decrease (P < 0.01). ④ The number of positive neurons in Groups D and E shows no significant difference (P > 0.05), as shown in Table 4.

4 DISCUSSION

4.1 Determination of behaviour by using a Morris water maze

The experiment showed that in Group C, compared with the previous modelling, we see an increased number of study for rats and prolonged escape latency on the seventh day. This indicates that rats from Group C suffered a decline in memory about the positioning of the underwater platform location, whilst looking for a means of escape in panic. The space exploration tests showed a reduced number of times of rats in Group C crossing the platform compared with that before modelling, indicating that the rats in Group C had a declined ability of visual space locating of the underwater platform and memory. In addition, it demonstrated that using a hippocampal stereotactic injection of A$\beta_{25\text{-}35}$ in rats to simulate cognitive and memory impairment caused by Aβ in AD, is a success.

Group D and Group E, compared with Group C, showed a significantly shortened escape latency of rats in the place navigation test on the 14th, 21st, and 28th day after modelling, and an increase in the number of times they crossed the platform. This indicated a less frequent behaviour for rats in the MCI-186 treatment group to blindly look for a means of escape, as well indicating enhanced abilities for memorizing the spatial location of the underwater platform and oriented exploration. Prompt treatment can improve the learning and memory obstacles of rats who are injected with Aβ25–35 into the hippocampus. Group E compared with Group D showed no significant difference in escape latency and the number of platform crossings, and increasing the dose does not achieve better results. The evidence above is preliminary proof of MCI-186's possible treatment of AD.

4.2 Neuronal inflammatory response and changes in apoptosis

To further examine whether NF-κB signalling pathways can play a role in the MCI-186 inhabitation of an inflammatory response excited by Aβ, we used an immunohistochemistry method to measure the NF-κB expression level of hippocampal neurons. Experimental results showed that the extent of the immune response of an NF-κB p65 subunit in the CA1 region of the hippocampus of Wistar rats, is sorted in the following order: Group C, Group E, Group D, Group B, and Group A. A prompt Aβ25–35 hippocampal injection, activated neuron NF-κB and nuclear transfer in the CA1 region of the hippocampus of rats, thus MCI-186 treatment can significantly reduce the above reaction.

109

The neuronal apoptosis is a universal phenomenon in the pathological process of a variety of neurodegenerative diseases[9]. Neuronal apoptosis is an important part in the AD pathogenesis process, as well as one of the main factors for reducing the number of neurons and causes of cognitive and memory dysfunction[10]. In spite of various reasons to trigger apoptosis, they will eventually undergo the caspase cascade reaction. Some studies suggest that caspase may decompose the family members of protection protein Bcl-2, damage the intermediate filament of the nucleation layer, lead to the disintegration of the solid structure, cause the nuclear membrane damage and chromosome condensation and result the inactivation of various intracellular enzyme for DNA repair, the decline of cell repair functions, finally inducing apoptosis of cells. Besides, through the signal channels of cell death, $A\beta$ and a variety of receptors interact with each other to generate casepase. The receptors of advanced glycation end-products of free radicals, apoptosis-inducing APP, and p75 neurotrophin receptors, are all involved in the generation of casepase. Activated casepase eventually leads to the apoptosis of AD neurones[11]. The immunohistochemistry is used to detect the expression variation of casepase-3 protein in the hippocampal CA1 region, as well as the expression of casepase-3 protein in the neuron plasma. An $A\beta25$–35 hippocampus injection causes the increase of casepase-3 expression of the enzyme associated with apoptosis in the hippocampal neurons of rats. MCI-186 treatment can inhibit the overexpression of casepase-3 caused by $A\beta_{25\text{-}35}$ in the hippocampal CA1 region, which is a mechanism of MCI-186 to inhibit the neuronal apoptosis. Besides, NF-κB could also inhibit caspase-3 expression, thus inhibiting apoptosis. This experiment also confirms that MCI-186 can inhibit the activation of NF-κB. Therefore, presumably, the MCI-186 inhibition for caspase-3 expression may also be related to the fact that it could inhibit NF-κB overexpression and prompt a decrease of the expression of inflammatory genes COX-2 and iNOS, thus reducing NO, prostaglandins, free radicals, and other inflammatory factors, and inhibit the inflammatory cascade.

ABOUT THE AUTHOR

Yang Xiaofan (1982–), male, Hailun, Heilongjiang Province, Master, an attending physician, mainly engaged in neurodegenerative diseases.

CORRESPONDING AUTHOR

Chen Pei (1981–), female, Shenyang, Liaoning Province, Master, lecturer, mainly engaged in medical education.

FUND PROJECT: Research Project of the Health Department in Heilongjiang Province, No. (2012-307).

Scientific Research Project of Mudanjiang Medical University, No. (2011-20).

REFERENCES

[1] Mullane K, Williams M. Alzheimer's therapeutics: continued clinical failures question the validity of the amyloid hypothesis-but what lies [J]. Beyond Biochem Pharmacol, 2013, 85: 289–305.

[2] Sosa- Ortiz AL. Acosta- Castillo I, Prince M J. Epidemiology of dementias and Alzheimer's disease [J]. Arch Med Res, 2012, 43 (8): 600–608.

[3] Ubhi K, Masliah E. Alzheimer's disease: recent advances and future perspectives. J Alzheimer's Dis, 2013, 33 (Suppl 1): S185–S194.

[4] Cuizeng Wei, Chen Jibin. *Research on the Epidemiologic Features and Risk Factors of Alzheimer's Disease* [J]. Journal of Chronic Diseases, 2014, 15 (1): 52–57.

[5] Du Wen, Chen Jianling. *Observation of the Curative Effect of Edaravone for the Vascular Dementia* [J]. China Modern Medicine, 2012, 19 (8): 60–62.

[6] Wang Zhongqing, Li Aili. *The Application of Edaravone in the Treatment of Neurological Diseases* [J]. Clinical drug world, 2013, 34 (8): 500–504.

[7] Jiang Xing, Zhou Yusheng, Pan Jie. *Research on the Mechanism of Edaravone and Its Clinical Application* [J]. Central South Pharmacy, 2013, 11 (8): 587–592.

[8] Liu Xuefeng, Li Yisong, Feng Liqian, and so on. *Research on the Duplicating of Alzheimer's Disease Rat Model* [J]. Journal of Chinese Medicine, 2012, 27 (12): 1606–1608.

[9] Pirat B, Muderrisoglu H, Unal MT, et al. Recombinant human-activated protein C inhibits cardiomyocyte apoptosis in a rat model of myocardial ischemia-reperfusion [J]. Coron Artery Dis; 2007, 18 (1): 61–66.

[10] Pallas M, Camins A. Molecular and biochemical features in Alzheimer's disease [J]. Curr Pharm Des; 2006, 12 (33): 4389–4408.

[11] Snigdha S, Smith E D, Prieto G A, et al. Caspase-3 activation as a bifurcation point between plasticity and cell death [J]. Neurosci Bull, 2012, 28 (1): 14–24.

Biomedical Engineering and Environmental Engineering – Chan (Ed.)
© *2015 Taylor & Francis Group, London, ISBN: 978-1-138-02805-0*

Expression of STIM1 in a rat with KA-causing epilepsy and the effect of edaravone

Zhaoping Zang, Zhiru Li, Zhongjin Liu, Weiwei Gao, Kexin Jing & Baokui Qi
Department of Neurology, Qiqihar Medical College, First Affiliated Hospital, Qiqihar, Heilongjiang Province, China

ABSTRACT: Purpose: By observing the expression of STIM1 in a rat with KA caused epilepsy and the expression change caused by the application of edaravone, the expression changes in the epilepsy process and after edaravone therapy were initially discussed, so as to define the effect of edaravone on epilepsy. Methods: 50 male Wistar rats were randomly divided into 5 groups, with 10 rats in each group. The 5 groups included the Normal Group, the PBS Contrast Group, the Epilepsy Group, the Therapy Group 1 (diazepam Therapy Group), and the Therapy Group 2 (diazepam+ edaravone Therapy Group). For the Epilepsy Group, KA was injected into the hippocampus on the right position of a rat, the dosage was $1.5\,\mu g/\mu l$ per kilogram of body weight; for the PBS Contrast Group, the same method was applied to inject the same dosage of PBS at the same position; for Therapy Group 1, after an epilepsy model was established, 10 mg/kg was used to terminate the epilepsy; for Therapy Group 2, on the basis of Therapy Group 1, one hour later, edaravone was used; the dosage was 30 mg, once per 12 hours. For the Normal Group, no medicine was used for the injection. Results: A laser confocal microscope was used for observation. It was found that the expression density of STIMI in the Epilepsy Group was markedly higher than that in the PBS Contrast Group and the Normal Group ($P < 0.05$), the expressions in the Therapy Groups were higher than the expression in the Normal Group, the PBS Contrast Group, and the Epilepsy Group, the difference has statistical significance ($P < 0.05$); the expression in Therapy Group 1 was higher than that in Therapy Group 2, the difference has statistical significance ($P < 0.05$). SDS-PAGE analysis shows that the relative molecular mass of STIM1 is 46KD. In addition, it was found that its expression in the Epilepsy Group was also higher than that in the Normal Group and the PBS Contrast Group; the expression in the Therapy Group was lower than that in the Epilepsy Group, and the difference has statistical significance; the result is the same as that of the laser confocal method ($P < 0.05$). Conclusion: The increase of STIMI expression has a cell protection effect on the nerve cell trauma of the hippocampus after epilepsy occurs.

1 INTRODUCTION

Epilepsy is a clinical syndrome caused by paroxysmal abnormal discharges of nerve cells, characterized by repetitive epileptiform outbreaks; its pathogeny is complex, and the pathogenesis is still not clear. Some scholars found that 30%–70% of epileptic patients suffer from cognitive disorder[1–2]. Epilepsy was written about in *Huangdi's Internal Classics* as follows: "All sudden muscular spasms and rigidities are ascribed to wind... All winds with vertigo and shaking are ascribed to the liver." In traditional Chinese medicine, the therapy for epilepsy is mostly initiated from "Discussing therapy from liver." However, due to the component complexity of Chinese herbal medicine, few epilepsy cases have applied Chinese herbal medicine for therapy, while there are fewer detections for the therapy effect of Chinese herbal medicine. Due to the singularity of the components of Western medicine, it is taken as the research object. Edaravone may clear away free radicals and restrain the release of the excitatory neurotransmitter glutamic acid. By using edaravone, we may research its

expression, and know further the pathogenesis and therapy of epilepsy. In addition, it may provide a new therapeutic target for the clinical medication of a clinical epilepsy patient, so to relieve the pain of the patient. The CRAC (Calcium Release-activated Calcium) passage is a slow calcium passage located in the plasma membrane; the STIM1 (Stromal Interaction Molecule 1) is one of the key proteins in the CRAC passage, mediating the formation and activation of the CRAC passage. The purpose of the experiment is to study the possible function of STIMI in the outbreak and therapy of temporal lobe epilepsy by investigating its expression change in kainic acid (KA, K-0250) in a rat hippocampus, and the expression change after therapy.

2 MATERIALS AND METHOD

2.1 *Materials*

A healthy, clean and mature male Wistar rat (with a body weight of 200–250 g). The environment temperature was proper and the experimental process obeyed the regulations of the Experiment Animal Management Byelaw promulgated by the Scientific & Technology Commission of the People's Republic of China. FITC (Beijing Zhongshan); PI propidium iodide (Beijing Zhongshan); confocal laser scanning biological microscope (LSM, Olympus510, META, USA); KA, Sigma, USA; Laborzentrifugen centrifuge (3K15, Sigma), and STIM1 rabbit monoclonal antibody IgG (Abcam, USA).

2.2 Establishment of an epilepsy animal model

Narcotize a Wistar rat by using 10% chloral hydrate with 1 ml (0.3–0.4 ml/kg body weight), then fix the rat onto a stereotaxic brain apparatus, shaving the skin, and cutting and stripping as usual. According to the Paxinos & Watson stereotaxic rat brain atlas (The Rat Brain in Stereotaxic Coordinates, 5th Edition), position with the bregma as the target spot (marking the point on the surface of the skull) (coordinates: X = 2.5 mm (right), Y = 3.0 mm, Z = 3.5 mm), drill holes in the hippocampus CA3 area. Under the conditions of an awake experimental animal and aseptic environment, use a lul microinjector to inject the liquid with concentration 1.5 μg/μl KA 1 μl prepared with PBS. The injection time continued for 10–15 min; the injector was kept inside for 5 min. According to the change of the ethology of the KA induced rat, the establishment of an epilepsy rat model was judged. In addition, it was contained into the experimental object. For the PBS Group, an equal dosage of PBS was injected within an equal time at the same position.

2.3 Standard of racine ethology

Level 0: normal behaviour state; Level I: convulsions of surface muscle such as mastication, blinking, whisker erection, and wet-dog type of tremors; Level II: convulsion of neck muscles, with nodding as the main expression; Level III: clonus and convulsion of front limbs; Level IV: front limbs on both sides straightening, accompanied with body erection; Level V: tumbling and full-body convulsions.

The behaviours below Level III belong to a complex partial outbreak; the behaviours at Levels IV and V are full-body intense clonus outbreaks. After medicine is injected, the ethological features are observed. For the rat group with KA partially injected into the hippocampus, if a behaviour change within level I-V appears within 1 hour, it may be regarded as an epilepsy model.

2.4 Preparation of specimen

According to experimental requirements, fix an animal with paraform, with paraffin embedding, cutting into slices with size 5 um, passing xylol for two times, dewaxing by applying gradient alcohols (100%, 95%, 80%, 70%), conducting hot recovery by microwave for 20 min, cooling to indoor temperature, applying PBS rinsing, restoring with antigen retrieval liquid

for 20 min, 37°C, then sequentially, closing with 5% BSA (Bovine Serum Albumin), hatching with rabbit antimouse STIM1 antibody monoclonal antibody (1–100)4°C overnight, recovering for 1 h at indoor temperature for No. 2d slice, applying PBS rinsing, adding (by dripping)goat anti-rabbit FITC mark—fluorescent second antibody (with concentration 1:50), dying with 2.5 μg/ml propidium iodide (PI) for 20 min, then rinsing the hydrated slice. It is necessary to avoid light for the double labelling experiment processes, starting from adding a fluorescent second antibody. The applied double labelling method referred to the introduction of Bernardini etc., In addition, proper improvement measurements were applied according to the requirements of the experiment on specimen.

2.5 Detecting by applying confocal LSM and data analysis

The applied instrument was an Olympus LSM 510 META confocal laser scanning biological microscope. For each slice, 10 view fields were selected at random and 10 layers were continually scanned. The digitized images were stored for the data analysis of the image.

2.6 Analysis on the print of protein:

Specimen treatment: Extracting the total protein of the cell, measuring the content of the total protein by microplate reader, adding 5 X loading buffer according to the proportion of 4: 1 (between specimen and 5 X loading buffer), boiling them for 5 min, to realize albuminous degeneration.

Steps: Preparing 12% SDS-PAGE gel, sampling 60 μg total protein, and applying electrophoresis for 45 min; under the conditions of indoor temperature and the constant voltage of 220 V, transferring membrane (PVDF membrane, Hybond, America) for 40 min by applying a wet transfer method. With the 5% solution of degreased milk powder prepared with PBST as the closing liquid, at 37°C, closing for 2 h; STIM1, antibody (1:50) 4°C, hatching for overnight; TBST rinsing for 3 times, 10 min/time; adding HRP-anti-rabbit IgG (1:8,000), hatching for 2 hours inside the shaker under indoor temperature condition for 2 h, TBST washing for 3 times, 5–10 min for every time; 30 min after ECL reaction, exposing X-ray plate inside darkroom; developing and fixing under indoor temperature conditions. Using GIS gel imaging system to scan the lamina density of film, recording the Integral Optical Density (IOD) of transmission light for reflecting the content of protein, β-actin was taken as internal reference.

2.7 Statistical analysis

The measured data were expressed by mean ± s.e.m. SPSS17·0 statistical analysis software was applied to carry out one-way analysis of variance and significance testing for data, etc. Each group of specimens was repeated three times, and the mean value was applied.

Table 1. Comparison on the STIM1 fluorescent intensity in the hippocampus CA3 area of five groups of brain epilepsy in model 1.

Fluorescent intensity of protein (means ± s.e.m)	
Group type	STIM1
Normal Group	1·01 ± 0·06*
PBS Group	1·07 ± 0·01*
Epilepsy Group	5·08 ± 0·10*
Therapy Group 1	3·11 ± 0·18*
Therapy Group 2	1·70 ± 0·28*

* There is significant difference, P < 0.

3 RESULTS

3.1 *Quantitative analysis for experimental animal*

For all fifty rats, the final data analysis was conducted.

3.2 *Expression of the protein in the hippocampus CA3 area for a rat suffering from epilepsy*

The experiment applied a confocal laser scanning microscope for detection. The instrument activates the laser source of visible light with wavelengths of 488 nm and 514 nm simultaneously; the emission wavelengths are 516 nm and 617 nm respectively, which have not crossed and overlapped the area. Such differentiation is the premise for double labelling and observation. When the open detection passage was 488 nm, only the facula (please refer to attached Figures 1, 4, 7, 10, and 13) dyed with green fluorescent light was displayed, whilst when the open detection passage was 514 nm, only the nucleic acid (please refer to attached Figures 2, 5, 8, 11, and 14) dyed with red was displayed. Thus, the contour of the cell nucleus was displayed more clearly. When the red nucleic acid, marked on the cell nucleus PI, was superimposed with the green of the fluorescent second antibody, marked on the FITC of the outside cell, the expression of the protein observed in the cytoplasm was seen more (please refer to attached Figures 3, 6, 9, 12, and 15). In the CA3 area, the expression density of the STIMI in the Epilepsy Group is higher than that in the Normal Group and the PBS Group, the expression density of the Therapy Group is lower than that in the Epilepsy, and the expression density of Therapy Group 1 is higher than that in Therapy Group 2. All differences have marked significance (P < 0.05).

3.3 *Expression of the protein in the hippocampus CA3 area for a rat suffering from epilepsy, when applying western blot*

SDS-PAGE analysis shows that the relative molecular mass of the STMI is 46kD. In addition, it finds that the change of the STIMI expression in the Epilepsy Group is higher than that in the Normal Group.

The expression of the STIMI in the hippocampus CA3 area of a rat:

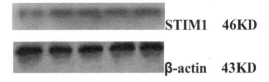

STIM1   46KD

β-actin   43KD

4 DISCUSSION

A KA epilepsy-causing model has been widely recognized at home and abroad. KA is an artificially-extracted matter, similar to glutamic acid. It has intense neurotoxicity, and it may be combined with the NMDA (N-methyl-D-aspartic Acid) receptor on the postsynaptic membrane of nerve cells inside the brain. It is deemed that the accumulation of EAA (Excitatory Amino Acid) and the up-regulation of the EAA receptor of the hippocampus nerve cell, are the main molecular mechanisms in the outbreak of the epilepsy induced by kainic acid[3]. The death of the nerve cell in the hippocampus CA3 area, the gliocyte proliferation, and the sprouting of a moss-shaped fibre, are one marked pathological characteristic of the kainic acid temporal lobe epilepsy model. Such pathological change, the pathological change of the temporal lobe epilepsy of humanity and the hippocampus hardening are consistent[4]. Therefore, in the experiment, the kinaic acid epilepsy-causing model was applied for research on the relevant expression of STIM1. A large amount of research shows that the calcium passage participates in the formation of epilepsy [6–8], the abnormality of the calcium passage causes the overmuch inward flow of Ca^{2+}, and many nerve cells synchronously discharge overmuch, leading to the outbreak. The CRAC passage is a slow calcium passage located in the plasma membrane, and the STIM1 is one of the key proteins in the CRAC passage, mediating the formation and activation of the CRAC passage. Therefore, the experiment research shows that its expression increased after the nerve cell had been damaged by the outbreak of epilepsy, while the expression decreased after edaravone therapy; it had a protection function for the trauma of the nerve cell caused by overmuch discharge after the outbreak of epilepsy. Therefore, the detection on the factor has profound significance for the research on the pathogenesis and therapy of epilepsy.

The chemical name of edaravone is 3-methyl-1-phenyl-2-pyrazoline-5-one; it is a kind of almost colourless and odourless liquid. Edaravone may restrain the release of glutamic acid and resist oxygen radicals, thereby playing its role of protection. Some other research deems that edaravone may realize its protection function by restraining lipid peroxidization. The molecular mass of edaravone is small; approximately 60% may penetrate the blood-brain barrier and reach the effective therapy concentration inside the brain. Intravenous administration may effectively clear

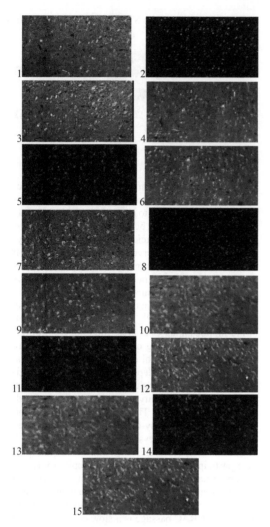

Figure 1. Expression of STIM1 in a rat hippocampus measured by the immunofluorescence double dyeing method. Expression of STIM1: Normal Group (1–3), PBS Group (4–6), Epilepsy Group (7–9), Therapy Group 1 (10–12), Therapy Group 2 (13–15).

hippocampus GFAP (Glial Fibrillary Acidic Protein) and interleukin-1β (IL-1β)and down- regulating the expression of C-Jun N end Kinase (JNK). It was found in research that edaravone may reduce the neuronal apoptosis of the rat hippocampus, after the continuation state of epilepsy, by lowering the expression of iNOS (inducible Nitric Oxide Synthase), thereby playing its role of protection. In the experiment, through the research on the correlation of the protein expression of a rat suffering from epilepsy, it was also verified that edaravone may prevent the neuron injury of a rat suffering from epilepsy. However, its concrete protection mechanism still awaits further research.

ACKNOWLEDGEMENT

Heilongjiang Province Natural Science Foundation-supported projects *Number H201498*.

REFERENCES

[1] Helmstaedter C, Kockelmann E. Cognitive outcomes in patients with chronic temporal lobe epilepsy. Epilepsia, 2006, 47: 96–98.

[2] Pulliainen V, Kuikka P, Jokelainen M. Motor Jor end and cognitive functions in newly diagnosed adult seizure patients before antiepileptic medication. Acta Neurol Scand, 2000, 101: 73–78.

[3] Loscher W, Lehmann H, Behl Bet al. A new pyrrolyl £quinoxalinedione serious of non NMDA glutamate receptor antagonists: pharmacological characterization and comparison with NBQX and valproate in the kindling model of epilepsy. Eur J Neurosci, 1999, 11: 250–262.

[4] Yang Zhongxu, Luan Guoming, Yan Li, etc. Establishment of a rat model suffering from temporal lobe epilepsy and research on the sensitivity of long-term epilepsy. Chinese Medical Journal, 2004, 84: 152–55.

[5] Yangeeisawa M, Kurihara H, Kimura S, et al. A novel potent vasoconstrictor produced by vascular endothelial cells. Nature, 1988; 332(3): 411.

[6] Zamponi GW, Lory P, Perez-Reyes E. Role of voltage-gated calcium channels in epilepsy [J]. Pflugers Arch, 2010, 460(2): 395–403.

[7] Li H, Graber KD, Jin S, et al. Gabapentin decreases epileptic form discharges in a chronic model of neocortical trauma [J]. Neurobiol Dis, 2012, 48(3): 429–438.

[8] Faria LC, Parada I, Prince DA. Interneuronal calcium channel abnormalities in posttraumatic epileptogenic neocortex [J]. Neuro biol Dis, 2012, 45(2): 821–828.

[9] Kamida T, Abe E, Abe T, et al. Edaravone, a free radical scavenger, retards the development of amygdala kindling in rats [J]. Neumsci Lett, 2009, 461 (3): 298–301.

[10] Miyamoto R, Shimakawa S, Suzuki S, et al. Edaravone prevents kainic acid-induced neur0MI death [J]. Brain Res, 2008, 13(1209): 85–91.

away the hydroxyl free radical (with high cytotoxicity) inside the brain [9–10]. As a kind of new free radical scavenger, the edaravone injection has been widely used in neurology. An edaravone injection may reduce the apoptosis of the nerve cell, alleviate brain damage and Kamida, etc. by clearing away free radicals, restraining the release of excitatory neurotransmitter glutamic acid, activating ERK1/2 (extra-cellular signal regulated kinase 1/2), up-regulating BDNF (Brain-derived Neurotrophic Factor) and the expression of Bcl-2 protein, down-regulating the expression of the

Effect of edaravone on ET-1 and survivin expression in patients with ischaemic stroke

Zhaoping Zang, Weiwei Gao, Ran Ma, Zhiru Li, Yanli Zhang, Wei Wang & Baokui Qi
Department of Neurology, Qiqihar Medical College, First Affiliated Hospital, Qiqihar, Heilongjiang Province, China

ABSTRACT: Purpose: Researching the effect of edaravone on ET-1 (ENDOTHELIN-1) and survivin of a patient suffering from ischaemic stroke. Method: 60 neurology in patients were randomly divided into the Routine Therapy Group (Contrast Group) and the Edaravone Therapy Group (Therapy Group), 30 cases respectively for each group. For the Contrast Group, Shuxuetong was applied by intravenous injection, and atorvastatin calcium tablets and enteric-coated aspirin tablets were used for oral medication. For the Therapy Group, on the basis of the therapy method of the Contrast Group, an additional 30 mg edaravone injection was applied to the intravenous drip, BID. The courses of treatment for both groups are all 14 days. Before therapy, on the 7th day and the 14th day during the treatment course, an assessment was conducted for each group respectively; the level of ET-1 and survivin were evaluated. Result: Before treatment, the difference between the level of ET-1 and survivin in serum between both groups had no statistical significance (P > 0.05). On the 7th day and the 14th day during the treatment course, the two groups were compared; the difference of the level of ET-1 and survivin between both groups had statistical significance (P < 0.05). Before treatment, the difference of NIHSS marks of both groups also had no statistical significance (P > 0.05). On the 7th day and the 14th day in the treatment course, the two groups were compared at the corresponding moment; the difference had statistical significance (for all, P < 0.05); the decrease degree of ET-1, survivin, and the mark improvement of NIHSS were apparent. Conclusion: For a patient suffering from ischaemic stroke, using the therapy of an edaravone injection, the level of ET-1 and survivin in serum will decrease respectively; the nerve functions of the patient suffering from ischaemic stroke may be effectively improved.

1 INTRODUCTION

Domestic and foreign research in recent years show that edaravone may restrain necrocytosis and lipid per-oxidation by a free radical scavenger, so to improve the cell functions in the process of cerebral ischemia and, after refilling when ischaemic stroke occurs, realize the effect of protecting the brain [1]. In the research, for a patient suffering from ischaemic stroke, before and after the therapy starts, NIHSS was assessed with mark, the change of survivin and ET-1 level were determined. The purpose of this is to know the nerve protection effect of edaravone on a patient suffering from ischaemic stroke, so to provide more perfect theoretical support for reasonable clinical diagnosis and therapy.

2 MATERIALS AND METHOD

2.1 Data and method

2.1.1 General data

60 inpatients (in the Neurology Department of our hospital) suffering from ischaemic stroke (internal carotid artery system) during the period of February 2011–February 2012 were selected; they all conformed to the diagnosis standards determined in the National 4th Cerebrovascular Disease Academic Conference [2], including 51 cases with a history of combined hypertension, 22 cases with a history of combined diabetes, and 12 cases with a history of atrial fibrillation. Exclusion standards: (1) acute infarction of the posterior circulation system and other combined CNS diseases; (2) with a history of a heart attack, serious trauma, or major operation within six months; (3) various acute and chronic inflammations, tumours, rheumatism, and connective tissue diseases; and (4) with apparent liver, kidney, and heart failure. The patients were randomly divided into the Routine Therapy Group (Contrast Group) and the Edaravone Therapy Group (Therapy Group). Therapy Group: 30 cases, including 18 male cases and 12 female cases; the mean age was (60.2 ± 8.4) years old. Before therapy, the NIHSS mark was (8.05 ± 2.20); Contrast Group: 30 cases, including 20 male cases and 10 female cases, the mean age was (64.2 ± 5.2) years old. Before therapy, the NIHSS mark was (7.03 ± 2.18). The disease time, age, sex, and NIHSS marks of two groups of patients were compared, the difference had no statistical

significance (P > 0.05), however, they did have comparability.

2.2 Method

On the 2nd, 7th, and 14th day from the commencement of the therapy, since all patients were hospitalized, they were kept under the state of empty stomach for over 8h since morning, then a 3 ml sample of venous blood at the elbow was taken and put into a heparin anti-coagulation tube, which was centrifuged for 5 min at 1500 r/min (with a centrifugal radius of 11 cm). After being centrifuged, the supermate was extracted, poured into the EP tube, and preserved in a refrigerator at 20°C for measurement.

2.3 Detection index of reagent

Survivin detection applied ELISA (Enzyme-linked Immunosorbent Assay); the kit was provided by the Wuhan Boshide Bioengineering Co. Ltd.; the measurement steps complied with the instruction manual of the kit; the ET-1 detection applied radio-immuno-assay, and the kit was provided by the Shanghai Kanu Biotechnological Co. Ltd.; all detection steps strictly complied with the operational manual. In the detection process, the quality was strictly controlled.

2.4 Therapy method

All patients were provided with enteric- coated aspirin tablets (produced by the German Bayer Company, 100 mg/time, QD) for the therapy of anti-platelet aggregation, Shuxuetong (Mudanjiang Youbo Co. Ltd., GYZZ: Z20010100, 6 ml, QD), in an intra-venous injection, for the improvement of blood circulation, and atorvastatin calcium (Youjia, Henan Tianfang Pharmaceutical, 20 mg, 1 time/d) for lowering blood grease. On basis of such therapy means, for the Therapy Group, for a patient with oncome time less than 24 h since hospitalized, edaravone (Jilin Huinan Changlong Biochemical Pharmaceutical Shareholding Co., Ltd., 30 mg+ saline with 100 ml, dripped off within 30 min, BID) was provided for scavenging free radicals and protecting the brain. 14d was a treatment course.

For the assessment of the neurologic impairment, the NIHSS (National Institute of Health Stroke Scale) method of the USA was applied. Before and after the therapy starting, the clinical degree of neurologic impairment was respectively assessed.

2.5 Statistical analysis

For the statistical treatment, SPSS17.0 was applied for statistical analysis, the measurement data applied (means ± s.e.m) for expression, the measurement data was inspected by t, and the enumeration data applied $X2$ for inspection; $P < 0.05$ was taken as the standard for difference significance.

Table 1. Comparison on the NIHSS marking results (points) of the two groups before and after therapy (means ± s.e.m).

Group type	before therapy	on the 7th day after therapy	on the 14th day after therapy
Therapy Group 30	8.05 ± 2.20	4.04 ± 2.17*	2.42 ± 1.43**
Contrast Group 30	7.03 ± 2.18	5.7 ± 2.06*#	3.54 ± 2.47**##

Note: * Compared with the same group before therapy P < 0.05; # Compared with the Therapy Group P < 0.05; ** Compared with the same group before therapy P < 0.05; ## Compared with the Therapy Group before therapy P < 0.05;

Table 2. Comparison on ET1 of two groups before and after therapy (means±s.e.m).

Group type	ET-1(ng/L)		
	before therapy	on the 7th day after therapy	on the 14th day after therapy
Contrast Group	75.49 ± 6.12	69.20 ± 6.41#	60.35 ± 6.91#
Therapy Group	75.89 ± 5.49	61.15 ± 5.63#*	52.54 ± 6.18*#

Note: * Compared with the Contrast Group P < 0.05; # Compared with the same group before treatment P < 0.05;

Table 3. Comparison on ET1 of two groups before and after therapy (means±s.e.m).

Group type	survivin (μmol/g)		
	before therapy	on the 7th day after therapy	on the 14th day after therapy
Contrast Group	11.65 ± 5.87	9.64 ± 3.75#	8.09 ± 5.62#
Therapy Group	11 .54 ± 5.42	8.36 ± 7.43#*	6.46 ± 5.38*#

Note: * Compared with the Contrast Group P < 0.05; # Compared with the same group before treatment P < 0.05;

3 RESULTS

On the 7th day since the therapy started, the two groups were assessed by NIHSS marking and compared. The NIHSS points of the two groups all apparently rose (P < 0.05); in addition, the difference of the two groups had statistical significance (P < 0.05), see Table 1; the survivin and ET-1 levels of the two groups were also compared, and the differences had statistical significance (P < 0.05), see Tables 1, 2, and 3.

4 DISCUSSION

The fatality rate and disability rate of a patient suffering from ischaemic stroke are extremely high, and until now, there has not been an effective therapy

116

method [3]. Therefore, emphasis has been placed on cerebral infarction therapy for the prevention of cerebrovascular disease [4–5]. The existing research shows the following basic pathological processes of ischaemic stroke: the cerebral vasospasm, thrombosis and blocking of the blood vessels causing ischemia, and the hypoxemia necrosis of brain tissue in the corresponding blood supply area. When ischaemic stroke occurs, many free radicals form in the brain cell, and these free radicals finally cause the death of the brain cell. Therefore, a free radical scavenger is a key link in the therapy process of ischaemic stroke. Edaravone is a type of powerful hydroxyl-radical scavenger; it may mitigate the cascading damage caused by free radicals mainly by scavenging the hydroxyl radical, so to prevent the peroxidized damage of the cerebrovascular endotheliocyte and nerve cell, and improve brain function [6] by improving the missing of nerve cell; it may also shrink the extension or infarction volume of the ischemic penumbra, and restrain and delay the death of the tardive nerve cell by lowering the concentration of the hydroxyl radical [7–8]. SUDA etc. [9] carried out magnetic resonance assessment for a patient suffering from ischaemic stroke, evaluated the anti-oedema effect of edaravone, found that edaravone may remedy the ischemic penumbra around the infarction nidus, and defined that edaravone is a kind of effective anti-oedema medicine with a cell protection function.

Survivin is a type of newly-found apoptosis inhibitor gene, which is not expressed in the brain tissue of a normal adult, but is expressed in most of the damaged brain tissue. ET-1 is the most powerful known peptide that shrinks the blood vessel; in addition, it is a kind of important neutrotransmitter and neuropeptide in the central nervous system. The cerebral artery is extremely sensitive to ET. Under the action of ET, the persistent cerebrovascular spasm is the important reason for causing the death of cells in the ischemic zone. Therefore, the selection of effective medicine to lower the level of ET in plasma will become an effective path for the therapy of ischaemic stroke [10]. The test verifies that edaravone may markedly lower the expression level of ET-1 and survivin, so to protect the brain. In addition, it may greatly improve the prognosis of a patient suffering from ischaemic stroke, therefore, it may be taken as a kind of popular medicine for a patient suffering from ischaemic stroke.

REFERENCES

[1] Wang GH, Jiang ZL, Li YC, et al. Free-radical scavenger edaravone treatment confers neuroprotection against traumatic brain injury in rats [J]. J Neurotrauma, 2011, 28 (10): 2123–2134.

[2] The National 4th Cerebrovascular Disease Academic Conference of Chinese Medical Association. Essentials of classified diagnosis for various cerebrovascular diseases [J]. Chinese Journal of Neurology, 1996, 29(6):379.

[3] Shao Rong, Zhou Gang, Liu Dinghua. Observation on the therapy effect of edaravone on massive cerebral infarction in the cerebral artery region [J]. Journal of Apoplexy and Nervous Diseases, 2007, 24(5): 614–615.

[4] Hankey G J, Eikelboom J W. Antithrombotic drugs for patients with ischaemic stroke and transient ischaemic attack to prevent recurrent major vascular events [J]. Lancet Neurol, 2010, 9 (3): 273-284.

[5] Dengler R, Diener H C, Schwartz A, et al. Early treatment with aspirin plus extended- release dipyridamole for transient ischaemic attack or ischaemic stroke within 24 h of symptom onset (EARLY trial): a randomised open-label blinded-endpoint trial [J]. Lancet Neurol, 2010, 9 (2): 159-166.

[6] Lee B J, Egi Yvan L, Eyen K, et al. Edaravone a free radical scavenger protects components of the neurovascular unit against oxidative stress in vitro [J]. Brain Res, 2010, 1307: 22-27.

[7] Wang Damo, Wen Shiquan, Feng Youjun, etc. Observation on the therapy effect of edaravone on acute cerebral infarction [J]. Chinese Practical Journal of Nervous Disease.

[8] Noor JI, Ikeda T, Ueda Y, et al. A free radical scavenger, edaravone, inhibits lipid peroxidation and the production of nitric oxide in hypoxic- ischemic brain damage of neonatal rats [J]. American journal of obstetrics and gynecology, 2005, 193(5): 1703–1708

[9] Suda S, Igarashi H, Arai Y, et al. Effect of edaravone a free radical scavenger on ischemic cerebral edema assessed by magnetic resonance imaging [J]. Neurol Med Chir (Tokyo), 2007, 47 (5): 197-202.

[10] Zhang Yanyan, Chen Xinkuan. Relation between the ET, Hcy level and the lipid peroxidization of a patient suffering from acute cerebral infarction [J]. Journal of Radioimmunology, 2008, 21(1): 23–24.

Biomedical Engineering and Environmental Engineering – Chan (Ed.)
© *2015 Taylor & Francis Group, London, ISBN: 978-1-138-02805-0*

Exploring the economical small-world characteristics of human brain functional networks based on electroencephalogram data

Lulu Wang & Wentao Huang
Department of Physics, Zhejiang Ocean University, Zhoushan, China

Lianhong Yu
School Hospital, Wuhan Textile University, Wuhan, China

Junfeng Gao & Chong Zhang
School of Life Science and Technology, University of Electronic Science and Technology of China, Chengdu, China

ABSTRACT: Exploring functional activities of the human brain will aid us to understand the functional patterns of the human brain in normal or disease states. Functional correlations between 238 scalp electrodes were calculated, and the thresholds were set to establish the simple undirected graphs, then some measurements of the brain functional networks were computed. The results confirmed that the human brain functional networks have both big clustering coefficients like regular networks and small characteristic path lengths, similar to random networks, and that the brain functional networks have a global efficiency greater than the regular but less than the random networks, and a local efficiency greater than the random but less than the regular networks. Taken together, the brain functional networks have economical small-world properties.

1 INTRODUCTION

The human brain has an estimated ~100 billion neurons and 100 trillion connections intricately linked; it is an open complex giant system of self-organization. The human brain, the most complex object in the universe, provides puzzling and challenging interdisciplinary research for people to explore and decode. Non-invasive technological methods, such as the Electroencephalogram (EEG), Magnetoencephalography (MEG), and functional Magnetic Resonance Imaging (fMRI), have been utilized to detect the signals of the brain under task or resting states, and to explore the functional organizational model of the human brain (Bullmore & Sporns, 2012). With more research and more in-depth understanding, people gradually realized that the sum of all its components are not the total of any complex system. There is a great need for the research paradigm to shift to system sciences, which may provide more additional abundant information to supplement or show differences from conventional views.

Complex networks theory is particularly appropriate for investigating the human brain functional connection (Bullmore & Sporns, 2009, Reijneveld et al., 2007). Since it was developed from social network problems, such as six degrees of separation and strength of weak ties, and once it was combined with graph theory, complex networks gradually formed into a new research field with the discovery of small-world and scale-free, and other important network characteristics, which differed with traditional regular lattice and random networks. Complex networks theory treats complex systems as collections of interconnected abstract nodes, explores the topological relationship between nodes and edges, and seeks to discover general principles, algorithms, and tools that govern real-world network behaviours. Nowadays, complex networks theory has become a new science across multiple scientific research fields (Albert & Barabási, 2002, Newman, 2003, Boccalettia et al., 2006, Costa et al., 2007, Fang et al., 2007), and is widely used in the biomedical field (Barabási & Oltvai, 2004, Mason & Verwoerd, 2007).

When complex network theory was applied to brain structural and functional research, many interesting results were acquired (Bullmore & Sporns, 2009, Biswal et al., 2010, Xue et al., 2013). When treating the cerebral cortices or scalp electrodes which are spatially separated as the nodes of the complex networks model, the temporal correlation between spatially remote regions of interest in the brain or on the scalp electrodes are regarded as edges in brain functional networks. For simplicity, brain functional networks are generally seen as undirected networks without loops. Based on EEG data, Stam et al. reported that the human brain has small-world characteristics (Stam et al., 2007). Researchers demonstrated that the functionally connected human brain has scale-free organization based on fMRI data (Eguíluz et al., 2005, van den Heuvel et al., 2008). Studies suggested that modularity plays a key role in the functional organization of the human brain during normal and pathological neural activities (Ahmadlou & Adeli, 2011, Chavez, 2010).

However, many early papers studying the functional brain networks based on EEG, have a relatively small number of scalp electrodes (usually 19-29 electrodes) (Stam et al., 2007, Peters et al., 2013, Kuhnert et al., 2012). Joudaki et al. demonstrated that the estimates of network metrics significantly differ depending on the network size (Joudaki et al., 2012). He, and others, suggested that it would be desirable to perform a more extensive reanalysis of the previously published data with complex networks theory, which would likely be able to offer new insights into the data (He et al., 2009). Thus, this article used theory of complex networks to reanalyse the publicly-available EEG data with 238 electrodes, to explore whether human brain functional networks, based on EEG data with a relatively larger number of electrodes, have economical small-world properties, just the same as the results from the fMRI data (Achard & Bullmore, 2007).

2 MATERIALS AND METHODS

2.1 EEG data

The EEG data are accessed from the project Head IT.org(http://sccn.ucsd.edu/eeglab/data/headit.html). Briefly, the (five-box) visuospatial selective attention tasks were conducted by one volunteer. The data were collected from 238-channel scalp electrodes from one subject with a 256 Hz sampling rate, using a BioSemi Active Two system. Input impedance was brought below 5 kΩ for all electrodes. Data were sampled with an analog pass band of 0.01–50 Hz. Data were minimized with line noise artefacts responses, and were digitally low-pass filtered below 40 Hz before data analysis. Data trials containing electrooculographic (EOG) potentials larger than 70 μV were rejected. Then, the conventional EEG frequency bands were filtered: delta, 0–4 Hz; theta, 4–8 Hz; alpha1, 8–10 Hz; alpha2, 10–13 Hz; beta, 13–30 Hz. The signal was filtered using the function eegfilt.m from the EEGLAB toolbox (Delorme & Makeig, 2004).

2.2 Construction of functional brain networks

For simplicity, brain functional networks are generally seen as undirected networks without loops. Electrodes were regarded as nodes, and the time correlations between nodes were treated as edges. Figure 1 illustrates the schematic representation of the human functional brain network construction, based on EEG data.

Achard and Bullmore's method has been used to obtained correlation matrices (one for each frequency interval in the study), which were then thresholded to produce a set of undirected graphs representing the

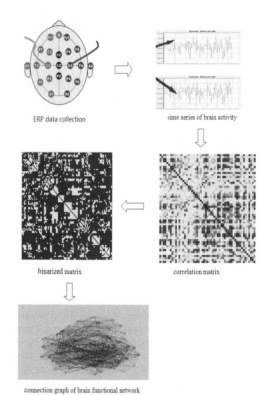

ERP data collection

time series of brain activity

binarized matrix

correlation matrix

connection graph of brain functional network

Figure 1. A flowchart for the construction of a functional network in the human brain, based on EEG data.

functional connections of the human brain (Achard & Bullmore, 2007). It is an important problem in the setting of threshold. It is reported that anatomical connections are sparse for non-human nervous systems, such as *Caenorhabditis elegans*, in which less than 6% of all possible unweighted synaptic connections exist between neurons. Evidence concerning anatomical connectivity in the human brain is sparse, based on both post-mortem observations and non-invasive mapping [24]. In order to avoid getting too sparse a network, numerous studies of brain functional networks have set the maximum correlation threshold, which means that the mean degree of the resulting network will be greater than the log of the number of nodes. However, minimum threshold usually have not been set based on some general accepted principle. Achard and Bullmore brought up using the cost (connection density) of networks as a criterion to set the threshold and demonstrated that human brain functional networks have economical small-world properties, comprising between 5% and 34% of the 4,005 possible edges in a network of 90 nodes. In the present study, we explore the human brain functional network in a similar range of 15% to 35% cost.

Complex network measurements mainly focused on computing the characteristic path length (L), clustering coefficients (C), the global efficiency (E_{global}), and the local efficiency (E_{local}). For a given graph G with N nodes, the characteristic path length of a network is defined as the average of the shortest path lengths l_{ij} between any two nodes i and j:

$$L = \frac{1}{N(N-1)} \sum_{i,j \in V, i \neq j} l_{ij}. \tag{1}$$

The clustering coefficient of a node C measures how close the local neighbourhood of a special node is to being part of a clique, defined as the ratio between the number of edges linking nodes adjacent to i, and the total possible number of edges among them, so that the C of the whole network is defined as:

$$C = <C_i> = \frac{1}{N} \sum_{i \in V} \frac{2e_i}{k_i(k_i-1)}, \tag{2}$$

where k_i named degree is the number of edges directly connected to the node i, and e_i is the corresponding element of the adjacent matrix between nodes i and j. To reduce sampling bias, the normalized characteristic path length $\lambda = L_{eeg}/L_{random}$ and the normalized clustering coefficient $\gamma = C_{eeg}/C_{random}$ were computed. A summary ratio scalar of small-worldness $\sigma = \gamma/\lambda$ was also calculated. The global efficiency is defined as the inverse of the harmonic mean of the minimum path length between each pair of nodes:

$$E_{global} = \frac{1}{N(N-1)} \sum_{i,j \in V, i \neq j} \frac{1}{l_{ij}}, \tag{3}$$

which means the ability of parallel information transfer in the network, or the ability of a network to combine the information of various parts. The local efficiency of a network is the mean of the global efficiency of each sub-network W:

$$E_{local} = \frac{1}{N} \sum_i E_{global}(W_i) \tag{4}$$

which reflects the tendency of a graph to form clusters. All the computations were performed by using R (http://www.r-project.org/) and Matlab; a connection graph was drawn with Pajek (http://vlado.fmf.uni-lj.si/pub/networks/pajek/).

3 RESULTS

The human brain functional networks from different bandwidth EEG data, were computed into a binarized adjacency matrix, which actually was an undirected simple graph comprising between 15% and 35% of the 31,878 possible edges in a network of 238 nodes. Subsequent results were all based on the symmetric matrixes from Target Box 1, which is similar to those from other Target Box data. The delta band brain network was presented with 7,905 connected edges, and thus with about 24.8% connection density (Fig. 2).

The average degree of human functional networks was 21.96, which was less than the number of nodes,

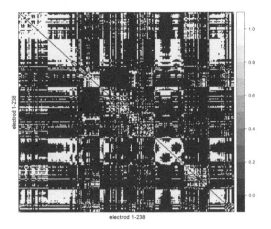

Figure 2. The heat map of the delta band brain network. The heat map is a 238×238 binarized matrix; each element of the map is either white (if there is significant correlation between electrodes) or black (if there is not).

Table 1. Characteristic parameters of the human brain functional networks for conventional EEG frequency bands.

	C_{eeg}	L_{eeg}	C_{rand}	L_{rand}	σ
Delta	0.68	2.10	0.25	1.75	2.30
Theta	0.77	1.85	0.25	1.75	2.86
Aph1	0.78	1.97	0.26	1.74	2.66
Aph2	0.80	2.24	0.26	1.74	2.38
Beta	0.70	2.18	0.24	1.77	2.41

however, it was greater than the natural logarithm of the number of 238 nodes. Measurements of the human brain functional networks are listed in Table 1. For comparison, characteristic path lengths and clustering coefficients of regular networks and random networks with corresponding number of nodes and edges were computed, respectively (Table 1).

For each respective frequency band, it could be found that the human brain functional network has both a large clustering coefficient, just like regular networks, and a small characteristic path length similar to random networks. $\gamma = 2.75 - 3.06$, $\lambda = 1.06 - 1.29$ was obtained and a summary ratio scalar of the small-world σ range from 2.30 to 2.86. It was obvious that $\gamma > 1$, $\lambda \approx 1$, and the summary ratio parameter $\sigma > 1$, which was consistent with the expected results of the theory of small-world characteristics (Watts & Strogatz, 1998).

Over a range of network costs in the delta frequency band, the global and the local efficiency of brain networks were respectively compared with the same parameters computed in a random network and a regular network. As shown in Figure 3, efficiency monotonically increased with cost in both cases; the random network had a higher global efficiency than the regular network, and the reverse for the local efficiency.

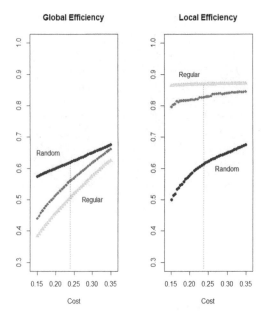

Figure 3. Economical characteristics of functional brain networks.

In both conditions, global and local efficiency increased with cost. The brain networks were intermediate between the random network and the regular network; the random network had greater global efficiency than the regular network (left); the regular network had greater local efficiency than the random network (right). The vertical dotted lines correspond with the maximum value of the difference between global efficiency and cost.

The efficiency curves of the brain networks were intermediate between the random network and the regular network: brain networks had global efficiency greater than the regular but less than the random networks, and local efficiency greater than random but less than regular. The economical small-world characteristics of the brain networks were clearly demonstrated in the cost range of 0.15 to 0.35, comprising between 15% and 35% of the 31,878 possible edges in a network of 238 nodes. Both the global and local efficiency of brain networks were greater than cost. The cost efficiency, which is the difference between the global efficiency (0.56 in the present study) and the cost, had a maximum value corresponding with the network, with 24% of the possible edges.

4 DISCUSSION

4.1 *Human functional networks have small-world properties*

It has been found that after being analysed by the complex network analysis approach, the Alzheimer's disease case-control data acquired by different means

of fMRI (Supekar et al., 2008), EEG (Stam et al. 2007), and MEG (Stam et al. 2009), show that the brain functional networks of diseases have easily damaged hubs and loss of small-world characteristics compared with those of the controls. Brain functional networks research about Down's syndrome (Ahmadlou et al., 2013), schizophrenia (Liu et al., 2008), and other brain diseases have reached similar conclusions. Based on the five-box EEG data with a large number of electrodes, the human brain functional networks are small-world networks in the present study, which imply that a small number of long-range connections not only help the differentiation of brain function and local connection cost constraints, but also benefit normal long-distance transmission and integration of information between different brain areas. The fact that the human brain turned into a disease state from a healthy state may be due to the destruction of high degrees of local integration and integrity of brain connectivity, so that different brain areas suffered from the hindered transmission and integration of information, which ultimately resulted in the development and formation of the "disconnection syndrome" of brain (Brier et al., 2014).

4.2 *Brain functional networks have economical characteristics*

When economical small networks, which are defined as having high efficiency with a low cost system, were first introduced by Latora et al. (Latora & Marchiori, 2001, Latora & Marchiori, 2003), human networks were also found to obey the rule (Bullmore & Sporns, 2012, Achard & Bullmore, 2007). Although human brain consume much of energy budget, 10 times that predicted by its weight alone, the energy consumption by human brain might trade off with brain size (Raichle, 2006, Navarrete et al., 2011). Then, the number and density of neurons, and the distance and cross-sectional diameters of axonal projections are severely constrained by brain volume. Economical brain networks may be naturally selected as a trade-off between minimizing the wiring cost and maximizing the adaptive value. How well information propagates over the brain network, measured by global and local efficiency, is of most importance for adaptation. Here, the present study demonstrated that the topological organization of human brain functional networks observe the economical selection pressure principle (about 56% of maximum efficiency for 24% of maximum cost), which is consistent with Achard's results based on fMRI data (Achard & Bullmore, 2007).

In conclusion, the complex network theory has been used to reanalyse and obtain brain functional networks based on the publicly-accessed EEG data with a large number of electrodes. Our results confirmed that the human brain functional networks have economical small-world characteristics as the early study, based on fMRI data. The results will also deepen the understanding for some disconnection syndromes, such as Alzheimer's disease.

ACKNOWLEDGEMENT

The work was supported by the Natural Science Foundation of China under Grant (31201001, 81271659), the Startup Project of scientific research, and the Young and Middle-aged Project of Zhejiang Ocean University (21065013413, 11062101712).

CORRESPONDING AUTHOR: Wentao Huang, EMAIL: ccnuhwt@aliyun.com.

REFERENCES

Achard S. & Bullmore E. 2007. Efficiency and cost of economical brain functional networks. *PLoS Comput Biol* 3(2): e17.

Ahmadlou M. & Adeli H. 2011. Functional community analysis of brain: a new approach for EEG-based investigation of the brain pathology. *Neuroimage* 58(2): 401–408.

Ahmadlou M., Gharib M., Hemmati S., *et al.* 2013. Disrupted small-world brain network in children with Down Syndrome. *Clin Neurophysiol* 124(9): 1755–1764.

Albert R., & Barabási A.L. 2002. Statistical mechanics of complex networks. *Rev Mod Phy* 74(1): 47–97.

Barabási A.L. & Oltvai Z.N. 2004. Network biology: understanding the cell's functional organization. *Nat Rev Genet* 5(2): 101–113.

Behrens T.E., Johansen-Berg H., Woolrich M.W., *et al.* 2003. Non-invasive mapping of connections between human thalamus and cortex using diffusion imaging. *Nat Neurosci* 6(7): 750–757.

Biswal B.B., Mennes M., Zuo X.N., *et al.* 2010.Toward discovery science of human brain function. *Proc Natl Acad Sci U, S A* 107(10): 4734–4739.

Boccalettia S., Latorab V., Morenod Y., *et al.* 2006. Complex Networks: Structure and dynamics. *Phys Rep* 424(4–5): 175–308.

Brier M.R., Thomas J.B., Ances B.M. 2014. Network dysfunction in Alzheimer's disease: refining the disconnection hypothesis. *Brain Connect* 4(5): 299–311.

Bullmore, E. & Sporns, O. 2009. Complex brain networks: graph theoretical analysis of structural and functional systems. *Nat Rev Neurosci* 10(4): 186–198.

Bullmore, E. & Sporns, O. 2012. The economy of brain network organization. *Nat Rev Neurosci* 13(5): 336–349.

Chavez M., Valencia M., Navarro V., *et al.* 2010. Functional modularity of background activities in normal and epileptic brain networks. *Phys Rev Lett* 104(11): 118701.

Costa L. DA. F., Rodrigues F.A., Travieso G., *et al.* 2007. Characterization of complex networks: A survey of measurements. *Adv Phys* 56(1): 167–242.

Delorme A. & Makeig S. 2004. EEGLAB: an open source toolbox for analysis of single-trial EEG dynamics including independent component analysis. *J Neurosci Methods* 134(1): 9–21.

Eguíluz V.M., Chialvo D.R., Cecchi G.A., *et al.* 2005. Scale-free brain functional networks. *Phys Rev Lett* 94(1): 018102.

Fang J.Q., Wang X.F., Zheng Z.G., *et al.* 2007. New interdisciplinary science: network science (I). *Prog Phys* 27(3): 239–343.

He Y., Chen Z., Gong G., *et al.* 2009. Neuronal networks in Alzheimer's disease. *Neuroscientist* 15(4): 333–350.

Joudaki A., Salehi N., Jalili M., *et al.* 2012. EEG-based functional brain networks: does the network size matter? *PLoS One* 7(4): e35673.

Kuhnert M.T., Geier C., Elger C.E., *et al.* 2012. Identifying important nodes in weighted functional brain networks: a comparison of different centrality approaches. *Chaos* 22(2): 023142.

Latora V. & Marchiori M. 2001. Efficient behavior of small-world networks. *Phys Rev Lett* 87(19): 198701.

Latora V. & Marchiori M. 2003. Economic small-world behavior in weighted networks. *Eur Phys J B* 32(2): 249–263.

Liu Y., Liang M., Zhou Y., *e t al.* 2008. Disrupted small-world networks in schizophrenia. *Brain* 131(4): 945–961.

Mason O. & Verwoerd M. 2007. Graph theory and networks in Biology. *IET Syst Biol* 1(2): 89–119.

Navarrete A., van Schaik C.P., Isler K. 2011. Energetics and the evolution of human brain size. *Nature* 480(7375): 91–93.

Newman M.E.J. 2003. The structure and function of complex networks. *SIAM Review* 45(2): 167–256.

Peters J.M., Taquet M., Vega C., *et al.* 2013. Brain functional networks in syndromic and non-syndromic autism: a graph theoretical study of EEG connectivity. *BMC Med* 11: 54.

Raichle ME. 2006. The brain's dark energy. *Science* 314(5803): 1249–1250.

Reijneveld, J.C., Ponten S.C., Berendse H.W., *et al.* 2007. The application of graph theoretical analysis to complex networks in the brain. *Clin Neurophysiol* 118(11): 2317–2331.

Stam C.J., Jones B.F., Nolte G., *et al.* 2007. Small-world networks and functional connectivity in Alzheimer's disease. *Cereb Cortex* 17(1): 92–99.

Stam C.J., de Haan W., Daffertshofer A., *et al.* 2009. Graph theoretical analysis of magnetoencephalographic functional connectivity in Alzheimer's disease. *Brain* 132(1): 213–224.

Supekar K., Menon V., Rubin D., *et al.* 2008. Network analysis of intrinsic functional brain connectivity in Alzheimer's disease. *PLoS Comput Biol* 4(6): 1–11.

van den Heuvel M.P., Stam C.J., Boersma M., *et al.* 2008. Small-world and scale-free organization of voxel-based resting-state functional connectivity in the human brain. *Neuroimage* 43(3): 528–539.

Watts D.J. & Strogatz S.H. 1998. Collective dynamics of 'small-world' networks. *Nature* 393(6684): 440–442.

Xue K.Q., Luo C., Yang T.H., *et al.* 2013. Disrupted structural connectivity of default mode network in childhood absence epilepsy. *Prog Biochem Biophys* 40(9): 826–833.

Biomedical Engineering and Environmental Engineering – Chan (Ed.)
© 2015 Taylor & Francis Group, London, ISBN: 978-1-138-02805-0

Computer modelling of Abdominal Aortic Aneurysms to predict the risk of rupture – the role of porosity of the thrombosis

Omar Altuwaijri
Department of Biomedical Technology, King Saud University, Saudi Arabia

ABSTRACT: Abdominal Aortic Aneurysm (AAA) is a cardiovascular disease occurring when the aorta becomes weak and develops a balloon expansion in its wall. This balloon diameter can reach sizes of up to four times the normal aortic diameter, with the diameter enlarging at rates of 0.2–1.0 cm/year. A ruptured aneurysm leads to death in 78%–94% of diseased aortas [1].

Aneurysm rupture is a biomechanical event that occurs when the mechanical stresses in the wall of the aorta exceed the failure strength of the aortic tissue [2].

In medical practice, when the maximum diameter of an AAA exceeds 5 cm it is considered at risk of rupture. Surgical repair is usually not considered until the diameter reaches at least 5 cm. However, it is frequently observed that AAAs with diameters of less than 4 cm can rupture, which raises the need for finding a more reliable method to assess rupture risk.

The role of the Intraluminal Thrombus (ILT), which exists in more than 75% of AAAs, was examined using variable thickness and material properties of the thrombus. For simplification purposes, it is assumed that ILT is a solid material, as sourced from previous studies, even though in reality ILT is a highly porous material with an average porosity of 80% [3]. The porosity of the ILT has been experimentally examined in a number of studies [4, 5, 6].

Two recent numerical studies have examined the porosity of ILT using finite element models [7, 8] which have provided useful information about the impact of thrombus porosity on AAA biomechanics, but both neglected the actual dynamics of the blood mass flow.

Keywords: Abdominal aortic aneurysm, Intraluminal Thrombus (ILT), Porosity of thrombosis

1 INTRODUCTION

Abdominal aortic aneurysm (AAA) is a cardiovascular disease occurring when the aorta becomes weak and develops a balloon expansion in its wall. This balloon diameter can reach sizes of up to four times the normal aortic diameter, with the diameter enlarging at rates of 0.2–1.0 cm/year. A ruptured aneurysm leads to death in 78% – 94% of diseased aortas [1].

Aneurysm rupture is a biomechanical event that occurs when the mechanical stresses in the wall of the aorta exceed the failure strength of the aortic tissue [2].

In medical practice, when the maximum diameter of an AAA exceeds 5 cm it is considered at risk of rupture. Surgical repair is usually not considered until the diameter reaches at least 5 cm. However, it is frequently observed that AAAs with diameters of less than 4 cm can rupture, which raises the need for finding a more reliable method to assess rupture risk.

The role of the intraluminal thrombus (ILT) which exists in more than 75% of AAAs was examined using variable thickness and material properties of the thrombus. For simplification purposes, it is assumed that ILT is a solid material, as sourced from previous studies, even though in reality ILT is a highly porous material with an average porosity of 80% [3]. The porosity of the ILT has been experimentally examined in a number of studies [4, 5, 6].

Two recent numerical studies have examined the porosity of ILT using finite element models [7, 8] which have provided useful information about the impact of thrombus porosity on AAA biomechanics, but both neglected the actual dynamics of the blood mass flow. There is a lack of knowledge on how modelling the porosity of the ILT may contribute to the understanding and the assessment of AAA.

In this study, the porosity of ILT was examined using a computational fluid dynamics model to investigate how blood flows within the thrombus, and how the information gained can be used to aid AAA diagnosis and give possible reasons for its growth.

2 METHOD

A simple fusiform axisymmetric three-dimensional (3D) aneurysm model was constructed with an inlet

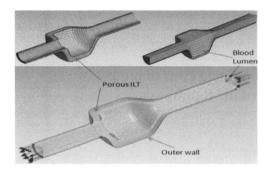

Figure 1. The porous ILT in grey, and the blood lumen domain in green mesh.

diameter of 2.0 cm and; maximum aneurysm diameter of 5.0 cm. An ILT layer with a 10 mm maximum thickness was added to the model (see Fig. 1).

ANSYS CFX computational fluid dynamics solution was used, an assumption of Newtonian flow. Steady flow with a velocity of 25 cm/s at the inlet and blood pressure of 120 mmHg at the outlet, were used with blood density, 1050 kg/m^3, and blood dynamic viscosity, 0.0035 Poiseuille (Pa.s).

In CFD, modelling porous material numerically involves applying Darcy's law which describes the flow of a fluid through a porous medium in the following form:

$$Q = \frac{kA}{\mu} \left(\frac{\partial P}{\partial L}\right)$$

where:
Q is the flow rate of the blood.
k is the permeability of the ILT.
A is the cross-sectional area of the aorta.
μ is the viscosity of the blood.
L is the length of the porous media the fluid will flow through.

$$\left(\frac{\partial P}{\partial L}\right)$$

represents the pressure change per unit length of the blood.

The interface between the blood lumen and the porous ILT material allows blood to flow inside the ILT. The permeability value of the ILT was taken from Adolph et al. [3], who reported that ILT of 0.91 0.54 mm4/N s. Three values of permeability will be examined to understand its effect.

For the fluid medium, a steady flow at a constant velocity of 25 cm/s at the inlet and blood pressure of 120 mmHg at the outlet was used, with a blood density of 1121 kg/m3 and blood dynamic viscosity of 0.0035 Poiseuille (Pa.s) [8].

3 RESULTS

Results from these analyses, illustrated in Figures 2 and 3, show that the flow inside the ILT has a much

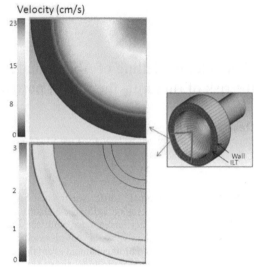

Figure 2. Velocity map on a cross-sectional view for blood flow on the whole aneurysm (top), and only in the ILT (bottom).

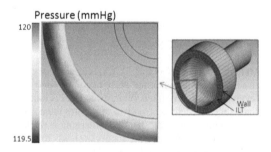

Figure 3. Velocity vectors of blood flow in the blood lumen (grey background) and the ILT (white background).

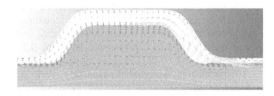

Figure 4. Pressure gradients of the cross-sectional view of the ILT in the middle of the aneurysm.

smaller velocity than the flow in the blood lumen. It also shows that the flow is more able to transmit to ILT at the distal region of the aneurysm.

Figure 4 shows the pressure gradients of the cross-sectional view of the ILT in the middle of the aneurysm, varying between 119 to 120 mmHg.

4 DISCUSSION

Figure 2 shows that the flow can actually move from the blood lumen to the ILT, and noticeable blood movement can be seen within the ILT, which agrees with the conclusion of Adolph et al. [3], who experimentally carried out a histologic analysis for the ILT and suggested that fluid and smaller molecules may pass freely through the thrombus via the canaliculi network. Another very interesting observation can be also seen at the distal end of the aneurysm, in that the flow is more able to transmit through the ILT in this region (recorded velocity in this region is the highest in ILT = 9.6 cm/s) compared with other locations of the ILT. This could be as a result of the perpendicular force of the flow and the continuous vortices' formation in this location, and could have implications for the rupture in these regions. This observation needs to be investigated clinically as it may explain why the aneurysm grows at the distal region of the aneurysm. It could be a result of the continuous movement of the blood and its smaller molecules in this region. Specifically, it could explain why the Matrix Metalloproteinases (MMPs) – which are highly linked with AAA pathology – are concentrated in a region of AAA which is at high risk of rupture, as concluded by Wilson et al. [9] in their clinical study, who reported that there was a localized increase in MMP-8 and MMP-9 concentrations at the site of the aortic rupture.

Another possible advantage of such modelling is that modelling the blood flow within the ILT will be important in future work for investigating drug delivery solutions in AAA, which is a significant potential new AAA treatment methodology as reported by Ayyalasomayajula et al. [10].

Figure 4 illustrates the pressure drop within the ILT, where it can be seen that the pressure decreases with the ILT depth, but that the total drop is very small (less than 0.5%). This is not a surprising observation because the ILT was modelled with high porosity (80%) and it will be expected that the pressure drop will be minimal. This is in strong agreement with the results reported by Schurink et al. [10], who found no reduction in blood pressure under the ILT. In their clinical study, Schurink et al. investigated nine patients who underwent operations for an AAA at the level of the thickest thrombus lining; the pressure within the aneurysmal thrombus (just inside the aneurysmal wall) was measured and compared with the systemic pressure and no deference was found. Similar results were reported by Takagi et al. [11] who conclude that the thrombus of an aneurysm does not significantly decrease the pressure on the aneurysmal wall. It is also in agreement with the recent blood pressure measurements of Polzer and Bursa [7], who numerically used poroelastic material for the ILT and reported no reduction of pressure between the blood lumen and underneath the ILT.

REFERENCES

[1] Finol, K., C. H. Amon. (2003) the effect of asymmetry in abdominal aortic aneurysms under physiologically realistic pulsatile flow conditions. Journal of Biomechanical Engineering 125: 207–217.

[2] Speelman A, Bosboom E, Schurink G, Vande vosse F, Makaorun M, Vorp D; (2008), Effects of wall calcifications in patient-specific wall stress analyses of abdominal aortic aneurysms. Journal of biomechanical engineering, 129(1): 105–109.

[3] Ashton J, Geest J, Bruce S, Haskett D; (2009), Compressive mechanical properties of the intraluminal thrombus in abdominal aortic aneurysms and fibrin-based thrombus mimics. J biomech, 42(3): 197–201.

[4] Adolph R, Vorp DA, Steed DL, Webster MW, Kameneva MV, Watkins SC; (1997) Cellular content and permeability of intraluminal thrombus in abdominal aortic aneurysm. J Vasc Surg.; 25(5): 916–926.

[5] Di Martino S, Guadagni G, Fumero A, Ballerini G, Spirito R, Biglioli P, Redaelli A; (2001), Fluid-structure interaction within realistic three-dimensional models of the aneurysmatic aorta as a guidance to assess the risk of rupture of the aneurysm. Medical engineering & physics, 23: 647–655.

[6] Gasser Christian, Giampaolo Martufi, Martin Auer, Maggie Folkesson, Jesper Swedenborg (2010), Micromechanical Characterization of Intraluminal Thrombus Tissue from Abdominal Aortic Aneurysms, Annals of Biomedical Engineering, volume (38) 2, 371–379.

[7] Polzer S. and Bursa J.; (2010) Poroelastic Model of Intraluminal Thrombus in FEA of Aortic Aneurysm, 6th World Congress of Biomechanics. August 1–6, 2010 Singapore, IFMBE Proceedings, Volume 31, Part 3, 763–767.

[8] Yu S; (2000), Steady and pulsatile flow studies in abdominal aortic aneurysm models using particle image velocimetry. International journal of heat and fluid flow, 21: 74–83.

[9] Wilson R, Anderton M, Schwalbe E, Jones L, Furness P, Bell P, Thompson M; (2006), Matrix Metalloproteinase-8 and -9 Are Increased at the Site of Abdominal Aortic Aneurysm Rupture. Circulation, 113: 438–445.

[10] Schurink G, Van Baalen J, Visser M, Van Bockel J; (2000), Thrombus within an aortic aneurysm does not reduce pressure on the aneurysmal wall. Journal of Vascular Surgery, 31: 501–506.

[11] Takagi H, Yoshikawa S, Mizuno Y, Matsuno Y, Umeda Y, Fukumoto Y, Mori Y; (2005) Intrathrombotic pressure of a thrombosed abdominal aortic aneurysm. Ann Vasc Surg. Jan; 19(1): 108–112.

Biomedical Engineering and Environmental Engineering – Chan (Ed.)
© 2015 Taylor & Francis Group, London, ISBN: 978-1-138-02805-0

One-way valve of the soft palate and Obstructive Sleep Apnoea Hypopnea Syndrome (OSAHS): A fluid-solid coupling study

L.J. Chen
School of Engineering, Sun Yat-Sen University, Guangzhou, Guangdong Province, P.R. China

X.M. Zhang
The Sixth Affiliated Hospital of Sun Yat-Sen University, Guangzhou, Guangdong Province, P.R. China

T. Xiao
Institute of Architecture and Civil Engineering, Guangdong University of Petrochemical Technology, Maoming, Guangdong Province, P.R. China

ABSTRACT: In this paper the upper airway and its surrounding tissues were reconstructed based on CT image data, and then a fluid-solid coupling analysis was conducted to simulate the entire collapse process of the soft palate by using a finite element software. The results show that when the respiration of patients with obstructive sleep apnoea hypopnea syndrome (OSAHS) temporarily halts during sleep, their soft palates serve the function of a one-way valve.

1 INTRODUCTION

The pathogenesis of obstructive sleep apnoea hypopnea syndrome (OSAHS) is hitherto unclear, but it is commonly believed that OSAHS is caused by abnormalities in the structure of the upper airway and its surrounding tissues, the pharyngeal muscle turning flaccid during sleep, and the collapse of airway soft tissue (Caples et al., 2005). Presently, the standard practice used in clinical medicine for the diagnosis of OSAHS is Polysomnography (PSG) (Tan et al., 2013, Li et al. 2010). Asides from PSG, techniques such as a portable home sleep apnoea monitoring instrument (Sun et al., 2011), upper airway and oesophageal manometry (Li et al. 2006), and 3D imaging CT and MRIs can also be adopted for the scanning of OSAHS and aid its diagnosis.

2 BIOMECHANICAL SIMULATIONS OF THE UPPER AIRWAY, USING A FLUID-SOLID COUPLING METHOD

The research presented in this article is based on the techniques of medical image and computer image processing, to construct a finite element model of the upper airway and related surrounding tissues of OSAHS patients. Supported theoretically by computational fluid mechanics and solid mechanics, finite element software analyses this mechanical model through fluid-solid coupling simulations, and retrieves data on the characteristics of the upper airway flow field, and the displacement pattern of the soft palate during temporary respiration halts that occur in sleep, which is valuable information for the study of the pathogenesis of OSAHS from a biomechanical perspective. The medical imaging files used in this research are taken from the CT scanning data of a 33-year-old male patient with a snoring history of 20+ years. In order to put the subject into a natural sleep condition and eliminate the influence of sedative drugs, the subject was required to stay awake for 24 hours prior to taking the medical images. Furthermore, the subject was required not to take any form of beverage or medication that is mentally stimulating, and before the imaging process, the subject was checked for sedative drugs such as tranquilizers. During the imaging process, when the patient was in a natural sleep condition, medical images were taken using a Toshiba Aquilion ONE 640 layer dynamics volume CT scanner, with the patient in a supine position.

In this work, some simplifications were made to the nasal tissues. The scanned data was firstly processed through the MIMICS (Materialise's Interactive Medical Image Control System) 14.0, and different anatomical modules were extracted using an image segmentation thresholding method, such as the point cloud spatial distribution of the airway, the soft palate, and the hard palate. The point cloud data of every anatomical structure were reverse engineered by a CAD software, Geomagic Studio 2013, and geometric models of the airway, the soft palate, and the hard palate were acquired from the NURBS curved surface fitting. Then, the geometric models were imported

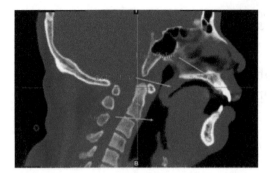

Figure 1. CT scanning image of an OSAHS patient's upper airway when sleeping in a supine position.

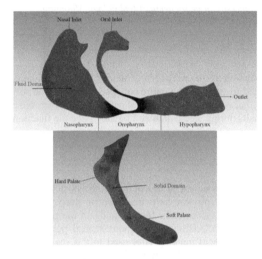

Figure 2. Finite element modelling of the flow field of the upper airway and the solid field of the soft and hard palates.

into Solidworks in order to be formatted so that they could be identified by related finite element models. Finally, the aforementioned models were imported into ANSYS 15.0 for finite element analyses, as shown in Figure 2. The numbers of nodes and elements produced in finite element mash generation are: 76,086 nodes and 127,712 elements for the flow field of the upper airway, and 85,631 nodes and 17,970 elements for the solid field of the hard and soft palates.

Linear constitutive relations were adopted for all tissues in these models. The soft palate was treated as incompressible material with a higher Poisson's ratio because its tissue contains a large quantity of water. The physical parameters used in this study are: density of 1000 and 1850 kgm^{-3}, Young's moduli of 1 MPa and 13.7 GPa, and Poisson's ratios of 0.4 and 0.3 for the soft and hard palates, respectively.

A realizable k-ε transient model for viscous incompressible turbulent flow was chosen in the airflow simulations. The parameters of the airflow are: density $\rho = 1.225 \, \text{kg/m}^3$, and dynamics viscosity index $\mu = 1.7894 \times 10^{-5} \, \text{kg/m·s}$. The upper airway was treated as a transient rigid body and the airflow as an

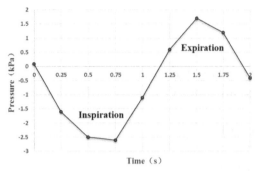

Figure 3. Curve of pressure when OSASH occurs in clinical trials.

incompressible fluid in simulations, while the change of temperature field was neglected. A whole process of respiration was simulated in consideration of the influence of weight on the soft palate when the patient sleeps in a supine position.

Considering the incompressible viscous airflow in the upper airway, the governing equations consist of the equation of mass conservation (Hahn et al. 1993):

$$\frac{\partial \rho_f}{\partial t} + \nabla \cdot (\rho_f v) = 0 \tag{1}$$

and Navier-Stokes equation (Martonen et al., 2002):

$$\frac{\partial \rho_f v}{\partial t} + \nabla \cdot (\rho_f vv - \tau_f) = f_f \tag{2}$$

where t denotes time, f_f volume vector, ρ_f flow density, v flow velocity vector, and τ_f shear tensor.

The equation of conservation for a solid field (Song, et al., 2012) can be derived from Newton's second law:

$$\rho_s \ddot{d}_s = \nabla \cdot \sigma_s + f_s \tag{3}$$

where ρ_s denotes solid density, σ_s Cauchy stress tensor, f_s vector of body force, and $\ddot{d}_s$ acceleration vector in the local solid field.

Meanwhile, the fluid-solid coupling of the upper airway with a soft or hard palate obeys stress and displacement conservations at the interface, which are written by the following equations, respectively (Song et al., 2012):

$$\tau_f \cdot n_f = \tau_s \cdot n_s \tag{4}$$

$$d_f = d_s \tag{5}$$

In this study the pharyngeal pressure measured by clinic PSG was used as the boundary load at the outlet of the flow field of the model. According to the pressure data intercepted for two seconds in a typical respiratory period when respiratory apnoea occurs, as shown in Figure 3, a distributed load was applied to the outlet of the airflow in the upper airway model, and an atmospheric pressure was applied at the nasal cavity, ie, the inlet of the airflow of the model. Only the mode of nasal inspiration and expiration was taken into account.

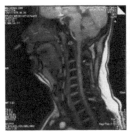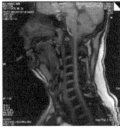

Figure 4. Actions of the one-way valve in an OSAHS patient's respirations. Left: inspiration. Right: expiration.

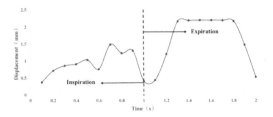

Figure 5. Curve of maximum displacement of nodes on an OSAHS patient's soft palate during the respiration period.

3 ONE-WAY VALVE STRUCTURE OF THE SOFT PALATE

Several parts in the pharynx have comparable anatomical structures of a one-way valve. The valve is a membranous structure which is capable of opening and closing, and which exists in the organs of human or certain other animals. The structural feather of a one-way valve is the existence of a device that is located somewhere in the fluid flow passage, and keeps fluid running merely in one direction. Such a device possesses one or several comparative fixed fulcrums or pedestals, with one or several flaps or valve membranes resting on them, alternating configurations when the fluid flow direction changes.

The structural fulcrum of the one-way valve of the soft palate is located at the joint between the soft and the hard palate. Its primary functions are to prevent the food flowing backwards into the nasopharynx when eating, and to control the airflow from the rhinolalia aperta when speaking.

OSAHS patients have a narrow pharyngeal cavity. When they sleep, the pharyngeal muscle sags, and their soft palates tend to form one-way valves which will affect respiration and cause hypopnea or apnoea. In nasal inspiration, air flows through the valve area and gives rise to snoring, while during expiration, the valve of the soft palate may close and air flows through the mouth instead of the nose, as shown in Figure 4. This phenomenon can be detected clinically through high-speed dynamic MR (Magnetic Resonance) detection.

4 BIOMECHANICAL SIMULATION RESULTS OF THE ONE-WAY VALVE OF THE SOFT PALATE

Most domestic and foreign researchers believe that the major and most direct factor of OSAHS pathogenesis is the morphological and functional abnormality of the upper respiratory anatomy. It is widely reported that the upper airway is an irregularly shaped pipeline with two openings at both ends. The ends are rigid but the middle part of the pipeline is easy to deform. OSAHS patients have narrow upper airways, which are muscular tubes and lack bony or cartilaginous support,

while respiration is an alternative process of positive and negative pressures. Theoretically, obstruction and an unsmooth airflow in the upper airway, arising from any cause, may initiate OSAHS. Through fluid-solid coupling, analysis of the upper airway, and the biomechanical simulation results demonstrate that the anatomical abnormality of OSASH patients' upper airways, in addition to soft palate hypertrophy, do affect the flow field of the upper airway, especially when the neurarchy weakens during sleep. Furthermore, an abnormal flow field would react against the soft palate, causing it to sag and collapse, resulting in airway obstruction and apnoea phenomenon. Figure 5 depicts the curve describing the maximum displacement of nodes on an OSASH patient's soft palate during his respiration period. It can be seen that the maximum displacement of nodes in the expiratory phase is much greater than that in the inspiratory phase. During time monitoring, the maximum displacement of 1.493 mm occurs at 0.7 s in the inspiratory phase, while that of 2.205 mm occurs at 1.5 s in the expiratory phase, where the displacement fluctuates round the peak value from 1.3 s to 1.8 s.

OSAHS patients usually have a hypertrophic soft palate and a narrow pharyngeal passage, which cause rapid change of pressure and unsmooth airflow in the upper airways during sleep. The soft palate becomes more deformed under a higher pressure difference in the flow field, resulting in a further decrease of internal pressure in the upper airways. The collapse of the soft palate leads to an incomplete or complete closure of the airways, resulting in snoring and apnoea. Figures 6 and 7 show the pressure distributions of the flow field in an OSASH patient's upper airway in the inspiratory and expiratory phase respectively, where the time history of 0.3 s, 0.5 s, and 0.8 s falls in the inspiratory phase. Due to some simplifications being made to the nasal structures, the highest value of pressure is located firstly at the anterior naris, and then through the whole nasal cavity during the inspiratory process, while the lowest value of pressure is located at the hypopharyngeal wall, near the soft palate. The time history on 1 s is the point when respiration alternates from inspiration to expiration, at the moment the minimum air pressure near the soft palate begins to rise. The time history on 1.3 s, 1.5 s, and 1.8 s is in the expiratory phase. When expiration begins, the highest value of air pressure in the upper airway starts to alternate between the hypopharynx and the oropharynx, while

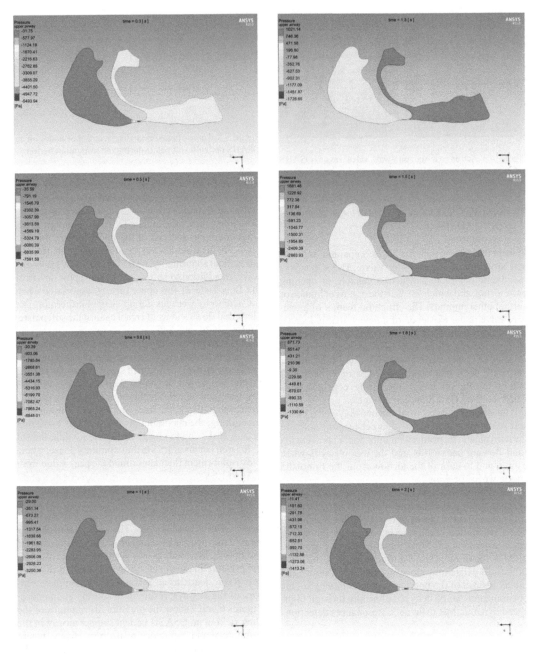

Figure 6.　Pressure distribution of an OSAHS patient's upper airway in the inspiratory phase.

Figure 7.　Pressure distribution of an OSAHS patient's upper airway in the expiratory phase.

the lowest value of air pressure is still located at the hypopharyngeal wall, near the soft palate.

Figures 8 and 9 show the displacement conditions of the OSAHS patient's soft palate in the inspiratory and expiratory phases, respectively. The simulation results show that the displacement of the soft palate in the expiratory phase is much larger than that in the inspiratory phase, and the airway closure, due to soft palate collapse, occurs in the expiratory phase. Moreover, the soft palate has already sagged to a certain extent owing

to its gravity when the patient lies in the supine position during natural sleep. In the inspiratory phase (from 0 to 1 s), the soft palate moves towards the retropharyngeal wall under the action of the flow field, reaches the maximum displacement at 0.7 s, and then returns back to the initial position when inspiration is ready to start. The inspiratory phase transforms into the expiratory phase at the time of 1 s. After that, the soft palate moves towards the retropharyngeal wall again, and as a result, the displacement increases gradually and reaches the

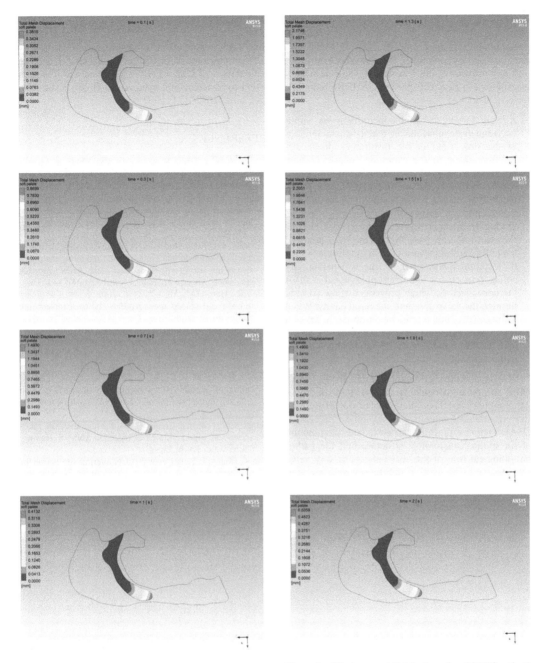

Figure 8. Displacement distribution of an OSAHS patient's soft palate in the inspiratory phase.

Figure 9. Displacement distribution of an OSAHS patient's soft palate in the expiratory phase and airway collapse.

maximum value at 1.5 s, and then returns back to the initial position when expiration is ready to start. It can be observed that the collapse of the soft palate occurs in the expiratory phase, which is consistent with our clinical observations by means of MR. Therefore, the airflow can easily pass through the valve area in inspiration, but the soft palate valve will collapse to close the airway in expiration.

5 SUMMARY

To date, biomechanical simulations of the upper airway, in combination with medical image processing and 3D reconstruction techniques, computational biomechanics, basic theories of modern otolaryngology, and advanced clinical detection means, are important means to explore OSHAS pathogenesis. The simulations would be closer to the real respiratory physiological processes in this paper because OSAHS patients' clinical monitoring data during sleep were used as boundary conditions. The analysis on the upper airway simulation results may shed light on the mechanism of upper airway collapse in OSAHS. In summary, the conclusions of this paper are:

1) A respiratory cycle of the upper airway was simulated and the results indicate that the characteristics of the flow field in the inspiratory phase differ from those in the expiratory phase. In both phases, the lowest pressure area is always located at the retropharyngeal wall of the soft palate, which means that this place is the position where collapse is most likely to occur.

2) During the upper airway respiratory phase, the airflow can smoothly pass through the valve area and the soft palate flaps at a certain amplitude. When the expiratory phase begins, the soft palate may collapse to obstruct the retropharyngeal wall of the upper airway, which prevents airflow passing through the valve area into the nasal cavity. When a low negative pressure in the retropharyngeal wall is not sufficient to reopen the soft palate valve, the patients may breathe with the mouth slightly open under neurarchy.

In this paper the phenomenon of the one-way valve of the soft palate is demonstrated by fluid-solid coupling simulations: The interaction of the airway and its surrounding tissues in the respiration process of OSAHS patients during sleep is studied; the function of the clinical anatomical structure and the pathophysiological role of the soft palate one-way valve are examined. This work is helpful not only for us to explore the pathogenesis of OSAHS, but also for the clinical therapy of OSAHS patients.

ACKNOWLEDGEMENTS

The work presented in this paper was supported by a research grant from the National Natural Science Foundation of China (11272360), and the Research Fund for the Doctoral Program of Higher Education of China (20120171110035).

REFERENCES

Caples, S.M., Gami, A.S. & Somers, V.K. 2005. Obstructive sleep apnea. *FOCUS: The Journal of Lifelong Learning in Psychiatry* 3(4): 557–567.

Hahn, I., Scherer, P.W. & Mozell MM. 1993. Velocity profiles measured for airflow through a large-scale model of the human nasal cavity. *Journal of Applied Physiology* 75: 2273–2273.

Li, S.Q., Ha, R.S., Qu, A.L., et al. 2010. Construction of the three-dimensional finite elemen model of the upper airway in patients with obstructive sleep apnea hypopnea syndrome. *Ningxia Medical Journal* 32(004): 314–316.

Li, Y.R., Han, D.M., Ye, J.Y., et al. 2006. Sites of struction in obstructive sleep apnea patients and their influencing factors an overnight study. *Chinese Journal of Otorhinolaryngology Head and Neck Surgery* 41(6): 437–442.

Martonen, T.B., Quan, L., Zhang, Z., et al. 2002. Flow simulation in the human upper respiratory tract. *Cell biochemistry and biophysics* 37(1): 27–36.

Sun, X.Q., Yin, M. & Cheng, L. 2011. The diagnostic application of portable sleep monitor in obstructive sleep apnea hypopnea syndrome. *Chinese Journal of Clinicians (Electronic Edition)* 5(6): 1695–1697.

Song, X.G., Cai, L. & Zhang, H. 2012. Fluid-structure interaction and engineering cases based on ANSYS. Beijing: China Water Power Press.

Tan, J., Huang, J., Yang J, et al. 2013. Numerical simulation for the upper airway flow characteristics of Chinese patients with OSAHS using CFD models. European Archives of Oto-Rhino-Laryngology. 270(3): 1035–1043.

Biomedical Engineering and Environmental Engineering – Chan (Ed.)
© *2015 Taylor & Francis Group, London, ISBN: 978-1-138-02805-0*

The zinc uptake system (*znuA* Locus) is important for bacterial adhesion and virulence in *Pseudomonas Aeruginosa*

Bo Li, Bobo Wang, Xiangli Gao, Lang Gao, Ying Liang, Kangmin Duan & Lixin Shen
Key Laboratory of Resources Biology and Biotechnology in Western China, Ministry of Education, Faculty of Life Sciences, Northwest University, Xi'an, China

ABSTRACT: ZnuA and its homologs have been reported to be a virulence factor and a zinc-binding periplasmic protein in different bacterial species. In this study, we characterized the ZnuA in the human pathogen *Pseudomonas aeruginosa*. It has three conserved histidine residues and a charged flexible loop which is rich in histidines and acidic residues. The expression level of *znuA* in *P. aeruginosa* PAO1 was regulated by ions of zinc, manganese as well as copper and the regulation is dependent on *zur*. The over expression of *znuA* increased the growth of PAO1 in zinc-restricted environment comparing to the level under zinc-sufficient conditions. These results suggest that *znuA* in PAO1 is involved in zinc uptake in *P. aeruginosa* as in other bacteria and there exist other zinc tansporter(s). Furthermore, we also found the relevance of *znuA* to the adhesion and virulence of PAO1 as both two abilities declined with the absence of *zunA*. The involvement of *zunA* on adhesion in PAO1 could depend on its role on zinc uptake.

1 INTRODUCTION

Pseudomonas aeruginosa (PA) is a major opportunistic human pathogen which is a leading cause of nosocomial infections. It plays a significant role in morbidity and mortality among burn victims, immunocompromised individuals and cystic fibrosis (CF) patients (Garau and Gomez, 2003).

The uptake and sequestration of transition metals, such as zinc, manganese etc, are important for bacterial survival in the environment and hosts (Banerjee et al., 2003). Zinc plays critically roles in both structural and catalytic aspects among all six classes of enzymes as well as the role in replication and transcription factors (Coleman, 1998). Since zinc is charged and hydrophilic, it cannot cross the bacterial membrane through passive diffusion but depends on the actions of specific or non-specific transport systems (Choudhury and Srivastava, 2001). Although zinc is essential, intracellular levels of zinc must be tightly regulated. Low level zinc does not support cellular growth, while excess zinc is toxic to cells because it competes with other metals for binding sites on enzymes (Costello et al., 1997).

A number of bacterial zinc transport systems have been characterized, including ZnuABC, ZupT and ZntA (Grass et al., 2005, Rensing et al., 1997). ZnuABC belongs to the family of ATP-binding cassette (ABC) transporters and consists of three proteins, ZnuA, ZnuB and ZnuC. ZnuB is a membrane permease, ZnuC is an ATPase and ZnuA is a periplasmic metallochaperone which captures zinc and then delivers it to ZnuB (Berducci, 2004., Chandra, 2007.). Two

regions in ZnuA are very important for zinc transport. One contains three conserved histidine residues which are crucial for zinc binding (Castelli et al., 2012). The other is a charged flexible loop, rich in histidines and acidic residues, which plays an important role in chaperoning zinc to the metal-binding cleft (Ilari et al., 2011). ZnuA is also reported to ensure the ability of adhesion and virulence in some species, such as *S. enteric*, *B. abortu* and *E. coli* CFT073 (Gunasekera et al., 2009, Ammendola et al., 2007, Kim et al., 2004).

In *P. aeruginosa* PAO1, two gene clusters of *znuA* and *znuCB* are separated by a 68bp region along with *zur*. The *zur* acts as a regulator of zinc transport system ZnuABC (Ellison et al., 2013). In our previous work, *znuA* was found to be involved in the virulence in PAO1(Δ2800) (Shen et al., 2012). In this study, we compared *znuA* with the homologs in other bacteria and investigated the role of *znuA* in zinc uptake, adhesion and virulence in PAO1.

2 MATERIALS AND METHODS

2.1 *Bacterial strains and growth conditions*

The bacterial strains and plasmids used in this study are listed in Table 1. *P. aeruginosa* PAO1 and *E. coli* DH10B strains were cultivated at 37°C on LB. Antibiotics were used at the following concentrations where appropriate, for *E. coli*, Kanamycin (Kan, 50 mg/mL), gentamicin (Gm, 15 mg/mL), and ampicillin (Amp, 100 mg/mL); and for *P. aeruginosa*, trimethoprim (Tmp, 300 mg/mL), gentamicin (Gm, 150 mg/mL).

Table 1. Strains and plasmids used in this study.

Strains or plasmids	Genotype or phenotype	References
E.coli		
DH10B	F *mcrA* Δ(*mrr-hsdRMS-mcrBC*) Φ80*lacZ*Δ*M*15 Δ*lacX74 recA1 ara*Δ139Δ(*ara-leu*)7697 *galU galK rpsL* (Str^R) *endA1 nupG*	Invitrogen
SM10-λ*pir*	*thi-1 thr leu tonA lacY supE recA::RP4-2-Tc:Mu,* Kan^R, λ*pir*	(R Simon, 1983)
P. aeruginosa		
PAO1	Wild type	(Holloway et al., 1994)
PAO1(Δ*znuA*)	*znuA* deletion mutant of PAO1	This study
PAO1(Δ*znr*)	*znr* replacement mutant of PAO1	This study
PAO1(Δ2407-10)	PA2407 replacement mutant of PAO1, Gm^R	This study
PAO1(Δ*znuA*Δ2407-10)	PA2407 replacement mutant of PAO1(Δ*znuA*), Gm^R	This study
pEX18Tc	Broad-host-range gene replacement vector; *sacB*, Tc^R	(Hoang et al., 1998)
pZ1918-*lacZ*	Source plasmid of Gm^R cassette	(Schweizer, 1993)
pRK2013	Broad-host-range helper vector; Tra^+, Kan^R	(Ditta et al., 1980)
pAK1900	*E. coli-P. aeruginosa* shuttle cloning vector carrying plac upstream of MCS, Amp^R, Cb^R	(Sharp et al., 1996)
pAK-*znuA*	pAK1900 with a 1.4 kb PCR fragment of gene *znuA*	This study
mini-CTX-*lacZ*	Integration plasmid, Tc^R	(Hoang et al., 2000)
pC*znuA*	miniCTX-lacZ with a 1.4 kb PCR fragment of *znuA* gene	This study
pMS402	Expression reporter plasmid carrying the promoterless *luxCDABE*, Kan^R,Tmp^R	(Bjarnason et al., 2003)
pKD-*znuA*	pMS402 containing *znuA* promoter region; Kan^R,Tmp^R	This study

All antibiotics used were purchased from Amresco (Solon, USA).

To remove metal contaminants, centrifuge tubes, 96-well plates and tips were treated over night with $10\,\mu M$ EDTA and then washed three times with ultra-pure water to eliminate EDTA traces.

2.2 Construction of mutant strains

The mutants PAO1(Δ*znuA*), PAO1(Δ2407-10) and PAO1(Δ*znuA*Δ2407-10) were constructed by allelic exchange as described previously (Hoang et al., 1998). Briefly, a 1403bp fragment of gene *znuA* was PCR amplified using primers 5′-TTAGAATTCCCGACGC TGTTCGCTCCG-3′ (*Eco*RI) and 5′-TATCTGCAGG GCAAATTCGCCTGCCAG-3′ (*Pst*I). The PCR product was digested with *Eco*RI and *Pst*I and then cloned into plasmid pEX18-Tc to generate plasmid pEX18Tc-*znuA*. pEX18Tc- *znuA* was then digested with *Xho*I and *Sal*I to get rid of a 541bp fragment in the amplified *znuA* DNA fragment. The remnant of pEX18Tc-*znuA* was self-ligated to form pEX18Tc-*znuA*-D.

znuA knockout was obtained by means of triparental mating. Briefly, over-night cultures of the donor strain DH10B containing the plasmid pEX18Tc-*znuA*-D, the recipient PAO1, and the helper strain containing pRK2013 were pelleted and resuspended in PBS respectively. The bacteria were mixed in a ratio of 1:1:1 and then spotted onto LB agar plates. After overnight growth at 37°C, the bacteria were scraped and resuspended in 1 mL PBS. The diluted suspensions were spread on PIA plates containing $300\,\mu g/mL$ tetracycline. The clones which had undergone a double crossover were selected on PIA plates containing 10% sucrose. The resultant *znuA* knockout mutant was verified by PCR and designated as PAO1(Δ*znuA*).

Similarly, PAO1(Δ2407-10) and PAO1 (Δ*znuA* Δ2407-10) was constructed in the background of PAO1 and PAO1(Δ*znuA*), respectively.

2.3 Complementation of the znuA knockout mutant

To complement *znuA* mutant, *znuA* gene together with its promoter was PCR amplified and integrated into the *attB* site on the chromosome using the mini-CTX1 system (Hoang et al., 2000). The primers used in the PCR amplification were 5′-TTAGAATTCCCGACG CTGTTCGCTCCG-3′ (*Eco*RI) and 5′-TATGGATCC GGCCATCAGTCACGGGTT-3′ (*Bam*HI). The product was ligated into mini-CTX-*lacZ* to generate pC*znuA* which was transferred into SM10. Transfer of pC*znuA* into PAO1(Δ*znuA*) was carried out by triparental mating. The resultant strain was designated as PAO1(Δ*znuA*)C.

2.4 Overexpression of znuA in PAO1

To overexpress *znuA* in PAO1, the gene *znuA*, together with its promoter, was integrated into pAK1900 (Sharp et al., 1996). The primers used in the PCR amplification were forward primer 5′-TATTCTAGAC CGACGCTGTTCGCTCCG-3′(*Xba*I) and reverse primer 5′-TATGGATCCGGCCATCAGTCACGGGTT- 3′(*Bam*HI). The DNA fragment was ligated into pAK1900 to form pAK-*znuA*. The plasmid was electroporated into PAO1 and the resulting strain was named PAO1(pAK-*znuA*).

2.5 Expression analysis of znuA

pMS402 carrying a promoterless *luxCDABE* reporter gene cluster was used to construct promoter-*luxCDABE* reporter fusion of the *znuA* as reported previously (Duan and Surette, 2007). This reporter fusion was transformed into PAO1 and PAO1(Δzur) The expression of *znuA* was monitored as counts per second (c.p.s.) of light production from the promoterless *luxCDABE* operon downstream of the gene promoter. Overnight culture of PAO1(pKD-*znuA*) was diluted 5% and used as inoculants after an additional 3 h incubation. The cultures were inoculated to parallel wells on a 96-well plate with a transparent bottom after being rediluted 1:20 in fresh LB, supplemented with or without 1 mM or 2 mM EDTA and 0.5 mM, 1 mM and 2 mM ZnSO$_4$. The expression of *znuA* were carried out in fresh M9, supplemented with or without 0.5 mM or 1 mM EDTA and 0.2 mM, 0.5 mM and 1 mM ZnSO$_4$, FeCl$_3$, CuSO$_4$, CdCl$_2$ and MnSO$_4$. The expression of *znuA* in PAO1(Δzur) were monitored similarly. Both absorbance (595 nm) and luminescence (c.p.s.) were measured every 30 min for 24 h in a Victor 3 Multilabel Counter(PerkinElmer, USA).

2.6 Growth analysis

Overnight cultures of PAO1, PAO1($\Delta znuA$), PAO1 ($\Delta 2407$-10) and PAO1($\Delta znuA\Delta 2407$-10) were respectively diluted to an OD$_{595}$ of 0.2 and cultivated for extra 3 h before being used as inoculants. Then the cultures were diluted 1:20 in fresh LB supplemented with or without 1 mM or 2 mM EDTA and 0.5 mM, 1 mM or 2 mM ZnSO$_4$. 300 μL aliquots were inoculated into 96-well plate and incubated at 37°C. Absorbance (595 nm) was measured. The growth of the *znuA* over-expression strain PAO1(pAK-*zunA*) and the control PAO1(pAK1900) was investigated in 500 μg/mL Cb added LB or in Cb added LB supplemented with 1 mM EDTA.

2.7 Adhesion assays of PAO1($\Delta znuA$)

Adhesion assays in polystyrene plates were performed according to the method in reference (Mulcahy et al., 2008). Overnight cultures of PAO1($\Delta znuA$), PAO1($\Delta znuA$)C and PAO1 were adjusted to an OD$_{595}$ of 0.1 with LB supplemented with or without 1 mM ZnSO$_4$ and then 100 μL dilutions were added to the wells of 96-well plates. Each group contained 10 wells and the experiments were repeated five times. The plate was incubated at 37°C for 8 h without shaking. The bacterial cells bound to wall of plate were stained with 0.1% crystal violet for 15 min and then washed several times with water. The attached cells were suspended in 1 mL of 75% ethanol. The absorbance at 595 nm was measured with a Victor 3 Multilabel Counter.

Adhesion ability to human cervical cancer (HeLa) cells was performed according to the report (Gabbianelli et al., 2011). HeLa cells were inoculated in D-MEM at 37°C. PAO1, PAO1($\Delta znuA$) and PAO1($\Delta znuA$)C were grown overnight in LB supplemented with 2 mM EDTA. The cultures were diluted to 10^6 cells/mL. 1 mL of the dilution was added to infect HeLa cells seeded on 24-well plate. After 2 h incubated at 37°C, each well was washed three times with PBS to remove the non-adherent bacteria. Then cold triton X-100 solution (0.5% in PBS) were used to lyse the HeLa cells. Serial dilutions of the cellular lysates were plated on PIA to enumerate adherent bacteria. Each group contained six wells and each experiment repeated three times.

2.8 Virulence assays of PAO1($\Delta znuA$)

The *Drosophila melanogaster* feeding assay was performed as described previously (Shen et al., 2012). PAO1($\Delta znuA$), PAO1($\Delta znuA$)C and PAO1 were cultivated overnight at 37°C in LB and the cultures were adjusted to an OD$_{595}$ of 2.0 with LB. The pellets from 1.5 mL of adjusted cultures were suspended in 100 μL 5% sucrose. The suspensions were spotted onto a sterile filter (Whatman GF/A 21 mm) which was placed on the surface of 5% sucrose agar in the wells of a 24-well plate (Falcon Cat No. 351147). 5% sucrose was used as negative controls. Ten male Canton S flies about 3–5 days old were added to each well of the 24-well plate after they had been starved for 3 h. Carbon dioxide was used to anesthetize flies for the sorting and transferring process. Each feeding group contained ten wells and the experiments were repeated five times. The number of live flies was counted at 24 h intervals.

3 RESULTS

3.1 Characterization of znuA

A ClustalX alignment of the amino acid of ZnuA from PAO1, PA14, PA7, *Pseudomonas protegens*, *Yersinia pestis* CO92, *Campylobacter jejuni* 11168 and *E. coli* K12 was performed (Figure 1). The ZnuA of PAO1 has high identity with the protein from other *Pseudomonas* strains, PA14 (99%), PA7 (97%), *P. protegens* (66%), but showing limited identity with the protein from *E. coli* K12 (36%), *Y. pestis* CO92 (39%), and *C. jejuni* 11168 (25%).

The sequence alignment revealed that ZnuA from PAO1 contains three conserved histidine residues (Figure 1). Those three conserved residues are crucial for zinc binding (Banerjee et al., 2003). Beginning at residue 117, 13 of the 15 amino acid residues are His (H) and acidic residues of Asp (D) or Glu (E) in the ZnuA of PAO1 (Figure 1). This domain has been proved to play an important role on the acquisition of periplasmic zinc in regulation of zinc uptake (Gabbianelli et al., 2011).

3.2 Activated expression of znuA under zinc-restricted conditions

To understand the activity of *znuA* in response to different concentrations of Zn(II), the expression of *znuA* was monitored using pKD-*znuA*.

The results indicated that the addition of 0.5 mM EDTA in LB increased the expression of *znuA* by

Figure 1. Alignment of ZnuA sequences from *P. aeruginosa* PAO1, PA14, PA7, *Pseudomonas protegens*, *E. coli* k12, *Yersinia pestis* CO92, *Campylobacter jejuni* 11168. The conserved zinc-binding histidines are boxed in black. A hyper-variable region rich in histidine and acidic residues Asp (D) or Glu (E), forming a putative charged flexible loop, is indicated by a horizontal bar.

nearly 3 folds. The expression of *znuA* restored to the level in LB when PAO1(pKD-*znuA*) was cultivated in LB supplied with 0.5 mM EDTA and 0.8 mM ZnSO$_4$ was added in presence of 0.5 mM EDTA (Figure 2A) (P > 0.05, Student's *t* test). Similar results were observed in M9 medium (data not shown). The results suggested that *znuA* expression correlates with Zn(II) concentrations in environment and zinc-restriction could activate *znuA* expression.

To evaluate the specificity of the response of *zunA* to metal ions, the effects of FeCl$_3$, CuSO$_4$, CdCl$_2$ and MnSO$_4$ were assessed. The results indicate that the expression of *znuA* was not affected by CdCl$_2$ or FeCl$_3$ but repressed by CuSO$_4$ and MnSO$_4$.

The expression of *zunA* was increased 2–3 folds in M9 supplied with EDTA compared with that in M9 or M9 supplied with EDTA and MnSO$_4$ (P < 0.0001, Student's *t* test). The same effect occurred in the presence of CuSO$_4$ (data not shown).

Zur is a repressor of ZnuABC in *E. coli* and other bacteria (Gaballa and Helmann, 1998, Patzer and Hantke, 1998). To further investigate the regulation of *znuA* by Mn(II) or Cu(II), the expression of *znuA* in PAO1(Δ*zur*) was monitored on the same conditions. The results indicated that the expression of *znuA* in PAO1(Δ*zur*) in LB with EDTA showed no obvious difference compared with that in LB with EDTA and Mn(II) (Figure 2B) (P>0.05, Student's *t* test). The situation for Cu(II) is similar. The results suggested that the regulation of *zunA* by Mn(II) or Cu(II) is dependent on *zur*.

3.3 Effect of znuA on PAO1 growth under zinc-restricted conditions

ZnuABC mutants of *Escherichia coli* O157:H7, *Salmonella enterica* and *Brucella abortu* were

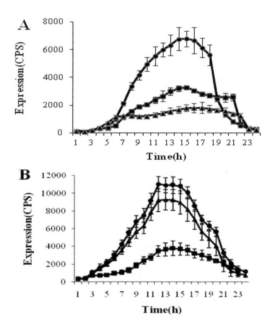

Figure 2. Influence of metal ions on *znuA* expression. Panel A: the expression of *znu*A in LB (squares), LB supplemented with 0.5 mM EDTA (circles) and additional 0.8 mM ZnSO$_4$ (triangles). Panel B: the expression of *znuA* in PAO1(Δ*zur*) in LB (squares), LB supplemented with 0.5 mM EDTA (circles) and additional 0.8 mM MnSO$_4$ (triangles).

observed dramatic growth retardation in zinc-deficient conditions (Gabbianelli et al., 2011, Ammendola et al., 2007, Kim et al., 2004). However, ZnuABC mutants in *Yersina ruckeri* and *E. coli* K-12 did not decrease the growth in zinc-restricted conditions (Sigdel et al., 2006, Dahiya and Stevenson, 2010).

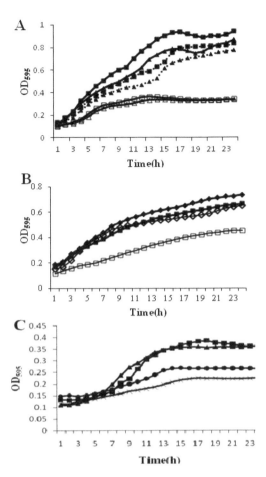

Figure 3. Growth curves. Panel A: growth curves of PAO1 (squares), PAO1(Δ*znuA*) (triangles) in LB medium (close symbols), in LB supplemented with 1 mM EDTA (open symbols) and 1.5 mM ZnSO$_4$ (dotted lines). Panel B: growth curves of wild type PAO1(pAK1900) (squares) and overexpression strain PAO1(pAK-*znuA*) (diamonds) in LB medium (close symbols), or in LB supplemented with 1 mM EDTA (open symbols). Panel C: growth of PAO1 (squares), PAO1(Δ*znuA*) (triangles), PAO1(ΔPA2407-10) (circle) and PAO1(Δ*znuA*ΔPA2407-10) (asterisk) in LB supplemented with 1 mM EDTA.

To reveal the role of *znuA* in zinc uptake in PAO1, growth of PAO1(Δ*znuA*) was compared with PAO1 and PAO1(pAK-*znuA*) in zinc-restricted conditions.

No obvious difference was observed between the growth of PAO1(Δ*znuA*) and PAO1 in both zinc-restricted and zinc-sufficient conditions ($P > 0.05$, Student's t test). However, the growth of these two strains was inhibited to the same extent in LB supplemented with 1 mM EDTA (Figure 3A). Because EDTA is known to chelate other metal ions besides zinc, the growth of PAO1 was measured with zinc added together with EDTA to determine if the growth inhibition was a result of chelation of other metal ions. The growth of both strains were recovered upon the supplementation of 1.5 mM ZnSO$_4$ to the LB with 1 mM EDTA (Figure 3A).The results demonstrated that the chelation of zinc by EDTA resulted in the growth inhibition.

Then we constructed the *zunA* overexpression strain PAO1(pAK-*znuA*) and compared its growth to PAO1(pAK1900) in both zinc-restricted and zinc-sufficient conditions. The results showed that the growth of PAO1 in zinc-restricted condition (LB supplemented with 1 mM EDTA) was increased to the level in zinc-sufficient condition (LB) by over expression of *znuA* on a plasmid pAK1900 (Figure 3B) ($P > 0.05$, Student's t test). Overexpression of *znuA* in PAO1 seems to offset the effect of EDTA. These results indicated that *znuA* in PAO1 is involved in zinc uptake as in other bacteria.

The *znuA* mutant had no marked growth effect compared to PAO1 in zinc-restricted conditions, indicating that there may be other zinc transporter(s) in PAO1. Using BLAST analysis, we found that PA2407, is homologous to *znuA* in PAO1. We constructed PAO1(Δ2407-10), PAO1(Δ*znuA*Δ2407-10). The growth of PAO1(Δ2407-10) decreased significantly compared with PAO1 and PAO1(Δ*znuA*) ($P < 0.01$, Student's t test) and the growth of PAO1(Δ*znuA*Δ2407-10) was slightly lower than PAO1(Δ2407-10) ($P > 0.05$, Student's t test) in zinc-restricted conditions (Figure 3C). These results support that other zinc transporter(s) exist in PAO1.

3.4 Involvement of znuA in adhesion

znuA is involved in the adhesion in *E. coli* O157:H7. Inactivation of *znuA* dramatically decreases its ability to adhere to Caco-2 cells (Gabbianelli et al., 2011).

To verify the role of *znuA* in the adhesion of PAO1, the ability of PAO1, PAO1(Δ*znuA*) and PAO1(Δ*znuA*)C to adhere to polystyrene plates and HeLa cells were analyzed.

Decreased adherence of PAO1(Δ*znuA*) to polystyrene plates was observed (Figure 4A). PAO1(Δ*znuA*) showed significant decrease in adhesion to polystyrene compared with PAO1 ($P < 0.05$, t test). The adhesion of PAO1(Δ*znuA*) was completely restored to PAO1 by *znuA* complementation.

The results from HeLa cells were similar to those from polystyrene plates (Figure 4C). The number of PAO1(Δ*znuA*) adhered to HeLa cells decreased dramatically compared with PAO1 ($P < 0.05$, t test). No difference was observed between PAO1 and PAO1(Δ*znuA*)C ($P > 0.05$, t test). The result indicates that *znuA* is involved in the adhesion of PAO1 to both bio and non-bio surfaces.

To verify whether *znuA* is directly or indirectly involved in adhesion, the ZnSO$_4$ in different amounts was added to the medium in the adhesion assays of polystyrene plates. The results indicated that the adhesion of PAO1(Δ*znuA*) was increased with the addition of zinc compared with the situation without zinc addition. Furthermore, the adhesion was increased in zinc concentration dependent manner

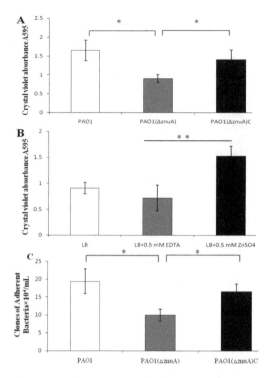

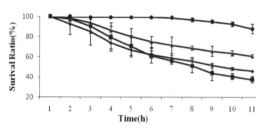

Figure 5. Survival rates of *Drosophila melanogaster* Canton S infected with PAO1 (squares), PAO1(Δ*znuA*) (triangles), PAO1(Δ*znuA*)C (diamonds). 5% sucrose was used as an uninfected control (circles). Each group contained 100 fruit flies and the experiments were repeated five times. Average numbers are shown in data.

Figure 4. Adhesion analysis of PAO1, PAO1(Δ*znuA*) and PAO1(Δ*znuA*) C. Panel A: adhesion of bacteria to polystyrene in LB medium. Panel B: adhesion of PAO1(Δ*znuA*) to polystyrene in LB supplemented with 0.5 mM EDTA or with 0.5 mM ZnSO₄. Panel C: adhesion of bacteria to HeLa epithelial cells. The data are averages of five independent experiments.

as shown in Figure 4B, PAO1(Δ*znuA*) showed significant increase in adhesion to polystyrene in LB supplied with 0.5 mM ZnSO4 compared with that in LB ($P < 0.05$, t test). This suggested that adhesion could probably be affected by ZnuA in a manner that is dependent on its role in zinc uptake.

3.5 Role of znuA in the virulence of PAO1

Our previous work suggested that the altered expression of *znuA* was involved in the increased virulence of a mutant PAO1(Δ2800) (Shen et al., 2012).

To confirm the role of *znuA* in *P. aeruginosa* virulence, a *Drosophila* feeding assay was carried out to compare the killing ability of PAO1(Δ*znuA*) with that of PAO1(Δ*znuA*)C and PAO1. As shown in Figure 5, the survival rate of fruit flies infected with PAO1(Δ*znuA*) was significantly higher than those infected with PAO1 ($P < 0.05$, Log-rank test) or PAO1(Δ*znuA*)C ($P < 0.05$, Log-rank test). The survival rates of flies infected with PAO1(Δ*znuA*)C showed no statistical difference from those infected with PAO1 ($P > 0.05$, Log-rank test). The results confirm that *znuA* plays an important role on the virulence of PAO1.

4 DISCUSSION

znuA is required for zinc transportation in bacteria including *B. abortus* and *S. enteric*. ZnuA contains three conserved histidine residues and a charged flexible loop, rich in histidines and acidic residue*s in Synechocystis* 6803 (Banerjee et al., 2003). T.J. Smith showed that this loop contains low-affinity zinc-binding sites but it does not directly affect the binding of zinc to the high-affinity site within the cleft of three hisdines (Smith and Pakrasi, 2009). It is suggested that this unusual loop region may play a role in the regulation of zinc uptake when the periplasmic concentration of zinc is high. ZnuA in PAO1 contains the three conserved histidines and the loop rich in histidines and acidic residues, indicating that *znuA* in PAO1 play a role in zinc transport. Our previous work suggested that the function of *znuA* is related to divalent cations (Shen et al., 2012).

The expression of *znuA* in PAO1 responded to Zn(II) concentrations. *znuA* expression was activated by 3-fold in zinc-restricted condition. Addition of Zn(II) decreased the expression of *znuA*. We also found that the expression of *znuA* is regulated by Cu(II) and Mn(II) through *zur*, which supports R. Gabbianelli's suggestion that properties of copper atoms is similar to zinc atoms and this could make Cu(II) bind in the zinc binding site of Zur (Gabbianelli et al., 2011). The repression of *znuA* by Zn(II) and Mn(II) is similar to that in *Y, ruckeri* (Dahiya and Stevenson, 2010).

Although no obvious growth difference was observed between PAO1(Δ*znuA*) and PAO1 under zinc-restricted conditions, overexpression of *znuA* in PAO1 facilitated the growth of PAO1 under the zinc-restricted conditions to a comparable level in zinc-sufficient conditions. These results suggest that *znuA* in PAO1 plays a role in zinc transportation or utilization. In terms of its importance for growth under zinc-restricted conditions, *znuA* in PAO1 is different from that in *E. coli* MC4100 (Berducci, 2004.) but similar to *znuA* in *Y. ruckeri* (Sigdel et al., 2006). In *E. coli* MC4100 there is only one high-affinity zinc transporter ZnuABC and the mutation of *znuABC* locus nearly abolished the growth in metal-restricted media. While in *Y. ruckeri*, deletion of *znuA* dose

140

not influence the growth under zinc-restricted conditions. In *E. coli* CFT073, the *zupT* mutant has no obvious growth inhibition, but in the background of *znuA* deletion, *zupT* mutation leads to an increased sensitivity to EDTA (Sabri et al., 2009). We have proved that *znuA* in PAO1 is involved in zinc uptake as in other bacteria and other zinc transporter(s) may exist in PAO1. PA2407, homologous to *znuA*, contains all the three conserved histidines involved in zinc binding according to Sequence alignment Analysis (Brillet et al., 2012). PA2407-10 is classified as a putative Mn(II)/Zn(II) transporter belonging to ABC family based on the TransportDB (Haritha et al., 2008). The mutants of PAO1(ΔznuAΔ2407-10) and PAO1(Δ2407-10) exhibited low level growth compared with PAO1 and PAO1(ΔznuA) in zinc-restricted conditions. These results support the notion that other zinc transporter(s) exists in PAO1 and the specific role of them needed further research.

znuA is involved in the adhesion in *E. coli* O157:H7 (Gabbianelli et al., 2011). We have also found that *znuA* plays a role on the ability of PAO1 to adhere both polystyrene and HeLa cells. The involvement of *znuA* on adhesion in PAO1 depends on its role on zinc uptake according our results.

A few recent reports suggest that several bacterial species require *znuA* for full virulence. The *znuA* knockout mutant in *S. enteric*, *Y. ruckeri*, *E. coli* K-12 and *Brucella abortus* 2308, showed attenuated virulence than wild-type in animal models (Ammendola et al., 2007, Dahiya and Stevenson, 2010, Sigdel et al., 2006, Yang et al., 2006). The *znuA* in PAO1 also plays a key role in the virulence because the survival rate of the flies infected with PAO1(ΔznuA) was significantly higher than those infected with PAO1 or PAO1(ΔznuA)C.

In conclusion, ZunA in PAO1 contains the three conserved histidine residues and the charged loop rich in histidines and acidic residues. The expression level of *znuA* in PAO1 is regulated by zinc, manganese as well as copper and the regulation is dependent on *zur*. *znuA* in PAO1 plays a role in zinc transportation or utilization and there exist other zinc transporter(s) in PAO1. *zunA* plays an important role in adhesion in PAO1 in zinc concentration depending manner and also plays an important role on virulence in PAO1.

ACKNOWLEDGMENTS

This work was supported by PCSIRT (IRT1174), NSFC (81171620, 1210063, 31370165) and the Western Resource Biology and Mordern Biological Technology Llab Open Foundation (ZS11007).

REFERENCES

Ammendola, S., Pasquali, P., Pistoia, C., Petrucci, P., Petrarca, P., Rotilio, G. & Battistoni, A. 2007. High-affinity Zn2+ uptake system ZnuABC is required for bacterial zinc homeostasis in intracellular environments and contributes to the virulence of *Salmonella enterica*. *Infect Immun*, 75, 5867–76.

Banerjee, S., Wei, B., Bhattacharyya-Pakrasi, M., Pakrasi, H. B. & Smith, T. J. 2003. Structural determinants of metal specificity in the zinc transport protein ZnuA from *synechocystis* 6803. *J Mol Biol*, 333, 1061–9.

Berducci, G., Mazzetti, A.P., Rotilio, G. and Battistoni, A. 2004. Periplasmic competition for zinc uptake between the metallochaperone ZnuA and Cu, Zn superoxide dismutase. *FEBS Lett.*, 289–292.

Bjarnason, J., Southward, C.M. & Surette, M.G. 2003. Genomic profiling of iron-responsive genes in *Salmonella enterica* serovar typhimurium by high-throughput screening of a random promoter library. *J Bacteriol*, 185, 4973–82.

Brillet, K., Ruffenach, F., Adams, H., Journet, L., Gasser, V., Hoegy, F., Guillon, L., Hannauer, M., Page, A. & Schalk, I.J. 2012. An ABC transporter with two periplasmic binding proteins involved in iron acquisition in Pseudomonas aeruginosa. *ACS Chem Biol*, 7, 2036–45.

Castelli, S., Stella, L., Petrarca, P., Battistoni, A., Desideri, A. & Falconi, M. 2012. Zinc ion coordination as a modulating factor of the ZnuA histidine-rich loop flexibility: a molecular modelling and fluorescence spectroscopy study. *Biochemical and biophysical research communications*.

Chandra, B.R., Yogavel, M. and Sharma, A. 2007. Structural analysis of ABC-family periplasmic zinc binding protein provides new insights into mechanism of ligand uptake and release. *J. Mol. Biol.*, 970–982.

Choudhury, R. & Srivastava, S. 2001. Zinc resistance mechanisms in bacteria. *Current Science*, 81, 768–775.

Coleman, J.E. 1998. Zinc enzymes. *Current opinion in chemical biology*, 2, 222.

Costello, L.C., Liu, Y., Franklin, R.B. & Kennedy, M.C. 1997. Zinc inhibition of mitochondrial aconitase and its importance in citrate metabolism of prostate epithelial cells. *Journal of Biological Chemistry*, 272, 28875–28881.

Dahiya, I. & Stevenson, R.M. 2010. The ZnuABC operon is important for *Yersinia ruckeri* infections of rainbow trout, Oncorhynchus mykiss (Walbaum). *J Fish Dis*, 33, 331–40.

Ditta, G., Stanfield, S., Corbin, D. & Helinski, D.R. 1980. Broad host range DNA cloning system for gram-negative bacteria: construction of a gene bank of *Rhizobium meliloti*. *Proc Natl Acad Sci USA*, 77, 7347–51.

Duan, K. & Surette, M.G. 2007. Environmental regulation of *Pseudomonas aeruginosa* PAO1 Las and Rhl quorum-sensing systems. *Journal of bacteriology*, 189, 4827–4836.

Ellison, M.L., IIII, J.M.F., Parrish, W., Danell, A.S. & Pesci, E.C. 2013. The Transcriptional regulator Np20 Is the Zinc uptake regulator in *Pseudomonas aeruginosa*. *PloS one*, 8, e75389.

Gaballa, A. & Helmann, J.D. 1998. Identification of a Zinc-Specific Metalloregulatory Protein, Zur, Controlling Zinc Transport Operons in *Bacillus subtilis*. *Journal of bacteriology*, 180, 5815–5821.

Gabbianelli, R., Scotti, R., Ammendola, S., Petrarca, P., Nicolini, L. & Battistoni, A. 2011. Role of ZnuABC and ZinT in *Escherichia coli* O157:H7 zinc acquisition and interaction with epithelial cells. *BMC Microbiol*, 11, 36.

Garau, J. & Gomez, L. 2003. *Pseudomonas aeruginosa* pneumonia. *Current opinion in infectious diseases*, 16, 135–143.

Grass, G., Franke, S., Taudte, N., Nies, D.H., Kucharski, L.M., Maguire, M.E. & Rensing, C. 2005. The metal permease ZupT From *Escherichia coli* is a transporter with a broad substrate spectrum. *Journal of bacteriology*, 187, 1604–1611.

Gunasekera, T.S., Herre, A.H. & Crowder, M.W. 2009. Absence of ZnuABC-mediated zinc uptake affects virulence-associated phenotypes of uropathogenic *Escherichia coli* CFT073 under Zn(II)-depleted conditions. *microblology letters*, 300, 36–41.

Haritha, A., Rodrigue, A. & Mohan, P.M. 2008. A comparative analysis of metal transportomes from metabolically versatile Pseudomonas. *BMC Res Notes*, 1, 88.

Hoang, T.T., Karkhoff-Schweizer, R.R., Kutchma, A.J. & Schweizer, H.P. 1998. A broad-host-range Flp-FRT recombination system for site-specific excision of chromosomally-located DNA sequences: application for isolation of unmarked *Pseudomonas aeruginosa* mutants. *Gene*, 212, 77–86.

Hoang, T.T., Kutchma, A.J., Becher, A. & Schweizer, H.P. 2000. Integration-proficient plasmids for *Pseudomonas aeruginosa*: site-specific integration and use for engineering of reporter and expression strains. *Plasmid*, 43, 59–72.

Holloway, B.W., Romling, U. & Tummler, B. 1994. Genomic mapping of *Pseudomonas aeruginosa* PAO. *Microbiology*, 140 (Pt 11), 2907–29.

Ilari, A., Alaleona, F., Petrarca, P., Battistoni, A. & Chiancone, E. 2011. The X-ray Structure of the Zinc Transporter ZnuA from *Salmonella enterica* Discloses a Unique Triad of Zinc-Coordinating Histidines. *Journal of molecular biology*, 409, 630–641.

Kim, S., Watanabe, K., Shirahata, T. & Watarai, M. 2004. Zinc uptake system (*znuA* locus) of *Brucella abortus* is essential for intracellular survival and virulence in mice. *J Vet Med Sci*, 66, 1059–63.

Mulcahy, H., O'Callaghan, J., O'Grady, E.P., Macia, M.D., Borrell, N., Gomez, C., Casey, P.G., Hill, C., Adams, C., Gahan, C.G., Oliver, A. & O'Gara, F. 2008. *Pseudomonas aeruginosa* RsmA plays an important role during murine infection by influencing colonization, virulence, persistence, and pulmonary inflammation. *Infect Immun*, 76, 632–8.

Patzer, S.I. & Hantke, K. 1998. The ZnuABC high-affinity zinc uptake system and its regulator Zur in *Escherichia coli. Mol Microbiol*, 28, 1199–210.

R. Simon, U.P, A Puhler 1983. A Broad Host Range Mobilization System for In Vivo Genetic Engineering: Transposon Mutagenesis in Gram Negative Bacteria. *Nature Biotechnology* 1, 784–791.

Rensing, C., Mitra, B. & Rosen, B.P. 1997. The *zntA* gene of *Escherichia coli* encodes a Zn(II)-translocating P-type ATPase. *Proc Natl Acad Sci U S A*, 94, 14326–31.

Sabri, M., Houle, S.B. & Dozois, C.M. 2009. Roles of the extraintestinal pathogenic *Escherichia coli* ZnuACB and ZupT zinc transporters during urinary tract infection. *Infection and immunity*, 77, 1155–1164.

Schweizer, H.P. 1993. Two plasmids, X1918 and Z1918, for easy recovery of the *xylE* and *lacZ* reporter genes. *Gene*, 134, 89–91.

Sharp, R., Jansons, I.S., Gertman, E. & Kropinski, A.M. 1996. Genetic and sequence analysis of the cos region of the temperate *Pseudomonas aeruginosa* bacteriophage, D3. *Gene*, 177, 47–53.

Shen, L., Gao, X., Wei, J., Chen, L., Zhao, X., Li, B. & Duan, K. 2012. PA2800 Plays an Important Role in Both Antibiotic Susceptibility and Virulence in *Pseudomonas aeruginosa*. *Current microbiology*, 65, 601–609.

Sigdel, T.K., Easton, J.A. & Crowder, M.W. 2006. Transcriptional response of *Escherichia coli* to TPEN. *Journal of bacteriology*, 188, 6709–6713.

Smith, T.J. & Pakrasi, H.B. 2009. Zinc-Binding Protein ZnuA. *Encyclopedia of Inorganic and Bioinorganic Chemistry*.

Yang, X., Becker, T., Walters, N. & Pascual, D.W. 2006. Deletion of znuA virulence factor attenuates *Brucella abortus* and confers protection against wild-type challenge. *Infect Immun*, 74, 3874–9.

Biomedical Engineering and Environmental Engineering – Chan (Ed.)
© 2015 Taylor & Francis Group, London, ISBN: 978-1-138-02805-0

The objectification of back scraping images of patients with acne, applied to a clinical study

Yifan Zhang, Guangqin Hu & Xinfeng Zhang
Beijing University Of Technology, Beijing, China

ABSTRACT: Purpose: Moving cupping therapy and scraping in Traditional Chinese Medicine (TCM) are therapeutically effective for acne. Colour and distribution of the scraping features on the back of patients after treatment can reflect different conditions of the patients. This paper describes a quantitative study on the scraping features of patients with acne. Methods: Firstly, scraping features including colour and texture of the whole back and related acupoints in the early and late stages of treatment, were extracted. Then, the difference in the scraping features of the whole back and at various acupoints between the earlier stage and later stage of treatment was analysed. Results: Red (R), Green (G), and Blue (B) values and the texture features of the overall scraping features of acne patients in the early stage of treatment were statistically significantly different from those in the late stage of treatment ($p < 0.05$ and $p < 0.01$). Conclusions: Characteristic parameters of colour and texture of scraping images in acne can provide objective evidence for auxiliary clinical diagnosis.

Keywords: Traditional Chinese Medicine; Acne; Scraping feature; Feature extraction

1   INTRODUCTION

Acne is a chronic inflammatory skin disease of the pilosebaceous units with a high incidence[1]. Scraping and moving cupping therapy is a method of Traditional Chinese Medicine (TCM) for treating acne with significant clinical effect[2]. The scraping features of patients after cupping can indicate the degree of related symptoms, thereby serving as the objective evidence for diagnosis and the basis of treatment. Traditional moving cupping therapy is based on clinical observation, and there has been no corresponding objective quantitative research. The scraping features of the back of the acne patients in the early stage and the late stage of moving cupping treatment were collected in this paper. Feature extraction and quantitative comparison were performed, and the performance of a number of parameters was analysed to provide a more scientific basis for the treatment of acne based on syndrome differentiation.

2   SUBJECTS AND METHODS

2.1   *Diagnostic criteria and exclusion criteria of acne*

Diagnostic criteria in TCM[3]: TCM's differential classification was performed according to the "Clinical guideline of new drugs for traditional Chinese medicine (Trial) 2002 edition." The acne was divided into the wind-heat syndrome, mainly including blackheads or whiteheads with red papules, the damp-heat syndrome, mainly featuring oily skin, painful papules, and pustules, the blood stasis and phlegm coagulation syndrome, involving skin lesions of nodules and cysts with a dark colour, as well as recurrent attacks, which are prone to scarring, and Chong and Ren disorders with all kinds of skin damage in female patients suffering from premenstrual exacerbation.

Exclusion criteria[3]: pregnant or lactating women, hyperergic patients, patients with severe primary diseases of the cardiovascular, cerebrovascular, liver, kidney, and hematopoietic systems, mental patients, cases with occupational acne induced by chemical substances, cases with drug-induced acne, and other cases which did not meet the inclusion criteria.

2.2   *Source of cases and collection methods*

All acne patients were confirmed by experts of TCM in the Hospital of the Beijing University of Technology. Thirty cases, 8 males and 22 females, aged 20 to 25 years old, were included in this article. The patients were treated with one course of moving cupping therapy (10 times). Three back scraping images of the early stage and the late stage of treatment were selected from each patient, respectively. A total of 180 images was obtained.

Image acquisition: Image acquisition was performed in the treatment room of the TCM Department of the Hospital of the Beijing University of Technology. The windows were covered with curtains, and an incandescent lamp was used as the light source to avoid the impact of weather on the images. A Canon

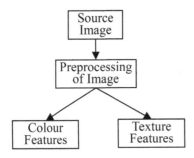

Figure 1. General framework for scraping feature extraction.

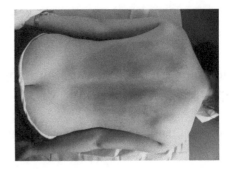

Figure 2. Original images.

PowerShot Pro1 camera was used to acquire the images in flash-off mode, non-macro mode, and automatic focus mode. The back scraping images of acne patients after moving cupping therapy were collected. Special glass jars of size 3 were used as the moving cupping tool. The image acquisition was performed with a vertical distance between the camera and human back of 60 cm to 80 cm. The shooting angle was vertical to the back to avoid image asymmetry.

Image requirements: Patients were requested to lie flat in the bed for receiving the moving cupping therapy, with their heads in the ventilation hole, their arms were relaxed and placed flat on the sides comfortably, and their cervical spines were straight. The back area was not covered by any object, such as the patient's hair and clothes. There were no jewellery or other obvious features on the neck. Dazhui and Changqiang acupoints on the human back were included in the images.

3 EXTRACTION OF SCRAPING FEATURES

A scraping feature refers to phenomena such as different colour, size, and groove depth on the skin after scraping or moving cupping, which reflects different symptoms, duration of the disease, and physical fitness. TCM mainly focuses on the colours and distribution of sand lumps and nodules. Specifically, a serious condition features more eruptions, more sand lumps, and a deeper colour. A mild condition features a lighter colour and less sand lumps. In this paper, R, G, and B values in the colour space[4] were used to characterize the colour of the scraping features. Texture features such as entropy, contrast, and correlation in the Grey Level Co-occurrence Matrix (GLCM)[5] were used to characterize the distribution of the scraping features. The colour and texture features of scraping were compared comprehensively. The block diagram is shown in Figure 1.

3.1 Image preprocessing

The source images are as shown in Figure 2. As shown in Figure 3, the source images were preprocessed by cutting and scale conversion, and transformed to

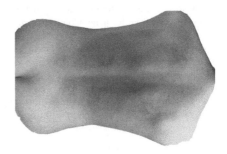

Figure 3. Back scraping images.

600×900 pixel images. The linea vertebralis was used as the horizontal median line to fix the positions of the Dazhui and Changqiang acupoints.

According to the description of each acupoint in TCM theory, Dazhui is located under the spinous process of the seventh cervical vertebra, Feishu is located 1.5 inches aside from under the spinous process of the third thoracic vertebra, Xinshu is located 1.5 inches aside from under the spinous process of the fifth thoracic vertebra, Geshu is located 1.5 inches aside from under the spinous process of the seventh thoracic vertebra, and Changqiang is located between the end of the coccyx and the anus. The specific locations are shown in Figure 4.

Therefore, the pixel point at the left upper corner of the back scraping image was used as the origin to create a coordinate system rightward and downward in the unit of pixel. Thus, the specific pixel locations of the above acupoints were obtained.

3.2 Colour feature extraction

Colour is one of the most important elements of a colour image and the most common bottom-layer feature. It has been widely used in image processing. Since TCM doctors diagnosed the disease condition by observing the colour of the scraping images, the colour features of the scraping images were extracted first. Colour moment[6] is a simple and effective indicator of colour features proposed by Stricker and Orengo. It is divided into first moment (mean), second moment (variance), and third moment (skewness). Because

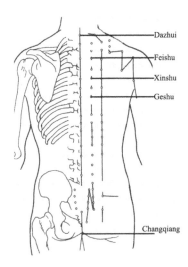

Figure 4. Locations of acupoints on the back.

colour information is mainly distributed in the lower-order moments, the colour moments can effectively indicate the colour distribution in the image. The mathematical definitions of the three colour moments are given as follows:

$$\mu_i = \frac{1}{n}\sum_{j=1}^{n} h_{ij} \qquad (1)$$

$$\sigma_i = \left(\frac{1}{n}\sum_{j=1}^{n}(h_{ij}-\mu_i)^2\right)^{1/2} \qquad (2)$$

$$s_i = \left(\frac{1}{n}\sum_{j=1}^{n}(h_{ij}-\mu_i)^3\right)^{1/3} \qquad (3)$$

where h_{ij} is the pixel value of the j-th pixel of the i-th colour channel component of the colour image; n represents the total number of pixels of the image.

In this paper, the Red Green Blue (RGB) colour space was used to calculate the colour moments of the R, G, and B colour components of the scraping image samples. The first moment reflects the average colour distribution of the scraping image; the second moment reflects the variance of the colour distribution of the scraping image. The significance of the third moment was not considered.

3.3 Extraction of texture features

Texture[5] is another important visual feature of the image. The statistical analysis is performed based on the description of the texture properties, thereby enabling the sparseness and smoothness of the image. The various forms of TCM scraping images are mainly shown as shade, density, or sparseness of the sand lumps. Therefore, GLCM was used in the statistical analysis to describe the texture features of the scraping images. The research on the texture feature extraction of the images has been performed based on GLCM[7][8]. According to the meaning of each parameter, entropy, correlation, and contrast were selected as the representatives to describe the texture features of the scraping images.

Entropy: A random measure indicating the information of the image. When all the elements in the co-occurrence matrix have the maximum randomness, and all the values in the spatial co-occurrence matrix are almost equal, the entropy is relatively large. It shows the degree of non-uniformity and complexity of the texture in the image. See, for example, Equation 4 below.

Correlation: The degree of similarity of GLCM elements in row direction with those in column direction in metric space. When the matrix element values are well-distributed and equal, the correlation value is large. Conversely, if the values vary greatly, the correlation is low. See, for example, Equation 5 below.

Contrast: It reflects the image clarity and the groove depth of texture. A deeper groove indicates greater contrast and more clear vision. Conversely, low contrast means a shallow groove and blurred vision. See, for example, Equation 6 below.

$$ENT = -\sum_i \sum_j p(i,j)\log(p(i,j)) \qquad (4)$$

$$COR = \frac{\sum_i \sum_j ijp(i,j) - \mu_x\mu_y}{\sigma_x\sigma_y} \qquad (5)$$

$$CON = \sum_i \sum_j (i-j)^2 p(i,j) \qquad (6)$$

Here $\mu_x, \mu_y, \sigma_x, \sigma_y$ is defined as:

$$\begin{cases} \mu_x = \sum_{i=0}^{N-1}\sum_{j=0}^{N-1} ip(i,j) \\ \mu_y = \sum_{i=0}^{N-1}\sum_{j=0}^{N-1} jp(i,j) \\ \sigma_x = \sum_{i=0}^{N-1}\sum_{j=0}^{N-1} (i-\mu_x)^2 p(i,j) \\ \sigma_y = \sum_{i=0}^{N-1}\sum_{j=0}^{N-1} (j-\mu_y)^2 p(i,j) \end{cases} \qquad (7)$$

N indicates the gray level of the image, i and j denote the gray values of the pixel, and P(i,j) is the matrix after GLCM normalization.

4 EXPERIMENTAL RESULTS AND ANALYSIS

In the experiments, the Matlab 2010a software was used to extract image colour and texture features. The obtained data were processed using SPSS17.0 software. Results were expressed as mean and standard deviation ($\bar{x} \pm s$). The means of the various groups were compared using t tests; $p < 0.05$ indicated a statistically significant difference.

Experiment 1. The back scraping images of 155×290 pixels showing seven acupoints namely, Dazhui, left Feishu, right Feishu, left Xinshu, right

145

Table 1. Comparison of colour features of the back scraping images ($\bar{x} \pm s$).

Groups n	Early stages of treatment 90	Late stages of treatment 90
R mean	188.59 ± 16.92	145.53 ± 18.65**
G mean	129.16 ± 14.43	96.74 ± 71.88**
B mean	112.80 ± 17.55	83.51 ± 13.71**
R variance	6.98 ± 0.39	8.07 ± 0.78**
G variance	7.528 ± 0.68	7.69 ± 0.43
B variance	7.57 ± 057	7.36 ± 0.42*

Comparison of colour features of back scraping images in the early stage and late stage of treatment,**p < 0.01, *p < 0.05

Table 2. Comparison of texture features of the back scraping images ($\bar{x} \pm s$).

Groups n	Early stages of treatment 90	Late stages of treatment 90
Entropy	1.98 ± 0.18	1.46 ± 0.19**
Correlation	0.74 ± 0.19	1.76 ± 0.68**
Contrast	2.16 ± 0.19	1.59 ± 0.21**

Comparison of colour features of back scraping images in the early stage and late stage of treatment, **p < 0.01.

Table 3. Comparison of colour features of scraping images near Dazhui ($\bar{x} \pm s$).

Groups n	Early stages of treatment 90	Late stages of treatment 90
R mean	184.13 ± 16.76	124.56 ± 14.54**
G mean	122.92 ± 12.87	78.81 ± 12.46**
B mean	109.16 ± 14.30	68.14 ± 12.78**
R variance	4.35 ± 1.10	3.39 ± 1.31*
G variance	5.06 ± 1.22	3.25 ± 0.94**
B variance	4.93 ± 1.16	2.71 ± 0.75**

Comparison of colour features of back scraping images in the early stage and late stage of treatment, **p < 0.01, *p < 0.05.

Table 4. Comparison of colour features of scraping images near Feishu ($\bar{x} \pm s$).

Groups n	Early stages of treatment 90	Late stages of treatment 90
R mean	189.72 ± 9.34	145.42 ± 7.66**
G mean	129.30 ± 9.38	95.59 ± 10.45**
B mean	113.54 ± 11.13	82.41 ± 15.90**
R variance	3.2 6 ± 0.81	2.83 ± 0.86*
G variance	3.83 ± 0.89	2.76 ± 0.63**
B variance	3.89 ± 0.97	2.53 ± 0.76**

Comparison of colour features of back scraping images in the early stage and late stage of treatment, **p < 0.01, *p < 0.05.

Xinshu, left Geshu and right Geshu were selected. Table 1 shows the colour contrasts of the scraping images in the early stage of treatment compared with those in the late stage of treatment. The colour contrast data showed that the R, G, and B colour characteristics of the scraping images of acne patients in the early stage of moving cupping treatment was statistically significantly different from those in the late stage (p < 0.01 and p < 0.05). The quantized values objectively indicated the difference.

Experiment 2. The same back scraping images of Experiment 1 were selected to compare the textures of scraping images in the early and late stages of treatment. The results are shown in Table 2.

Texture contrast data showed that the texture features in terms of entropy, correlation, and contrast of the back scraping images of the patients with acne in the early stage of treatment, were highly statistically significantly different from those in the late stage of treatment (p < 0.01).

Experiment 3. Scraping images of 20 × 20 pixels near Dazhui, left and right Feishu, left and right Xinshu, and left and right Geshu were selected, respectively. As shown in Tables 3 to 6, the colour features of the acupoints were compared.

The colour contrast data of the acupoints showed that the R, G, and B values of all selected acupoints, especially Dazhui and Feishu, in the early stage of treatment, were statistically significantly different

Table 5. Comparison of colour features of scraping images near Xinshu ($\bar{x} \pm s$).

Groups n	Early stages of treatment 90	Late stages of treatment 90
R mean	185.50 ± 6.73	159.85 ± 9.15**
G mean	124.35 ± 5.99	107.20 ± 8.40**
B mean	106.69 ± 7.50	92.31 ± 9.75*
R variance	3.09 ± 0.77	2.48 ± 0.92*
G variance	3.72 ± 0.90	3.01 ± 0.85*
B variance	3.83 ± 0.87	2.74 ± 0.54*

Comparison of colour features of back scraping images in the early stage and late stage of treatment, *p < 0.05.

from those in the late stage of treatment (p < 0.01 and p < 0.05).

Experiment 4. As shown in Tables 7 to 10, the same scraping images of acupoints from Experiment 3 were selected, and the texture features of various acupoints were compared.

The texture contrast data of the acupoints showed that the texture features of the acupoint scraping images in the early stage were statistically different from those in the late stage of treatment (p < 0.01 and p < 0.05).

146

Table 6. Comparison of colour features of scraping images near Geshu ($\bar{x} \pm s$).

Groups n	Early stages of treatment 90	Late stages of treatment 90
R mean	178.44 ± 8.58	172.01 ± 7.70*
G mean	120.54 ± 8.60	121.35 ± 11.62
B mean	99.16 ± 9.15	104.42 ± 9.70
R variance	2.88 ± 0.75	2.56 ± 0.79
G variance	3.08 ± 0.85	3.01 ± 0.85*
B variance	3.07 ± 0.81	2.93 ± 0.77*

Comparison of colour features of back scraping images in the early stage and late stage of treatment, *p < 0.05.

Table 7. Table 7. Comparison of texture features of Dazhui scraping images ($\bar{x} \pm s$).

Groups n	Early stages of treatment 90	Late stages of treatment 90
Entropy	1.04 ± 0.23	0.78 ± 0.14*
Correlation	2.65 ± 0.73	4.69 ± 0.91**
Contrast	1.22 ± 0.23	0.90 ± 0.12*

Comparison of texture features of back scraping images in the early stage and late stage of treatment, *p < 0.05, **p < 0.01.

Table 8. Comparison of texture features of Feishu scraping images ($\bar{x} \pm s$).

Groups n	Early stages of treatment 90	Late stages of treatment 90
Entropy	0.85 ± 0.23	0.75 ± 0.18
Correlation	4.08 ± 0.76	5.73 ± 0.99*
Contrast	1.01 ± 0.32	0.88 ± 0.12*

Comparison of texture features of back scraping images in the early stage and late stage of treatment, *p < 0.05.

Table 9. Comparison of texture features of Xinshu scraping images ($\bar{x} \pm s$).

Groups n	Early stages of treatment 90	Late stages of treatment 90
Entropy	0.86 ± 0.22	0.86 ± 0.18
Correlation	3.48 ± 0.72	3.75 ± 0.69
Contrast	1.22 ± 0.30	1.02 ± 0.15*

Comparison of texture features of back scraping images in the early stage and late stage of treatment, *p < 0.05.

5 DISCUSSION

In TCM practice, diagnosis and differentiation of diseases were performed based on comprehensive

Table 10. Comparison of texture features of Geshu scraping images ($\bar{x} \pm s$).

Groups n	Early stages of treatment 90	Late stages of treatment 90
Entropy	0.91 ± 0.17	0.60 ± 0.12**
Correlation	4.08 ± 0.82	9.01 ± 1.98**
Contrast	1.07 ± 0.21	0.71 ± 0.14**

Comparison of texture features of back scraping images in the early stage and late stage of treatment, **p < 0.01.

analysis with four examination methods – watching, smelling, asking, and sphygmopalpation. Observations on the colour and texture distribution of the scraping images help to determine the condition and duration of the disease. Currently, all research on scraping images by TCM experts was performed based on clinical observations, and no objective study on the corresponding features have been carried out. In this paper, the scraping images of acne patients were extracted and quantitatively analysed to provide an objective basis for the treatment of acne, based on syndrome differentiation.

According to TCM clinical experience, the disease condition and physical fitness of patients are reflected by the pattern of scraping images of the corresponding acupoints. The overall trend is that the darkest colour and most sand lumps are found in Dazhui and Feishu acupoints, followed by Xinshu and Geshu acupoints. In this article, the colour and texture features of the scraping images of various acupoints in the early stage of treatment were compared with those in the late stage of treatment. The results showed that the difference was statistically significant (p < 0.01 and p < 0.05). Dazhui and Feishu acupoints underwent the most significant changes after moving cupping therapy.

ACKNOWLEDGEMENT

This work was supported by the National 973 Project of 2010: Multidimensional health status identification of traditional Chinese medicine theory and method research. No: 2011CB505406.

REFERENCES

[1] Fabbrocini G, Staibano S, De Rosa G, et al. Resveratrol-Containing Gel for the Treatment of Acne Vulgaris[J]. American journal of clinical dermatology, 2011, 12(2): 133–141.
[2] Wang Q, Wang G. Therapeutic effect observation on treatment of acne with acupuncture plus moving cupping and blood-letting [J]. Journal of Acupuncture and Tuina Science, 2008, 6: 212–214.
[3] Zheng XY. Clinical guideline of new drugs for traditional Chinese medicine. Beijing: Medicine Science and Technology Press of China. 2002: 292–295. Chinese

[4] Naoki T, Seo M, Igarashi T, et al. Optimal color space for quantitative analysis of shinny skin[C]//Biomedical Engineering and Informatics (BMEI), 2013 6th International Conference on. IEEE, 2013: 836–840.

[5] Haralick R M, Shanmugam K, Dinstein I H. Textural features for image classification [J]. Systems, Man and Cybernetics, IEEE Transactions on, 1973 (6): 610–621.

[6] Stricker M A, Orengo M. Similarity of color images[C] //IS&T/SPIE's Symposium on Electronic Imaging: Science & Technology. International Society for Optics and Photonics, 1995: 381–392.

[7] Xiaoyi Song, Yongjie Li, Wufan Chen, A Textural Feature Based Image Retrieval Algorithm, in Proc. of 4th International conference on Natural Computation, Oct. 2008.

[8] Song X, Li Y, Chen W. A Textural Feature-Based Image Retrieval Algorithm[C]//Natural Computation, 2008. ICNC'08. Fourth International Conference on. IEEE, 2008, 4: 71–75.

Biomedical Engineering and Environmental Engineering – Chan (Ed.)
© 2015 Taylor & Francis Group, London, ISBN: 978-1-138-02805-0

Screening and preliminary identification of a new type of lactic acid bacterium producing a broad spectrum active antibacterial ingredient

Caiwen Dong, Mengmeng Duan, Shuaihong Li, Chunsheng Chen & Yongfang Zhou
College of Food and Biological Engineering, Zhengzhou University of Light Industry, Zhengzhou, China

ABSTRACT: One unknown lactic acid bacterium was isolated from fresh meat stored at a low temperature. The supernatant of this strain could inhibit the growth of the indicator strain strongly after excluding hydrogen peroxide and organic acid. The inhibitory activity had no obvious change after treatment with pepsin and trypsin, which indicated that this inhibitory material could not obtain the feature of protein. The test for bacterial inhibition spectrum proved that the supernatant of this strain could inhibit not only gram-positive bacteria but also gram-negative bacteria and yeast. Therefore, this strain produced a bacteriocin with a wide inhibition spectrum. Through the detection of its appearance, and physiological and biochemical characteristics, it was not in accordance with known *Lactobacillus*, and it could be a new type of *Lactobacillus*.

Keywords: Lactic acid bacterium; Antibacterial ingredient; Screening

1   INTRODUCTION

Bacteriocins from Lactic Acid Bacteria (LAB) are kinds of natural and safe antimicrobial peptides or proteins and attracted more interest than those from other resources (Cleveland, 2001). There are now four groups of bacteriocins produced by LAB according to their biochemical characteristics (Klaenhammer, 1993). Class I bacteriocins (lantibiotics) includes small and heat-stable peptides containing thioether amino acids, and Class II bacteriocins are small, heat-stable, non-lantibiotic peptides. Some large, heat-stable proteins are classified as Class III bacteriocins and Class IV are large, complex bacteriocins which contain lipid or carbohydrate groups. Among these, Class I and Class II bacteriocins are more suitable to be used as bio-preservatives in the food industry as substitutes for some chemical preservation. However, presently industrialized bacteriocins exhibit certain limitations which limit their application range. At the present, research is focused on finding new sources of lactic acid bacteria which can produce an antibacterial ingredient with a broad antibacterial spectrum.

In this paper, we report on the screening and preliminary identification of new types of lactic acid bacterial-producing novel bacteriocins with a broad antimicrobial spectrum in meat.

2   MATERIALS AND METHODS

2.1   *Bacterial strains and preliminary screening*

Thirty-six strains of lactic acid bacteria (LAB), isolated from fresh meat during storage and identified by phenotypic tests, were used in this study (Dong CW,

2009). In this work, one strain of lactic acid bacteria from meat was first determined by physiological tests.

For the test for an antibacterial ingredient, the selected strain was inoculated in tubes containing 10 ml MRS (de Man, Rogosa and Sharp) broth each and statically incubated at 37°C for 18 h. Then 1% (v/v) of those cultures were inoculated in 100 ml MRS broth individually and incubated at 37°C for 24 h without agitation. Bacterial cells were removed by centrifugation (6000 g, 10 min, 4°C) and supernatants were examined by the diameters of the inhibition zones using the agar diffusion assay method with *Escherichia coli* (ATCC 53323) as indicator strains (Ennahar, 2000). Briefly, 200 μl of cell-free supernatants were placed into wells (8.00 mm in diameter) on MRS agar plates seeded with the above indicator strain. After incubation at 37°C for 12 h, the diameters of the inhibitory zones were determined.

2.2   *Eliminating the effect of low pH*

After the first screening, another screening was done according to the next step. The pH value of the fermented supernatants was adjusted to 6.0 with NaOH (5 mol /l) to eliminate the effect of low pH value, and the pH of the MRS broth was also adjusted to the same value using either lactic acid or acetic acid as a control, respectively. Then, supernatants were examined by the diameters of the inhibition zones according to the above method.

2.3   *Eliminating the effect of hydrogen peroxide*

In order to eliminate the effect of hydrogen peroxide, after adding a catalase (50 mg/ml, Sigma), the cell-free supernatants (pH 6.0) were incubated at 37°C

Table 1.　Inhibition zones of the supernatant before and after pH adjustment.

Specimen	Inhibition zone (mm)
Supernatant	33.5 ± 0.4
Supernatant (pH 6.0)	30.6 ± 0.3
Lactic acid (pH 6.0)	8.0 ± 0.1
Acetic acid (pH 6.0)	8.0 ± 0.1

Table 2.　Inhibition zones of the supernatant before and after hydrogen peroxide elimination.

Specimen	Inhibition zone (mm)
Supernatant (pH 6.0)	30.5 ± 0.3
Supernatant (catalase treatment)	29.4 ± 0.2
Activity decreasing rate	3.6%

for 2 h to eliminate the effect of hydrogen peroxide, and the same supernatants without a catalase were used as a control. The diameters of the inhibition zones were measured again.

2.4 Effect of proteolytic enzymes on the antibacterial ingredient

To determine the possible nature of the antibacterial ingredient, active isolates were grown in MRS broth at 37°C for 24 h, the cells were harvested by centrifugation (6000 g, 10 min, 4°C), and the pH of the cell-free supernatant was adjusted to 6.0 with NaOH. Aliquots of these samples were treated with the enzymes trypsin and pepsin (1 mg/ml) and incubated for 2 h at 37°C. Antimicrobial activity was monitored by using the agar diffusion assay method as described above.

2.5 Antimicrobial spectrum

The indicatory LAB was statically incubated in MRS broth at 37°C for 24 h. All the other indicatory bacteria were cultivated overnight in the nutrient broth at 37°C, with gentle agitation for 12 h. After eliminating hydrogen peroxide and low pH effects, the bacterial cell-free supernatants (pH 6.0) were obtained through centrifugation (6000 g, 10 min, 4°C). The diameters of the inhibition zones were recorded using Staphylococcus aureus (ATCC 29740), Escherichia coli (ATCC 53323), and Saccharomyces cerevisiae (ACCC 20036) as indicators.

2.6 Preliminary identification of the indicatory LAB

Preliminary identification was first determined by phenotypical and physiological tests including Gram staining, cell morphology, and carbohydrate fermentation patterns.

3 RESULTS

3.1 Antibacterial effect after eliminating the effect of low pH

In Table 1, the values were the mean of three independent experiments carried out in duplicate; the diameter of the inhibition zone including that of the Oxford cup, was 8.0 mm. After eliminating the effect of low pH, the supernatant obtained good antibacterial effect.

Table 3.　Inhibition zones of the supernatant after proteolytic enzymes treatment.

Specimen	Inhibition Zone (mm)	Activity Decreasing rate
Supernatant (pH 6.0)	30.4 ± 0.3	0
Supernatant (pepsin treatment)	30.0 ± 0.2	1.6%
Supernatant (trypsin treatment)	29.8 ± 0.2	2.3%

Table 4.　Inhibition zones of the supernatant against different indicatory bacterial strain.

Indicatory bacterial strain	Inhibition zone (mm)
Escherichia coli	30.3 ± 0.4
Staphylococcus aureus	16.5 ± 0.2
Saccharomyces cerevisiae	15.0 ± 0.2

3.2 Antibacterial effect after eliminating the effect of hydrogen peroxide

In Table 2, the values were the mean of three independent experiments carried out in duplicate; the diameter of the inhibition zone, including that of the Oxford cup, was 8.0 mm. It showed that there is little difference in the inhibition zone of the supernatant before and after catalase treatment. The activity decreasing rate is only 3.6%.

3.3 Antibacterial effect after proteolytic enzymes treatment

As shown in Table 3, after treatment by proteolytic enzymes, the antibacterial activity of the supernatant decreased slightly. This result showed that the antibacterial ingredient was not a type of peptide.

3.4 Antimicrobial spectrum

The antibacterial ingredient exhibited a broad antimicrobial activity (Table 4). It could significantly inhibit Escherichia coli, though less so against Staphylococcus aureus and Saccharomyces cerevisiae.

Table 5. Carbohydrate fermentation results of the strain.

Item	Result	Item	Result
Arabinose	−	Mannitose	+
Cellobiose	+	Melezitose	−
Esculin	+	Melibiose	+
Fructose	+	Raffinose pentahydrate	−
Galactose	+	Rhamnose	−
Glucose	+	Ribose	+
Sodium gluconate	−	Sorbitol	−
Lactose	+	Sacrose	+
Maltose	+	Fucose	+
Mannitol	−	Xylose	−

3.5 Strain identification tables

Cells of this strain were Gram-positive, rod-shaped, and without spore formation. They could grow at 10°C and 15°C but not at 45°C. The carbohydrate fermentation results are shown in Table 5 (+ (positive), − (negative)).

4 CONCLUSIONS

The antibacterial ingredient produced by this strain is not a peptide. Further separation and purification is needed to study its property. The phenotypical and physiological characteristics of this strain are also not in accordance with any known lactic acid bacterial strain. So, it is possible that it belongs to a new kind of lactic acid bacterial strain. Other method of identification is needed for its final identification.

ACKNOWLEDGEMENT

This research was funded by the Doctorate Research Funding of Zhengzhou University of Light Industry (000393).

REFERENCES

Cleveland, J., Montville, T. J., Nes, I. F., et al. 2001. Bacteriocins: Safe, natural antimicrobials for food preservation. International Journal of Food Microbiology, 71: 1–20.

Dong CW, Bai YH, Cao GS, et al. 2009. Isolation and identification of corrupted lactic acid bacteria from fresh meat during storage in low temperature. Food research and development, 30(7): 117–120.

Ennahar, S., Sashihara, T., Sonomoto, K., et al 2000. Class IIa bacteriocins: Biosynthesis, structure and activity. FEMS Microbiology Reviews, 24: 85–106.

Klaenhammer, T. R. 1993. Genetics of bacteriocins produced by lactic acid bacteria. FEMS Microbiology Reviews, 12: 39–85.

Biomedical Engineering and Environmental Engineering – Chan (Ed.)
© 2015 Taylor & Francis Group, London, ISBN: 978-1-138-02805-0

The effect on fermentation products by engineering *Saccharomyces Cerevisiae* GY-1, which can express an α-acetolactate decarboxylase gene

Fang-Yi Pei, Shou-Feng Huang, Yuan Gao & Jing-Ping Ge
Key Laboratory of Microbiology, College of Life Science, Heilongjiang University, Harbin, China

ABSTRACT: α-Acetolactate decarboxylase has been widely applied in the beer brewing industry due to its influence on flavour compounds produced in the brewing process, which can significantly affect the taste of the beer. In this paper, comparison of six flavour compounds (acetaldehyde, ethyl formate, ethyl acetate, n-propanol, iso-butanol, and isoamyl acetate) in the fermentation broth between the original Saccharomyces cerevisiae HDY-01 and Saccharomyces cerevisiae GY-1 (an engineered strain which is transformed by an α-acetolactate decarboxylase gene) was done to analyse the effect on fermentation products of the α-acetolactate decarboxylase gene expression. The results illustrated that the transformed GY-1 strain showed a faster decreasing rate of the sugar than the HDY-01 strain, and throughout the fermentation process, the six flavour substances in the transformed strain were in accordance with the standards of beer production, and were lower than the original strain as well. The highest contents of the six flavour compounds of the transformed strain GY-1 strain were 3.45747 ppm (acetaldehyde), 2.85376 ppm (n-propanol), 1.89048 ppm (iso-butanol), 0.91249 ppm (ethyl formate), 1.35292 ppm (ethyl acetate), and 1.19908 ppm (isoamyl acetate), respectively.

Keywords: α-Acetolactate decarboxylase, Beer brewing, Flavour compounds, Fermentation products

1 INTRODUCTION

α-Acetolactate decarboxylase (α-ALDC, EC.4.1.1.5, 33 kD) can quickly decompose α-acetolactate to acetoin, while α-acetolactate is the precursor of a few flavour substances in the beer production process[1–3], thereby the contents of the flavour substances will be rapidly decreased, which helps to promote beer to mature, shortens the fermentation cycle, and the beer will taste better.

The taste of beer can be affected by different kinds of factors, most of which are flavour substances[4–5]. Reduction of alcohol will lead to formation of aldehydes in the beer fermentation process, and oxidation of most unsaturated aldehydes will age the beer flavour and produce a variety of bad tastes. Acetaldehyde has an important influence on beer flavour and quality under the flavour threshold of 10 ppm[6–8], thus the contents of acetaldehyde in mature beer should be less than 10 ppm, while the contents in high-quality beer should be less than 6 ppm[9]. Otherwise, an unpleasant rough bitterness with a spicy grassy rot will be produced if the content is too high.

Esters are significant influence factors as well[10]. The total amount of esters in beer are normally in the region of 15~50 ppm[11–12] and the fruity esters can enhance the flavour of beer. In general, the higher the esters' flavour index, the more effective the flavour composition is in the beer. Isoamyl acetate is the most important component of esters, followed by ethyl acetate, hexane, ethyl acetate, and ethyl benzene respectively[13–14]. Thus, moderate amounts of esters should be obtained in the beer to ensure the fragrance is pleasant, otherwise an unpleasant scent will be formed if the contents excess the threshold. Therefore, it is of great importance to control the concentration of flavour compounds within a standard scope in the practical beer brewing industry.

A genetically engineered strain, *Saccharomyces cerevisiae* GY-1, was obtained by transforming the exogenetic acetolactate decarboxylase gene into *Saccharomyces cerevisiae* HDY-01, which can catalytically transfer α-acetolactatethe to acetoin, decreasing the concentration of aldehydes and eaters that are not necessary to much in beer[15–16]. Results obtained from our earlier experiments demonstrated that exogenetic acetolactate decarboxylase gene was stably expressed in *Saccharomyces cerevisiae* GY-1, and the α-acetolactatethe content was lower than that before constructed (data was not shown here), which was strong evidence that the gene was transformed successfully.

In this paper, cell numbers, the degree of sugar, and the contents of six flavour compounds were determined by Gas Chromatography (GC) in the fermentation broth by the *Saccharomyces cerevisiae* HDY-01 strain and *Saccharomyces cerevisiae* GY-1. According to the results, the influence of the α-acetolactate decarboxylase gene expression on the fermentation products of beer was determined, which will lay the foundation for the detection of beer fermentation and provide theoretical support for controlling a good flavour of the beer.

2 MATERIALS AND METHODS

2.1 Microorganisms

Saccharomyces cerevisiae HDY-01 was kept in storage by the Key Laboratory of Microbiology of Heilongjiang University, China. *Saccharomyces cerevisiae* GY-1 is a genetically engineered yeast strain harbouring an α-acetolactate decarboxylase gene. It was also constructed and preserved by the Key Laboratory of Microbiology of Heilongjiang University, China.

2.2 Medium and cultivation

YEPD medium was used for yeast culture, which contained: 1% yeast extract, 2% peptone, and 2% glucose. Wort medium was used for yeast fermentation, which contained barley and water at the ratio of 1:4(v/v), then kept in 60°C water for 4 h and filtered 2 times through gauze. In general, the cultivation was maintained at 30°C at a shaking speed of 140 rpm.

2.3 Cell count in yeast suspension

The HDY-01 strain and the GY-1 strain were cultured on the YEPD medium at 30°C under agitation at 140 rpm for 36 h, and subsequently inoculated into 700 mL of wort medium at 5% (v/v). The fermentation process was static cultivated at 13°C and sealed with parafilm in strictly anaerobic conditions. Samples were taken every other day, the fermentation suspension was measured by a blood counting chamber, and the number of suspended yeast cells were calculated.

2.4 Sugar degree measurement

The fermentation process was carried out with identical conditions as above. Samples were taken every other day, and a saccharometer (WYT-10-32%) was applied to determine the sugar degree (%) in the fermentation broth.

2.5 Detection of six flavour compounds

Samples were taken every other day in the fermentation process, and samples from the HDY-01 strain were utilized as the control. Acetaldehyde, ethyl formate, ethyl acetate, n-propanol, iso-butanol, and isoamyl acetate in the fermentation broth were measured by gas chromatography (GC) equipped with an Agilent Us5337511H capillary column (60*0.53ID, 1.5 μm). Nitrogen, hydrogen, and air were used as the carrier gases, and the flow rate was 30 mL/min, 40 mL/min, and 400 mL/min, respectively. Using a split injection with split ratio of 6:1, the column temperature was controlled by the following gradient program: start at 40°C for 3 min; increase at a rate of 5°C/min; isotherm at 90°C for 1 min.

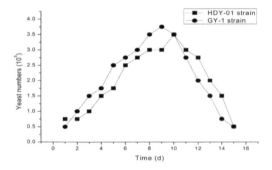

Figure 1. The yeast numbers of HDY-01 and GY-1 strains.

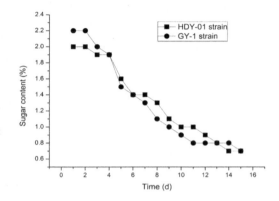

Figure 2. The sugar degree change in the fermentation liquid.

3 RESULTS AND DISCUSSION

3.1 Cell count in yeast suspension

The number of cells in the suspension of the HDY-01 strain and the GY-1 strain demonstrated trends of increasing at first and subsequently decreasing during the fermentation process (Figure 1); the maximum number in the HDY-01 strain and the GY-1 strain were 3.5×10^5/mL and 3.75×10^5/mL respectively. The number of cells in the suspension of the GY-1 strain accumulated and decreased faster than the HDY-01 strain, indicating that the GY-1 strain could accumulate more biomass than the HDY-01 strain. Thus the GY-1 strain is more suitable for beer production.

3.2 Sugar degree measurement

The sugar degree of the GY-1 strain exceeded 0.2% more than the HDY-01 strain during the fermentation time (Figure 2), but the trend was not significantly different between the two strains. From 1 to 4 days, the GY-1 strain displayed a higher sugar degree than the HDY-01 strain, while in the next 9 days, the GY-1 strain sugar degree was lower than that of the HDY-01 strain. Therefore, the sugar content of the GY-1 strain visibly decreased faster than that of the HDY-01 strain, which indicates that the GY-1 strain is more suitable for industrial beer production.

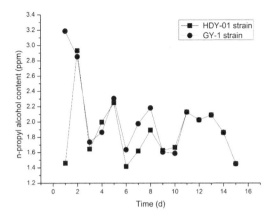

Figure 3. The change of acetaldehyde content.

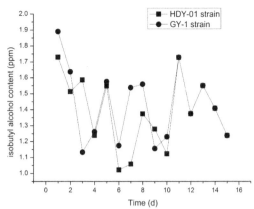

Figure 5. The change of isobutyl alcohol content.

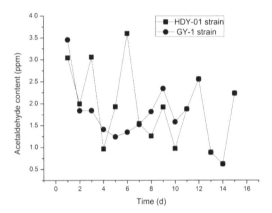

Figure 4. The change of n-propyl content.

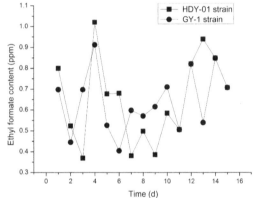

Figure 6. The change of ethyl formate content.

3.3 Detection of six flavour compounds

The content of acetaldehyde in the HDY-01 strain and the GY-1 strain was all less than 10 ppm (Figure 3). The acetaldehyde contents of the GY-1 strain was lower than that of the HDY-01 strain in the first 7 days, but it was contrary after the seventh day. Comparison of acetaldehyde contents throughout the fermentation process by the GY-1 strain and HDY-01 strain illustrated that the maximal values were 3.6009 ppm and 3.45747 ppm, respectively.

Figure 4 suggests that the HDY-01 strain and the GY-1 strain contained an almost identical concentration of n-propanol throughout the fermentation process. The maximal n-propanol amounts of the GY-1 strain and the HDY-01 strain were 2.9321 ppm and 2.85376 ppm, respectively.

The HDY-01 strain contained more iso-butanol than the GY-1 strain in the fermentation broth during the first 11 days (Figure 5). In addition, the results were opposite in the following fermentation process. The maximal iso-butanol concentration of the HDY-01 strain was 1.73 ppm, and the minimal value was 1.02209 ppm, while the maximal and minimal iso-butanol concentrations of the GY-1 strain were 1.89048 ppm and 1.1329 ppm, respectively.

As can be seen in Figure 6, the ethyl formate contents in the initial stage of the fermentation broth of the HDY-01 strain was slightly higher than that of the GY-1 strain. In the following fermentation process, both strains fluctuated from higher to lower contents of ethyl formate after 5 days of fermentation, and the contents were almost identical. The maximal values of ethyl formate in the HDY-01 strain and the GY-1 strain were 1.35292 ppm and 0.91249 ppm, respectively.

From Figure 7 we find that the variation trends of the ethyl acetate contents in the HDY-01 strain and in the GY-1 strain were very similar. However, from 3 to 7 days during the fermentation, the ethyl acetate contents of the GY-1 strain was a little higher than that of HDY-01 strain. From 7 to 9 days, the ethyl acetate contents took on the opposite trends. The maximal value of ethyl acetate was identical.

The changes in the isoamyl acetate contents in the GY-1 strain and the HDY-01 strain are shown in Figure 8, which implies that in the earlier stages, during 1 to 6 days, the HDY-01 strain produced lower levels of isoamyl acetate than that of the GY-1 strain, whilst the contents tended to be similar from 12 to 15 days. Finally, the GY-1 strain produced lower contents of isoamyl acetate than that of the HDY-01 strain.

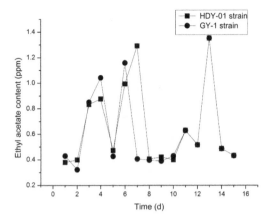

Figure 7. The change of ethyl acetate content.

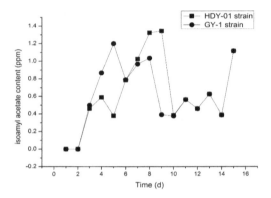

Figure 8. The change of isoamyl acetate content.

Furthermore, the GY-1 strain obtained 1.1991 ppm of isoamyl acetate at the highest point while the HDY-01 strain produced 1.3423 ppm of isoamyl acetate.

4 CONCLUSIONS

The results demonstrated the feasibility of using *Saccharomyces cerevisiae* GY-1 for the beer fermentation, not only for the good taste it produces, but also for shortening the fermentation period. Acetaldehyde, ethyl formate, ethyl acetate, n-propanol, iso-butanol, and isoamyl acetate were the major flavour substances under lower levels of the GY-1 strain, thus the beer could be made to taste better. This significantly reduced the production costs, and laid the foundation for the detection of beer fermentation and for providing theoretical support for controlling its good flavour.

ACKNOWLEDGEMENTS

The research was supported by The National Natural Science Foundation of China (Grant No. 31270534), The National Natural Science Foundation of China (Grant No. 31270143), The National Natural Science Foundation of China (Grant No. 31470537), and the National Science Foundation for Distinguished Young Scholars of China (31300355).

REFERENCES

[1] Y. L. Yao. Talking Beer Flavor [J]. Technology Exchange. 2010, 5: 37–38.

[2] C.H. Zhao, Z.R. Han. A preliminary study of α-acetolactate decarboxylase fermentation conditions [J]. Food technology, 2005, (4): 5–8.

[3] H.Z. Wei, S.M. Ma, C.Y. Wei, application of α-acetolactate decarboxylase for beer fermentation production diacetyl [J]. Guangxi Light Industry, 2011, 8: 17–18.

[4] H. Yan, Y.J. Zhang, T.L. Zhao, *et al*. Cloning and Expression Methods of α-acetolactate decarboxylase Gene and Detection Strategy[J]. Food Research and Development. 2014, 3(35): 130–133.

[5] Y.J. Qin, W.F. Liu, D. Gao, *et al*. Containing α-acetolactate decarboxylase gene recombination *Saccharomyces cerevisiae* beer fermentation[J]. Shandong University (Natural Science). 2000, 35(2): 235–240.

[6] Sone H, T Fujii, K Kondo, *et al*. Nucleotide sequence and expression of the α-acetolactate decarboxylase gene in brewer's yeast [J]. Appl. Environ. Microbrol. 1988, 54(1): 38–42.

[7] Godtfredsen S E, H Lorcck and P Sigsgaard. On the occurrence of α-acetolactate decarboxylases among microorganisms [J]. Carlsberg Res Commun. 1983, (48): 239–247.

[8] X.L. Song, H.Y. Zhu, Gas chromatography – mass spectrometry Beer Flavor [J], Food Research and Development. 2006, 5(127): 115–116.

[9] Sone H, T Fujii, K Kondo, *et al*. Nucleotide sequence and expression of the α-acetonlactate decarboxylase gene in brewer's yeast[J]. Appl Environ Microbrol, 1988, 54(1): 38–42.

[10] X.R. Zhao, Y.W. Xia, Y.B. Zhao. A new progress of *Saccharomyces cerevisiae* Fermentation [J]. Wine, 2007, 34 (1): 59–62.

[11] Suihko ML, K Blomqist, M Penttila, *et al*. Recombinant brewer's yeast strains suitable for accelerated brewing [J]. Biotechnol. 1990, (14): 285–300.

[12] Goossens D. Decreased diacetyl production in lager brewing yeast by integration of the ILVS gene [J]. l. Inst Brew.1993, 99(3): 208.

[13] Yamauchi Y, Okamoto T, Murayama H, *et al*. Rapid maturation of beer using an immobilized yeast bioreactor: heat conversion of α-acetolactate [J]. Biotechnology. 1995, 38(2): 101–116.

[14] Shindo S, Sahara H and Koshino S. Lupptession of α-acetolactate formation in brewing with immobilized yeast [J]. Inst.Brew. 1994 (100): 69–72.

[15] Olsen F, K Aunstrup. α-acetolactate decarboxylase. Eur Pat Appl[C]. 1984, EP 128714.

[16] Y. Gao. Flavor formation and change by Beer fermentation process [J]. Beer Technology. 2005: 5–10.

Biomedical Engineering and Environmental Engineering – Chan (Ed.)
© 2015 Taylor & Francis Group, London, ISBN: 978-1-138-02805-0

Propolis suppresses the transcription of proinflammatory-related factors in HeLa cells

K. Chen
College of Pharmaceuticals, Liaocheng University, Shandong, China

S.C. Li
Department of Gynaecology and Obstetrics, Liaocheng People's Hospital, Shandong, China

N.L. Liang
Medical Institute, Liaocheng Vocational and Technical College, Shandong, China

ABSTRACT: Previous studies have shown that propolis can inhibit inflammatory responses, but very little is known about the mechanisms involved. In this study, the transaction and expression of three preinflammatory factors, iNOS, PTGS-2, and IL8, which are closely related to inflammatory response, were investigated by luciferase reporter assay and RT-PCR in HeLa cells treated with propolis. The results showed that propolis sharply down-regulated the transcription of these three genes. Then, the effects of NF-κB on the transaction of three preinflammatory factors were further addressed. As was expected, NF-κBs remarkably up-regulated the activity of the reporter gene, but had no effect on IL-8 promoter transaction was found after the NF-κB binding site mutation This work may set a foundation for further study of propolis and related functional products that provide health benefits to patients suffering from inflammatory diseases.

Keywords: Propolis; NF-κB; Proinflammatory-related factors; HeLa cells

1 INTRODUCTION

Inflammation is a beneficial host response to external challenges or cellular injury that leads to the release of a complex array of inflammatory mediators, finalizing the restoration of tissue structure and function. However, prolonged inflammation can be harmful, contributing to the pathogenesis of many diseases. A number of cytokines are involved in the inflammation response. iNOS (inducible Nitric Oxide Synthase) (Vane et al., 1994), PTGS-2 (Prostaglandin endoperoxide Synthase 2, also called cyclooxygenase 2) (Hwang, 1992; Masferrer et al., 1994), and IL-8 (Interleukin 8) are among the most important proinflammatory-related factors. It has been hypothesized that inhibition of high-output NO, prostaglandins, and IL-8 production in macrophages, by blocking iNOS (Titheradge, 1999; Xie & Nathan, 2014), PTGS-2 and IL-8 (Hu et al., 1993) expressions or their activities, could serve as the basis for potential development of anti-inflammatory drugs.

NF-κB (Nuclear Factor-κB) is an important transcription factor complex that regulates the expression of many genes involved in immune and inflammatory responses (Baeuerle & Baltimore, 1996; Kopp & Ghosh, 1995). Following activation, a NF-κB heterodimer is rapidly translocated to the nucleus, where it activates the transcription of target genes, including the genes encoding the proinflammatory cytokines, adhesion molecules, chemokines, and inducible enzymes such as PTGS-2 and iNOS. Because NF-κB plays such a pivotal role in the amplifying loop of the inflammatory response, it has become a logical target for new types of anti-inflammatory treatment.

Propolis, a resinous material collected by honeybees (*Apis mellifera L.*) from leaf buds and cracks in the bark, and mixed with pollen, wax, and bee enzymes, possesses various physiological activities (Premratanachai & Chan, 2014). It has been used in traditionalmedicine since ancient times in many countries. Recently, it was reported to possess various biological activities, such as being antibacterial, anti-inflammatory, regulating immunity etc, which has been well known by many researchers (Zhang et al., 2013; Khacha et al., 2013, & Wang et al., 2014). The anti-inflammatory effects of propolis solution extracted by means of modern pharmacological methods has been studied (de Groot, 2013). Some experiments and human studies have been carried out in order to clarify how propolis regulates inflammatory responses (Lima et al., 2014; Rajendran et al., 2014). However, the mechanisms underlying the effects of propolis on inflammatory response and the transcriptional regulation of inflammation factors, are still being outlined. In this paper, the effects of propolis on the regulation of proinflammatory factors in HeLa cells were investigated.

2 MATERIALS AND METHODS

2.1 *Preparation of ethanol extract from propolis*

Raw Chinese propolis samples were collected from *A. mellifera* colonies in Shandong province and stored at $-20°C$ until used. The main plant origin of the propolis samples collected was the poplar tree (*Populus sp.*). Ethanol extract from propolis was prepared according to a published technical paper (Wang, 2014). 100 gm of the propolis sample was weighed and broken into a powder with a grinder. Then the sample was extracted by 95% (v/v) ethanol (1L), and sonicated at 37°C for 4 h. After sonication, the supernatant was filtered with Whatman No.4 filter papers to remove the residues. The residues were collected and then extracted and sonicated with 95% ethanol for another 3 h. The raw propolis was extracted three times. Thereafter, all of the supernatants were collected together and evaporated in a rotary evaporator under a reduced pressure at 50°C. Finally, the extract was dried in the oven until it reached a constant weight, and stored at $-20°C$ until it was used further. During the cell experiments, the final extract was weighed and redissolved in 100% ethanol. The ethanol extract of the propolis solution was filtered with 0.22 μm filters (Pall) to make 20 mg/mL stock, and stored at $-20°C$ in the dark. The final concentration of ethanol in the cell culture medium did not exceed 0.1% (v/v).

2.2 *Cell culture*

The HeLa cells were obtained from the American Type Culture Collection and grown in DMEM (Dulbecco's Modified Eagle's Medium, Sigma, USA), and supplemented with 10% (v/v) foetal bovine serum at 37°C in a 5% CO2 humidified incubator. For each experiment, the HeLa cells were collected by dissociation of a confluent stock culture with 0.25% trypsin and 1 mmol/L EDTA(ethylenediaminetetraacetic acid). The cells used for all the experiments were cultured in triplicate at a density of 5×105 cells/mL in 24-well tissue culture plates (Corning) with various propolis concentrations. LPS(Lipopolysaccharide) (100 ng/mL) (E. coli O55:B5, Sigma, USA) was used as a positive control. PDTC (Pyrrolidine Dithiocarbamate) was used as an NF-κB inhibitor (Liu et al., 1999).

2.3 *Plasmids, transient transfections, and luciferase reporter gene assays*

The human IL-8 promoter reporter construct (IL-8-Luc-promoter plasmid wild type (-162 to $+44$ bp) and NF-κB site mutation construct (IL-8mutNFκB-Luc) were gifts from Brasier (University of Texas Medical Branch, Galveston, TX, USA). The human iNOS promoter reporter plasmid (iNOS-Luc) and the PTGS-2 promoter reporter plasmid (PTGS-2-Luc) were from this laboratory. pNFκB-Luc (Clontech) was a commercial vector. pCMV2 constructs, containing cDNAs for human NF-κB1 (p50) and RelA/p65, were generous gifts from Dr. Marty Mayo of the University of North Carolina (Madrid et al., 2000). All constructs were verified by DNA sequencing (Invitrogen, Shanghai, China). DNA transfections of cells were carried out in 24-well plates by using TurboFectTM (Fermentas) according to the manufacturer's instructions. pEGFP-N1, a plasmid containing a promoter-driven green fluorescent protein gene, was added to each well to control for transfection efficiency between the groups. Cells were lysed with reporter lysis buffer (Promega). Luciferase activity was assayed with Biotek Gen5 Microplate Reader using a commercially available kit (Promega).

2.4 *RT-PCR analysis*

5 μg of the total RNA (ribo nucleic acid) was reverse transcribed with Moloney murine leukaemia virus reverse transcriptase (Promega) by olio(dT) primers for 90 min at 42°C in 20 μL of reaction mixtures, and then further diluted to 100 μL with water for the subsequent procedures. The resulting cDNA was amplified by RT-PCR using the following: human GAPDH (glyceraldehyde-3-phosphate dehydrogenase) primers (sense, 5′-ATTCAACGGCAC AGTCAAGG-3′, and anti-sense, 5′-GCAGAAGGGG CGGAGATGA-3′); human hiNOS primers (sense, 5′-AGAGTGGAGAGTCCAGCC-3′, and anti-sense, 5′-AGGCACACGCAATGATGG-3′); human PTGS-2 primers (sense, 5′-TTCAAATGAGATTGTGGGAAA ATTGCT-3′, and anti-sense, 5′-AGATCATCTCTGCC TGAGTATCTT-3′); and human IL-8 primers (sense, 5′-CTTGGCAGCCTTCCTGATTTCT-3′, and anti-sense, 5′-CGCCTTTACAATAATTTCTGTGTTG GCG-3′). The protocol consisted of 30 cycles of incubation at 94°C for 30 s, 58°C for 30 s, and 72°C for 1 min, followed by extension for 5 min at 72°C. The amplified products were analysed by 2% agarose gel electrophoresis and visualized by EB staining under UV(ultraviolet) light.

3 RESULTS

3.1 *Effect of propolis on the transcription of proinflammatory-related genes*

The propolis was examined for its ability to influence the proinflammatory-related gene promoter construct in HeLa cells.

To identify the various proinflammatory-related gene promoter constructs in response to propolis, the HeLa cells transfected with iNOS-Luc, PTGS-2-Luc, IL-8-Luc, and IL-8mutNFκB-Luc were treated with propolis at a concentration of 4, 8, 16, and 25 μg/mL, respectively. LPS was used as positive control in this experiment (Spitzer et al., 2002). As can be observed (Fig. 1a), in HeLa cells transfected with

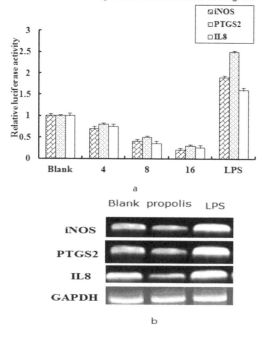

a

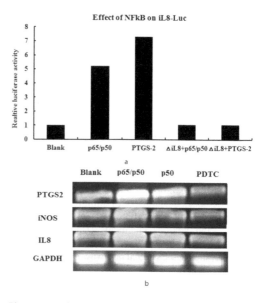

Figure 2. Effects of NF-κB on the promoter-Luc activity. a. Effects of NF-κBs on IL8-Luc promoter b. RT-PCR analysis.

Figure 1. Effects of propolis on the promoter-lLuc activity. a. Effects of propolis on the promoter-luc activity; b. RT-PCR analysis of the effects on transcriptional level of proinflammatory-related genes.

iNOS-Luc, PTGS-2-Luc, or IL-8-Luc, the expression of the luciferase gene was dramatically suppressed by propolis. Considering that the NF-κB binding site was located on the promoter region of iNOS, PTGS-2, and IL8, we deduced that NF-κBs might be involved in the regulations.

To test the effects of propolis on the endogenous expression of iNOS, PTGS2, and IL-8, we used a semi-quantitative RT-PCR to evaluate the levels of these proinflammatory-related gene mRNAs after the propolis treatment. In agreement with the luciferase promoter detection results (Fig. 1b), the mRNA levels of iNOS, PTGS-2, and IL-8 were also down-regulated by propolis.

3.2 NF-κBs directed the regulation of proinflammatory-related genes

To test the role of NF-κBs in the transcription of proinflammatory-related genes, HeLa cells were transfected with iNOS-Luc, PTGS-2-Luc, or the IL-8-Luc vector, respectively, along with p65 and p50 expression plasmids. From the luciferase activity measured (Fig. 2a), the heterodimmer p65-p50 remarkably up-regulated the expression of the reporter genes. In contrast, in the HeLa cells transfected with IL-8mutNF-κB-Luc, the expression of the luciferase gene almost had no response to various forms of NF-κB dimmers. PDTC significantly attenuated the expression of the reporter genes.

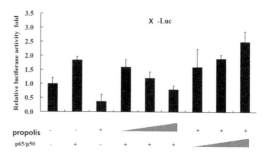

Figure 3. Effects of propolis on the NF-κB promoter-Luc.

Furthermore, the effects of NF-κBs on the endogenous transcription of iNOS, PTGS-2, and IL-8 were detected by RT-PCR, and the results also showed that iNOS, PTGS-2, and IL-8 were all up-regulated by NF-κBs (Fig. 2b).

To investigate the role of propolis in NF-κBs regulation, luciferase reporter assay and RT-PCR analysis were performed. As shown in Figure 3, the expression of the reporter gene was markedly suppressed dose-depended.

4 DISCUSSION

Purified propolis is being used in clinical and experimental studies for its healthy properties (Bankova et al., 2014). Based upon animal experiments and human studies (Khacha et al., 2013; Wang et al., 2014, & Rajendran et al., 2014), there is a potential therapeutic role for propolis in patients with inflammatory bowel disease. However, large, double-blinded controlled trials are needed to confirm efficacy and to document dosage and treatment parameters. Whether these positive results were, in part, due to patient selection or the use of propolis, remains to be determined in a larger patient study. Suffice it to say, the role of propolis in treating or preventing inflammation has not yet been clearly proven.

In the present study, the mechanisms of propolis regulation of proinflammatory genes iNOS, PTGS-2, and IL-8 in HeLa cells were discussed. It was demonstrated that propolis down-regulates the transcription of proinflammatory genes. Genetic studies have revealed an important role: the unique κB site 5'-GGGRNWYYCC-3' (Ulrich et al., 1994), through which NF-κBs can bind to drive the transcription of them, was located. A variety of stimuli or inhibitors, such as microbial and viral products, cytokines, DNA damage, and noxious chemicals (Bharrhan et al., 2012), have been shown to have abilities to regulate the activity of the NF-κBs signal pathway. Here, we found that propolis could also act as an inhibitor and down-regulated iNOS, PTGS-2, and IL-8 via the NF-κB cell signal pathway.

ACKNOWLEDGEMENTS

This work was supported by the doctorate scientific research fund of *Liaocheng* University.

REFERENCES

Baeuerle P.A. & Baltimore D. 1996. NF-kappa B: ten years after. *Cell*, 87: 13–20.

Bankova V., Popova M. & Trusheva B. 2014. Propolis volatile compounds: chemical diversity and biological activity: a review.Chem Cent J. 2; 8: 28.

Bharrhan S., Chopra K., Arora S.K., Toor J.S. & Rishi P. 2012. Down-regulation of NF-κB signalling by polyphenolic compounds prevents endotoxin-induced liver injury in a rat model. *Innate Immun* 18: 70–79.

de Groot A.C. 2013. Propolis: a review of properties, applications, chemical composition, contact allergy, and other adverse effects. Dermatitis 24(6):263–82.

Hu D.E., Hori Y., & Fan T.P. 1993. Interleukin-8 stimulates angiogenesis in rats. *Inflamm.* 17: 135–43

Hwang D. 1992. Selective expression of mitogen-inducible cyclooxygenase in macrophages stimulated with lipopolysaccharide. *J. Biol. Chem.* 267: 25934–38.

Khacha-ananda S., Tragoolpua K., Chantawannakul P. & Tragoolpua Y. 2013. Antioxidant and anti-cancer cell proliferation activity of propolis extracts from two extraction methods. Asian Pac J Cancer Prev. 14(11): 6991–5.

Kopp E.B. & Ghosh S. 1995. NF-kappa B and Rel proteins in innate immunity. *Adv. Immunol.*58: 1–27.

Lima L.D., Andrade S.P., Campos P.P., Barcelos L.S., Soriani F.M., Moura S.A., & Ferreira M.A. 2014. Brazilian green propolis modulates inflammation, angiogenesis and fibrogenesis in intraperitoneal implant in mice. Eur J Med Chem. 86: 103–12.

Liu S.F., Ye X., & Malik A.B. 1999. Inhibition of NF-kappaB activation by pyrrolidine dithiocarbamate prevents in vivo expression of proinflammatory genes. *Circulation*, 100: 1330–37.

Madrid L.V., Wang C.Y., Guttridge D.C., Schottelius A.J., Baldwin A.S. Jr., & Mayo M.W.. 2000. Akt Suppresses Apoptosis by Stimulating the Transactivation Potential of the RelA/p65 Subunit of NF-κB. *Mol. Cell. Biol.* 20: 1626–38.

Masferrer J.L., Zweifel B.S., Manning P.T., Hauser S.D., Leahy K.M., Smith W.G., Isakson P.C., & Seibert K. 1994. Selective inhibition of inducible cyclooxygenase 2 in vivo is anti-inflammatory and nonulcerogenic. *Proc. Nati. Acad. Sci. USA* 91: 3228–32.

Pahl H.L. 1999. Activators and target genes of Rel/NF-κB transcription factors. *Oncogene* 18: 6853–66.

Premratanachai P., & Chanchao C. 2014. Review of the anti-cancer activities of bee products. Asian Pac J Trop Biomed. 4(5): 337–44.

Rajendran P., Rengarajan T., Nandakumar N., Palaniswami R., Nishigaki Y., & Nishigaki I. 2014. Kaempferol, a potential cytostatic and cure for inflammatory disorders. BMC Complement Altern Med. 14:177.

Spitzer J.A., Zheng M., Kolls J.K., Vande Stouwe C., & Spitzer J.J. 2002. Ethanol and LPS modulate NF-kappaB activation, inducible NO synthase and COX-2 gene expression in rat liver cells in vivo. *Front. Biosci.* 7: 99–108.

Titheradge M.A. 1999. Nitric oxide in septic shock. *Biochem. Biophys. Acta*, 1411: 437–55.

Ulrich S., Guido F., & Keith B. 1994. Structure, regulation and function of NF-κB. *Annu. Rev. Cell Biol.* 10: 405–55.

Vane J.R., Mitchell J.A., Appleton I., Tomlinson A., Bishop-Bailey D., Croxtall J., & Willoughby D.A. 1994. Inducible isoforms of cyclooxygenase and nitric oxide synthase in inflammation. *Proc. Nati. Acad. Sci. USA* 91: 2046–50.

Wang K., Zhang J., Ping S., Ma Q., Chen X., Xuan H., Shi J., Zhang C., & Hu F.. 2014. Anti-inflammatory effects of ethanol extracts of Chinese propolis and buds from poplar (Populus×canadensis).J Ethnopharmacol. 155(1): 300–11.

Xie Q. & Nathan C. 1994. The high-output nitric oxide pathway: role and regulation. J. Leukoc. Biol., 56: 576–82.

Zhang J.L., Wang K, Hu F.L. 2013. Advance in studies on antioxidant activity of propolis and its molecular mechanism. 38(16): 2645–52.

Biomedical Engineering and Environmental Engineering – Chan (Ed.)
© *2015 Taylor & Francis Group, London, ISBN: 978-1-138-02805-0*

The current situation of community home care services in Beijing

Na Yu

School of Management and Economics, Beijing Institute of Technology, Beijing, China

ABSTRACT: From 2010, community home care services were started in Beijing. Based on the analysis of the survey data and the research of the status quo, problems with community home care services were raised and recommendations were given for future work.

1 INTRODUCTION

At the end of 2008, the Beijing Municipal Government drew up a new pension model called '9064.' By 2020, 90% of the elderly will have home care services with assistance by social services; 6% of the elderly will have care services through community services purchased by the government; the balance of 4% will be covered by the elderly care service agencies. In order to make sure 90% of the elderly enjoy care service at home, the Civil Affairs Bureau in Beijing, together with another thirteen government departments, jointly issued the 'Approach for Beijing Citizens Home Care Service' (including 'Help to the Handicapped'), effective from the 1st of January, 2010.

Five years have passed since then. How are the Beijing community home care services being carried out? What is the current status and what are the needs of the elderly? In this paper, we will analyse the needs, situations, and existing problems of the community home care services through the survey data, and try to show and analyse a clear picture of the community home care service in Beijing.

2 METHOD

2.1 Research objects

This research distributed 986 questionnaires to the elderly who are 60 years old and above, and collected 986 valid questionnaires. Among the 986 elderly, 36% (355) are male and the balance is female. The interviewees are between 60 to 94 years old. Most are from 60 to 80 years old, and the interviewees above 80 are minimal. 34% of the interviewees are senior high school graduates; 28% have junior high school education, 15% have primary or lower education, and very few have master's degrees or above; 22% graduated from university or polytechnics. The main source of income of the 96.5% is their own retirement salary, 1.2% receive alimony, 3% have savings and investment income, and a few receive government grants

or other sources of income. Approximately 36% of the surveyed said their monthly income was between 1,000 to 2,000 yuan, nearly 50% have between 2000–4000 yuan, and a few are lower than 1,000 yuan or higher than 4,000 yuan.

2.2 Research methods

The questionnaire contains questions concerning: basic personal information, pension, will, and care services, financial security in the status quo, life care status and demands, health care status and demands, solace and legal protection status and demands, and volunteer service.

3 RESULTS

3.1 Analysis of care service willingness

67% of the elderly prefer staying at home with their family; 18% are willing to stay with the care service agencies, to avoid being a burden to the family; and the remaining 15% are willing to try a new of community home care service. This reflects Chinese people's preference of staying at home.

3.2 Demands on material security

Most of the surveyed elderly live with their spouses and/or children, and most of them receive an occasional allowance from their children, although they mainly rely on their own retirement pay. In their answers about financial situations, over 50% of the elderly reckon that their financial conditions are more or less sound; approximately 23% of them have some difficulties; 19% consider their own life to be well off; only a few respondents consider themselves very poor or very rich. 75 respondents receive the subsidies for being very senior citizens, 5 respondents receive the subsidy for the handicapped, 8 respondents receive the subsidies for very low-income families, 30 respondents get food vouchers, and 4 respondents enjoy some other subsidies, while most of the respondents do not benefit from any kind of government subsidies.

What troubles the elderly most? 33% of the respondents agree that their life is very comfortable; while among the troubled elderly, 30% are worried because of their low income, and over 50% encountered medical treatment issues. For them, daily chores or loneliness are not the main difficulties. Besides the above problems, poor quality of marriages, housing, and their ID registration etc. also troubled some of them. This shows that our community home care service should pay more attention to improve the living conditions of the elderly with low incomes, and their concerns about the difficulties and expenses of medical treatment.

About 33% of the respondents indicated their communities did not provide any services, and 66% indicated the community provided certain services, mainly medical care, solace, and life care.

3.3 Demands on life care

Over 90% of the respondents are able to take care of themselves completely, only 5% can take care of themselves partly, and about 1% of them are completely dependent. Out of the 14 selections, most of the elderly can eat, walk, dress, and take medicines independently. Noticeably, more respondents cannot do the housework, cook, go shopping, or wash independently. These services may be needed more by the oldest people and the elderly who are ill.

More than 80% of the respondents want the community to open restaurants and laundries for the elderly. 65% of the respondents want hourly home care services. These two demands are significantly more than the other two demands of the day-care centres and volunteers. Over 70% of the respondents need 1-2 hours of service at home, which is troublesome and labour demanding for the elderly, and they hope the hourly service price is as low as possible. Nearly 90% of respondents said they can take care of themselves; 19% are taken care of by their spouses; 13.5% by their children; another 3% by a domestic helper; and almost no one by hourly workers, neighbours, or community service staff. About 80 % of respondents do not know much about the life care services provided by the community, or some communities do not provide any life care services at all. We are somehow surprised by the low coverage of the community life care service. In addition, the services of maintenance, house cleaning, and food delivery are more known compared to assistance with taking medicine, going to the toilet, and paying bills As not many people receive or even know about the community life care service, less than 40% of the respondents are satisfied. Instead, most of the respondents are not happy with the service, which tells us that there is still much room for the community service to improve.

3.4 Demands on life security

More than 25% of the respondents do not have any chronic diseases; the most common diseases are hypertension, osteoarthritis, diabetes, and heart disease. Due to the features of long duration, slow recovery, and non-obvious symptoms of these diseases, the sick elderly require long-term and sustainable health services. Over 60% of the respondents are expecting the community health stations to provide regular medical screening; the elderly also like to be attended by the doctors and nurses at home.

More than 75% of the respondents pay less than 500 yuan in medical expenses. Over 30% of the respondents mention that it is difficult to bear such an amount for medical expenses. For medical insurance, 66% of respondents have the basic medical insurance for urban residents; 26% enjoy free medical services; and some of the elderly, whose ID is registered in rural areas, have the new rural cooperative medical insurance. For the most visited medical organizations, more than 60% of the respondents choose public hospitals, mainly due to the better quality of service and closer proximity to their homes; 16% frequently go to the doctor in the community health stations, because health stations are normally in the community and the drugs there are cheaper. Less than 30% go to the community health stations often or occasionally. Poor service quality and fewer services available are the main reasons for not visiting the community health station by the remaining respondents.

3.5 Demands on friendship services

15% of the respondents are very happy with their life and 24% of them are happy; 60% feel it is just so-so; 1% are not satisfied at all. About 80% of the respondents never feel lonely; 15% feel lonely occasionally; nearly 4% feel lonely often. Nearly 20% of the respondents hope the volunteer can have a chat, or walk with them, or do some exercise with them. More than 90% of the respondents have excellent or good relations with their family and/or neighbours. 13% of the respondents said their community does not have this solace service and 28% of them said they do not know if their community has this service or not. This ratio is lower than that of the life care services. So the solace and legal services have been carried out more efficiently and effectively, with more publicity. People are more satisfied by this type of service than the life care service.

4 CONCLUSION

From a practical point of view, the family pension is still dominating the ideology of the social-supporting system for the elderly, and children are still the main supporters of the elderly in terms of finance, health care, and spiritual solace.

4.1 Current situations and issues of material security

The most prominent problem in the material security services, is that few respondents receive such care services provided by the community; some of the elderly just hear about this service, while many people do not know of this service or their community does not have this service. In contrast, almost all community social workers or persons in charge, said their communities have carried out many such services for the elderly. This shows that the communities are not giving enough publicity to this service, or many potential clients are not aware of the existence of the service they need. It also tells us that the community home care services have been promoted by government policies in Beijing, but they have had little implementation.

During the survey, we also found that the elderly from places outside of Beijing are not protected at all because of the difference in policies between different cities or urban and rural areas. Some of them were working outside of Beijing before retirement. Although they can receive their retirement pay through their bank account on time, they might not be able to get their new or festival allowances on time. Though they have medical insurance, they cannot go to their nominated hospital and claim the expenditures. They can pay for some minor ailments, but not for a sudden and serious disease, in which case they will have to take care of the charges themselves. They are worried about this most of all. Furthermore, even if the communities in Beijing provide perfect life care and solace services, the elderly from places outside of Beijing cannot enjoy the services because their contacts have not been registered, and the community workers are too busy to attend to them. The demarcation and difference in the local welfare system and resources between urban and rural areas caused by the urban-rural dual structure, has excluded the farmers from the social welfare and security system. Although the demarcation, caused by the urban-rural dual structure or similar isolation between cities in China, has existed for many years, the elderly from the places outside of Beijing should also be taken care of in our community home-stay pension system, which is a people-oriented highly efficient new system.

4.2 Current situations and problems of life care

Community home care services in Beijing are very limited and the services are normally very simple. Almost no single community could provide all the types of services to the elderly, and a lot of the elderly said that the services which are provided by the community, are not required by them. The services they require, like chatting or shopping assistance, are not provided.

In addition, we found during the conversations in the same survey, that there is also a difference in the people with low satisfaction of the community care services: those received the service are holding a positive attitude towards the community work; however the others with unsolved difficulties do not like the services at all. So it does not suit everyone. Some respondents also mentioned that the workers in some communities are not well trained and professional, and might not

be able to handle the household chores. This reminds us that the care services in the community need to further improve the quality of their manpower, build sufficient reserves of professional people, and create specialized community home care service teams and volunteer teams, in order to achieve professionalism and specialization in the whole team. At the same time, a standardized assessment and regulatory system shall be established, so that the elderly can build trust in the community service.

4.3 Current situations and problems of life protection

The health level of the elderly population is lower than others, and it deteriorates with an increase in age. Medical charges are one of the important expenditures for the elderly. Many of the respondents mentioned that they tried to avoid the best medicine in order to save money, and they could not afford to have a serious disease, though some minor ones do not cost a lot. Medical care in the chronic phase of a disease is particularly important for the elderly with chronic disease. Therefore, the big expense and financial difficulties in medical treatment are still the general problems faced by the elderly. As for many respondents, the Beijing rule of 85% reimbursement over an amount of 1300 yuan is not good because the elderly are normally suffering from a chronic or major disease, costing around 1000 yuan a month. So they have to pay the entire charges. Meanwhile, many specialized drugs are not within the scope of reimbursement, so the elderly cannot benefit much from their medical insurance.

The coverage of the community health station for the elderly is not great enough. Most of the stations are only able to prescribe and dispense medicines, with little day-to-day diagnosis and follow-up for the elderly, and fewer medical treatment services and lack of equipments make the elderly not willing to visit the health station. Moreover, due to limited funds and venue issues, the commuting time between some health stations and residential areas are over half an hour. The elderly will choose the hospital nearby due to convenience and concerns over the quality of the medical treatment, which is the main reason that the health station cannot replace the large hospital.

4.4 Current situations and problems of solace

Currently, there are some issues with the solace service provided by the community. There are not enough professional psychological counselling and legal advice services. The community could not handle most of the cases and they have to refer it to the concerned government agencies when receiving the inquiries of legal issues, which is tedious and inefficient. Situation permitting, each community should have at least one professional legal adviser and one psychological counsellor. Otherwise, other viable channels should be provided to the elderly.

The elderly normally have a strong sense of loneliness. They hope to be cared for by their children and the whole society. Currently, there are very few one-to-one forms of services, such as accompanied meals or chats in the community. Most communities provide an activity centre and organize collective entertainment activities. It is easy for the elderly with outgoing personalities to participate in singing and dancing activities, and their mental status gets better and better. However, the unsociable elderly will not take the initiative to participate in collective activities, staying lonely and not concerned with others. The community should provide a solace service as per the different situations of the elderly. Of course, the capability of the community is very limited. The key is to form a multiple supply system, led by the government, with participation from market and social organizations, which would then be able to integrate the social resources and provide better services for the elderly.

REFERENCES

Cheng Haijun, (2012), The Plight and Countermeasures of China's Current Social Welfare for the Elderly, Journal of Capital Normal University (Social Sciences Edition), (1) pp. 123–129.

Cheng Shuling & Zang Xiaoying & Zhao Yue (2012), The Investigation of the Current Status and Demand of the Nursing for Chronic Disease of the Elderly in China's Urban and Rural Area and the Analysis of the Factors, Chinese Journal of Practical Nursing, 28(23), pp. 66–70.

Lao Yinqian & Cao Weigu & Li Yingdan, (2011), How the Medical Insurance System to Deal with the Impact of the ageing society, Union Expo: theoretical research (7) pp. 133–134.

Li Fengqin & Chen Quanxin (2012), Exploring the Urban Community Home care service Model—Example of Purchase of Service by Government of Gulou District, Nanjing City from the Heart-to-heart Service Center for the Elderly, Northwest Population Journal, 33(1) pp. 46–50.

Ma Yuqin & Dong Gang & Xiong Linping & Teng Haiying & Zhao Xiaojun, (2012), Research on China's Medical Treatment Services for the Elderly: Based on the Demand of Medical Treatment Services, Chinese Health Economics, 31(7), pp. 20–22.

Wang Shibin & Shen Qunxi (2012), The Main Psychological Problems of the Urban Elderly People and the Countermeasures – Cases of 878 Urban Elderly people in Guangdong, Chinese Journal of Gerontology, 32(2), pp. 361–363.

Zhu Jige & Liu Xiaoqiang, (2008), the Progress and Implications of the Medical Security System in Foreign Countries, Foreign Medical, Health and Economics Volume, 25(3) pp. 97–102.

Biomedical Engineering and Environmental Engineering – Chan (Ed.)
© 2015 Taylor & Francis Group, London, ISBN: 978-1-138-02805-0

The demand for home care services for the elderly in Beijing

Na Yu
School of Management and Economics, Beijing Institute of Technology, Beijing, China

ABSTRACT: The ageing issue in Beijing is getting more serious day by day. To meet the demand for care services for the elderly, a community home-based care service is being promoted in Beijing. To improve community home care services and better understand the demands of the elderly, an analysis has been made on different demands, based on a questionnaire conducted in nine communities in Beijing, from the aspects of care workers hours, door-to-door services, regular medical screening, door-to-door diagnosis, and door-to-door care.

1 INTRODUCTION

Effective from the 1st of January, 2010, the 'Approach for Beijing Citizens Home Care Service' (including 'Help to the Handicapped') was jointly issued by the Civil Affairs Bureau in Beijing, together with another thirteen government departments. In this Approach, all communities were specifically requested to establish the rules of care service tickets, set up canteens for the elderly, and build a community old people's' home. These old people's' homes were to be built in each and every community or village in Beijing within three years [1].

To carry out professional and personalized community home care services, we had to find out the demand for door-to-door services from the elderly. Thereafter, we will discuss it from the aspects of the cost of an hourly maid worker, door-to-door services, regular health check-ups, door-to-door diagnosis, and door-to-door care [2].

2 METHOD

2.1 Research objects

This research distributed 986 questionnaires to the elderly who are 60 years old and above, and collected 986 valid questionnaires. Among the 986 elderly, 36% (355) are male and the balance is female. The interviewees are from 60 to 94 years old. Most are from 60 to 80 years old, and the interviewees above 80 are minimal. 34% of the interviewees are senior high school graduates; 28% have a junior high school education, 15% have a primary or lower education, and very few have master's degrees or above; 22% graduated from university or ploytechnics. The main source of income of the 96.5% is their own retirement salary, 1.2% receive alimony, 3% have savings and investment income, and a few receive government grants

or other sources of income. Approximately 36% of the surveyed said their monthly income was between 1,000 to 2,000 yuan, nearly 50% have an income between 2,000–4,000 yuan, and a few have an income lower than 1,000 yuan or higher than 4,000 yuan.

2.2 Research methods

The questionnaire contained questions concerning: basic personal information, pension, will, and care services, financial security in the status quo, life care status and demands, health care status and demands, solace and legal protection status and demands, and volunteer services.

3 RESULTS

3.1 Analysis of the cost of an hourly maid worker

On the wish list of services to be carried out by the community, the elderly with low incomes place high demands on various services, in particular the pension services provided by the community stations and volunteers. Their spending on expensive care services is quite limited due to their financial situation [3].

Another obvious finding is that the low-income elderly are only willing to accept the lowest cost of an hourly maid worker, while the high-income elderly are willing to pay the hourly maid workers as per the market wages. So, we shall consider the affordability of the elderly [4] and provide different types of services at different prices, for them to choose.

3.2 Analysis of the demand of to-door care services

When the elderly are becoming older year by year and the losing their self-care capability, the demands

Table 1. The acceptance of the price of an hourly maid worker from the elderly with different incomes [n(%)].

Monthly Income	Below RMB 10	RMB 10–15	RMB 16–20	Above RMB20	Not Consider	Not require
Below RMB1000	83.6	7.3	0.0	0.0	7.3	1.8
RMB1000–2000	78.3	12.9	3.5	0.9	2.9	1.5
RMB2000–4000	63.0	21.3	5.0	3.6	3.9	2.9
Above RMB4000	45.4	23.7	21.6	2.1	1.0	6.2
Grand Total	68.0	17.7	5.9	2.2	3.4	2.7

Table 2. The demand for door-to-door care services from the elderly of different ages [n(%)].

Ages	Yes	No	Total
60–70 YO	305(51.6)	286(48.4)	591(100)
70–80 YO	150(52.1)	138(47.9)	288(100)
Above 80 YO	68(70.1)	29(29.9)	97(100)
Grand Total	523(53.6)	453(46.4)	976(100)

Table 3. The demand for regular medical screening from the elderly of different ages [n(%)].

Ages	Yes	No	Total
60–70 YO	392(66.3)	199(33.7)	591(100)
70–80 YO	196(68.1)	92(31.9)	288(100)
Above 80 YO	67(69.1)	30(30.9)	97(100)
Grand Total	655(67.1)	321(32.9)	976(100)

Table 4. The demand for door-to-door diagnosis from the elderly with different health statuses [n(%)].

Health Status	Yes	No	Total
Excellent	77(45.1)	90(53.9)	167(100)
Good	275(46.9)	265(49.1)	540(100)
Average	91(47.4)	101(52.6)	192(100)
Bad	47(68.1)	22(31.9)	69(100)
Very Bad	8(100)	0(0)	8(100)
Grand Total	498	478	976(100)

Table 5. The demand for door-to-door care services from the elderly with different health statuses [n(%)].

Health Status	Yes	No	Total
Excellent	78(46.4)	90(53.6)	168(100)
Good	276(51.1)	264(48.9)	540(100)
Average	114(59.1)	79(40.9)	193(100)
Bad	49(71.0)	20(29.0)	69(100)
Very Bad	8(100)	0(0)	8(100)
Grand Total	525	453	978(100)

for pension services is increasing tremendously. The same results are evident for door-to-door care services, community station services, and volunteer services.

3.3 Analysis of the demand for regular health check-ups

Most of the elderly have difficulty in walking and they have very high expectations of the door-to-door care services offered by the community health stations. As shown in Table 2, most of the elderly above 80 years old wish the community health stations could provide door-to-door care services, as well as door-to-door diagnosis and health guidance, with the result of $x2 = 11.832$ and $P = 0.003 < 0.05$. There is no big difference in demand for the services to be done at the station, such as the periodical medical screening, between the elderly of different ages, with the result of $x2 = 0.449$ and $P = 0.799$. This reminds us to consider the walking difficulties of the elderly when providing the care services, especially the medical service.

3.4 Analysis of the demand for door-to-door diagnosis

The difference in the health status is reflected more in the demands of the medical service. As Table 4 shows, the elderly who think they are not healthy enough place more demands on the door-to-door diagnosis from the community health station, with the result of $x2 = 18.374$ and $P = 0.001$. It is also true in the

demand for the door-to-door care service, as shown in Table 5, with the result of $x2 = 22.481$ and $P = 0.000$.

4 CONCLUSION

On the one hand, the low-income people welcome the low-priced, or even free, care services provided by the community. On the other hand, the high-income people, who are normally more knowledgeable and experienced, are more likely to opt for a quality service at a higher price [5]. The low-income elderly have a higher requirement for the various life care services provided by the community, especially in the community station services and volunteer services. Their expenditure on the expensive care services is limited by the economic burden. So, we shall consider the affordability of the elderly and provide different types of services at different prices, for them to choose [6].

When the elderly are becoming old, their capability of taking care of themselves is also declining. So the demand for door-to-door care services from the elderly increases tremendously. Most of the elderly have difficulty in walking and they have very high expectations for the door-to-door services, nearby health check-ups, door-to-door diagnosis, and door-to-door care provided by the community. Like most of the elderly above 80 years old, the elderly with a bad health status require more door-to-door medical services.

REFERENCES

[1] Documents of China National Working Committee on Ageing and National Development and Reform Commission etc, forwarded by the General Office of the State Council of the People's Republic of China: "Notice on Quickening the Development of the Care service" (GOSC-2006-6).

[2] Lao Yinqian, Cao Weigu, Li Yingdan, "How the Medical Insurance System to Face the Impact of the Ageing Society", A Review of Labor Unions: Theory and Reserch Beijing, vol.7, 2011, pp. 133–134.

[3] Kathleen Melnnis-Dittfich, Sui Yujie, "The Evaluation and Intervention of Physiology, Psychology and Social Aspect in Ageing Society Work", China Renmin University Press, Beijing, 2008, page 91.

[4] Pei Xiaomei, Fang Lijie, "Introduction of Long-term Care for the Elderly", Social Science Academic Press, Beijing, 2010, pp. 48–49.

[5] Dong Yahong, "The Analysis and Reform of the Social Care service System in China", Social Science, Beijing, vol.3, 2012, pp. 68–75.

[6] Zhuang Qi, " The Current Status and Solution for the Urban Community Home Care service in China", Journal of Beijing Vocational College of Labor and Social Security, Beijing, vol.4, 2008.

Biomedical Engineering and Environmental Engineering – Chan (Ed.)
© 2015 Taylor & Francis Group, London, ISBN: 978-1-138-02805-0

A comprehensive assessment of laboratory teaching of haematology and haematological tests: Research and practice

Li jun Gao, Wei Xia, Ya li Zhang & Ming Sun
Medical Laboratory Institute, Beihua University, Jilin City, Jilin Province, China

ABSTRACT: Haematology and haematological tests are one of the most important main courses in laboratory medicine. They also include laboratory classes which cover 60% of the total teaching hours. Over the years of teaching process using the assessment methods of haematology and haematological tests, the Teaching and Research Office of Clinical Blood and Body Fluid Examination at Medical Laboratory Institute of Beihua University believes that the comprehensive assessment method for laboratory evaluation can improve not only the initiative and effects on students of learning haematology and haematological tests, but can also enhance the students' practical ability and comprehensive capability of analysing and solving problems.

1 INTRODUCTION

The objective of Laboratory Medicine Institute at medical colleges and universities is to train laboratory medicine undergraduates to be the senior specialists who not only master the theoretical knowledge of basic and clinical medicine but are capable of working in medical testing laboratories in hospitals of various levels, blood banks and epidemic prevention stations after graduation[1]. To achieve this objective, the teaching quality and the students' practical capability must be greatly improved through the "knowledge focused but ability neglected" teaching thought[2]. The blood and hematopoietic tissues, as well as the specimens from the secondary clinical diseases that are derived from blood and hematopoietic tissue diseases are the main study objects of clinical haematology and haematological tests. In performing these tests, various laboratory testing technologies and methods are utilized, in order to analyse and study the pathological changes concerning hematopoietic tissues and blood, to clarify the mechanism of blood diseases and to coordinate the diagnosis, treatment and prognosis of blood disorders. The course, which includes haematology and haematological tests, establishes a new system which combines and links haematological theories, laboratory methods from the laboratory science, and clinical blood diseases[3]. This system thus is called the theory-testing-disease system. The laboratory diagnosis of various blood diseases is critical in laboratory teaching. Therefore, the contents and styles of laboratory assessment have to be changed from the original traditional examination methods to the comprehensive assessment, according to the subject characteristics and syllabus requirements in order to enhance the training of comprehensive capability, a problem-analysing capability, and the hands-on capability of all the students, enabling students to qualify for clinical work.

2 NECESSITY AND IMPORTANCE OF THE IMPLEMENTATION OF COMPREHENSIVE ASSESSMENT METHOD

Haematology and haematological tests is one of the most specialized courses in medical laboratory. It is not only a practical course but also a technical course; the laboratory classes cover 60% of the entire curriculum. A laboratory class is mostly divided into morphology and techniques. For example, it has always been difficult for students to identify bone marrow cells, especially abnormal cells. Thus, identifying cell morphology is also an important aspect of teaching.

Previously, the laboratory assessment focused mostly on identification of 10 normal bone marrow cells, which is not applicable to the training of students' clinical diagnosis capacity. In order for the students to adapt to clinical work as soon as possible, the comprehensive assessment method is used for the laboratory class assessment, i.e., students are encouraged to design their own diagnosis programs based on a given case and available specimens, such as peripheral blood, marrow and other laboratory instruments and agents that they might need. Finally the students are required to make laboratory diagnosis as per observation and analysis.

The comprehensive assessment method for a medical laboratory examination enables the students to truly realize the importance of this course and cultivate their practical ability, enabling them to better adapt to any future clinical work.

3 AN IMPLEMENTATION PLAN OF THE COMPREHENSIVE ASSESSMENT METHOD

3.1 Data collection and clinical case arrangements

3.1.1 Clinical hospitals should be visited for collection of the complete data of cases including medical history, physical examination, relevant auxiliary examination and necessary clinical specimens such as blood, urine and bone marrow.

3.1.2 The data of cases should be well prepared by means of keeping records on case data and specimens and numbering them by rules.

3.1.3 Clinical specimens should be well preserved. Blood and bone marrow should be made into smears and fixed with alcohol for long-term preservation.

3.2 Preparation of a specific implementation plan of comprehensive assessment for medical laboratory teaching

3.2.1 In daily medical laboratory teaching, students should deign and perform the tests based on the knowledge they have acquired whilst being provided with typical case data as per teaching contents. Thus, the comprehensive capacity of self-study, self-knowledge and self-assessment can be developed.

3.2.2 All members of a subject group should be assigned respective tasks and should cooperate with each other in pre-testing prior to an assessment. The teachers should be fully aware of the testing method used for each case and the testing results that may come up.

3.2.3 In assessment, there are four groups of students and three to four students in each group. Different case data should be distributed to the students randomly. The test should be finished within 240 minutes. In the end the laboratory diagnosis must be made.

3.2.4 Teachers in each lab should observe the on-going tests of students in each group and record problems in the testing process. According to the performance of each student, the laboratory assessment scores should be given. The assessment contents include the following three parts; 40 points are for testing design, 40 points are for testing operation and 20 points are for laboratory diagnosis.

3.2.5 After the assessment, the teacher should make conclusions and point out the problems that the students have during the entire laboratory testing process. Finally the correct diagnosis must be presented.

4 THE PRACTICAL CONDITIONS AND PRACTICAL EFFECTS OF THE COMPREHENSIVE ASSESSMENT

4.1 Practical conditions

The initiatives of the students were significantly improved after this assessment method was applied to the course of hematology and hematology tests in Grade 2006. Their capability to combine theories with clinical practice was greatly strengthened and the teaching effects were so good that the students had been well received by many internship organizations such as Beijing Friendship Hospital, Jilin University Affiliated Hospital, and Hanna Provincial Hospital.

4.2 Good teaching effectiveness achieved

4.2.1 The students were trained to be innovative. Thirty students of class of 2006 majoring in laboratory medicine participated in the College Students Innovative Experiments. They won one national class project and five college class projects. Forty students of class of 2007 attending the experiments won two national class projects and four college class projects. Both the number of participants and the projects they won increased. Twenty students majoring in laboratory medicine of class of 2006 and class of 2007 joined teachers' scientific research activities successively.

4.2.2 The ability to combine theory with practice was improved. In the laboratory assessment of this course in class of 2006 majoring in laboratory medicine, the diagnostic accuracy rate for blood diseases reached 92% accuracy rate compared to 79% accuracy rate in the past, using 10 normal bone marrow identifications as standard.

4.2.3 The students' academic performance was improved. In daily assessment and final exams, the students' scores were improved. The average score of the final exam for the course of laboratory medicine of class of 2004 was 78.9 points and that for class of 2006 was 85.8 points. Their academic performance has been much improved.

4.2.4 The students' interest in learning was enhanced. There were four post graduates who majored in laboratory medicine of class of 2006 in haematology tests subject. There were six post graduates of class of 2007.

4.3 An improvement in students' clinical working ability

Investigations and visits assisted by the college teaching affairs department have been paid to internship hospitals in Beijing, Shanghai, Shenzhen, and Guangzhou. The students' ability to use professional knowledge during their internship and their postgraduate clinical work has been well appreciated.

According to information feedback, the students' learning initiatives have been enhanced and their clinical thinking ability has been much improved compared to previous years.. They are able to master professional knowledge and use it in their clinical work.

4.4 *The Completion of the reform of laboratory assessment methods*

Laboratory assessment in the past was mostly about the identification of 10 normal bone marrow cells for laboratory medicine majors before Grade 2004, which was irrelevant to the clinical diseases and therefore did not train students' comprehensive thinking ability. Since Grade 2005, some classes have utilized a comprehensive assessment method for laboratory medicine assessment. The students' design diagnosis programs observe, analyze, and make final laboratory diagnosis according to the case data, which develop the students' clinical diagnosis ability, as well as clinical thinking ability. At present, this method has been undergoing a full implementation for the laboratory medicine major.

4.5 *Enhancing the influence of this course in the country*

4.5.1 This assessment concept has been integrated into the teaching materials by editing multiple textbooks, in particular the laboratory textbooks. It has nationally obtained a unanimous acknowledgement in the discussion and exchanges with several universities in the field of laboratory testing.

4.5.2 In the inter-college meeting of the national laboratory testing in 2010, the subject group made a thematic report on the assessment method of this course. This was highly praised by experts. Many medical colleges came to visit our university to learn how to use this method.

4.5.3 The course of haematology and haematology tests was awarded a quality course by Jilin Province in 2009. During an adjudication process, this assessment method was approbated and highly acclaimed by several experts in 2010. The Ministry of Education of Jilin Province has recommended it to compete for national quality course selection.

4.5.4 The network course of haematology and haematology tests won the first prize for the year 2010 Educational and Technological Achievement Award.

5 AN APPLICATION EXPERIENCE OF THE COMPREHENSIVE ASSESSMENT METHOD

At present, almost all the laboratory medicine majors apply the traditional assessment method. The establishment of the comprehensive assessment will break the traditional assessment method in motivating the students' learning enthusiasm and developing their initiative thinking ability. This assessment method focuses on students in the teaching process, guiding them to use the learnt knowledge in practice comprehensively.

The comprehensive assessment method not only assesses the students' judgment on laboratory testing results, but inspects their abilities to test the design, the operations, and the usage of instruments and materials in filling in the bone marrow reports. It avoids the misjudgement resulting from previous carelessness in cells identification. Arbitrary judgments are also prevented from happening.

Through the assessments, the existing problems have been exposed as well, such as non-standard or unskilful operations, inaccurate testing results, carelessness or lack of confidence etc. Therefore, in the future laboratory teaching, we should focus on the development of students' practical ability and problem analysing capability. We should further optimize teaching methods to improve the laboratory teaching quality. We should shorten the distance from academic teaching to practical clinical work, so that the students are able to adapt themselves to the clinical testing work as soon as possible.

ACKNOWLEDGEMENTS

This work was supported by Teaching and Research Subject of Beihua University Medicine Faculty in 2009.

REFERENCES

[1] Xiao Huixiang, Liu Baowen, Zhou Xin, Tentative plan of teaching reform on undergraduate medical laboratory education [J]. Higher Medical Education in China, 1995, 23(2): 20–21.

[2] Lu Jun, Teaching practice of clinical haematology and haematology laboratory testing [J]. Shanxi Medical University Journal: basic medical education version, 2009, 11(4): 463–465.

[3] Xu Wenrong, Clinical haematology and testing {M}, the 4th version, Beijing: People's Health Publishing House, 2007: 2.

Environmental science and technology

Biomedical Engineering and Environmental Engineering – Chan (Ed.)
© *2015 Taylor & Francis Group, London, ISBN: 978-1-138-02805-0*

Molecularly modelling the effect of citrate on the removal of phosphate by ferric chloride

Yanpeng Mao
Institute of Energy and Environment, School of Energy and Power Engineering, Shandong University, Jinan, PR China

Liwei Zhan
Shandong Environmental Audits and Reception Center of Construction Projects, Jinan, PR China

Wenlong Wang & Chunyuan Ma
Institute of Energy and Environment, School of Energy and Power Engineering, Shandong University, Jinan, PR China

ABSTRACT: The extent of phosphate removal by ferric iron salts is strongly dependent upon the pH and concentration of competing organic matter. The phosphate removal capacity was optimized at the pH range of 6.0–8.0 in the absence of citrate and at the pH range of 7.0–8.0 in the presence of citrate. A molecular model, based on mass balances, which takes $Fe_{1.6}H_2PO_4(OH)_{3.8}$ ($\log_{10}K_{sp} = -69.6$) as a chemical formula for precipitation, has been developed and provides a good description of phosphate removal behaviour over a wide range of conditions. Through this model, the speciation for phosphate in the presence of citrate has been successfully calculated.

Keywords: phosphate, citrate, mass balances model, pH

1 INTRODUCTION

The removal of phosphate from wastewater is a major concern, especially in areas that are sensitive to eutrophication. Chemical precipitation of phosphorus with salts of aluminium, iron (e.g. ferric chloride) or lime are used in order to achieve satisfying phosphate removal providing the advantages of significantly smaller equipment footprints, which are being easier and often more reliable to operate over the enhanced biological phosphorus removal. For the chemical phosphate removal by Fe(III) at the circumneutral pH, many researcher hypothesized that the isolated $FePO_4(s)$ (Luedecke, 1989; Stumm and Sigg, 1979; Zhang et al., 2010) would form, or at least be involved in Fe(III) precipitation with the empirical chemical formula $Fe_{1.6}H_2PO_4(OH)_{3.8}$($\log_{10}Ksp = -67.2$) (Fytianos et al., 1998; Szabo et al., 2008). However, the presence of varied organic matters in natural water or wastewater will complex with iron (III), which will compete with the production of iron (III) phosphate, and hence inhibit the chemical removal of phosphate by Fe(III) salts by consuming large quantities of Fe. Historically, investigators have investigated the effect of Soluble Microbial Produces (SMP) (Stuckey and Barker, 1999) and other natural organic matters on phosphate removal by Fe(III) salts (Qualls et al.,

2009; Weng et al., 2008) and shown that the presence of hydroxyl-carboxylate functional groups in organic matters would significantly decrease the removal efficiency of phosphate by ferric chloride though these investigators did not attempt to model this effect.

Therefore, in this study, the removal of phosphate by ferric chloride over a range of pH in the absence and presence of a model organic compound (citric acid) with functional groups similar to those found on naturally occurring organic compounds was examined. A molecular model based on mass balances was investigated to describe the data obtained.

2 MATERIALS AND METHODS

2.1 *The preparation of solutions*

All solutions were prepared using analytical grade reagents in 18 MΩ cm Milli-Q water and stored in the dark at 4°C in the fridge. Only glassware was used in the experiment to avoid the absorption of phosphorus by plastic ware. The glassware was soaked in HCl basin for several days and washed with MQ water before use. A 100 mmol/L Fe(III) stock solution was prepared by dissolving $FeCl_3$ in 0.5 mol/L HCl.

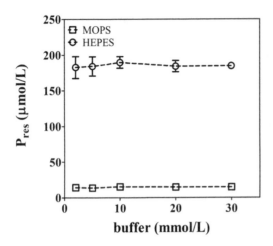

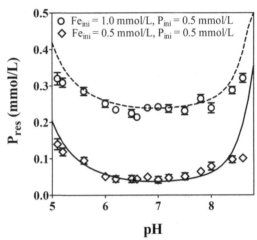

Figure 1. Effect of concentrated buffer on the removal of 0.5 mmol/L phosphate by 0.5 mmol/L (○), and 1.0 mmol/L ferric chloride (□).

Figure 2. The residual phosphate at a varying pH 5.0–9.0 and at $I = 0.1$ by Fe/P molar ratios of 1:1 (○) and 1:2 (◇). The lines are from the mass balances model.

A 1 mol/L NaH$_2$PO$_4$ stock solution was prepared by dissolving NaH$_2$PO$_4$ in Milli-Q water.

Municipal wastewaters may contain from 5 to 20 mg/L of total phosphorous, of which 1–5 mg/L are organic and the rest in inorganic (4–15 mg/L, namely 0.1–0.5 mmol/L). In this experiment, 0.5 mmol/L NaH$_2$PO$_4$ with the electrolyte of 2 mmol/L NaHCO$_3$ and 0.1 mol/L NaCl to adjust the solution alkalinity and ionic strength and with the 10 mmol/L buffer to control pH was used as phosphate solution to represent the inorganic phosphorous in municipal wastewaters. All the experiments were carried out at the lab temperature which was controlled to $22 \pm 1°C$.

2.2 The controlled conditions of experiments

The pH of all solutions used was carefully controlled with an appropriate biological buffer. Both of MOPS and HEPES are non-complexing agents, and are unlikely to influence the behaviour of other solution constituents (Kandegedara and Rorabacher, 1999; Mao et al., 2012). To confirm this, an additional set of experiments of phosphate removal by ferric chloride using different concentration of MOPS or HEPES (2 mmol/L, 5 mmol/L, 10 mmol/L, 20 mmol/L and 30 mmol/L) was performed at a pH = 7. As shown in Figure 1, both MOPS and HEPES had little effect on the removal of phosphate by ferric chloride, and therefore are suitable for controlling the pH in this study.

The hydraulic condition may affect the floc produced, and hence inhibit the removal of phosphate. The reaction solution was agitated 10 min (at 120 rpm) for coagulation and 20 min (60 rpm) for flocculation, whilst the pH was readjusted at a given experimental level during the process. The formed precipitate kept static to settle in the reaction solution for 30 min. Therefore, 1 hour was used as a Hydraulic Retention Time (HRT), which is enough to guarantee the reaction equilibrium.

2.3 The removal of phosphate by ferric chloride

Different dosages of FeCl$_3$ (Fe$_{ini}$ = 0.5 and 1.0 mmol/L) were added into the phosphate solutions for the removal of NaH$_2$PO$_4$ (P$_{ini}$ = 0.5 mmol/L) at a pH = 5.0–9.0. Then 1 mmol/L FeCl$_3$ was used to remove 0.5 mmol/L NaH$_2$PO$_4$ at a pH 6.0 and 7.0 in the presence of different concentrations of citrate from 0 to 1 mmol/L.

Without disturbing the precipitate within the bottles, the samples for PO$_4^{3-}$ analysis were pipetted from a depth of 1 cm below the surface of the solution. The samples were immediately filtered through 0.45 um size filter to be analysed for the concentration of residual phosphate (P$_{res}$) in the solution by using the ascorbic acid molybdate blue method (A.A.P.H., 2005) with a 1 cm cell.

The thermodynamic modelling of phosphate removal by FeCl$_3$ was investigated by the Visual MINTEQ program (Gustafsson, 2004), based on a chemical equilibrium.

3 RESULTS

3.1 Ph on phosphate removal by FeCl$_3$

The experimental data for the phosphate removal by ferric chloride at varying pH 5.0–9.0 are presented in Figure 2. As can be seen, the phosphate removal efficiencies at a pH 5.0–9.0 are at a range of 35.8%–57.2% and a range of 72.0%–91.7% for Fe/P molar ratios of 1:1 and 1:2, respectively. The optimum pH for obtaining the best phosphate removal efficiency in this study is a pH range of 6.0–8.0, which is substantially accordant with the previous studies (Fytianos et al., 1998). Additionally, an increase in the pH from 5.0 to 6.0 caused a significant decrease in P$_{res}$, whilst with an increase in the pH from 8.0 to 9.0, the P$_{res}$ gradually increased. The lower the pH, the more arduous it is to

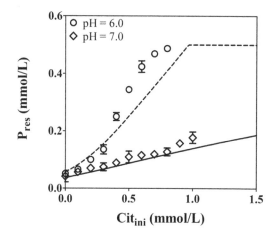

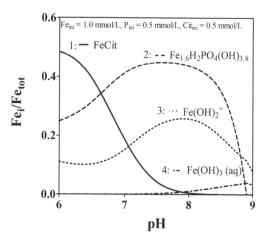

Figure 3. The residual phosphate with initial citrate concentration of 0–1.0 mmol/L at pH = 6.0 (○) and 7.0 (◇) by Fe/P molar ratio of 1:2. The lines are from the mass balances model.

Figure 4. Ferric iron species at the pH range of 6.0–9.0 at $I = 0.1$.

form the species PO_4^{3-} by releasing H^+ from H_2PO4^-. Whilst the higher the pH, the easier it is for Fe^{3+} to form Hydrous Ferric Oxides (HFO) through the OH^- compete with PO_4^{3-}.

3.2 Effects of citrate on phosphate removal by FeCl₃

Results of phosphate removal by Fe/P molar ratio of 1:2 at pH 6.0 and 7.0 with initial citrate concentration of 0–1.0 mmol/L are shown in Fig. 3.

The presence of citrate significantly reduced the extent of phosphate removal by FeCl₃ at each pH with citrate clearly able to outcompete phosphate for Fe^{3+} under the conditions used in this study. At pH = 6.0, the activity of ferric ions for phosphate removal could be totally blocked by dosage of 0.8 mmol/L citrate into solutions. While at pH = 7.0, with increasing the dosage of citrate from 0 to 1.0 mmol/L, the phosphate removal efficiency was only reduced from 91.6% to 64.5%, which is less than the drop range at pH = 6.0.

It could be explained by its influence on the speciation of ferric iron. As shown in Figure 4, 4 ferric iron species (i.e. FeCit, $Fe_{1.6}H_2PO_4(OH)_{3.8}$, $Fe(OH)_2^+$, $Fe(OH)_3(aq)$) were dominating at the pH range of 6.0–9.0. With an increase in pH, the radios of $Fe_{1.6}H_2PO_4(OH)_{3.8}$, $Fe(OH)_2^+$, $Fe(OH)_3(aq)$ raised first and then declined, and the radio of FeCit decreased. Therefore, the ability of citrate complexation with ferric iron is declined with the increase of pH. It implied that more citrate is needed to inhibit the phosphate removal ability of the same amount of ferric iron at a higher pH.

4 DISCUSSION

4.1 The mass balances model

Results of the mass balances model in the experimental data for P_{res} in the absence and presence of citrate over

Table 1. Key reactions and associated constants for the mass balances model.

Reactions[a]	$\log_{10} K$ ($I = 0$)	Refs.[b]
$H^+ + PO_4^{3-} \Leftrightarrow HPO_4^{2-}$	12.4	(1)
$2H^+ + PO_4^{3-} \Leftrightarrow H_2PO_4^-$	19.6	(1)
$3H^+ + PO_4^{3-} \Leftrightarrow H_3PO_4$	21.7	(1)
$Fe^{3+} + H_2O \Leftrightarrow Fe(OH)^{2+} + H^+$	−2.02	(2)
$Fe^{3+} + 2H_2O \Leftrightarrow Fe(OH)_2^+ + 2H^+$	−5.75	(2)
$Fe^{3+} + 3H_2O \Leftrightarrow Fe(OH)_3(aq) + 3H^+$	−15	(2)
$Fe^{3+} + 4H_2O \Leftrightarrow Fe(OH)_4^- + 4H^+$	−22.7	(2)
$Fe^{3+} + PO_4^{3-} + H^+ \Leftrightarrow FeHPO_4^+$	22.3	(2)
$Fe^{3+} + PO_4^{3-} + 2H^+ \Leftrightarrow FeH_2PO_4^{2+}$	23.9	(2)
$cit^{3-} + H^+ \Leftrightarrow Hcit^{2-}$	6.40	(3)
$cit^{3-} + 2H^+ \Leftrightarrow H_2cit^-$	11.2	(3)
$cit^{3-} + 3H^+ \Leftrightarrow H_3cit$	14.3	(3)
$Fe^{3+} + cit^{3-} \Leftrightarrow Fecit$	13.1	(2)
$Fe^{3+} + cit^{3-} + H^+ \Leftrightarrow FeHcit^+$	14.4	(2)
$Fe^{3+} + cit^{3-} + H_2O \Leftrightarrow FeOHcit^- + H^+$	1.79	(2)
$Fe_{1.6}H_2PO_4(OH)_{3.8}(s)$	−69.6[c]	this study

[a] Some reactions involving Na^+ and Cl^- were included in the model but not shown in the table. [b] (1) (Stumm and Morgan, 1981), (2) (Gustafsson, 2004), (3) (Morel and Hering, 1993). [c] $\log_{10} K_{sp}$ ($I = 0$) = −69.6.

a range of pH are plotted in Figures 2 and 3 (lines). It was shown that the model used in this study could successfully describe the experimental data except the data at a higher pH in the absence of citrate where the data was less than the predicted data from this model, as underestimating the adsorption of phosphate onto the HFO. The reactions assumed to be involved and the equilibrium constants assigned for these reactions are presented in Table 1.

In this model, we assumed the precipitate consisted of the isolated FePO₄(s) and Fe(III) precipitation (such as the HFO) with the empirical chemical formula $Fe_{1.6}H_2PO_4(OH)_{3.8}$. Furthermore, the solubility product constant ($\log_{10} Ksp$) was estimated to be -69.6 by optimizing the fit of the mass balances model to the

177

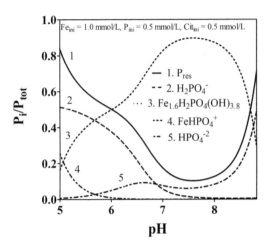

Figure 5. Phosphorus species at the pH range of 6.0–9.0 at $I = 0.1$.

obtained data, which is a little higher than the data from literatures (Fytianos et al., 1998; Szabo et al., 2008). In modelling the effects of the presence of citrate, three Fe(III)-citrate species (i.e. FeCit, FeHCit$^+$, FeOHCit$^-$) were involved and the species FeCit was dominating under the conditions used in this study (Figure 4).

4.2 The fate of phosphorus in the presence of citrate

The speciation variations of phosphate at a different pH (Figure 5) showed that the dominating fates for phosphorus in the presence of citrate were $H_2PO_4^-$, $Fe_{1.6}H_2PO_4(OH)_{3.8}$, $FeHPO_4^+$, HPO_4^{2-}. At a lower pH 6.0 and at a higher pH 9.0, more that 50% of phosphate was present as $H_2PO_4^-$ and HPO_4^{2-}, respectively. At the pH 7.0–8.0, more than 80% of phosphate was transferred to a solid phase as $Fe_{1.6}H_2PO_4(OH)_{3.8}$, and hence, a pH range of 7.0–8.0 would be controlled to obtain an optimum phosphate removal in the presence of citrate.

5 CONCLUSIONS

In this study, a molecular model based on mass balances has been developed and used to describe the phosphate removal by ferric iron salt in the absence and in the presence of citrate at circumneutral pH. It was concluded that the best phosphate removal efficiency by ferric iron was obtained at a pH range of 6.0–8.0 in the absence of citrate and at a pH range of 7.0–8.0 in the presence of citrate. The solubility product constant ($\log_{10} Ksp$) for $Fe_{1.6}H_2PO_4(OH)_{3.8}$ used as chemical formula for precipitate was re-estimated to be −69.6.

The results obtained here emphasize the likely competition for ferric iron between phosphates and dissolved organic compounds, which represents a limitation to the iron-mediated removal of phosphate from the solution. Possible approaches to overcome this problem include removing dissolved organics and the addition of ferrous rather than ferric iron with precipitation of vivianite.

ACKNOWLEDGEMENT

We gratefully acknowledge the financial support from the Postdoctoral Science Foundation of China (2013M531602) and Shandong Post-Doctoral Science Foundation (201202017).

REFERENCES

A.A.P.H. 2005. Standard methods for examination of water and wastewater. APHA, Washington.

Fytianos K., Voudrias E., Raikos N. 1998. Modelling of phosphorus removal from aqueous and wastewater samples using ferric iron. Environmental Pollution 101:123–130.

Gustafsson J.P. (2004) Visual MINTEQ.

Kandegedara A., Rorabacher D.B. 1999. Noncomplexing tertiary amines as "better" buffers covering the range of pH 3–11. Temperature dependence of their acid dissociation constants. Analytical Chemistry 71:3140–3144.

Luedecke C. 1989. Precipitation of Ferric Phosphate in Activated-Sludge – a Chemical-Model and Its Verification – Reply. Water Science and Technology 21:1564–1565.

Mao Y., Ninh Pham A., Xin Y., David Waite T. 2012. Effects of pH, floc age and organic compounds on the removal of phosphate by pre-polymerized hydrous ferric oxides. Separation and Purification Technology 91:38–45.

Morel F.M.M., Hering J.G. 1993. Principles and applications of aquatic chemistry Wiley, New York.

Qualls R.G., Sherwood L.J., Richardson C.J. 2009. Effect of natural dissolved organic carbon on phosphate removal by ferric chloride and aluminum sulfate treatment of wetland waters. Water Resources Research 45:W09414.

Stuckey D.C., Barker D.J. 1999. A review of soluble microbial products (SMP) in wastewater treatment systems. Water Research 33:3063–3082.

Stumm W., Sigg L. 1979. Colloid-Chemical Basis of Phosphate Removal by Chemical Precipitation, Coagulation and Filtration. Zeitschrift Fur Wasser Und Abwasser Forschung-Journal for Water and Wastewater Research 12:73–83.

Stumm W., Morgan J.J. 1981. Aquatic Chemistry, An Introduction Emphasizing Chemical Equilibria in Natural Waters 2nd ed. Wiley, J. & Sons Inc., New York.

Szabo A., Takacs I., Murthy S., Daigger G.T., Licsko I., Smith S. 2008. Significance of design and operational variables in chemical phosphorus removal. Water Environment Research 80:407–416.

Weng L.P., Van Riemsdijk W.H., Hiemstra T. 2008 Humic Nanoparticles at the Oxide-Water Interface: Interactions with Phosphate Ion Adsorption. Environmental Science & Technology 42:8747–8752.

Zhang T., Ding L.L., Ren H.Q., Guo Z.T., Tan J. 2010. Thermodynamic modeling of ferric phosphate precipitation for phosphorus removal and recovery from wastewater. Journal of Hazardous Materials 176:444–450.

Biomedical Engineering and Environmental Engineering – Chan (Ed.)
© 2015 Taylor & Francis Group, London, ISBN: 978-1-138-02805-0

A review of accelerated ageing and the composition analysis of aviation kerosene

Liu-yun Wang, Xuan-jun Wang, Xu Yao & Jing-yue Li
Xi'an High Tech. Institute, Xi'an, China

ABSTRACT: Aeronautics kerosene accelerated aging tests can expose weak links of kerosene in advance and they will give service information before reaching the aeronautics kerosene shelf life. This paper focuses on the accelerated ageing of aviation kerosene and the methods of component analysis, pointing out the characteristics and shortcomings of various analytical methods. In summing up the experience of domestic and foreign advanced testing through improved technology, we deduce facts about kerosene's life.

Keywords: Aeronautics kerosene; Ageing; Life evaluation

1 INTRODUCTION

As one of the petroleum products, aviation kerosene has attracted attention because of its ageing problems in recent years. Aviation kerosene used by the militaries all around the world has its serial number which is called a JP series. The US military jet fuel starts with JP-1, JP-2 and JP-3, which is mostly extracted from gasoline or kerosene. The JP-1 kerosene-type fuel was developed in 1944 which contained water more than others, JP-2 has not been adopted widely because of the high cost of crude oil during the refining process. JP-3 has a low flash point (40°C) and is easily volatilized. JP-4 and JP-5 have excellent overall performances. RJ-4, RJ-4I, JP- 9, JP-10, RJ-5, and RJ-7 are a series of synthetic fuels which contain one or more compounds. Some types of them are almost the same as domestic heating oil in spite of the additive content[1]. Military jet fuel is a type of a highly specialized product. It is made for special uses, and we need to inhibit its oxidation because of the wide use of hydrotreating fuel which is susceptible to oxidation[2]. Unstable aeronautics kerosene cannot guarantee the reliable use and life of the aeronautic technical equipment[3]. The contact between aeronautics kerosene and the surface of metal materials in storage, transport and the process of use may change the technical indicators in varying degrees. The composition of aeronautics kerosene have a significant impact on its performance and reliability levels, and the comparing of analysis of the kerosene composition is very important in order to improve the production process, quality control, etc.[4]. With the extension of storage time of the aeronautics kerosene, the problem of how to improve storage conditions has gradually stated to show.

2 THE ACCELERATED AGING METHOD

Because the natural ageing of petroleum products takes too long, we usually make high temperature accelerated ageing tests to estimate its storage life. The work in this area has been carried out abroad in the 1960s[5]. The earliest method used is to put the petroleum product in a hot environment for some time and measure the change of the components to determine the degree of reliability of petroleum products. The accelerated ageing test has been used for 40 years. In 1957, *Accelerated Life Testing of Capacitors* published by Levenbach was regarded as the first paper which wrote about the accelerated ageing test[6].

According to the research status of kerosene, the ageing test gradually formed two types of method: one is the natural ageing test method, which directly uses the natural environment to perform the ageing test; the other is the artificial accelerated aging test method, which uses an ageing box to simulate certain ageing factors of natural environmental conditions in the laboratory. Because of the diversity and complexity of the ageing factors, the natural aging is undoubtedly the most important and reliable aging test method. However, the natural aging takes too long and the different years, seasons and climatic conditions of regions make the test results incomparable. The artificial accelerated ageing test strengthens some certain important factors of the natural storage, such as light, temperature, humidity, etc. shortens the cycle of the ageing test, and the reproducibility of the experimental results which is high because of the controllability of the testing conditions.

3 THE RESEARCH ON AGEING METHODS

Because the natural ageing takes too long, the artificial accelerated ageing test strengthens some certain important factors in the natural storage. The accelerated ageing test methods include: artificial climate tests, thermal ageing tests (anaerobic, hot air, thermal oxidation tests, etc.), ozone aging tests, UV-induced tests, etc.

3.1 *The artificial climate accelerated aging tests*

The regular changes in the atmospheric temperature reflect the changes in solar radiation intensity that the rotating Earth has accepted. The effect from the atmospheric temperature on a metal tank is not restricted by the day, and it always affects the temperature of the oil and gas in the tank. Vishnu, et al.[10] studied the effects of the solar radiation on the spherical tank gas space. Yang Hong-wei, et al.[11] measured and analysed the value and gradient changes of temperature in the upper, middle and lower apace of a tank in continuous 24 hours. The factors simulated in the artificial climate accelerated ageing tests are light, oxygen, heat, humidity, and rain. Among these, light is the most important one. The similarity between the sunlight spectrum and the artificial light spectrum will directly affect the reliability of the results of the ageing tests. The test's acceleration rate depends on a radiation intensity of the light source. Under the condition of the same degradation mechanisms, a choice of the ideal light source not only shows the validity of the test method, but also builds the correlation between the natural exposed ageing and the artificial climate accelerated exposed. Currently the light source used in artificial climate accelerated aging tests includes a carbon arc lamp, UV lamps, a xenon arc lamp, a high pressure mercury lamp, et al. Among these, the UV ageing instrument QUV and a Xenon arc lamp illumination are the most popular methods[12–14].

3.2 *Thermal ageing tests*

Thermal aging tests accelerate the process of the ageing of kerosene in oxygen and heat, in order to reflect the kerosene's anti-ageing properties in the heat oxygen. The oven ageing test method is a common method for the heat resistance tests, which is widely used in petrol, diesel, lubricating oil and other oil products. The test temperature is determined for the requirements of kerosene and experimental purposes. The upper temperature limit can be confirmed according to the relevant technical specifications, and we can select the appropriate temperature according to the highest heat resistance temperature of kerosene. If the oil temperature is too high, the viscosity of the oil will decline

too much to guarantee the required normal bearing oil film thickness, and the normal lubrication and bearing capacity can be broken, the oil aging deterioration accelerated, the asphalt jelly produced to clog the filter, so that the oil changeover cycle is reduced[15]. A sample is placed within the selected criteria heat oven, and periodical checks should be taken to figure out the change of the appearance and performance of the test sample in order to evaluate its heat resistance.

3.3 *The ozone ageing tests*

As the ozone content in the atmosphere is small, the ozone aging tests simulate and enhance the ozone conditions, using the strong oxidation of ozone to oxidate the sample, and under the circumstances of not affecting the continuity of ozone oxidation conditions, we can take the method of arranging the samples in series in order to analyse the density, viscosity and distillation range of the sample in different time periods by using the national standard. In the process of continuous oxidation, the colour of kerosene turns from transparent to light yellow, and finally yellow, as the time passed by the colour change becomes more and more obvious. By researching the rules of the ozone effects on kerosene, the rapid identification and evaluation of the anti-ageing properties of ozone and ozone-resistant protective efficacy can be made, and more effective measures can be taken to improve the life of kerosene.

4 COMPONENT ANALYSIS

Currently, the measuring of the quality of petroleum products mostly includes viscosity, flash point, acid value, etc. but there is little domestic study of component analysis. Reviewing the relevant literature, the main analytical methods at home and abroad include a fluorescent indicator adsorption, chromatography, spectrometry and gas chromatography – mass spectrometry.

High efficiency liquid chromatography, supercutical fluid chromatography, and mass spectrometry are mostly used to determine the group composition of oil, which is the distribution of compounds in oil. GB/T 11132-2002 adopts ASTM D1319 "the measurement of hydrocarbon in liquid petroleum products (fluorescent indicator adsorption)" equivalent, specifies the method that shows up hydrocarbons in petroleum products on the silica gel adsorption column with a fluorescent indicator, and calculates the volume fraction. This standard suits the determination of the petroleum hydrocarbons whose boiling point are less than 315°C.

Since 1996 ASTM regarded the supercritical fluid chromatography as a standard method to measure the single-ring aromatic hydrocarbons and polycyclic aromatic hydrocarbons in diesel and aviation fuel. Many scholars have conducted the studies following this direction, thinking that the supercritical fluid chromatography had excellent prospects in the composition analysis of diesel and the heavy distillate above it[17]. Wu Di et al.[18] have summarized the application of supercritical fluid chromatography analysis in petrochemical products, pointing out that as a chromatogram between gas chromatography and HPLC, the supercritical fluid chromatography can solve the problem that the gas chromatography and HPLC cannot solve, that is the separation of low polarity high-boiling organic compounds. However, as the supercritical fluid chromatography instruments are expensive, and the technology is not mature yet, they have not been widely used.

The infrared spectroscopy has been widely used in oil analysis. The near-infrared spectroscopy is a fast analytical technique, and more suitable for the needs of production and control for the advantages of fast, low energy consumption, low pollution and online analysis. There are many reports about the near-infrared spectroscopy used for the determination of gasoline[19] and diesel[21] fuel composition, but none for the kerosene reported. Wang Shao-jun, et al.[21] rapidly determined of the aromatic content of diesel with the mid-infrared spectroscopy, whilst Wang Li-jun, et al.[22] determined the oxygenates content of gasoline with mid-infrared spectroscopy. In addition, the ultraviolet spectroscopy can also be combined with the chemometrics component analysis. The UV spectrophotometry is mostly used for the determination of aromatic hydrocarbons in the sample, but it cannot get the information of other components (the ASTM D1840 determines the naphthalene hydrocarbons in the aviation fuel by using UV spectrophotometry, China's petrochemical industry standard SH/T0181-2005 modified method with the ASTM D1840 standard. This method has cumbersome operations, and its preparatory work is more complex, which extends the products analysis time)[23].

The hydrocarbons' isomers in kerosene have a large variety and a very complex composition. When the gas chromatography – mass spectrometry is used for the analysis of the composition of kerosene, the problem of lacking the sufficient capacity of the column arises. Therefore, this is mostly used for the main components' analysis, otherwise the components to be analysed should be separated at first. Liu Mao-lin, et al.[22] analysed kerosene lamps with gas chromatography – mass spectrometry. Nine types of aromatic hydrocarbons such as the straight-chain C7-C14 alkanes, benzene, and naphthalene, etc. were detected, as well as isoparaffins, cycloalkanes, and a very small amount of multi-carbonic acid and ketone. Bernabei, et al.[25] determined the total naphthalene in aeronautics kerosene with the ultraviolet spectrum. The method of the determination of the total aromatics content with the fluorescent indicator adsorption cannot measure every trace level of aromatics and PAHs prescribed by exposure limits. We can adopt a SIM to measure over 60 types of aromatic compounds and more than 30 types of PAHs separately in the 35 min and 40 min chromatographic processes. However, the GC has difficulties in qualitative issues, and the use of standard material presents the problems of having difficulties in obtaining all the standard material,

higher cost, and the long process analysis in characterizing every component in petroleum products. The use of gas chromatography – mass spectrometry can greatly solve the problems of qualitative issues in chromatography.

5 THE RELEVANCE AND LIFE ASSESSMENT OF AVIATION KEROSENE ACCELERATED AGEING AND NATURAL AGEING

Accelerated ageing is generally expressed as the ratio of the time used in a natural environment for the accelerated conditions to cause the same characteristics of the fuel changes[26]. In the artificial accelerated tests, only with the intensity of climate correctly determined, can the life of fuels forecast be carried out. The excessive exposure will reduce the relevance between the accelerated ageing and natural ageing fuels, and the following factors will reduce this relevance[27]:

(1) Shortwave light (outside the solar spectrum)
(2) High intensity of light (especially artificial light)
(3) Abnormally high temperature of the sample
(4) The difference temperature between the samples in different colour depth
(5) The absence of contaminants or other biological factors

As the ageing time passes by, oxidative degradation occurs to the aviation kerosene, resulting in a polymer and an acidic substance. Moreover, the acid value and viscosity of the oil will increase, which can lead to the oil deterioration. Study of the oil oxidation stability is more, but less on the study of storage and life assessment in different temperatures. The ageing studies of aviation kerosene in the future may use the combining methods of numerical simulation and experimental study, and the prediction of the service life of aviation kerosene under the conditions of different environmental factors will gradually be taken seriously. The method that is widely used now is the high-temperature accelerated ageing tests method[28]. The Arrhenius or Berthelot equations still should be used in a life prediction process. Weibull distribution is carried out by a Swedish physicist Weibull, and as one of the important life distributions it has been widely used in electronic devices life tests and mechanical products fatigue life tests, and their statistical analysis (tests, assessments, etc.) have been widely studied, provides many valid statistical inference methods[29].

6 CONCLUSION

In summary, under natural ageing conditions, the cycle of aviation kerosene changes in various physical and chemical conditions takes a long time and can have high costs. With the accelerated ageing test method, the ageing process of aviation kerosene is simulated, the changes of kerosene parameters in the ageing process analysed, the law of the ageing process is mastered, and the unstable fuel composition and the changes of acid degree value are tracked. In order to get information on oil's ageing deterioration and provide the basis for aviation kerosene's reliability, security, and some other assessments, and on the basis of this, make the implementation of life extension measures for aviation kerosene. With the update of basic research progress and technology, ideas regarding the problem of acceleration in the ageing of kerosene are put forward, based on the advanced test experience abroad, the management of tests on the information about storage can be strengthened and we can finish the test work on the life prediction for aviation kerosene storage.

REFERENCES

[1] Jiao Yan, Feng Lili, Zhu Yuelin, et al. Review of American military jet fuels development [J]. Journal of Rocket Prop-ulsion, 2008, 34(1): 30–35.

[2] Du Bofujin, Changruji (Translation). Aviation kerosene performance manual [M], Beijing: Aviation Industry Press, 1990

[3] Denisov E T, Changruji (Translation). The oxidation and suppress of jet fuel [M], Beijing: Hydrocarbon Processing Press, 1987.

[4] Liu Guozhu, Wang Li, QU Haijie, et al. Artificial neural network approaches on composition-property relationships of jet fuels based on GC-MS [J]. Fuel, 2007, 86(16): 2551–2559.

[5] Holl G. Former and modern method for the determination of the service life of rocketed propellant[R]. AD-A330303.

[6] Levebbach G J. Accelerated life testing of capacitors IRA-trans on reliability and quality control [J]. PGRQC, 1957, 10(1): 9–20.

[7] Wang Lingling. Accelerated life test [M], Beijing: The Science Press, 1997.

[8] Li Xiaojun, Chen Xinwen. Study on composites mechanical properties under accelerated ageing conditions [J]. Journal of Aeronautical Materials, 2003, 21(23): 286–288.

[9] Zhang Hairen, He Wensheng, Lin Mali. Domestic and foreign laboratories light source accelerated aging test equipment [J]. Synthetic Materials Aging and Application, 2007, 36(4): 47–49.

[10] Vishnu Verma, Ghosh A K, Kushwaha H S. Temperature Distribution and Thermal Stress Analysis of Ball-Tank Subjected to Solar Radiation [J]. Journal of Pressure Vessel Technology, 2005, 127(5): 119–122.

[11] Yang Hongwei, Yang Shiliang, Fei Yiwei, et al. Temperature gradient of hydrocarbon space in fixed roof tank and measures of reducing consumption by breathing loss [J]. 2011, 30(4): 314–315.

[12] Wang Chunchuang. Overview of artificial accelerated light aging test methods [J]. Electronic Product Reliability and Environmental Testing, 2009, 27(1): 65–69.

[13] ASTM D6152-2006, Standard Practice for Operating Open flame Carbon Arc Light Apparatus for Exposure of Non-metallic Materials.

[14] ASTM D6155-2005x, Operating Xenon Arc Light Apparatus for Exposure of Non-Materials.

[15] Jiang Tong. Analysis and research on reducing the oil temperature of diesel engine [J]. Shandong Internal Combustion Engine, 2004, 3: 39–43.

[16] Xu Yongye. The study of gasoline and poly tetrahydrofuran with near-infrared spectra [D], Nanjing: Nanjing University of Technology, 2005.

[17] Adam F, Thiébaut D, Bertoncini F, et al. Supercritical fluid chromatography hyphenated with twin comprehensive two-dimentional gas chromatography for ultimate analysis of middle distillates [J]. Journal of Chromatography A, 2010, 1217(8):1386–1394.

[18] Wu Di, Qi Bangfeng, Cheng Zhongqian. Applications of supercritical fluid chromatography in petrochemical analysis [J]. Chemical Analysis and Meterage, 2008, 17(5): 74–77.

[19] He Yuguang. Application of the near infrared spectroscopy in analysis of gasoline [J]. Liaoning Chemical Industry, 2008, 37(11): 787–789.

[20] Xu Guangtong, Liu Zelong, Yang Yuxin, et al. Determination of diesel fuel composition by near infrared spectroscopy and its application [J]. Acta Petrolei Sinica (Petroleum Processing Section), 2002, 18(4): 65–71.

[21] Wang Shaojun, Wu Hongxin, Ling Fengxiang. Determination of aromatics in diesels by MIR [J]. Journal of Analytical Science, 2003, 19(5): 437–438.

[22] Wang Lijun, Cheng Zhongqian, Qi Bangfeng. The Determination of oxygenates in clean gasoline by mid-infrared spectroscope [J]. Contemporary Chemical Industry, 2005, 34(5): 359–361.

[23] Zhao Hongmei, Liu Tiancai. Ultraviolet-spectrophotometric determination of aromatic in specialty solvents [J]. Guangdong Chemical Industry, 2002, 29(6): 31–33, 19.

[24] Liu Maolin, Xu Lilian, Wang Kunhua, et al. Analysis of aromatie components in the domestic lamp kerosene [J]. Petrochemical Technology, 1995, 24(12): 888–890.

[25] Bernabei M, Reda R, Galiero R, et al. Determination of total and polycyclic aromatic hydrocarbons in aeronautics jet fuel [J]. Journal of Chromatography A, 1995, 985(1/2): 197–203.

[26] Zhang Chunhua, Wen Xiseng, Chen Xun. A Comprehensive Review of Accelerated Life Testing [J]. Acta Armamentarii, 2004, 25(4): 485–490.

[27] George Wypych. Acceleration, Correlation, and Service Life Prediction [J]. Environmental Technology 2001, (4): 23–26.

[28] Zhang Zhihua. Accelerated life testing and statistical analysis [M]. Beijing: Beijing Industrial University Press, 2002.

[29] GB2689.4-81. Life testing and accelerated life testing best linear unbiased estimator (for the Weibull distribution), People's Republic of China National Standrad, 1981.

Biomedical Engineering and Environmental Engineering – Chan (Ed.)
© *2015 Taylor & Francis Group, London, ISBN: 978-1-138-02805-0*

The use of diagnostic ratios in Polycyclic Aromatic Hydrocarbons (PAHs) as a method for the identification of pollution sources in the Moravian-Silesian region of the Czech Republic

M. Kucbel, B. Sykorova, J. Kucharova, H. Raclavska & K. Raclavsky
ENET – Energy Units for Utilization of non Traditional Energy Sources, Ostrava, Czech Republic

ABSTRACT: Polycyclic Aromatic Hydrocarbons (PAHs) analytes were investigated in order to identify their sources of emission using the diagnostic ratio method. The characterization was conducted in 40 sites of the Moravian-Silesian region of the Czech Republic. The sites were divided into five groups depending on the level of their pollution. In addition, the particulate matter with a diameter of 10 mm or less (PM_{10}) and soot from diesel engines were analyzed. It was found that (i) phenantrene (Phe), fluoranthene (Flu), fluorene (Fl), and pyrene (Pyr) were present in the highest amounts, (ii) the ratio of benzo[b]fluoranthene (BbF) and benzo[k]fluoranthene (BkF) was a good marker of traffic emissions, (iii) the ratio of Benzo[a]Pyrene (BaP) and Benzo[ghi]Perylene (BghiP) as well as the ratio of Flu/Flu + Pyr could be a suitable marker for emissions from coal combustion, but (iv) coronene (Cor) is not a reliable marker of traffic emissions. Furthermore, the diagnostic ratio of PAHs present in PM_{10} were dependent on size distribution, whereas the soot from diesel engines was affected the age of the car.

Keywords: polycyclic aromatic hydrocarbons, diagnostic ratios, pollution sources, Moravian-Silesian Region, coronene

1 INTRODUCTION

Polycyclic aromatic hydrocarbons (PAHs) are universal environmental pollutants that consist of two or more condensed benzene rings. They are formed at high temperatures during the incomplete combustion of carbonaceous materials (Ravindra et al., 2008). The presence of PAHs in the air of most urban and rural areas is mostly associated with emissions from cars, burning fossil fuels in residential furnaces (Singh et al., 2008) and industrial activities involving combustion processes (Omar et al., 2002). Depending on their origin, PAHs can be divided into two groups: pyrogenic PAHs and petrogenic PAHs (Hylland, 2006). Pyrogenic PAHs are formed during the incomplete combustion of organic material, while petrogenic PAHs are present in crude oil and in some petroleum products such as kerosene, gasoline, diesel, fuel, lubricating oils, etc. The main sources of pyrogenic PAHs are forest fires, the incomplete combustion of fossil fuels (especially coal) and wood, waste incineration and to some extent the combustion of gasoline and diesel fuels from motor vehicles (Mohammadi et al., 2010). PAHs analytes with a Higher Molecular Weight (HMW) are typical for pyrogenic sources, whereas PAHs analytes with a Lower Molecular Weight (LMW) are common for petrogenic sources (Neff et al., 2003). The analytes ratio of LMW/HMW is often used to distinguish pyrogenic and petrogenic sources of the PAHs. Specifically, a higher concentration of PAHs with LMW indicates a higher content of unburned petroleum products, while the presence of PAHs with HMW is typically associated with pyrogenic products released mainly during burning of fossil fuels (e.g. coal) (Hassanien & Abdel-Latif, 2008). Ravindra et al. (2006) state that the main source of PAHs with 3 and 4 rings (LMW) is coal, whereas the main source of PAHs with HMW such as benzo[a]pyrene (BaP), benzo[b]fluoranthene (BbF), benzo[ghi]perylene (BghiP) and indeno[1,2,3-cd]pyrene (IND) are gasoline vehicles. In addition, the PAHs with LMW are more volatile than the PAHs with HMW. Therefore, the gaseous PAHs with LMW are often subjected to a long-range transport and allow measurements of their higher concentrations in remote areas (Mazquiarán & Pinedo, 2007). The sum of fluoranthene (Flu), pyrene (Pyr), chrysene (Chr), BaP, IND, BghiP, benzo[k]fluoranthene (BkF) and benzo[a]anthracene (BaA) analytes (i.e., the typical markers for combustion) is referred to as COMB PAHs (Kong et al., 2013; Wu et al., 2014).

The most commonly used technique for the qualitative determination and characterization of sources of PAHs is a method utilizing diagnostic ratios of PAHs; their analytes are characteristic for certain emission sources (Galarneau, 2008; Ravindra et al., 2008). The principle of this method is based on calculating the ratio of concentrations of specific pairs of PAHs that are characteristic for a particular source or group of sources of pollution, and then comparing these ratios with the values reported in the literature. However, this method is not unrestricted from some drawbacks such as (i) change in the concentration of analytes used for diagnostic ratios due to their reactivity with other atmospheric agents (e.g. ozone and/or nitrogen oxides), (ii) considerable variability of the PAHs concentrations within seasons (Tobiszewski & Namiśnik 2012), and (iii) significant lack of diagnostic ratios that is primarily attributed to the wide and variable range of ratios of the individual PAHs and depends on the source and author of the study (Liu et al., 2013) Specifically, according to Hu et al (2012) the ratio of BaP/BghiP should be a good indicator of the emission of PAHs from transport (0.3–0.4 range) and coal (0.9–6.6 range) Whereas Katsoyiannis et al. (2007) report that the sources of the ratio of BaP/BghiP with value <0.6 are non-traffic emissions and with a value >0.6 the source is a traffic emission. However, some ratios represent well the emission of PAHs; for example, the ratios of BaA/(BaA + Chr) and IND/(IND + BghiP) which indicate the sources of emission as transport or home heating. (Tobiszewski & Namiśnik 2012)

The purpose of this study is to determine the potential sources of 18 PAHs analytes using diagnostic ratios In particular the sum of 16 major PAHs defined by the USEPA [acenaphtylene (AcPy) acenaphtene (Acp) anthracene (Ant) benzo[a]anthracene (BaA) benzo[a]pyrene (BaP) benzo[b]fluoranthene (BbF) benzo[g,h,i]perylene (BghiP) benzo[k]fluoranthene (BkF) dibenzo[a,h] anthracene (Dia,hA) fluoranthene (Flu) fluorene (Fl) chrysene (Chr) Indeno[1,2,3-cd]pyrene (IND) naphthalene (Nap) phenanthrene (Phe) pyrene (Pyr)]; coronene (Cor), and benzo[e]pyrene (BeP) are investigated. The PAHs are extracted from the fraction of PM_{10} at 40 selected locations of the Moravian-Silesian Region (MSR), in the Czech Republic (CZ) in the winter season.

2 MATERIALS AND METHODS

2.1 Sampling sites

The map of 40 sampling sites of MSR, CZ are presented in Figure 1. The sites were grouped according to the prevailing sources of PAHs into 5 groups:

Background and mountain sites without contamination: 10, 23, 30, 31, and 35.
Sites with majority of residential heating: 7, 8, 9, 11, 12, 13, 14, 15, 19, 26, 27, 28, 29, 33, 36, 38, and 39.
Urban agglomerations: 1, 2, 3, 5, 6, 17, 20, and 21.
Sites with cities without significant industrial activity: 22, 24, 25,and 37.
Sites with a high traffic load: 4, 16, 18, 32, 34, and 40.

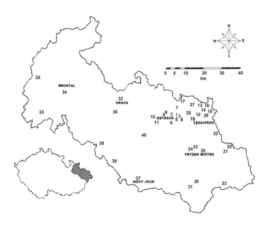

Figure 1. Map of the sampling sites.

* 1. Ostrava-Radvanice, 2. Ostrava-Marianske Hory, 3. Ostrava-Radvanice OZO, 4. Ostrava-Poruba, 5. Ostrava-Radvanice church, 6. Ostrava-Dubina, 7. Ostrava-Hermanice, 8. Senov, 9. Ostrava-Hostalkovice, 10. Klimkovice-Sanatoria, 11. Klimkovice, 12. Karvina Doly, 13. Karvina Stare Mesto, 14. Karvina Frystat, 15. Karvina-Raj,16. Karvina, 17. Bohumín-Skrecon, 18. Havirov, 19. Havirov-Sumbark, 20. Trinec, 21. Trinec-Oldrichovice, 22. Cesky Tesin, 23. Horni Lomna, 24. Frydek-Mistek A, 25. Frydek-Mistek B, 26. Nosovice, 27. Orlova, 28. Petrvald, 29. Stonava, 30. Ostravice, 31. Celadna, 32. Opava, 33. Rymarov, 34. Bruntál, 35. Karlova Studanka, 36. Hradec nad Moravici, 37. Novy Jicin, 38. Vitkov, 39. Odry, and 40. Bilovec.

2.2 Sampling

The samples were collected during the winter season 2013/2014. Sampling was conducted at each site for 24 hours using a high-volume sampler Digittel MD 05. Concentrations of PAHs analytes were also monitored by direct analysis of PM_{10} samples that were collected from major energy sources (i.e., power plant Tøebovice and heating plant Teplárna Přívoz, operated by DALKIA a.s.). The sampling of PM_{10} particulate was conducted using a tapping impactor Kalman KS 404 according to the standard test method EN ISO 23210 "Stationary Source Emissions – Determination of $PM_{10}/PM_{2.5}$ Mass Concentration in Flue Gas – Measurement at Low Concentrations by Use of Impactors". Sampling of soot from six cars (Skoda Superb and Skoda Octavia) was performed on a station of technical inspection.

2.3 Analytical procedure

The separation and identification of PAHs were conducted using the following methods (i) ISO 11338-2:2003 "Stationary Source Emissions – Determination of Gas and Particle-Phase Polycyclic Aromatic Hydrocarbons – Part 2: Sample Preparation, Clean and

Destination" using a HPLC – PDA system and (ii) iso-topic dilution method using a HRGC – HRMS system. The analyses were performed in the ALS laboratories, CZ. The analysis of PM_{10} emission was conducted using py-GC/MS in the IGI laboratories at VŠB – Technical University of Ostrava, CZ.

3 RESULTS AND DISCUSSION

3.1 *Diagnostic ratios*

Identification of PAHs sources in the air of MSR was based on the selected diagnostic ratios and the reference values that are listed in Table 1. Table 2 presents the diagnostic ratios of PAHs obtained from MSR, CZ. The ratio of ΣCOMB/ΣPAHs was not unambiguous for any of the sites and the average value for all locations was in the amount of 0.47. Thus it was not possible to determine the pyrogenic or petrogenic source of the PAHs. The ratio of IND/(IND + BghiP) was in the range of 0.35–0.44 in all sites and was equivalent to the emissions from diesel transport. The values of the diagnostic ratios accounting for emission from coal should be higher than 0.56 albeit this value was not achieved in either case in this study demonstrating that this ratio does not have an indicative value for the MSR. The ratio of BaP/(BaP + Chr) was approximately 0.47 for all sites. This value corresponds well with the combustion of coal but also transport emissions and can be used to distinguish emission sources. Diagnostic ratio of BbF/BkF was in the range of 1.62 1.84 which is higher than the diagnostic ratio reported for the combustion of fossil fuels and biomass and for emissions from diesel engines (>0.5) and can be used to identify the sources of contamination. The ratio of BaP/BghiP was in the range of 1.17–1.87 and it is two to three times higher than values reported for non-transport and transport emissions in the literature (Katsoyiannis et al., 2007). The ratio of Flu/(Flu + Pyr) in the range of 0.54–0.66 refers to the combustion of coal and biomass. The value of the BaA/(BaA + Chr) ratio in the amount of 0.46 was observed only for 10 sites and was identified as a source coal combustion. In addition this value does not match any other values reported in the literature.

The best determination of pollution sources is provided by the ΣLMW/ΣHMW ratio which divides the sites into sites affected by transportation and by coal combustion. For sites with majority of residential heating the ΣLMW/ΣHMW ratio corresponded predominantly with petrogenic sources (sites 7, 8, 9, 11, 19, 26, 27, 29 36, and 38), whereas for the remaining sites this ratio was mostly pyrogenic (sites 12, 13, 14, 15 and 39). A clear determination of the emission source was not possible only in two cases, i.e., for sites 27 and 33. A pyrogenic origin of emissions was therefore detected for all locations within the Karvina agglomeration although based on the remaining diagnostic ratios the coal combustion origin which is typical for this area was not confirmed. The values of the ΣLMW/ΣHMW ratio for the background

Table 1. Reference values of PAHs diagnostic ratios.

Ratio	Source of emission	Value
Fl/(Fl + Pyr)	Diesel	>0.5[a]
	Gasoline	<0.5[a]
	Coal/coke	0.53[b]
	Coal combustion	0.57[c]
	Iron smelt	0.4[d]
	Coke manufacture	0.5[d]
BbF/BkF	Diesel	>0.5[e]
IND/(IND + BghiP)	Petrogenic sources	<0.2[f]
	Oil combustion	0.2–0.5[f]
	Grass, wood, and coal combustion	>0.5[f]
	Coal combustion	0.56[g]
	Iron smelt	0.37[g]
	Coke manufacture	0.90[g]
BaP/(BaP + Chr)	Diesel	<0.5[h]
	Gasoline	>0.5[h]
	Coal combustion	0.46[c]
	Vehicles with diesel engine	0.65[c]
	Vehicles with gasoline engine	0.5[c]
	Wood combustion	0.59[c]
BaA/(BaA + Chr)	Coal/coke	0.5[g]
	Coal combustion	0.46[g]
	Iron smelt	0.48[g]
	Coke manufacture	0.30[g]
Flu/(Flu + Pyr)	Petrogenic sources	<0.4[i,j]
	Fossil fuels combustion	0.4–0.5[i]
	Grass, wood, and coal combustion	>0.5[i]
BaP/BghiP	Non-transport emissions	<0.6[j]
	Transport emissions	>0.6[j]
ΣLMW/ΣHMW	Petrogenic sources	>1[k]
	Pyrogenic sources	<1[k]
ΣCOMB/ΣPAHs	Petrogenic sources	0.3[i,k]
	Pyrogenic sources	0.7[i,k]

ΣLMW: PAHs with 2–3 benzene rings, ΣHMW: PAHs with 4–6 benzene rings, [a]*Mandalakis et al., 2002;* [b]*Kong et al., 2010;* [c]*Galarneau, 2008;* [d]*Manoli et al., 2004;* [e]*Pandey et al., 1999;* [f]*Yunker et al., 2002;* [g]*Kong et al., 2013;* [h]*Teixeira et al., 2012;* [i]*Wu et al., 2014;* [j]*Katsoyiannis et al., 2007;* [k]*Jamhari et al., 2014.*

and mountain sites ranged from 0.95 to 1.93. This is a further confirmation that PAHs with LMW are more volatile than PAHs with HMW, so the gaseous PAHs with LMW have a greater potential for a long-range transport. This is also the reason why the concentration of PAHs with LMW is higher than the concentration of PAHs with HMW in remote areas.

The comparison of the median of diagnostic ratios of 40 sites with diagnostic ratios determined from the coal combustion emissions in large power plants (PM_{10} and $PM_{2.5-10}$) and soot from diesel engines is shown in Table 3. The differences between the ratios in each class (PM_{10} and $PM_{2.5-10}$) are significant which

Table 2. PAHs diagnostic ratios in the MSR.

	Site	Fl/(Fl+Pyr)	IND/(IND+BghiP)	BaP/(BaP+Chr)	BbF/BkF	BaP/BghiP	BaA/(BaA+Chr)	Flu/(Flu+Pyr)	ΣLMW/ΣHMW	COMB/ΣPAHs
Urban agglomerations	1	**0.46**	0.45	0.48	1.81	1.60	0.47	0.66	0.72	0.58
	2	0.22	0.44	0.45	1.84	1.36	0.38	0.62	1.24	0.45
	3	**0.43**	0.44	0.50	1.69	1.87	0.53	0.56	0.84	0.54
	5	**0.40**	0.40	0.48	1.78	1.73	0.50	0.59	0.82	0.54
	6	0.51	0.35	0.50	1.62	1.40	0.50	0.54	0.97	0.50
	17	**0.41**	0.39	0.43	1.83	1.17	0.47	0.65	0.95	0.51
	20	0.57	0.44	0.47	1.75	1.39	0.50	0.59	0.94	0.51
	21	0.64	0.40	0.46	1.72	1.39	0.52	0.58	1.16	0.46
Sites without urban acitivies	22	0.41	0.42	0.49	1.67	1.28	0.53	0.63	1.16	0.46
	24	0.50	0.42	0.54	1.80	1.46	**0.48**	0.64	1.35	0.42
	25	0.51	0.45	0.49	1.69	1.62	0.49	0.60	1.17	0.46
	37	0.59	0.43	0.43	1.80	1.00	0.40	0.65	1.36	0.42
Transport sites	4	0.36	0.44	0.50	1.78	1.32	0.52	0.61	1.03	0.49
	16	0.55	0.46	0.42	1.72	1.27	0.42	0.64	0.90	0.52
	18	0.38	0.39	0.46	1.73	1.59	**0.48**	0.60	0.85	0.54
	32	0.42	0.44	0.41	1.59	1.29	0.52	0.56	0.94	0.51
	34	0.56	0.45	0.43	1.70	1.40	0.50	0.58	1.06	0.48
	40	0.69	0.56	0.41	1.63	1.92	0.47	0.64	1.43	0.41
Sites with residential heating	7	0.51	0.41	0.48	1.65	1.56	0.45	0.61	1.27	0.45
	8	0.44	0.42	0.47	1.65	1.67	0.45	0.56	1.51	0.40
	9	0.66	0.46	0.43	1.78	1.19	0.46	0.62	1.37	0.42
	11	0.42	0.42	0.46	1.62	1.54	0.44	0.64	1.33	0.46
	12	0.58	0.41	0.41	1.75	1.34	0.46	0.66	0.98	0.50
	13	0.36	0.39	0.48	1.78	1.49	0.53	0.65	0.83	0.54
	14	0.49	0.43	0.41	1.67	1.50	0.52	0.58	0.67	0.59
	15	0.43	0.38	0.47	1.65	1.37	0.52	0.59	0.80	0.55
	19	0.64	0.44	0.46	1.60	1.32	0.54	0.63	1.14	0.46
	26	0.58	0.42	0.43	1.71	1.44	0.47	0.57	1.41	0.41
	27	0.50	0.44	0.54	1.60	1.95	0.53	0.64	1.03	0.45
	28	0.53	0.40	0.55	1.74	1.85	0.54	0.66	1.71	0.36
	29	0.53	0.42	0.50	1.64	1.26	0.49	0.56	1.19	0.45
	33	0.56	0.45	0.45	1.77	1.61	**0.48**	0.59	1.04	0.49
	36	0.62	0.43	0.49	1.60	1.38	0.55	0.59	1.23	0.44
	38	0.61	0.45	0.43	1.65	1.58	0.46	0.61	1.45	0.41
	39	0.33	0.45	0.47	1.60	1.34	0.46	0.60	0.82	0.55
Background and mountain sites	10	0.65	0.44	0.45	1.78	1.34	0.38	0.65	1.93	0.37
	23	0.53	0.45	0.48	1.95	1.13	0.33	0.61	1.75	0.38
	30	0.42	0.46	0.44	1.58	1.33	**0.48**	0.60	0.95	0.52
	31	0.46	0.44	0.47	1.81	1.57	0.47	0.44	1.26	0.44
	35	0.57	0.48	0.44	1.69	1.47	0.45	0.60	1.00	0.49

PAHs from combustion in diesel engines	Coal combustion	**Iron smelt**	Pyrogenic sources	No clear determininion
PAHs from combustion in gasoline engines	Transport emissions	Coal and biomass combustion	Petrogenic sources	

indicates that the diagnostic ratios are influenced by the particle size distribution of PM$_{10}$ The analytes from diesel engines soot are affected by the age of the vehicle. The comparison of the median values of diagnostic ratios with newly defined ratios show that the ratio BbF/BkF may be a good indicator of the air pollution from traffic whereas the ratio of BaP/BghiP and Flu/Flu + Pyr may indicate the emissions from coal combustion.

3.2 Coronene as a marker of transport emissions?

There is a wide range of molecular markers used to identify substances of pyrogenic or petrogenic origin the source of transport or non-transport emissions or to detect the type of fuel used during various combustion processes. One of these markers is Cor which is considered as a typical molecular marker

of transport emissions The average value of Cor for the sites with a high load of traffic was the greatest (2.93 ng/m^3), although even for the sites characterized by industrial activities and the sites with majority of residential heating the values of Cor were also high (2.03 ng/m^3 for the sites with industrial activities and. 2.00 ng/m^3 for the sites with majority of residential heating). Lower average values of Cor were obtained for the sites without industrial activities (1.34 ng/m^3) and very low average concentrations of Cor were observed for the background and mountain sites (0.77 ng/m^3).

The average value of the sum of 16 PAHs for the sites with a high traffic load was 400.97 ng/m^3 for the industrial sites 575.75 ng/m^3 for the sites without industrial activity 218.83 ng/m^3 for the sites with a predominance of residential heating 442.41 ng/m^3 and finally the lowest average value of the sum of 16 PAHs

Table 3. Comparison of the median of diagnostic ratios of 40 sites with diagnostic ratios determined from the coal combustion emissions in large power plants and soot from diesel engines.

	Fl/ (Fl + Pyr)	IND/ (IND + BghiP)	BaP/ (BaP + Chr)	BbF/ BkF	BaF/ BghiF
Median	0.51	0.44	0.45	1.69	1.40
PM_{10}	0.50	0.50	0.20	0.40	1.00
$PM_{2.5-10}$	0.20	1.00	0.30	0.81–0.90	0.28
Soot	0.22	0.34	0.37	1.41	0.39

	BaA/ (BaA+ Chr)	Phe (Phe+ Ant)	Flu/ (Flu+ Pyr)	$\Sigma LM/$ ΣHMW Σ PAHs	COMB/
Median	0.48	0.48	0.60	1.23	0.46
PM_{10}	0.39	0.39	0.50	0.29	0.23
$PM_{2.5-10}$	0.86	0.86	0.21	0.88	0.56
Soot	0.37	0.37	0.22	0.25	0.69

Table 4. Ratio of BaP and Cor for fuel and combustion processes (Shen et al., 2014).

Combustion processes	Ratio of BaP and Cor
Residential coal combustion	2.0
Residential wood combustion	2.5
Combustion of pellets (biomass)	1.8
Gas transport	3.2
Diesel transport	4.2

showed that the ratios which varied between the grain size classes and their value were related to a particle sizes

In addition, it was found that the ratios of BbF/BkF were good indicator of the air pollution from traffic and that the ratio of BaP/BghiP and Flu/Flu + Pyr could indicate coal combustioTable 2. PAHs diagnostic ratios in the MSR Coronene was not found to be a unique marker of traffic emissions as described in the prior literature. Furthermore, significant correlations were observed between Cor and the sum of 16 PAHs and other analytes. The high concentrations of Cor were obtained mainly from industrial sites of the MSR and sites characterized by prevailing residential heating. Finally, high concentrations of Cor were observed because of the combustion of fossil fuels and biomass, which in some cases even exceeded the values obtained from traffic emissions.

was obtained for the mountain and background sites $(172.87\,\mathrm{ng/m^3})$.

Of the ten sites with the highest concentrations of Cor only two sites were included in the category influenced by a high load of traffic. Specifically, site 32 for which a value of $4.19\,\mathrm{ng/m^3}$ of Cor was observed and site 4 for which a value of $2.73\,\mathrm{ng/m^3}$ of Cor was observed. Four other sites fell into the category of urban-industrial sites without industrial activities: site 3 for which a value of $6.28\,\mathrm{ng/m^3}$ of Cor was observed, site 5 with a value of $4.94\,\mathrm{ng/m^3}$ site 21 with a value of 3.3 $\mathrm{ng/m^3}$ and site 20 with values of $2.79\,\mathrm{ng/m^3}$ The remaining four sites were in the category with a predominance of residential heating: site 12 with a Cor concentration of $4.91\,\mathrm{ng/m^3}$ site 15 with $3.87\,\mathrm{ng/m^3}$ site 13 with $3.88\,\mathrm{ng/m^3}$ and site 27 with $3.22\,\mathrm{ng/m^3}$ Since Cor is being emitted in greater amounts from the combustion of solid fossil fuels or biomass, from the residential heating and during various industrial activities, it cannot be considered as a product only delivered by transport emissions, and therefore it cannot be considered as a characteristic indicator of PAHs emissions from transport. In addition, positive correlations were found not only with BaP ($R = 0.897, p < 0.05$) but with a sum of 16 PAHs ($R = 0.917, p < 0.05$) were investigated as well. Therefore the determination of a clear origin for Cor was not possible. This fact was also confirmed by Shen et al. (2014) who showed the mean ratio of BaP and Cor for different types of fuels and various combustion processes.

4 CONCLUSIONS

Among the 18 PAHs investigated the highest concentrations were observed for phenantrene, fluorahnthen, fluorene, and pyrene. The calculation of the diagnostic ratios of PAHs analytes in the high pollution sources

ACKNOWLEDGEMENT

This paper was supported by research projects of the Ministry of Education, Youth and Sport of the Czech Republic: Modelování a měření termických a energetických procesů SP2014/59 and the Technology Agency of the Czech Republic TA02020004 The Research of Chemical and Physical Character of Particles in Emissions.

REFERENCES

Galarneau, E. 2008. Source specificity and atmospheric processing of airborne PAHs: Implications for source apportionment. *Atmospheric Environment* 42: 8139–8149.

Hassanien, M.A. & Abdel-Latif, N.M. 2008. Polycyclic aromatic hydrocarbons in road dust over Greater Cairo, Egypt. *Journal of Hazardous Materials* 151: 247–254.

Hu, J., Liu, C.Q., Zhang, G.P. & Zhang, Y.L. 2012. Seasonal variation and source apportionment of PAHs in TSP in the atmosphere of Guiyang, Southwest China. *Atmospheric Research* 118 (15): 271–279.

Hylland, K. 2006. Polycyclic aromatic hydrocarbon (PAH) ecotoxicology in marine ecosystems. *Journal of Toxicology and Environmental Health* Part A 69:109–123.

Jamhari, A.A., Sahani, M., Latif, M. T., Chan, K. M., Tan, H. S., Khan, M. F. & Tahir, M. 2014. Concentration and source identification of polycyclic aromatic hydrocarbons (PAHs) in PM_{10} of urban, industrial and semi-urban areas in Malaysia. *Atmospheric Environment* 86: 16–27.

Katsoyiannis, A., Sweetman, A.J. & Jones, K.C. 2011. PAH molecular diagnostic ratios applied to atmospheric sources: a critical evaluation using two Decades of source

Inventory and air concentration data from the UK. *Environmental Science and Technology* 45 (20): 8897–906.

Kong, S., Ding, X., Bai, Z., Han, B., Chen, L., Shi, J. & Li, Z. 2010. A seasonal study of polycyclic aromatic hydrocarbons in PM$_{2.5}$ and PM$_{2.5-10}$ in five typical cities of Liaoning Province, China. *Journal of Hazardous Materials* 183 (1–3): 70–80.

Kong, S., Ji, Y., Li, Z., Lu, B. & Bai, Z. 2013. Emission and profile characteristic of polycyclic aromatic hydrocarbons in PM$_{2.5}$ and PM$_{10}$ from stationary sources based on dilution sampling. *Atmospheric Environment* 77: 155–165.

Liu, J., Li, J., Lin, T., Liu, D., Xu, Y., Chaemfa, C., Qi, S., Liu, F. & Zhang, G. 2013. Diurnal and nocturnal variations of PAHs in the Lhasa atmosphere, Tibetan Plateau: Implication for local sources and the impact of atmospheric degradation processing. *Atmospheric Research* 124: 34–43.

Mandalakis, M., Tsapakis, M., Tsoga, A. & Stephanou, E.G. 2002. Gas-particle concentrations and distribution of aliphatic hydrocarbons, PAHs, PCBs and PCDD/Fs in the atmosphere of Athens (Greece). *Atmospheric Environment* 36: 4023–4035.

Manoli, E., Kouras, A. & Samara, C. 2004. Profile analysis of ambient and source emitted particle-bound polycyclic aromatic hydrocarbons from three sites in northern Greece. *Chemosphere* 56: 867–878.

Mazquiarán, M.A.B. & Pinedo, L.C.O. 2007. Organic composition of atmospheric urban aerosol: variations and sources of aliphatic and polycyclic aromatic hydrocarbons. *Atmospheric Research* 85: 288–299.

Mohammadi, C., Adeh, Z., Saify, A. & Shalikar, H. 2010. Polycyclic Aromatic Hydrocarbons (PAHS) along the Eastern Caspian Sea Coast. *Global Journal of Environmental Research* 4(2): 59–63.

Neff, J.M., Boehm, P.D. Kropp, R., Stubblefield, W.A. & Page, D.S. 2003. Monitoring recovery of Prince William Sound, Alaska, following the Exxon Valdez oil spill: Bioavailability of PAH in offshore sediments. *In Proceedings of the international oil spill conference Publication No. I 4730 B* Washington, DC: American Petroleum Institute.

Omar, N.Y.M.J, Abas, M.R.B., Ketuly, K.A. & Tahir, N.M. 2002. Concentrations of PAHs in atmospheric particles (PM-$_{10}$) and roadside soil particles collected in Kuala Lumpur, Malaysia. *Atmospheric Environment* 36 (2): 247–254.

Pandey, P.K., Patel, K.S. & Lenicek, J. 1999. Polycyclic aromatic hydrocarbons: need for assessment of health risks in India? Study of an urban-industrial location in India. *Environmental Monitoring and Assessment* 59: 287–319.

Ravindra, K., Bencs, L., Wauters, E., Hoog, J., Deutsch, F., Roekens, E., Bleux, N., Berghmans, P & Grieken, R. 2006. Seasonal and site-specific variation in vapour and aerosol phase PAHs over Flanders (Belgium) and their relation with anthropogenic activities. *Atmospheric Environment* 40: 771–785.

Ravindra, K., Sokhi, R. & Grieken, R. 2008. Atmospheric polycyclic aromatic hydrocarbons: Source attribution, emission factors and regulation. *Atmospheric Environment* 42: 2895–921.

Shen, G., Chen, Y, Wei, S., Fu, X., Ding, A., Wu, H. & TAO, S. 2014. Can Coronene and/or Benzo(a)pyrene/Coronene ratio act as unique markers for vehicle emission? *Environmental Pollution* 184: 650–653.

Singh, K.P., Malik, A., Kumar, A, Saxena, P. & Sinha, S. 2008. Receptor modeling for source apportionment of polycyclic aromatic hydrocarbons in urban atmosphere. *Environmental Monitoring and Assessment* 136: 183–196.

Teixeira, E.C., Agudelo-Castañeda, D. M., Guimarães Fachel, J. M., Leal, K. A., Oliveira Garcia, G. & Wiegand, F. 2012. Source identification and seasonal variation of polycyclic aromatic hydrocarbons associated with atmospheric fine and coarse particles in the Metropolitan Area of Porto Alegre, RS, Brazil. *Atmospheric Research* 118 (15): 390–403.

Tobiszewski, M. & Namieśnik, J. 2012. PAH diagnostic ratios for the identification of pollution emission sources. *Environmental Pollution* 162: 110–119.

Wu, D., Wang, Z, Chen, J., Kong, S., Fu, X., Deng, H., Shao, G. & Wu, G. 2014. Polycyclic aromatic hydrocarbons (PAHs) in atmospheric PM$_{2.5}$ and PM$_{10}$ at a coal-based industrial city: Implication for PAH control at industrial agglomeration regions, China. *Atmospheric Research* 149: 217–229.

Yunker M.B, Macdonald R.W, Vingarzan R, Mitchell R.H, Goyette D & Sylvestre, S. 2002. PAHs in the Fraser River basin: a critical appraisal of PAH ratios as indicators of PAH source and composition. *Organic Geochemistry* 33: 489–515.

Biomedical Engineering and Environmental Engineering – Chan (Ed.)
© *2015 Taylor & Francis Group, London, ISBN: 978-1-138-02805-0*

Use of zebrafish (danio rerio) embryos to assess the effect of α-Naphthoflavone on the toxicity of M-methylphenol

Jianming Yang
CAS Key Lab of Bio-Medical Diagnostics, Suzhou Institute of Biomedical Engineering and Technology, Chinese Academic of Sciences, Suzhou, Jiangsu, China
Changchun Institute of Optics, Fine Mechanics and Physics, Chinese Academy of Sciences, Changchun, China
University of Chinese Academy of Science, Beijing, China

Pengli Bai, Huancai Yin, Ying Lin, Mingli Chen & Jian Yin
CAS Key Lab of Bio-Medical Diagnostics, Suzhou Institute of Biomedical Engineering and Technology, Chinese Academic of Sciences, Suzhou, Jiangsu, China

ABSTRACT: Zebrafish early life stage (ELS) assays were used to assess the effect of α-Naphthoflavone (ANF) on the toxicity of m-cresol. Treatment of m-cresol (5–250 mg/L) alone resulted in both lethal and sub-lethal toxic effects. While in the previous experiment, ANF (0.1–3 mg/L) only caused sub-lethal effects including cardiac edema and delayed hatching Based on the endpoints of no blood circulation and slower heart rate at 48 hpf, toxicity of m-cresol was significantly enhanced by ANF. Furthermore, corresponding oxidative stress responses of zebrafish embryos after 48 h treatment were evaluated and the results showed that the mixture of m-cresol and ANF caused a much more significant decrease of reduced glutathione (GSH) content and superoxide dismutase (SOD) activity, as well as a more obvious increase of malondialdehyde (MDA) level than m-cresol alone. Thus, ANF significantly enhanced the embryotoxicity of m-cresol toxicity, which should be due to the modulation of detoxification pathways in zebrafish embryos.

Keywords: M-methylphenol; α-Naphthoflavone; Zebrafish embryo; Oxidative stress.

1 INTRODUCTION

The composition of polluted aquatic environment is usually very complicated, which contains a wide variety of heavy metal ions, organic pollutants, bacteria, virus, parasite eggs as well as other poisons. Thus, studies on the joint toxicity are urgently needed in assessing the aquatic pollution levels, which unfortunately are still insufficient (Duan *et al.*, 2008).

Typical cresols include p-cresol, m-cresol and o-cresol, and all of them are methylated derivatives of phenol (Dietz, 1991). Due to their large range of applications, such as being solvents and intermediates in compositing disinfectants, fragrances, pesticides, dyes and explosives, cresols are widely distributed in the environment and are harmful to the aquatic environment as well as aquatic organisms (Alva and Peyton, 2003; Kumar *et al.*, 2005). It has been reported that the LC_{50} of cresols for fish was 2.3–29.5, 6.4–24.5 and 4.0–21.2 mg/L for o-cresol, m-cresol and p-cresol, respectively (Post, 1987). And m-cresol causes developmental toxicities on D. magna embryo and mainly affects the second antennae, malpighian tubes well as rostrum (Ton *et al.*, 2012). In addition, the toxicity of cresols was attributed to the oxidative stress induced by their metabolites (Schepers *et al.*, 2007).

On the other hand, polycyclic aromatic hydrocarbons (PAHs) are emitted from all kinds of sources like firewood, straw, domestic and coking in the energy supply. Most of them are stored in freshwater sediment (Dai *et al.*, 2013) and cause functional disorder of fish immune responses as well as immune gene expression (Hur *et al.*, 2013). More importantly, it has been reported that toxicity of various chemicals would be modulated by PAH. For example, by inhibiting CYP1A, α-naphthoflavone (ANF) potentiated (+)-usnic acid -induced inhibition of cellular respiration and exacerbated cellular ATP depletion by usnic acid (Shi *et al.*, 2013). Besides, ANF was reported to enhance the toxicity of cadmium by blocking its efflux and inducing the production of oxidative stress (Yin *et al.*, 2014).

Zebrafish (Danio rerio) has gained merits as a model species over the past few years owing to its similar cellular and physiological characteristics with higher vertebrates (Gestri *et al.*, 2012; Vatine *et al.*, 2013), Therefore, the early life stage test using zebrafish embryos is widely used in investigating the toxicity and teratogenicity of chemicals known to cause significant impacts on environmental and human health (Berry *et al.*, 2007; Weil *et al.*, 2009).

Due to the wide use of cresols and PAHs, they often co-exist in the aquatic environments together for a long time (Renoux et al., 1998). However, the effect of PAH on the toxicity of m-cresol is still unknown. In the present study, we used ANF together with one of cresols, m-cresol to examine their joint effects on the development of zebrafish embryos. To further investigate the role of oxidative stress, we assayed the reduced glutathione (GSH) content, malondialdehyde (MDA) level as well as superoxide dismutase (SOD) activity.

2 MATERIALS AND METHODS

2.1 Chemcals

α-Naphthoflavone (ANF, $C_{19}H_{12}O_2$, purity > 99%) and m-cresol (C_7H_8O) were purchased from Sigma Chemical Co. (US). Their molecular structure is shown below:

α-Naphthoflavone   m-cresol

2.2 Maintenance of Zebrafish and embryo harvesting

The wide type adult zebrafish were purchased from a local fish aquarium (Suzhou, China). The male and female zebrafish were kept separately in the 10-L tanks with carbon-filtered system and every 50 zebrafish were distributed in a tank. The zebrafish were acclimated to the new environment for a month at the water temperature of $28 \pm 0.5°C$ with a 14:10 h photoperiod cycle, and fed twice a day with commercial fish food. The zebrafish used for breeding were separated by sex at the end of the light cycle in hatching tanks. At the beginning of the next light cycle, we collected embryos within 30 min after natural mating (Si et al., 2013). The embryos were washed with Holt buffer (3.5 g/L NaCl, 0.05 g/L KCl, 0.1 g/L CaCL$_2$ and 0.025 g/L NaHCO$_3$, pH 7.5) three times and kept in Holt buffer. Then the unfertilized and abnormal embryos were discarded under a stereomicroscope. The remained embryos would be used for toxic exposure experiment.

2.3 Exposure experiment

Acetone was used as cosolvent of m-cresol and ANF and its final concentration was no more than 0.1% which had no adverse impacts on embryos. At 4 hpf, 10 embryos were randomly distributed in one well of the 24-well plate (Corning, USA) which contained 2 mL exposure solution (m-cresol, ANF or m-cresol/ANF solutions). Six replicates for each concentration were

performed in one experiment. The 24-well plates were covered with preservative film and put in an incubator chamber at 28°C with a 14:10 h photoperiod cycle. Every 24 h we replaced half of the exposure solutions of each well.

In the previous experiment, ANF (0.1–3 mg/L) only caused sub-lethal effects including 48 hpf cardiac edema and 72 hpf delayed hatching (Yin et al., 2014). This paper mainly focused on the toxicity of m-cresol (5–250 mg/L, plus the control and solvent control). The joint exposure concentrations were selected based on the individual exposure effects. Differential toxicity endpoints for assessing developmental toxicity were observed and representatively photographed at different developmental phases of zebrafish with an Axio observer A1 microscope (Carl Zeiss, Inc., Oberkochen, Germany). Mortality was calculated at 24 hpf. At 48 hpf, embryos with no blood circulation (the embryo had heart beat, but the blood stopped flowing in the vessels), cardiac edema and slower heart beat (beats/min) were recorded. And we measured the heart rate of each group using the method described by Shi Du et al. (2008).

2.4 Biomarkers Assay

60 embryos treated by toxic solutions or vehicle alone were washed with Holt buffer and transferred to a 1.5 ml centrifuge tube containing 300 ul PBS buffer (pH 7.4). The embryos were then "flash frozen" in liquid nitrogen and stored at $-80°C$ until biomarkers extraction (Jaja-Chimedza et al., 2012). The embryo samples were defrosted and homogenized by ultrasonic on ice. The supernatant was collected after the contrifugation of the embryo homogenate (8000g, 4°C, 10 min). The protein content of the samples was subsequently determined using the bicinchoninic acid (BCA) protein assay kit (Beyotime Institute of Biotechnology, Jiangsu, China). GSH content was evaluated based on the conjugation of GSH to 5,5-dithiobis (2-nitrobenzoic acid) and the spectrophotometric of production was measured at 412 nm (Khan et al., 2011). The level of MDA was detected based on spectrophotometric measure (532 nm) of the production of thiobarbituric acid (TBA) (Liu and Lei, 2012). The activity of SOD was assayed using NBT/Riboflavine methodand determinedby the increase in the absorbance of formazan at 560 nm (Chaabane et al., 2012). The absorbance was measured by microplate reader (Biotech Instrument, USA) using 96-well plates.

2.5 Statistical analysis

All the experiments were repeated at least 3 times ($n = 3$). Statistical Package for Social Science (version 15.0, SPSS Inc., Chicago, Illinois, USA) was used for statistical analyses. All the results were reported as mean $\pm$ SD. Tukey's HSD test was used to compare results for two different treatments for statistical comparisons. And a p-value <0.05 was considered to be significant.

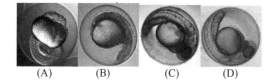

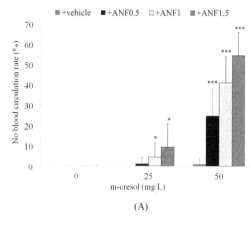

Figure 1. Representatives of normal or abnormal zebrafish embryo and larva fish. (A) Normal embryos, 24 hpf; (B) stagnant development of the embryo, 24 hpf, 150 mg/L m-cresol; (C) normal embryo, 48 hpf; (D) lack of eye and body pigmentation in the embryo, 48 hpf, 10 mg/L m-cresol.

(A)

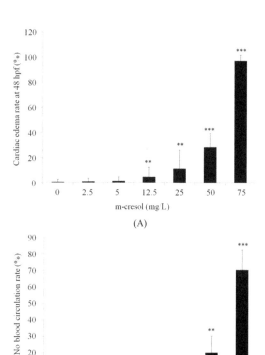

(A)

(B)

Figure 2. Concentration -response relations for m-cresol. (A) Cardiac edema rate in zebrafish embryos exposed to m-cresol, 48 hpf. (B) No blood circulation in zebrafish embryos exposed to m-cresol, 48 hpf. *p < 0.05, **p < 0.01, ***p < 0.001 compared with untreated control.

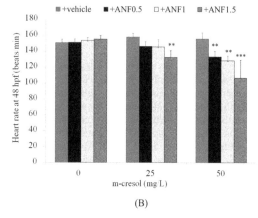

(B)

Figure 3. (A) Percentage of no blood circulation in zebrafish embryos at 48 hpf upon exposure to the combination of m-cresol and ANF. (B) Heart rate in zebrafish embryos at 48 hpf upon exposure to the combination of m-cresol and ANF. *p < 0.05, **p < 0.01, ***p < 0.001 compared with groups treated by m-cresol only.

3 RESULTS

3.1 Toxicity of m-cresol alone

At 4 hpf, zebrafish embryos were exposed to various concentrations of m-cresol, and the exposure lasted for 72 h. Normal and abnormal individuals were photographed and shown in Fig. 1. Treatment of m-cresol (5–250 mg/L) resulted in both lethal and sub-lethal toxic effects, such as death at 24 hpf, cardiac edema,
lack of eye and body pigmentation, slower heart rate, no blood circulation at 48 hpf, delayed hatching and complete hatching failure at 72 hpf Dose-response toxic effects of m-cresol such as no blood circulation were obtained and shown in Fig. 2. For example, after exposure to 50 mg/L m-cresol for 48 hpf, embryos with cardiac edema and no blood circulation were 28.33% and 0.83% respectively. While, almost all the embryos got cardiac edema and 20.42% of them got no blood when the concentration increased to 75 mg/L.

3.2 Effect of ANF on the toxicity of m-cresol

In the joint exposure experiment, the concentration of 25 and 50 mg/L for m-cresol as well as 0.5–1.5 mg/L for ANF which had no lethal effect on embryos were used. No blood circulation and slower heart rate at 48 hpf were selected as two representative endpoints. Blood circulation of the embryos was generally normal

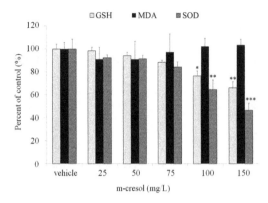

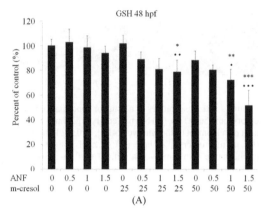

(A)

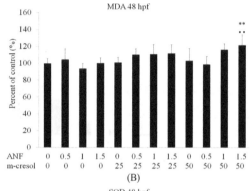

(B)

Figure 4. Effects of m-cresol on glutathione (GSH) content, malondialdehyde (MDA) level and superoxide dismutase (SOD) activity in 48 hpf zebrafish embryos. Values are presented as means ± standard deviation (n = 3). *p < 0.05, **p < 0.01, ***p < 0.001 compared with untreated control.

when treated with ANF alone. But the toxicity of m-cresol was significantly enhanced when it was used together with ANF. For example, the percentage of embryos getting no blood circulation was only 0.83% when they were treated with 50 mg/L m-cresol alone, but it increased to 30.83%, 51.67% and 72.08% when co-treated with 0.5, 1 and 1.5 mg/L ANF, respectively (Fig. 3A). Besides, the heart rates of embryos treated with m-cresol groups were significantly reduced when high dose of ANF (1.5 mg/L) was added. While, no appreciable change was observed when m-cresol and ANF were used separately (Fig. 3B).

3.3 Effects of m-cresol on detoxification pathways in the Zebrafish embryo

The content of GSH and MDA and the activity of SOD in zebrafish embryos were examined after 48 h exposure and illustrated in Fig. 4. As a result, the presence of m-cresol caused a gradual decrease of GSH content and SOD activity in a dose-dependent manner, but it caused no alteration of MDA (Fig. 4A).

3.4 Joint effects of m-cresol and ANF on detoxification pathways in the Zebrafish embryo

As shown in Fig. 5, when the concentration of m-cresol was constant, oxidative stress responses were enhanced by ANF in a dose-dependent manner. For example, after the co-treatment of 50 mg/L m-cresol and 1.5 mg/L ANF for 48 h GSH content and SOD activity in the embryos were reduced by 48.1% and 60% respectively, but the reductions were only 11.5% and 0 when the embryos were treated by m-cresol alone. Compared with the individual treatment, a much more significant increase of MDA levels was observed in the embryos which were exposed to a co-treatment of m-cresol with 1 and 1.5 mg/L ANF mixtures.

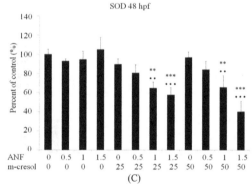

(C)

Figure 5. GSH level (A), MDA content (B) and SOD activity (C) in zebrafish embryos at 48 hpf upon exposure to the combination of m-cresol and ANF. Values are shown as means ± S.D. (n = 3). ·p < 0.05, ··p < 0.01, ···p < 0.001 compared with groups treated by m-cresol only; *p < 0.05, **p < 0.01, ***p < 0.001 compared with groups treated by ANF only.

4 DISCUSSION

Due to their wide use, cresols and PAH often co-exist in the aquatic environments for a long time. However, studies on the effects of PAH on the toxicity of cresols to zebrafish embryo are still insufficient. In this paper, m-cresol and ANF were selected as the representatives of cresols and PAHs, and their joint toxicity as well as the corresponding mechanism was studied.

In the study of individual toxicity of m-cresol, we found that both m-cresol exhibited significant effects on the development of zebrafish embryos. Liu *et al.* (2011) demonstrated that after a treatment of 25 mg/L m-cresol, the embryos exhibited malformation including 24 h incomplete eyespots development, 20 s no active movement and 48 h lack of eye and body pigmentation. In our study, only the lack of eye and body pigmentation at 48 hpf were observed when the embryos were treated with 25 mg/L m-cresol. Such a phenomenon could be explained by the different hours post fertilization when m-cresol was added to the culture medium.

As shown in Fig. 4, a linear decrease of GSH content caused by m-cresol treatment was observed, which corresponded well with the previous reports that m-cresol depleted intracellular glutathione levels (Thompson *et al.*, 1994) Such a finding could be explained by the fact that major metabolites of m-cresol might firstly be conjugated with GSH by GST (Yan *et al.*, 2005). Under the treatment of 100 and 150 mg/L m-cresol, the activity of SOD decreased significantly. This result could be explained by previous research, which found that reactive oxygen species (ROS) level could be increased by P-cresylsulphate, the main in vivo metabolite of p-cresol (Schepers *et al.*, 2007). Therefore, due to the inhibition of SOD activity, the concentration of ROS might increase significantly, exceeding the capacity of the cellular antioxidant system (Shao *et al.*, 2012). However, the results didn't show any significant alteration of MDA levels in the embryos exposed to m-cresol. Previous studies showed that the MDA level did not always correlate with the cellular antioxidants (Nigam *et al.*, 1999), which might be a possible explanation for our phenomenon.

In the joint exposure experiments, the percentage of no blood circulation treated with m-cresol was significantly enhanced by ANF (Fig. 3A). Similarly, the heart rates of embryos become much slower with the co-treatment of m-cresol and ANF (Fig. 3B). Both of these results indicated an enhanced toxicity of m-cresol by ANF, which corresponded well with the oxidative stress responses in the zebrafish embryos (Fig. 5). Although ANF didn't induce any oxidative stress at the selected concentrations, they did enhance the oxidative stress caused by m-cresol. Enhanced m-cresol toxicity by ANF could be explained by its inhibition of CYP1A activity and the modulation effects on the process of oxidative stress induced by m-cresol (Hodson *et al.*, 2007; Timme-Laragy *et al.*, 2007).

In addition, ANF might have an interactive mechanism with m-cresol and enhance the damage to the antioxidant system when they were used together. Furthermore, it has been reported that PAH can increase the sensitivity of the cell to chemical agents by blocking the function of transporters (Epel *et al.*, 2008; Yin *et al.*, 2014) which might be another possible explanation for the joint toxicity of m-cresol with ANF.

In conclusion, M-cresol caused oxidative stress in zebrafish embryos and the response was significantly affected by the co-treatment of ANF, which might be the reason for the enhanced toxicity of m-cresol by ANF. Beside, zebrafish could be an ideal model for such investigation.

ACKNOWLEDGEMENT

This work was supported by grants from the National Natural Science Foundation of China (No. 21307154) and partly by the National Key Instrument Developing Project of China (No. ZDYZ2013-1) and Hi-Tech Research and Development (863) Program of China (No. 2014AA020905).

REFERENCES

Alva, V.A., Peyton, B.M., 2003. Phenol and catechol biodegradation by the haloalkaliphile Halomonas campisalis: influence of pH and salinity. Environ Sci Technol, 37(19), 4397–402.

Berry, J.P., Gantar, M., Gibbs, P.D., Schmale, M.C., 2007. The zebrafish (Danio rerio) embryo as a model system for identification and characterization of developmental toxins from marine and freshwater microalgae. Comp Biochem Physiol C Toxicol Pharmacol, 145(1), 61–72.

Chaabane, F., Boubaker, J., Loussaif, A., Neffati, A., Kilani-Jaziri, S., Ghedira, K., Chekir-Ghedira, L., 2012. Antioxidant, genotoxic and antigenotoxic activities of daphne gnidium leaf extracts. BMC Complement Altern Med, 12153.

Dai, J.F., Jiang, M., Qu, L.L., Sun, L., Wang, Y.Y., Gong, L.L., Gong, R.J., Si, J., 2013. Toxoplasma gondii: enzyme-linked immunosorbent assay based on a recombinant multi-epitope peptide for distinguishing recent from past infection in human sera. Exp Parasitol, 133(1), 95–100.

Dietz, D., 1991. NTP technical report on the toxicity studies of Cresols (CAS Nos. 95-48-7, 108-39-4, 106-44-5) in F344/N Rats and B6C3F1 Mice (Feed Studies). Toxic Rep Ser, 91–128.

Duan, Z., Zhu, L., Kun, Y., Zhu, X., 2008. Individual and joint toxic effects of pentachlorophenol and bisphenol A on the development of zebrafish (Danio rerio) embryo. Ecotoxicol Environ Saf, 71(3), 774–80.

Epel, D., Luckenbach, T., Stevenson, C.N., Macmanus-Spencer, L.A., Hamdoun, A., Smital, T., 2008. Efflux transporters: newly appreciated roles in protection against pollutants. Environ. Sci. Technol, 42(11), 3914–20.

Gestri, G., Link, B.A., Neuhauss, S.C., 2012. The visual system of zebrafish and its use to model human ocular diseases. Dev Neurobiol, 72(3), 302–27.

Hodson, P.V., Qureshi, K., Noble, C.A., Akhtar, P., Brown, R.S., 2007. Inhibition of CYP1A enzymes by alpha-naphthoflavone causes both synergism and antagonism of retene toxicity to rainbow trout (Oncorhynchus mykiss). Aquat Toxicol, 81(3), 275–85.

Hur, D., Jeon, J.K., Hong, S., 2013. Analysis of immune gene expression modulated by benzo[a]pyrene in head kidney of olive flounder (Paralichthys olivaceus). Comp Biochem Physiol B Biochem Mol Biol, 165(1), 49–57.

Jaja-Chimedza, A., Gantar, M., Mayer, G.D., Gibbs, P.D., Berry, J.P., 2012. Effects of cyanobacterial lipopolysaccharides from microcystis on glutathione-based detoxification pathways in the zebrafish (Danio rerio) embryo. Toxins (Basel), 4(6), 390–404.

Khan, H., Khan, M.F., Jan, S.U., Ullah, N., 2011. Effect of aluminium metal on glutathione (GSH) level in plasma and cytosolic fraction of human blood. Pak J Pharm Sci, 24(1), 13–8.

Kumar, A., Kumar, S., Kumar, S., 2005. Biodegradation kinetics of phenol and catechol using Pseudomonas putida MTCC 1194. Biochemical Engineering Journal, 22(2), 151–59.

Liu, C.Z., Lei, B., 2012. [Effect of acupuncture on serum malonaldehyde content, superoxide dismutase and glutathione peroxidase activity in chronic fatigue syndrome rats]. Zhen Ci Yan Jiu, 37(1), 38–40, 58.

Liu Zaiping, Zhang Songlin, Yang Jinghui, Tang Rui, Bijie, L., 2011. Effect of M-cresol on Embryonic Development of Zebrafish Environmental Science and Technology 24(2), 1–3.

Nigam, D., Shukla, G.S., Agarwal, A.K., 1999. Glutathione depletion and oxidative damage in mitochondria following exposure to cadmium in rat liver and kidney. Toxicol Lett, 106(2–3), 151–7.

Post, G., 1987. Textbook of fish health. 2ndEd. TFH Publ. Inc., pp. 259.

Renoux, Y, A., Millette, D., Tyagi, R.D., Samson, R., 1998. Detoxification of fluorene, phenanthrene, carbazole and p -cresol in columns of aquifer sand as studied by the Microtox assay. Water Res, 33(9), 2045–52.

Schepers, E., Meert, N., Glorieux, G., Goeman, J., Van der Eycken, J., Vanholder, R., 2007. P-cresylsulphate, the main in vivo metabolite of p-cresol, activates leucocyte free radical production. Nephrol Dial Transplant, 22(2), 592–6.

Shao, B., Zhu, L., Dong, M., Wang, J., Xie, H., Zhang, Q., Du, Z., Zhu, S., 2012. DNA damage and oxidative stress induced by endosulfan exposure in zebrafish (Danio rerio). Ecotoxicology, 21(5), 1533–40.

Shi, Q., Greenhaw, J., Salminen, W.F., 2013. Inhibition of cytochrome P450s enhances (+)-usnic acid cytotoxicity in primary cultured rat hepatocytes. J Appl Toxicol.

Shi, X., Du, Y., Lam, P.K., Wu, R.S., Zhou, B., 2008. Developmental toxicity and alteration of gene expression in zebrafish embryos exposed to PFOS. Toxicol Appl Pharmacol, 230(1), 23–32.

Si, J., Zhang, H., Wang, Z., Wu, Z., Lu, J., Di, C., Zhou, X., Wang, X., 2013. Effects of (12)C(6+) ion radiation and ferulic acid on the zebrafish (Danio rerio) embryonic oxidative stress response and gene expression. Mutat Res, 745-74626-33.

Thompson, D.C., Perera, K., Fisher, R., Brendel, K., 1994. Cresol isomers: comparison of toxic potency in rat liver slices. Toxicol Appl Pharmacol, 125(1), 51–8.

Timme-Laragy, A.R., Cockman, C.J., Matson, C.W., Di Giulio, R.T., 2007. Synergistic induction of AHR regulated genes in developmental toxicity from co-exposure to two model PAHs in zebrafish. Aquatic Toxicology, 85(4), 241–50.

Ton, S.S., Chang, S.H., Hsu, L.Y., Wang, M.H., Wang, K.S., 2012. Evaluation of acute toxicity and teratogenic effects of disinfectants by Daphnia magna embryo assay. Environ Pollut, 16854–61.

Vatine, G.D., Zada, D., Lerer-Goldshtein, T., Tovin, A., Malkinson, G., Yaniv, K., Appelbaum, L., 2013. Zebrafish as a model for monocarboxyl transporter 8-deficiency. J Biol Chem, 288(1), 169–80.

Weil, M., Scholz, S., Zimmer, M., Sacher, F., Duis, K., 2009. Gene expression analysis in zebrafish embryos: a potential approach to predict effect concentrations in the fish early life stage test. Environ Toxicol Chem, 28(9), 1970–8.

Yan, Z., Zhong, H.M., Maher, N., Torres, R., Leo, G.C., Caldwell, G.W., Huebert, N., 2005. Bioactivation of 4-methylphenol (p-cresol) via cytochrome P450-mediated aromatic oxidation in human liver microsomes. Drug Metab Dispos, 33(12), 1867–76.

Yin, J., Yang, J., Zhang, F., Miao, P., Chen, M., Lin, Y., 2014. Individual and joint toxic effects of cadmium sulfate and α-Naphthoflavone on the development of zebrafish embryo. J Zhejiang Univ Sci B, 15(9), 766–75.

Biomedical Engineering and Environmental Engineering – Chan (Ed.)
© 2015 Taylor & Francis Group, London, ISBN: 978-1-138-02805-0

Numerical analysis of impact of land reclamation on the residual current and water exchange of Bohai Bay in China

X.B. Li, B.H. Zheng & L.J. Wang

State Key Laboratory of Environment and Criteria and Assessment, Chinese Research Academy
of Environmental Sciences, Beijing, China
State Environmental Protection Key Laboratory of Drink Water Resource Protection,
Chinese Research Academy of Environmental Sciences, Beijing, China

ABSTRACT: An integrated 2D hydrodynamic-dispersion numerical model was used to investigate the impact of land reclamations on the water exchange of Bohai Bay. This analysis was mainly done under three cases, corresponding to the different coastlines of Bohai Bay at year 2003, 2010 and 2020. With the whole Bohai Bay divided into four sub-areas, the water exchange amongst these areas was investigated by the water exchange matrices, the residual current and the average sub-area exposure times. The results show that: 1) The water exchange matrices under the different coastlines demonstrate that the water exchange between the four sub-areas has a significant relationship with the change of Eulerian residual current; 2) The water exchanges of Tianjin port were affected seriously by the land reclamations. The variation of average sub-area exposure times show that water exchange between sub-areas changes more noticeably from year 2003 to 2010 than 2010 to 2020.

1 INTRODUCTION

The coastal and estuarine ecosystems are mainly affected by physical, chemical and biological processes. As a physical process, the water exchange between a semi-enclosed bay and the open sea has been widely studied recently (Andrejev et al. 2004, Liu et al. 2004, Ribbe et al. 2008) since it has important implications for the fates of introduced substances and the primary productivity in estuaries. Because the land reclamation changes the coastline of the semi-enclosed bay, it will obviously make the hydrodynamic process change. Therefore, study on the water exchange before and after the land reclamations can be helpful in illustrating the variation of the ecological environment in the semi-enclosed sea bay.

The most widely used characteristic timescales to quantify the renewal of water in sea bays are the age (Bolin & Rodhe 1973, Deleersnijder et al. 2001), the residence time (Takeoka 1984, Zimmerman 1976) and the exposure time (Monsen et al. 2002). The age of a particle or water parcel is defined as the time elapsed since it entered the region of interest, while the residence time of a particle or water parcel is the time it needs to leave the study area. A significant drawback of the residence time is its inability to consider water parcels which re-enter the region of interest. Therefore, considering the re-entrance of the water parcels in the studied area, the exposure time is a more realistic timescale which considers the total time spent by the water parcel. Obviously, the residence time will significantly overestimate the water exchange ability in the area of interest.

For a small water body, it is usually reasonable to assume that the mass is well mixed in the body (Guo & Lordi 2000), and it is not very important to know the details of the water exchange amongst different parts of the water body. Thus, it is enough to evaluate the water exchange between the study water body and its surrounding water bodies with the residence time of the whole water body. However, if the area of the body of water is large, it is important to understand not only the water exchange between the study water body and its surrounding water bodies but also the water exchange amongst different parts of the study water body (Bilgili et al. 2005, Sun & Tao 2006, Thompson et al. 2002).

The area of interest in this study is Bohai Bay (Fig. 1). The effects of its land reclamation on the ecological environment along the coastal areas have been studied extensively (Li et al. 2010, Nie & Tao 2009, Yang et al. 2011). Nie (2010) analyzed the variation of eco-environment near the Tianjin coastal area and found that the land reclamation changed the current and weakened the tidal prism in this area. By analyzing the Shannon-Weaver diversity index on three investigations in Tianjin Harbor Industrial Zone from 2006 to 2008, Li (2010) found that the reclamation projects have significantly affected the growth, reproduction and distribution of phytoplankton, thus decreasing biodiversity and changing the structure of the community.

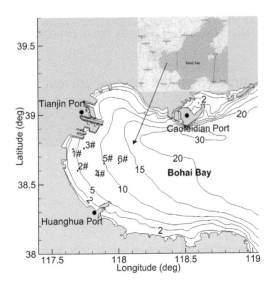

Figure 1. Topography and coastline of Bohai Bay at year 2010. The sites numbered from 1# to 6# are the positions where field data was collected in July 2008.

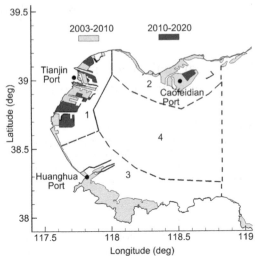

Figure 2. Land reclamations and the division of sub-areas. The outmost line is the land boundary around Bohai Bay at year 2003, the filled light gray area is the land reclamation that had been carried out from 2003 to 2010, and the filled dark gray area is the planned reclamation at year 2020.

However, an exact and detailed interpretation of the land reclamation's effect on the hydrodynamic and water exchange processes in Bohai Bay was not studied in depth, and providing this interpretation was the aim of this paper.

In this paper, brief introductions of the numerical model and timescales used to illustrate the water exchange in sea bays are given in Sections 2 and 3, respectively. Section 4 presents the model verification results for hydrodynamic parameters. Section 5 outlines the model simulation results and provides a discussion of these results. Finally, the main conclusions drawn from this investigation are given in Section 6.

2 STUDY AREA

Bohai Bay is a semi-enclosed interior sea bay in the west part of Bohai Sea. The bay covers an area of about 14700 km² and has an average depth of about 10 m (Fig. 1). It has a wide, mild slope mud beach and takes a large amount of waste water from the surrounding Tianjin city and Hebei province and even the waste water from Beijing city. At present, there are mainly three large ports located at the west, north and south part of the bay which are the Tianjin port, Caofeidian port and Huanghua port, respectively.

Since the demand of the social and economic development of the surrounding coastal cities, a lot of land reclamations have been or will be carried out around all three large ports in Bohai Bay beginning in 2003 to continuing to the year 2020 (Fig. 2).

In order to form a comprehensive understanding of the transport and fate of the contaminants in Bohai Bay,

the whole bay was divided into four sub-areas (Fig. 2). The purpose in the division of these four sub-areas was to represent the coastal area around the three different ports and the sea area in the center of Bohai Bay. Therefore, sub-area 1 and sub-area 3 are comprised of sea area where the depth is below 10 m, while sub-area 2 was defined as the sea area whose distance from the shoreline is less than 20 km. Then, an integrated hydrodynamic-dispersion numerical model was used to study the water exchange between each sub-area in this sea bay.

3 MODEL DESCRIPTION

Since Bohai Bay is a sea bay with shallow water depth, the 2D depth-integrated hydrodynamic and dispersion models are adequately precise and were applied to predict the flow fields and distributions of the passive dissolved conservative matter (PDCM). The concepts of water exchange matrix and average sub-area exposure time were applied to illustrate the effect of land reclamations on the water exchange in Bohai Bay.

3.1 Hydrodynamic model & Dispersion model

A 2D depth-integrated hydrodynamic model is used in the flow field simulation (Falconer 1993) and a 2D depth-integrated advection-diffusion equation is used to predict the distribution of PDCM here. These models had been used a lot in coastal hydrodynamic simulation, and also be proved to be robust in the study of Bohai Bay (Yuan et al. 1999, Sun & Tao 2006).

3.2 Definition of water exchange timescales

Following the definition given by Zimmerman (1976), Takeoka (1984) defined the average residence time of a water body as following:

$$\tau_r = \int_0^\infty \frac{R(t)}{R_0} dt = \int_0^\infty r(t) dt \tag{1}$$

where $R(t) =$ the amount of the PDCM which still remains in the water body at time t, R_0 represents the amount of the PDCM in the water body at the initial time. The function $r(t) = R(t)/R_0$ is called the remnant function (Takeoka 1984). When considering the return of PDCM at the boundary of studied area, equation (1) is also the average exposure time.

The remnant function can be obtained by using the integrated hydrodynamic-dispersion model with proper initial and boundary conditions as follows:

$$\begin{cases} C(t_0, x \in \Omega_i) = 1 \\ C(t_0, x \notin \Omega_i) = 0 \end{cases} \tag{2}$$

where $t_0 =$ the initial time when PDCM was released. Equation (2) presents that the concentration of PDCM at t_0 is set to be 1 in domain Ω_i and 0 elsewhere. Mostly, Ω_i represents an entire sea area when the water exchange of this area with outer sea is focused. It should also be noted that the boundary of the dispersion model where zero concentration is set up will have a great effect on the result of simulation and exposure time.

In order to determine the water exchange amongst sub-areas of an entire water body, the initial conditions for the PDCM in these sub-areas are also set up according to equation (2) with Ω_i representing the ith sub-area of the studied water body. The definition of the remnant function and average sub-area exposure time for each sub-area is:

$$\begin{cases} r_{ij}(t) = R_{ij}(t)/R_0^i \\ \tau_r^i = \sum_{j=1}^n \tau_{ij} = \sum_{j=1}^n \int_0^\infty r_{ij}(t) dt \end{cases} \tag{3}$$

where $R_{ij}(t) =$ the amount of the PDCM in sub-area j which is initially in sub-area i and R_0^i represents the amount of the PDCM in sub-area i at the initial time. Thus, $r_{ij}(t)$ can be considered as the fraction of PDCM which transports from sub-area i to sub-area j at time t. Function $r_{ij}(t)$ is also called the water exchange matrix. It is obvious that this matrix allows the identification of water exchange between all the sub-areas. This function paints a rough picture of where water parcels released at different places spend most of the time on their journey out of the study area without necessitating knowledge of the complex circulation and transport patterns. Also, it can be easily seen that $r_i(t) = \sum r_{ij}(t)$ is the remnant function of sub-area i in the whole study area, and τ_r^i is the average sub-area exposure time sub-area i resides in the whole study area.

4 MODEL CALIBRATION AND VERIFICATION

In this study, the Bohai Sea is meshed within a rectangular area, and the open boundary was set at the position of the solid line in Figure 1. The numerical model used in this study was calibrated and verified against observed harmonic constants and also with measured data for Bohai Bay collected in July 2008 (Fig. 1), respectively.

4.1 Harmonic constants

In this study, the M_2, S_2, K_1, O_1, N_2, P_1, K_2 and Q_1 tides of Dalian station and Yantai station are selected to drive the hydrodynamic model at the open boundary because they are the largest tide components in Bohai Sea. Thus, further verification of the hydrodynamic model is made for these eight tide constituents. The tidal harmonic constants of the eight tide constituents are calculated and compared against the observed data. The predicted harmonic constants of M_2, S_2, K_1 and O_1 tides at ten gauge stations in Bohai Sea are in good agreement with the corresponding observed data (Huang 1991, Lv & Fang 2002). The absolute average error of amplitude of M_2, S_2, K_1 and O_1 was 7.9 cm, 3.2 cm, 3.1 cm and 3.8 cm, while the absolute average error of phase lag of M_2, S_2, K_1 and O_1 was 5.08 degree, 7.21 degree, 9.2 degree and 10.4 degree, respectively.

4.2 Water elevations and velocities

The main parameter needed to be calibrated in the hydrodynamic model is the Manning's roughness coefficient. The value of this coefficient is suggested as 0.017 by many researchers (Huang 1991, Wang et al. 1984). In order to keep the parameters in the model up-to-date and ensure the accuracy of the numerical modeling results, the hydrodynamic model is calibrated and verified against field measured data that was collected from 11 July 2003 to 15 July 2003. The predicted water elevations are in good agreement with the field measured data at three stations (Sun & Tao 2006). This means that the roughness coefficient is still valid in this area at year 2003.

In order to take into account the variation of hydrodynamics affected by shoreline change and to check if the roughness coefficient is still valid under the new topography, another series of field measured data was collected in July 2008. The velocities were measured at six stations located around Tianjin port (Fig. 1). Figure 3 illustrates the comparisons of the numerical predicted depth-averaged velocities and their corresponding measured data at all these six stations.

Figure 3 shows that the predicted velocity amplitude and velocity direction with the measured data at the number 1 to number 6 sites. It can be seen that simulation results at all the stations had a preferable agreement with the field data except for site 1. The discrepancy between the predicted and measured results

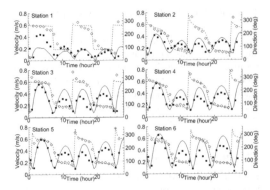

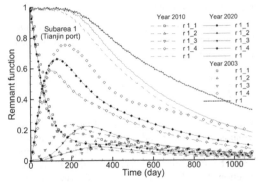

Figure 3. Predicted results and measured velocity amplitude and velocity direction at six stations in July 2008 (Solid line: predicted velocity amplitude; Dashed line: predicted velocity direction; Filled square dots: measured velocity amplitude; Square dots: measured velocity direction).

Figure 4. The water exchange matrix of Tianjin Port in Bohai Bay at year 2003, 2010 and 2020, with the PDCM released at high water level time.

at site 1 is most likely due to the effect of the short wave and wave-induced near-shore current. This means that the calibrated model is still robust for the simulation in the next steps.

5 RESULTS AND DISCUSSION

5.1 Water exchange matrix amongst sub-areas

In order to get a comprehensive understanding of the water exchange amongst different parts of Bohai Bay, the whole Bohai Bay was divided into four sub-areas (Fig. 2) and the water exchange amongst these sub-areas was studied. Twenty-four cases were carried out in total with six cases corresponding to each sub-area. The six cases of each sub-area were combinations of water exchange under two different water levels (high tide level and low tide level) with the three different coastlines (year 2003, 2010 and 2020). As illustrated by equation (2), in each case PDCM was released in the domain with the initial concentration set as unity in current sub-area and zero in the other part of the domain. By analyzing the predicted concentration of PDCM in each case according to equation (3), the water exchange between the studied sub-area and other sub-areas could be understood.

Figure 4 shows the water exchange of Tianjin port (sub-area 1) among other sub-areas, and the meaning of water exchange matrix defined in equation (3) is shown clearly in this figure. Only the cases for the PDCM being released at high water level are illustrated in Figure 4 because the results of the cases for the PDCM being released at low water level are similar. It can be seen that water exchanges between sub-area 1 and sub-area 4 were strong (the values of r_{14} was large), and its water exchanges with sub-area 2 and sub-area 3 were weak (the values of r_{12} and r_{13} were small). Similarly, the water exchange amongst the other sub-areas can be understood by analyzing the other curves (other elements in matrix). Thus, the water exchange

matrix can be used to represent the water exchange amongst the sub-areas.

Figure 4 also shows the comparison of water exchange between sub-areas at year 2010 and year 2020. Comparing the value of r_{14} at year 2010 to its value at year 2020 shows an increase in the water of sub-area 1(Tianjin port) transferred through sub-area 4. The value of r_{13} at year 2010 and 2020 shows that less water in sub-area 1 will pass through sub-area 3 at year 2020. The water of sub-area 1 transferred through sub-area 2 will hardly change between these two years. The variation of r_1 shows that the water of sub-area 1 will remain in Bohai Bay longer after the land reclamation at year 2020. This is mainly because the water of this sub-area will be transferred more through sub-area 4 (Bohai Bay center), in which the flow is mainly a rectilinear current and the residual current is very weak.

5.2 Residual current

In order to figure out the reason for the variation in the water exchange between these sub-areas, the Eulerian residual current was calculated as following:

$$\mathbf{u}_R = \frac{1}{T}\int_{t_0}^{t_0+T}\mathbf{u}(x_0,t)dt \qquad (4)$$

where $\mathbf{u}_R$ = the residual current, t_0 = the moment starting to statistic the flow filed, T = the time period of the model, $\mathbf{u}(x_0,t)$ = the velocity at place x_0 and time t.

Figure 5 shows the distribution of residual current in Bohai Bay at year 2003, 2010 and 2020. It can be seen from Fig. 5(a) that the spatial distribution of residual current in Bohai Bay at year 2003 is mainly of two kinds, the clockwise current in the north part and the counterclockwise current in the south part which are located along the shoreline with the Tianjin port as the divide for the current. Figure 5(b) shows that the residual current at the sea area between northern Bohai Bay and Caofeidian port has a value between 1 cm/s and 10 cm/s with the low speed area mostly near the

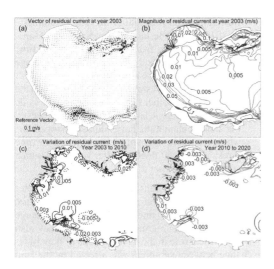

Figure 5. The Eulerian residual current in Bohai Bay at year 2003, 2010 and 2020. (where, (c) is the difference of residual current magnitude between year 2010 and 2003, and (d) is the difference of residual current magnitude between year 2020 and 2010).

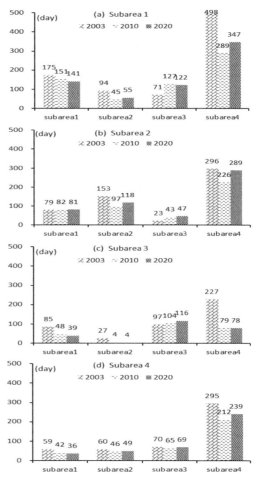

Figure 6. Average sub-area exposure time of each sub-area in other sub-areas at year 2003, 2010 and 2020. (The PDCM in all these sub-areas were released at a high water level time).

Tianjin port and the high speed area near the Caofeidian port. The residual current at the sea area between the south part of Tianjin port and the north part of Huanghua port has a value between 3 cm/s and 5 cm/s. At the sea area in the south part of Huanghua port, the residual current is about 5 cm/s to 10 cm/s, and the east part around Caofeidian port also has a residual current between 5 cm/s and 10 cm/s. The residual current in the center of Bohai Bay is below 1 cm/s, which is much smaller than the residual current of the other parts of Bohai Bay. The spatial distribution of residual current in Bohai Bay indicates that the contaminants in Bohai Bay were mainly transported out by the residual current along the shoreline.

The variation of residual current from year 2003 to 2010 is shown in Fig. 5(c). According to the land reclamation carried out till year 2010, the coastal area around Tianjin port with low residual current was eliminated and the sea area of this sub-area became small while the residual current became stronger along the Tianjin port at year 2010. So, it can be illustrated that the water exchange in sub-area 1 became stronger from year 2003 to 2010, which is also shown in Fig. 4 by analyzing the values of $r_{1j}(t)$ and $r_1(t)$ at these two years. The land reclamation taken around Caofeidian port makes the residual current in this place weaker. However, this sea area had a very shallow water depth and the residual current here was a ring current which made the PDCM stay longer in this area at year 2003. We can conclude that the reclamation project broke the ring current in Caofeidian port at year 2003 thereby making the water exchange in this area stronger. The land reclamation taken at the south part of Bohai Bay (sub-area 3) made the sea area in sub-area 3 much smaller, which also decreased the residual current that flows into sub-area 4 (Bohai Bay center). This combined

action led to the improvement of total water exchange in sub-area 3. It can also be seen from Fig. 7(c) that the residual current in Bohai bay center changes slightly from year 2003 to year 2010, except for part of its area near sub-area 1 and sub-area 3.

The variation of residual current from year 2010 to year 2020 is shown in Fig. 7(d). The further land reclamation of Tianjin port makes the northern Bohai Bay too narrow, and the decrease of residual current in this area is about 0.3 cm/s to 0.5 cm/s. Also, residual current in the sea area near the mid of Tianjin port has an average decline of about 0.3 cm/s. The time cost of the water that will flow from sub-area 1 to sub-area 2 (Caofeidian port) will be larger. Additionally, more water will flow through the center of Bohai Bay and it will also take a longer time than at year 2010. The residual current around sub-area 2 (Caofeidian port) also has a decline about 0.3 cm/s to 0.5 cm/s. This will result in much more of the water of sub-area 2 flowing through sub-area 4 which will increase the time cost since this area has a low residual current. Because of this the water exchange of sub-area 2 will also became weaker. The residual current of sub-area 3 (Huanghua port) will change slightly from year 2010 to year 2020, and it can also be seen from Fig. 7(c) that the water exchange of sub-area 3 hardly changes.

5.3 *Average sub-area exposure time*

Sometimes, it is important to form a comprehensive understanding of the water exchange amongst different parts of Bohai Bay, and in order to give out a more accurate and detailed explanation about changes in water exchange between these sub-areas, average sub-area exposure times between all the sub-areas were calculated according to equation (3). The average sub-area exposure time of all these sub-areas at high water level are presented in Fig. 6.

From Fig 6(a), it can be seen that the water of sub-area 1 (Tianjin port) spent about 175 days in itself, 94 days in sub-area 2, 71 days in sub-area 3 and 498 days in sub-area 4 at year 2003. While at year 2010, this situation changed. The water spent less time in sub-area 1, sub-area 2 and sub-area 4, but more time in sub-area 3. This indicates that the land reclamation will make the water from sub-area 2 influences the water quality in sub-area 3 to a greater degree, and the water of this sub-area will be exchanged out quicker through sub-area 2 and sub-area 4. Other similar findings can also be seen by comparison of the average sub-area exposure time at different years.

The exposure time usually depends on the initial time that the PDCM is released and it should therefore be studied. In our cases, a common trend in the variation of the water exchange between these sub-areas is that the water exchange was not affected significantly by the initial conditions but was mainly affected by the shape of the coastline. Therefore, the average sub-area exposure time of all these sub-areas at low water level are not presented here. Some common findings can be illustrated as follows:

(1) Water from sub-area 1 stayed in sub-area 2 and sub-area 4 for a shorter period of time after the land reclamation of year 2010, but will again stay slightly longer in these two sub-areas at year 2020. Water from this sub-area stayed in sub-area 3 longer after the land reclamation of year 2010 which will also stay unchanged after the land reclamation of year 2020.

(2) Water from sub-area 2 stayed significantly less time in the sub-area 3 at year 2003, and the time becomes longer after year 2010 and 2020. Most of the time, it resides in itself and sub-area 4. It stayed for a shorter time in these two sub-areas at year 2010, but increased slightly in length again at year 2020. The time spent in sub-area 1 rarely changed.

(3) Water from sub-area 3 stayed for a shorter time in sub-area 1 and sub-area 4 after the land reclamation of year 2010, while the time resided in itself increased slightly. Water of sub-area 3 rarely flowed into sub-area 2 after the land reclamation of 2010. Since the land reclamation of sub-area 3 did not change from 2010 to 2020, the time it spent in all the sub-areas will hardly change between these two years.

(4) Water of sub-area 4 stayed mostly in itself, and the time it stayed in other sub-areas is almost the same under the different land reclamations. However, the land reclamation have almost changed the whole residual current of the sea bay, and the time it resided in itself declined after year 2010 with a slight increase after 2020.

The above discussions illustrate how the water exchange matrix and the average sub-area exposure time can be used to interpret the water exchange between different sub-areas in Bohai bay. They also played a critical role in the interpretation of how water exchange between sub-areas was affected by the land reclamations along the coastline of Bohai Bay. It was found that Eulerian residual current has a significant relationship with the inner water exchange of Bohai Bay, but it cannot provide much detailed information for understanding the variation of water exchange in different sub-areas. However, the average sub-area exposure time method can presents a more detailed and precise value to figure out the impact of land reclamations on water exchange in Bohai bay. It should be noted that this study does not consider the wind's non-trivial effect on the water exchange of the sea bay.

6 CONCLUSIONS

The water exchange in a semi-enclosed sea bay (Bohai Bay) is investigated in this paper using an integrated hydrodynamic-dispersion model, with particular attention being paid on the water exchange amongst different sub-areas of Bohai Bay. The main conclusions drawn from the model predictions can be summarized as follows:

1) Calibration and verification of the hydrodynamic model against the newly measured data shows that the roughness coefficient in Bohai Bay has not changed substantially.

2) By analyzing the Eulerian residual current under the three different cases, it was found that the water exchange has a great deal of spatial variation of residual current in Bohai Bay, especially from year 2003 to year 2010. The water exchange matrices under these three different cases show that the water exchanges of Tianjin port were affected seriously by the land reclamations.

3) The average sub-area exposure time was used to give a more accurate and detailed explanation for the variation of water exchange among these sub-areas under the different land reclamation projects. All the sub-areas have a stronger water exchange with the outside sea after 2010 and will be a slightly weaker after 2020. Water exchange between sub-areas changes noticeably

from 2003 to 2010, but not as obviously from 2010 to 2020.

ACKNOWLEDGMENTS

This research was supported by the Chinese National Special Science and Technology Program of Water Pollution Control and Treatment (2012ZX07503-002) and National Basic Research Program of China (2012CB417004).

REFERENCES

Andrejev, O., Myrberg, K. & Lundberg, P.A. 2004. Age and renewal time of water masses in a semi-enclosed basin–application to the Gulf of Finland. *Tellus A* 56: 548–558.

Bilgili, A., Proehl, J.A., Lynch, D.R., Smith, K.W. & Swift, M.R. 2005. Estuary/ocean exchange and tidal mixing in a Gulf of Maine Estuary: A Lagrangian modeling study. *Estuarine, Coastal and Shelf Science* 65: 607–624.

Bolin, B. & Rodhe, H. 1973. A note on the concepts of age distribution and transit time in natural reservoirs. *Tellus* 25: 58–62.

Deleersnijder, E., Campin, J.M. & Delhez, E.J. 2001. The concept of age in marine modelling: I. Theory and preliminary model results. *Journal of Marine Systems* 28: 229–267.

Falconer, R.A. 1993. An introduction to nearly horizontal flows. *In*: Michael B Abbott, W.A.P. (ed.) *Coastal, Estuarial and Harbour Engineers' Reference Book*.

Guo, Q. & Lordi, G.P. 2000. Method for quantifying freshwater input and flushing time in estuaries. *Journal of environmental engineering* 126: 675–683.

Huang, Z.K. 1991. Tidal waves in the Bohai Sea and their variations. *Journal of Ocean University of Qingdao* 21: 1–12 (in Chinese).

Li, K., Liu, X., Zhao, X. & Guo, W. 2010. Effects of Reclamation Projects on Marine Ecological Environment in Tianjin Harbor Industrial Zone. *Procedia Environmental Sciences* 2: 792–799.

Liu, Z., Wei, H., Liu, G. & Zhang, J. 2004. Simulation of water exchange in Jiaozhou Bay by average residence time approach. *Estuarine, Coastal and Shelf Science* 61: 25–35.

Lv, X. & Fang, G. 2002. Numerical experiments of the adjoint model for M2 tide in the Bohai Sea. *Acta Oceanologia Sinica* 24: 17–24 (in Chinese).

Monsen, N.E., Cloern, J.E., Lucas, L.V. & Monismith, S.G. 2002. A comment on the use of flushing time, residence time, and age as transport time scales. *Limnology and Oceanography* 47: 1545–1553.

Nie, H.T. & Tao, J.H. 2009. Eco-Environment status of the Bohai Bay and the impact of coastal exploitation. *Marine Science Bulletin* 2: 81–96.

Ribbe, J., Wolff, J.O., Staneva, J. & Gräwe, U. 2008. Assessing water renewal time scales for marine environments from three-dimensional modelling: A case study for Hervey Bay, Australia. *Environmental Modelling & Software* 23: 1217–1228.

Sun, J. & Tao, J.H. 2006. Relation matrix of water exchange for sea bays and its application. *China Ocean Engineering* 20: 529–544.

Takeoka, H. 1984. Fundamental concepts of exchange and transport time scales in a coastal sea. *Continental Shelf Research* 3: 311–326.

Thompson, K.R., Dowd, M., Shen, Y. & Greenberg, D.A. 2002. Probabilistic characterization of tidal mixing in a coastal embayment: a Markov Chain approach. *Continental Shelf Research* 22: 1603–1614.

Wang, Z., Liu, Z., Shan, G. & Xu, H. 1984. Numerical simulation of the tidal mixing in Bohai Sea, II: treatment of problems in simulating numerically the principal semidiurnal constitudent in Bohai Sea. *Oceanologia Et Limnologia Sinica* 15: 58–66 (in Chinese).

Yang, H.Y., Chen, B., Barter, M., Piersma, T., Zhou, C.F., Li, F.S. & Zhang, Z.W. 2011. Impacts of tidal land reclamation in Bohai Bay, China: ongoing losses of critical Yellow Sea waterbird staging and wintering sites. *Bird Conservation International* 21: 241–259.

Yuan, D., Lin, B., Tao, J. & Falconer, R.A. Verification of a numerical model using field monitoring data for modelling Bohai Bay. Proceedings of 28th IAHR Congress, Graz Austria, Theme D, 1999.

Zimmerman, J.T.F. 1976. Mixing and flushing of tidal embayments in the western Dutch Wadden Sea part I: Distribution of salinity and calculation of mixing time scales. *Netherlands Journal of Sea Research* 10: 149–191.

Biomedical Engineering and Environmental Engineering – Chan (Ed.)
© 2015 Taylor & Francis Group, London, ISBN: 978-1-138-02805-0

A design of a three phase AC sampling module for an electricity information acquisition device

H.L. Sun, R. Zhang, K.J. Zhou & J. Ye
State Grid Chongqing Electric power Research Institute, Chongqing, China

ABSTRACT: The paper analysed the design status of a three-phase AC sampling function in electric energy data acquisition equipment and pointed out its shortcomings. At the same time, according to the related standards of electric energy data acquisition equipment formulated by the State Grid Corporation of China (SGCC), the paper puts forward the requirements of the function and performance of a three-phase AC sampling module and proposes the specific hardware design scheme from the aspects of the composition of the module, the working principle, the size, the interface definition, the component selection, and the circuit design. The module can not only be used in a variety of electric energy data acquisition devices, but can also be used in other equipment that needs a three-phase AC sampling function. The module can be used universally in the transverse, longitudinal series and even cross product series and has a wide application prospect.

Keywords: AC sampling; module; data acquisition equipment; concentrator; special transformer.

1 INTRODUCTION

According to the related standards of the electric energy data acquisition equipment, formulated by the SGCC, type I concentrator, type I and type III terminals of special transformer and other electrical data acquisition equipment all require a three-phase AC sampling function. Therefore, introducing a modular design idea into the designing process of electric data acquisition equipment and designing a type of the three-phase AC sampling module with the characteristics of relatively independent, convenient installation, and a high ratio of performance to price, can not only reduce a repeat word of the design of the same function the performance of different products, and the waste of resources, but can also improve the efficiency of a product design.

2 ANALYSIS OF THE PRESENT STATUS

With the development of power user electric energy data acquire system construction continuously, electric energy data acquisition equipment (hereinafter referred to as the acquisition device), including type I concentrator, type I and type III terminals of special transformer required for three-phase AC sampling (hereinafter referred to as the sampling) will be a large number of applications. The manufacturers are designing the sampling part of acquisition equipment on the motherboard respectively, and not the realizing the modular design of the sampling part, which is not conducive to the same circuit function and

performance in the transverse and longitudinal series and even cross series of products to achieve universal.

Due to the modular design with the characteristics of "relative independence, interchangeability and general", introducing the module design concept into the design of sampling part of acquisition equipment and designing a sampling module with the characteristics of "relative independence, interchangeability and general" and wide application is necessary and important. The work is conducive to the three-phase sampling module in the transverse, longitudinal series and even cross series of products to achieve universal. The repeated design work of the circuit and the waste of resources can be reduced, and the efficiency of a product design can be improved.

3 DESIGN REQUIREMENTS

The there-phase AC sampling module is needed to finish the measurements of the there-phase voltage, current, phase and total reactive power, phase and total reactive electric energy, frequency, power factor, which is actually equivalent to a three-phase electric energy meter metering function. According to the relevant technical specifications of the SGCC, the design requirements of the three-phase AC sampling module are given as follows in 3.1 and 3.2.

3.1 *Performance requirements*

1) Operating temperature range: $-40°\sim+70°$;
2) Working voltage: DC, 5 V;

3) AC voltage: rated value is 57.7 V/100 V, 220 V/380 V, the input voltage range: (0~120%) Un. Un is the rated voltage;
4) AC current: the input AC current rating of 5 A (or 1.5 A), the input current range: 0~6 A, can withstand 1.2 times the Imax (the maximum value of the input current), and at least 4 hours of continuous overload; the tolerance 20 times rated current overload 5S does not damage;
5) The voltage class of accuracy and error limits: 0.5S, 0.5%;
6) The current class of accuracy and error limits: 0.5S, 0.5%;
7) The active power class of accuracy and error limits: 0.5S, 0.5%;
8) The reactive power class of accuracy and error limits: 2S, 2%;
9) The power factor class of accuracy and error limits: 2S, 2%.

3.2 Function requirements

1) Has the advantages of small volume, convenient installation;
2) Has the reactive pulse outputs (for calibration) interface;
3) With the SPI interface and the MCU external communication;
4) With voltage, and the current transformer input interface;
5) Can measure three-phase voltage, current;
6) Can measure the three-phase and the total/reactive power, and three-phase and the total/reactive electric energy;
7) Can measure the frequency, and the power factor;
8) Can measure 2~31 times of the voltage and current harmonic component;
9) Phase voltage and current measurement angle;
10) Can measure the zero sequence current;
11) Can be calibrated to a separate module; the calibration parameters are stored in a module.

4 THE DESIGN SCHEME

4.1 The composition and principle diagram

The there-phase AC sampling, including the whole circuits except the voltage transformer, the current transformer and the MCU, is mostly composed of voltage in the current sampling circuits, the three-phase measurement chip, the EEPROM memory, voltage and the current sampling electrostatic protection circuit, the input and the output interface. The module communicates with the host through the SPI interface processor. The current sensor connected is the isolating current transformer and the voltage sensor is the isolating voltage transformer ; with and without pulse output opt coupler isolation circuit in the main machine plate, as close as possible to the whole output interface is beneficial to improving the electromagnetic compatibility of insulating performance and safety performance.

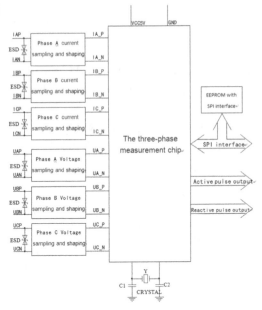

Figure 1. The principle diagram.

The principle diagram of three-phase sampling module is shown in Figure 1.

4.2 The working principle

As shown in Figure 1, the output of external voltage current transformer, which is proportional to the electric voltage and current, is transported to the input of sampling shaping circuit thorough the interface connector to the module. Connecting an ESD protection diode in shaping circuit input, to suppress the electrostatic and group pulse interference came from the transformer induction and improve the performance of electromagnetic compatibility. The AC sampling signal shaped by a shaping circuit is input to a low noise, and a high gain differential amplifier. The three-phase voltage, current, phase and total reactive power and the three-phase and the total measurement reactive energy, frequency and power factor are measured after amplification and A/D transformation in the interior of the three-phase measurement chip. The measurement data is read by the external MCU through the SIP interface. The module is very convenient and can be calibrated through software. The calibration parameters can be written by an external microcontroller through the SPI to the EEPROM memory module. The parameters can be read by the external MCU and written into the three-phase measurement chip.

4.3 The structure design

4.3.1 Outline dimension

Consider the three-phase AC sampling module to be built into the inside of the product, the outline dimensions should be small and easy to install, as shown in Figure 2.

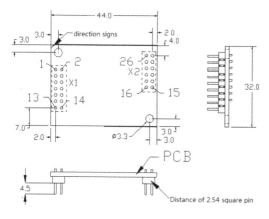

Figure 2. The outline dimension of the three-phase sampling module.

Description: X1 is the input interface connector of voltage and current transformer and X2 is the output interface connector. The two connector pin number designs are not the same, and asymmetric distribution, avoiding installation inserted in the wrong direction. The two mounting holes are set for fixing the module, the available PCB column spacing or strut screw module and the host substrate are fixed, in order to increase the reliability of the product.

4.3.2 The input and output interface pin definition
The module's input and output interface pins distribution and arrangement directions are shown in Figure 2 and the definitions are shown in Table 1 (Appendix 1).

4.4 Components selection

The components selection of the three-phase AC sampling module should follow the following principles:

1) In order to meet the requirements of module miniaturization and easy mass production, resistive and capacitive components should be chosen as the SMD device 0603 or a smaller size, the SMD integrated circuit device of a small size SOP, the QFP package, etc.
2) The working voltage, the device operating temperature range, the function and performance parameters are the key considerations to meet the requirements of the whole module and the corresponding circuit.

In accordance with the above selection principle, the selection suggestions of the main devices of the three-phase are given separately.

4.4.1 The sampling shaping circuit device
The voltage and the current sampling shaping circuit are mostly composed of the ESD protection diodes, resistors, capacitors and other devices. This part of the circuit is a low noise high gain measurement chip difference input circuit amplifier. Precision and

Device Type	Device Marking Code		Breakdown Voltage $V_{(BR)}$ at I_T[1] (V)		Test Current I_T (mA)	Stand-off Voltage V_{WM} (V)	Maximum Reverse Leakage at V_{WM} I_D (μA)[3]	Maximum Peak Pulse Surge Current I_{PPM} (A)[2]	Maximum Clamping Voltage at I_{PPM} V_C (V)
	UNI	BI	Min	Max					
SMAJ6.5A	AK	WK	7.22	7.98	10	6.5	500	35.7	11.2

Figure 3. Performance parameters of SMAJ6.5A.

consistency requirements of the device is high. The suggestions are as follows:

1) ESD protection diode: ESD protection diode is connected to the output ends of the external voltage, the current transformer, the most front-end module voltage, and the current input. The main role is to improve the group pulse voltage and the current loop antistatic ability. The normal working voltage of two ends of the maximum value is very low, generally less than 3 V. Therefore, SMAJ6.5A bidirectional ESD diode is a recommended choice. The performance parameter is shown in Figure 3.
2) Resistance: a high precision chip resistor, such as 0603 or 0402 package, are suggested choices, the requirement of error is less than 0.5%.

4.4.2 The measurement chip
4.4.2.1 The measurement chip section
A measurement chip is the major and key device of the three-phase sampling module. Zhuhai fire force of ATT7022B, ATT7022C, ADE7758, ADE7878 of ADI Company and Shenzhen Ruining micro RN8302 single chip are generally used as the core chips in an AC sampling device. These chips are integrated with a special DSP, can measure three-phase voltage, current, phase and total reactive power, three-phase and total reactive electric energy, frequency, power factor, has the characteristics of high integration, high performance price ratio, high reliability. However, the ATT7022B and the ADE7758 do not have a sub harmonic measurement function. To realize the sub harmonic measurement, additional high-speed A/D converter chips are needed, which not only increases the cost but also cannot meet the requirements of the module miniaturization. The ATT7022C, ADE7878, RN8302 are the three measurement chips which are integrated single chips in the sub harmonic measuring circuit solution; they can meet the requirement of miniaturization and a high performance to price ratio module. However, the RN8302 is just listed soon scheme, practical performance is not clear. Therefore, it is not currently recommend.

Judging by the above, the ATT7022C and the ADE7878 can best meet the requirements of the single chip solution, and are therefore recommended. In the design of the programmer, ATT7022C are selected from the prices point of view.

4.4.2.2 A performance of the measurement chip
The ATT7022C is an upgraded version of the ATT7022B with a compatible hardware pin. Based on retains all the features of ATT7022B, ATT7022C are increased the ADC sampling data cache opening function, does not need frequent interrupt to read the

real ADC data. The ATT7022C is an upgraded version of the ATT7022B with a compatible hardware pin. Based on retains all the features of ATT7022B, ATT7022C are increased the ADC sampling data cache opening function, does not need frequent interrupt to read the real ADC data. The 16bit data can be read by an external CPU through the SPI interface in each cycle of the output of the original sampling data. Using the Lagrange interpolation and the FFT algorithm, the harmonic component and the phase can be calculated.

The main performance characteristics are as follows:

1) Working voltage is 5 V;
2) The operating temperature range: $-40°\sim+85°$;
3) The package: 44Pin, QFP;
4) This is a high reliability, high precision, high stability of fundamental wave and harmonics of the three-phase electric energy metering chip;
5) The Accuracy better than 0.1% in the range of 1000:1
6) Effective value measurement accuracy is better than 0.5% of the current and voltage;
7) Using the standard SPI interface, speed is up to 2 Mbps, and provides a four way flexible pulse output;
8) Fundamental wave energy can be separately metered;
9) 16 high precision ADC the internal integration of 7 channel and a 24 bit high-speed DSP, seventh road ADC can be used for preventing electricity stealing;
10) Giving the total active reactive power, fundamental active reactive power and harmonic active reactive power and apparent energy parameters at the same time;
11) Compared with 7022B, the small signal current effective value accuracy obtained the promotion;
12) This has 240 bit ADC sampling data cache function, that can be used as a su harmonic analysis;
13) Synchronous sampling that supports a single channel and a dual channel or three channels, can be obtained indirectly by voltage included an angle with high accuracy (better than 0.5 degrees);
14) Software calibration, convenient and quick

4.4.3 The EEPROM memory chip

The EEPROM memory chip in the three-phase AC sampling module is mostly used to form the calibration parameter storage. The section is considered from the aspects of communication interface operating voltage, operating temperature range, storage capacity, and the external processor and metering chip.

In this paper, the total voltage of the working power supply module is 5 V, the operating temperature range of requirements is $-40°\sim+70°$, communication interface and the external processor and the metering chip is the SPI and, the interface storage amount is 1Kbit. According to these requirements, the selection of ATMEL AT25010N-10SI can meet the requirements

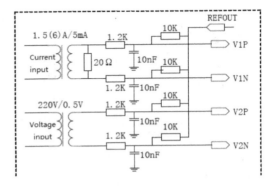

Figure 4. AC analogue input circuit.

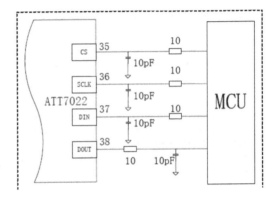

Figure 5. A typical SPI interface circuit diagram.

of the module design, and the main performance parameters are as follows:

1) Working voltage: 2.7 V~5.5 V;
2) The operating temperature range: $-40°\sim+85°$;
3) Communication interface: SPI;
4) Storage capacity of 1Kbit (128 * 8);
5) Package: SOIC8

4.5 Circuit design of module

4.5.1 Input circuit of the AC analogy signal

ATT7022C is within the integrated Road 7 16 bit A/D converter. The effective value of current channels is in the 2 mV to 1 V range and the linear error is less than 0.5%; the effective value of voltage channel is in the 10 mV to 1 V range and linear error is less than 0.5%; voltage values is at 0.2 to 0.6 V (amplified voltage value, advice voltage sampling the signal in the 0.1 V, the voltage amplification multiple channel selected 4 times), the current value is in the 2 mV to 1 V, and the electric energy of linear error is less than 0.1%. AC analogue input circuit of a typical circuit is shown in Figure 4.

4.5.2 SPI interface circuit

Considering the SPI interface transmission rate is high and signal line may have interference or jitter, a suggestion of seriating a small resistance in the SPI signal line is given. Typically, this resistor value is in the 10 to 100 ohms and can eliminate any oscillation on an SPI interface. If the parasitic capacitance of the input to that end of the IC is not large enough, it also can be connected with a capacitance of about 10P. A typical SPI interface circuit is shown in Figure 5.

5 CONCLUSION

The scheme proposed in this paper is with characteristics of a high precision measurement, complete functions, a wide working temperature range, good electromagnetic compatibility, high reliability, small size, and a convenient installation. The AC sampling module can not only be used in type I concentrator, type I and type III terminals of special transformer, but also can be used in other devices requiring three-phase AC sampling function. The module is conducive to the three-phase sampling module in the transverse, longitudinal series, and even cross series of products to achieve universal application prospects.

REFERENCES

Chen, A.G., Meng, Y. Q/GDW 1376.2-2013 power user information acquisition system communication protocol-Part 2: concentrator local communication module interface protocol [S], 2013(4): 1–337, Beijing: China Electric Power Press.

Liu, X., Chang, L. Q/GDW 1374.2-2013 power user information acquisition system technical specification-Part 2: automatic meter reading terminal technical specification [S], 2013(4): 1–384, Beijing: China Electric Power Press.

Liu, X., Dong, L.J. Q/GDW 1375.2-2013 power user information acquisition system type specification-Part 2 concentrator type specification [S], 2013(4): 1–347, Beijing: China Electric Power Press.

Liu, X., Chen, A.G. Q/GDW 1376.3-2013 power user information acquisition system communication protocol-Part 3: acquisition terminal remote communication module interface protocol [S], 2013(4): 1–354, Beijing: China Electric Power Press.

Xiao, H.Y. A Survey on measuring method for harmonics in power system[j] Power System Technolog, 2002, 26.

Zhang, X., Liu, X. Q/GDW 1374.1-2013 power user information acquisition system technical specification-Part 1: designed variable terminal technical specification [S], 2013(4): 1–378, Beijing: China Electric Power Press.

Zhang, X., Tang, Y. Q/GDW 1375.1-2013 power user information acquisition system type specification-Part 1: designed variable terminal type specification [S], 2013(4): 1–388, Beijing: China Electric Power Press.

Appendix

Table 1. Pin definitions of three-phase sampling module

Pin number	Signal type	Signal name	Signal direction for the module	Description
1	signal	IAP	input	A phase current differential input +, connecting A phase current transformer input.
2	signal	IAN	input	A phase current differential input −, connecting A phase current transformer output.
3	signal	IBP	input	B phase current differential input +, connecting B phase current transformer input.
4	signal	IBN	input	B phase current differential input −, connecting B phase current transformer output
5	signal	ICP	input	C phase current differential input +, connecting C phase current transformer input
6	signal	ICN	input	C phase current differential input +, connecting C phase current transformer output
7	Null	NC	/	NO
8	Null	NC	/	
9	signal	UAP	input	A phase voltage differential input +, connecting A phase voltage transformer input.
10	signal	UAN	input	A phase voltage differential input −, connecting A phase voltage transformer output.
11	signal	UBP	input	B phase voltage differential input +, connecting B phase voltage transformer input.
12	signal	UBN	input	B phase voltage differential input −, connecting B phase voltage transformer output.
13	signal	UCP	input	C phase voltage differential input +, connecting C phase voltage transformer input.
14	signal	UCN	input	C phase voltage differential input −, connecting C phase voltage transformer output.
15	Power input	VCC	input	The total power input module of the DC 5 V/30 mA, by the host supply module.
16	Power ground	GND	input	
17	SPI export signal	SPI_DI	input	SPI interface, single chip computer using the interface and communication module, at a 5VTTL level.
18	SPI export signal	SPI_DO	output	
19	SPI export signal	SPI_CLK	input	
20	SPI export signal	SPI_CS	input	
21	SPI export signal	SPI_RST	input	
22	signal	SIG	output	Measurement chip state signal output
23	signal	MOD_SEL	input	Mode choice, controlled by an external microcontroller. Set high, works in the three-phase four wire mode, set low works in the three phase three wire mode.
24	Power ground	GND	input	
25	signal	CF1	output	Active electric energy pulse output, no opt coupler isolation. Opt coupler design in the host near the leading out terminal, in order to improve the electromagnetic compatibility.
26	signal	CF2	output	Reactive electric energy pulse output, no opt coupler isolation. Opt coupler design in the host near the leading out terminal, in order to improve the electromagnetic compatibility.

Biomedical Engineering and Environmental Engineering – Chan (Ed.)
© *2015 Taylor & Francis Group, London, ISBN: 978-1-138-02805-0*

The modular design of an electricity information acquisition device

H.L. Sun, J. Ye, K. Zheng & R. Zhang
State Grid Chongqing Electric Power Research Institute, Chongqing, China

ABSTRACT: This paper researched I a concentrator and III designed a variable terminal which is widely used in the electricity information acquisition system, analysed their functions and applications combined with the relevant enterprise standards of the State Grid Corporation, considered the characteristics and requirements of the electronic product modular design, and then planned the main module of two types of an acquisition device. Furthermore, given their physical structure, the interface definition and communication protocol so as to achieve the two acquisition device shared host and uplink communication module, modularized the uplink and downlink communication channel. The research results show that the two types of acquisition device can fully share the same host. If change the local communication module (downlink communication channel) into the control module, I concentrator can be changed to III designed variable terminal in the same host, and vice versa. The local communication module can be freely converted in a narrowband power line carrier, a broadband power line carrier, a micro power wireless and other communication means while a remote communication module (uplink communication channel) can be converted freely in the GPRS, the CDMA, 3G, the PSTN, a micro power wireless, and a 230 MHz network. Thus satisfy customer needs diversity using in the most economic method.

Keywords: modular, acquisition device, host, local communication module, remote communication module.

1 INTRODUCTION

The electricity information acquisition system is an important part of a smart grid construction, and the electric energy data acquisition device (hereinafter referred to as the acquisition device) is the key equipment of the electric information acquisition system. As the grid environment is complex and changeable, the power line channel interference, and the terrain can also make a difference. Therefore the acquisition device communication channel requires multiple communication ways in order to select suitable uplink and downlink communications according to the actual communication environment by the user. The modular design is the best choice to meet the user's diverse application requirements.

2 THE PRESENT SITUATION AND PROBLEMS

The State Grid Corporation has issued a series of relatively complete enterprise standards, composed of functional specifications, technical specifications, type specifications, communication protocol, security standard, inspection specifications and other six categories of 15 standards in electricity information acquisition system and its device.

Although there are requirements in the related standards for acquisition device standardization and modularization, the research of the related specifications found that the following problems still exist:

(1) Specific provisions and requirements for a hardware connector between a remote or a local communication module and the host have not been provided. The specification only stipulate the module uses pin header and the host uses female header, but there are no specific instructions on which module should use a specific pin, and which type of row should the host use. If this is not clearly defined, it can cause problems for the exchange amongst modules.

(2) There did not make clear pin number orientation the remote or local communication module and host, so it may cause ambiguity that lead to the modules are not interchangeable.

(3) The four-way remote signal/pulse acquisition interface of two types of acquisition device do not stipulate the common power supply (+) or the common power supply (GND).

(4) The host of the two acquisition device is still required, and it should be designed independently rather than as a common module. Therefore, it cannot achieve the purpose of the two types of acquisition devices sharing the same host.

3 THE CHARACTERISTICS AND ADVANTAGES OF THE DESIGN MODULE

A modular electronic product is an effective method used for a single piece production in large quantities. An electronics module is a standard component which has an independent function and an input output. The principle of a modular product design is that it makes function analysis for products with different or same functions in a certain range, divides and designs a series of functional modules, through the selection and combination of modules constitute different users to customize products, to meet the different needs of users. This is the application of similarity principle in the product function and structure, and an organic combination of standardization and diversification. A module is a functional unit of a modular design and manufacturing, which has three major characteristics:

(1) Relative independence. A module can be separately designed, manufactured and debugged, modified, and stored, which facilitate product respectively by different professional enterprises.
(2) Interchangeability. The standardization of a module interface structure, size and parameters makes the interchange between modules easy, so that a greater number of modules meet the requirements of different products.
(3) Universal property. It facilitates to achieve the modules become universal between transverse series of products and vertical, achieve the modules become universal among the across series of products.

4 THE MODULAR DESIGN OF THE ACQUISITION DEVICE

4.1 *Analysis of function differences between a I concentrator and a III designed variable terminal*

Analysing the technical specification and type specification of a I concentrator and a III designed variable terminal, we can find that the host of both acquisition device has the same shape structure and function requirements:

(1) With the same shape structure and installation size.
(2) With the same liquid crystal display size and position.
(3) The position and definition of a status indicator lamp, a key and the external interface are the same, or compatible.
(4) The uplink traffic channel includes: the 230 MHz network, the GPRS, the CDMA, the micro power wireless, the Ethernet, the PSTN and so on.
(5) The AC sampling function is exactly the same as the requirements.
(6) The requirements of power supply are the same, or compatible.

The difference between the two acquisition devices is that the I concentrator is mostly used for residential users' electricity information collection; the downlink communication mode includes a micro power wireless, a power line carrier, and a wired network, achieved by a local communication module. While the III designed variable terminal is mostly used in a transformer substation, industrial enterprises and other users of special transformers for the electrical energy information acquisition and load control, through the control module to realize. However, the shape, size and interface definition of the two modules can be designed to be the same or compatible, in order to achieve the aim of two devices sharing the same host.

4.2 *Module planning*

According to the features of the modular design, the I concentrator and the III designed variable terminal could be planned for the following modules: the host, the remote communication module, the local communication module (the III designed variable terminal for a control module).

4.3 *The host*

According to the State Grid related specifications, the most functions of the host of the I concentrator and the III designed variable terminal are the same therefore this can be planned as a relatively independent module and a unified design for the shape structure, the input/output interface, the status indication, the display, etc.

4.3.1 *The host interface design*
The host interface includes the interface between a host and a remote communication module, the weak current interface between a host and a local communication module (a designed variable terminal for the control module), the coupling interface between a host and a local communication module, the main terminal and the auxiliary terminal interface and so on. Here is given its interface type, the interface direction, the interface definition, and a communication protocol.

4.3.1.1 *Interface type*
The State Grid type specification of Q/GDW 1375.1-2013 and Q/GDW 1375.2-2013 do not specify the specific requirements for each hardware interface connector on the host side, such as connector types and pins spacing. It must be determined, otherwise different manufacturers' products would not be able to interchange with each other. Now put forward the following solutions:

(1) The interface between the host and a remote communication module
  Adopt a double rows female header suitable for a 2*15 square pin-header with a 2.54 mm spacing as a connector and a bottom empty centipede feet female header.

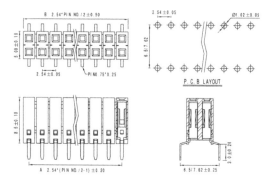

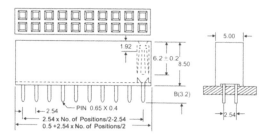

Figure 1. The centipede feet female header structure size.

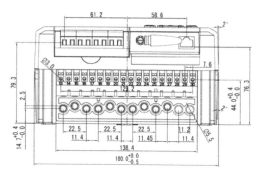

Figure 3. The main terminal and the auxiliary terminal positions and sizes.

Figure 2. The female header structure size of the local communication module coupling interface.

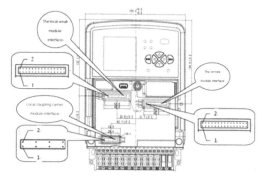

Figure 4. The direction definition of pins for each host interface.

(2) The weak current interface between the host and a local communication module/control module

Adopting a double rows female header suitable for a 2*13 square pin-header with a 2.54 mm spacing as a connector and a bottom empty centipede feet female header.

The above two interfaces are weak current interfaces, and the reason for using a hollow centipede feet female header is that a centipede feet female header can tolerate more length errors of the communication module corresponding pin-header. The Communication module can still be inserted smoothly when the pin-header is longer, otherwise it may be inserted in a bad position. The size and the type of a centipede feet female header is shown in Figure 1.

(3) The coupling interface between the host and a local communication module

The interface is only used by I concentrators, but the host must design it. While used as III designed variable terminal without an electrical connection so that it does not produce any effect, so the host can be done in two public acquisition device, to achieve interoperability, interchangeability objective.

The interface adopts a double rows female header suitable for a 2*10 square pin-header with a 2.54 mm spacing as a connector. Because the interface is an AC 3×220 V voltage access, it should use an ordinary female header that does not allow the module corresponding pin header

Figure 5. The local communication module/control module interface host side pin-out.

through the PCB from the safety point of view; its structure is shown in Figure 2.

(4) The main terminal and the auxiliary terminal interface

The position and size of the main terminal and the auxiliary terminal interface for the I concentrator and the III designed variable terminal are exactly the same, as shown in Figure 3.

4.3.1.2 *The direction definition of interface pins*
The direction definition of pins for each host interface is shown in Figure 4.

4.3.1.3 *The interface pin signal definition*
(1) The interface between the host and the remote communication module

The interface pin-out, pin signal definitions of the I concentrator and the III designed variable terminal are identical. They are designed according to the State Grid related type specification, specifically referring to the terminals of the special transformer type specification Q/GDW 1375.1-2013

Table 1. Description of the weak interface module definition of local communication module

The weak interface pin of the local communication module		Implication
NO.20 STATE0	NO.14 STATE1	
1	X	Module is not inserted
0	1	Insert the local communication module, run the concentrator program
0	0	Insert the control module, run the III designed variable terminal program

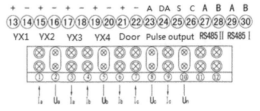

Figure 6. The main terminal and auxiliary terminal of I concentrator.

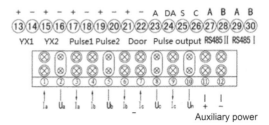

Figure 7. The main terminal and the auxiliary terminal of the III designed variable terminal.

and the concentrator type specification Q/GDW 1375.2-2013.

(2) The weak current interface between the host and the local communication module/control module

The pins of the I concentrator and the III designed variable terminal have the same sequence, as shown in Fig 5.

Pin signal definition of the interface can be compatible for an organic whole and designed based on relevant type specification of State Grid, except that use the NO.14 pin which reserve for disuse combine with the NO.20 as module type identification signal to identify the control module is inserted or local communication modules, achieve that I concentrator and III designed variable terminal share the same host. Specifically refer the terminals of special transformer type specification Q/GDW 1375.1-2013 and concentrator type specification Q/GDW 1375.2-2013.

Here, only explain the difference from the module type definition in State Grid specification, see Table 1.

(3) The coupling interface between the host and a local communication module

The interface only used by I concentrators, although the host must design it. While used as a III designed variable terminal without an electrical connection so that it does not produce any effect, so the host can do it in two public acquisition devices, in order to achieve the interoperability, and the interchangeability objective.

The pin definition of the interface is completely based on the State Grid type specification for designs, see the specific concentrator type specification Q/GDW 1375.2-2013.

(4) The main terminal and auxiliary terminal interface

The order of the interface pin-out and signals definition are shown in Figure 6 and Figure 7.

From the above Figure 6 and Figure 7 it can be seen that the pin-out sequence of the main terminals and the auxiliary terminals for two acquisition devices are the same as well as their pin definitions and application.

To sum up, the main internal design of the main terminal and the auxiliary terminal for the I concentrator and the III designed variable terminal can be completely combined to one, then the purpose of sharing, interoperability, and interchangeability can be achieved.

4.3.1.4 The auxiliary terminal remote communication/pulse input port design

Since the two acquisition devices remote signal/pulse input interface have two more roads, there must be a common power anode (common anode) or a common power ground (common cathode) problem. However, the related specifications of the State Grid only provide a collection of nature or pulse switches, i.e. a passive switch or a passive pulse volume, rather than a common cathode or a common anode. There is the hardware circuit design using a common cathode, for the following reasons:

When it comes to using the device for a meter to collect an active pulse and a reactive power pulse, the pulse output terminal of the installed meter has only three leading out terminals: active positive, reactive positive, and public. The public end of the optocoupler emitter, is namely a common cathode, as shown in Figure 8. The pulse acquisition interface design of the device is a common anode, as shown in Figure 9.

In this case, the acquisition device will not be able to collect the meter for the active pulse and the reactive pulse at the same time, or either the active pulse or the reactive pulse. Because the electric meter is an emitter (common cathode), it could not be connected to a common end, a 12 V (common anode) of the acquisition device, so that the pulse acquisition circuit could not work, but this can also cause the meter optocoupler pulse to output the optocoupler damage due to the reverse power up to the electric meter.

212

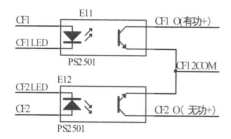

Figure 8.   The electric meter pulse output circuit.

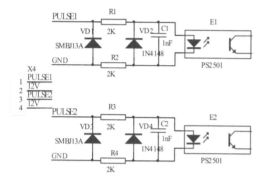

Figure 9.   The pulse acquisition input circuit of an acquisition device.

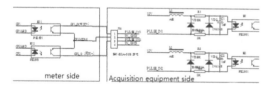

Figure 10.   The pulse acquisition device circuit.

If the design is a common cathode, it can solve this problem, as shown in Figure 10.

4.3.1.5   The interface communication protocol

The remote communication module interface and the weak local communication module interface of the host involves the problem of communication protocols, respectively as follows.

The communication protocol between the host and the remote communication module according to the State Grid Specification Q/GDW 1376.3-2013 power user information acquisition system communication protocol – Part 3: The acquisition terminal remote communication module interface protocol.

The communication protocol is between the host and the local communication module according to the State Grid Specification Q/GDW 1376.2-2013 power user's information acquisition system communication protocol – Part 2: the concentrator local communication module interface protocol.

The communication protocol is between the host and the control module according to the State Grid

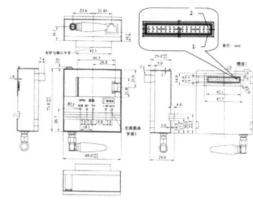

Figure 11.   The structure size and the interface pin orientation of the remote communication module.

Specification C.12 communication protocol between the control module and the terminal of Q/GDW 1375.1-2013 power user's information acquisition system type specification – Part 1: the designed variable terminal type specification.

4.4   The remote communication module

4.4.1   The structure size and the interface pin orientation

The structure size and the interface pin orientation of the remote communication module is shown in Figure 11.

4.4.2   The interface type

Adopting a double rows 2*15 square pin-header with a 2.54 mm spacing as connector; the grounding pin is longer than other 0.5 mm. The reason for suggesting the square pin is that the structure of the square pin is a stronger anti break than the round pin, the long grounding pin facilitates the module's hot plug.

4.4.3   The interface definition

The pin definition of the interface is completely based on the State Grid type specification for the design. See the specific designed variable terminal type specification for the Q/GDW 1375.1-2013, or the Q/GDW 1375.2-2013 concentrator type specification.

4.4.4   The communication protocol

The communication protocol between the remote communication module and the host based on the Q/GDW 1376.3-2013 power user's information acquisition system communication protocol – Part 3: The acquisition terminal remote communication module interface protocol.

4.5   The local communication module/control module

4.5.1   The structure size and the interface pin orientation

The structure size and the interface pin orientation of the local communication module/control module shown in Figure 12 and Figure 13.

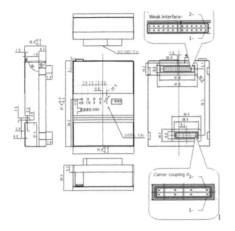

Figure 14. The pin-out of local communication module weak interface/control module interface.

Table 2. Description of the interface module definition of the local communication module

The weak interface pin of local communication module		Implication
NO.20	NO.14	
STATE0	STATE1	
0	1	The local communication module: No.20 grounded, No.14 pull up
0	0	Control module: No. 20 and No. 14 grounded

The pin signal definition of the interface can be compatible for an organic whole and designed based on relevant type specification of the State Grid, except that it uses the No.14 pin which is reserved for disuse combined with the No.20 pin, as a module type identification signal in order to determine the control module is inserted or local communication modules. This specifically refers to the terminals of special transformer type specification Q/GDW 1375.1-2013 and the concentrator type specification Q/GDW 1375.2-2013.

Here, only explain the difference from the module type definition in State Grid specification, see Table 2.

4.5.4 The communication protocol

The local communication protocol based on the Q/GDW 1376.2-2013 power user's information acquisition system communication protocol – Part 2: concentrator local communication module interface protocol.

The control module communication protocol based on the State Grid Specification C.12 communication protocol between the control module and the terminal of Q/GDW 1375.1-2013 power user's information acquisition system type specification – Part 1: the designed variable terminal type specification.

5 CONCLUSION

It can be seen from the above design that the two kinds of acquisition devices can fully share the same host. If changing the local communication module into the control module, the I concentrator can be changed to III designed variable terminal in the same host, and vice versa. The local communication module can be converted freely in the narrowband power line carrier, the broadband power line carrier, micro power wireless and other communication means while remote communication module can be converted freely in the

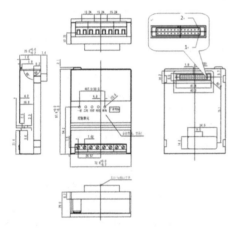

Figure 12. The structure size and the interface pin orientation of the local communication module.

Figure 13. The structure size and the interface pin orientation of the control module.

4.5.2 The interface type

The local communication module weak interface and the control module interface adopt a double rows 2*15 square pin-header with a 2.54 mm spacing as connector; the grounding pin is longer than other 0.5 mm. The reason for suggesting the square pin is that the structure of the square pin is stronger anti break than the round pin; the long grounding pin facilitates the module's hot plug.

4.5.3 Interface definition

(1) The local module carrier coupling interface
    The pin definition of the interface is completely based on the State Grid type specification for the design; see the specific concentrator type specification Q/GDW 1375.2-2013.
(2) The local communication module weak interface/control module interface
    The local communication module weak interface/control module interface uses the same connector type and pin-out, as shown in Figure 14.

GPRS, CDMA, 3G, PSTN, micro power wireless, and 230 MHz network. It can be matched into more than ten types of different products by replacing the module, the users only need to select the appropriate function module according to their requirements, thereby reducing the cost and avoiding the waste of resources. Thus, it satisfies customer needs' diversity using the most economic method.

REFERENCES

Chen, A.G. Meng, Y. Q/GDW 1376.2-2013 power user information acquisition system communication protocol-Part 2: concentrator local communication module interface protocol [S], 2013(4): 1–337, Beijing: China Electric Power Press.

Liu, X., Chang, L. Q/GDW 1374.2-2013 power user information acquisition system technical specification-Part 2: automatic meter reading terminal technical specification [S], 2013(4): 1–384, Beijing: China Electric Power Press.

Liu, X. Dong, L.J. Q/GDW 1375.2-2013 power user information acquisition system type specification-Part 2 concentrator type specification [S], 2013(4): 1–347, Beijing: China Electric Power Press.

Liu, X., Chen, A.G. Q/GDW 1376.3-2013 power user information acquisition system communication protocol-Part 3: acquisition terminal remote communication module interface protocol [S], 2013(4): 1–354, Beijing: China Electric Power Press.

Zhang, X., Liu, X. Q/GDW 1374.1-2013 power user information acquisition system technical specification-Part 1: designed variable terminal technical specification [S], 2013(4): 1–378, Beijing: China Electric Power Press.

Zhang, X., Tang, Y. Q/GDW 1375.1-2013 power user information acquisition system type specification-Part 1: designed variable terminal type specification [S], 2013(4): 1–388, Beijing: China Electric Power Press.

Road traffic noise prediction model in Macao

N. Sheng, Z.R. Xu, M. Li & Q.W. Hong
Department of Decision Sciences, School of Business, Macau University of Science and Technology, Macao

ABSTRACT: As the largest gambling hub in the world, Macao faces serious traffic noise pollution. This study applied the Calculation of Road Traffic Noise (CRTN) model to predict traffic noise in Macao. The validity of the CRTN model was assessed by comparing the CRTN predictions and the measurements. It was found that the predictions correlated well with the measurements with an R^2 of 0.832 and a mean difference of +0.52 dB(A).

1 INTRODUCTION

Macao is a high-density city with 607,500 people living in a land area of only 30.3 km² (DSEC 2013). The population density is as high as 20,049 inhabitants/km² and the traffic density is as high as 541 vehicles/km. Up to now there is no generally accepted model for traffic noise prediction in Macao. Traffic noise prediction models are essential for traffic noise pollution assessment and sound environmental management in urban areas. Many traffic noise prediction models have been developed for traffic noise assessment in different countries (Delany et al. 1976; CRTN 1988; EMPA 1987; RLS-90 1990; Tachibana and Sasaki 1994; Anderson et al. 1996; Bendtsen 1999; Suksaard et al. 1999; Steele 2001; Li et al. 2002; To et al. 2002; Tang and Wang 2007; Murphy et al. 2009; Petraitis et al. 2011). Among these, the Calculation of Road Traffic Noise (CRTN) model (Delany et al. 1976; CRTN 1988) developed by the Department of Transport of the United Kingdom is one of the popular ones. Many researchers have studied the reliability of traffic noise prediction using the CRTN model in different urban areas (Lam and Tam 1998; Leung and Mak 2008; Mak et al. 2010; Mak and Leung 2013). This study will assess the applicability of the CRTN model to Macao.

2 METHODOLOGY

In this study, 31 sites were selected in the Macao Peninsula and data were collected during the morning and evening peak hours between October and December 2013. The data collected included traffic noise level, traffic composition, traffic volume, vehicle speed and road characteristics. The traffic noise measurements of sound pressure level were conducted by using noise statistics analyzer HS6298A. In each traffic noise measurement, traffic characteristics including traffic composition, traffic volume and speed of vehicles

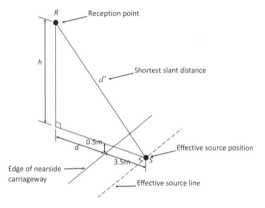

Figure 1. An illustration of the positions of source and reception point.

on the road were recorded simultaneously for noise prediction purposes.

The traffic noise prediction was based on the CRTN model which originated in the United Kingdom and has been officially adopted in the Hong Kong Planning Standard and Guidelines for traffic noise assessment (Lam and Tam 1998). The calculation of the CRTN model assumes the source of traffic noise to be a line 0.5 m above the carriageway and 3.5 m in from the nearside carriageway edge, see Figure 1.

The calculation of the CRTN model starts with a basic noise level determined by the traffic flow, traffic speed, traffic composition and road gradient. At a reception point with a reference distance of 10 m away from the nearside carriageway edge, the basic noise level can be calculated by (CRTN 1988):

$$L_{A10} = 10\log_{10} Q + 33\log_{10}(V + 40 + 500/V)$$
$$+ 10\log_{10}(1 + 5P/V) + 0.3G - 26.6 \qquad (1)$$

where Q is the hourly traffic volume of all heavy and light vehicles, V is the traffic speed, P is the percentage of heavy vehicles, G is the road gradient, and L_{A10} is the basic noise level exceeded for just 10% of the time over a period of one hour.

The corrections are then added to the basic noise level to take into account the effects of distance from the source line, the nature of the ground surface, and reflections from buildings and façades.

Figure 1 shows the positions of source and reception point. The distance correction in the CRTN method is calculated as:

$$\Delta L_d = -10\log_{10}(d'/13.5) \qquad (2)$$

where d' is the shortest slant distance from the source position given by $d' = \sqrt{(d+3.5)^2 + h^2}$, d is the shortest horizontal distance between the nearside carriageway edge and the reception point, and h is the vertical distance between the source position and the reception point. The shortest horizontal distance d is assumed to be not less than 4 m.

In the CRTN method, the reflection correction is calculated by:

$$\Delta L_f = 2.5 + 1.5(\theta'/\theta) \qquad (3)$$

where the correction of 2.5 dB(A) is to take into account the reflection of noise from façade adjacent to the reception point (or on the nearside of the reception point), $1.5(\theta'/\theta)$ dB(A) is the correction for reflection from opposite façade facing the reception point, θ' is the sum of the angles subtended by all the reflecting façades on the opposite side of the road facing the reception point, and θ is the total angle subtended by the source line at the reception point.

The validation of CRTN model was carried out by comparing the on-site traffic noise measurements and the corresponding CRTN predictions.

3 RESULTS AND DISCUSSION

In this study, Macao Peninsula was selected as the study area. On-site traffic noise measurements at 31 sites showed that the average traffic noise at the 31 sites was 77.16 dB(A). All traffic noise levels exceeded 70 dB(A). It should be noted that until now, there has been no road traffic noise standard in Macao. This study referred to the Hong Kong Planning Standards and Guidelines in which a standard of 70 dB(A) (LA10 in 1 hour) has been set for road traffic noise. The results showed that about 80% of traffic noise levels at the sites investigated in this study were above the benchmark of 70 (A) by 5 dB(A), which indicate that the Macao Peninsula has fallen into a situation of serious traffic noise pollution.

By comparing the on-site traffic noise measurements and the corresponding CRTN predictions, it was found that the deviations between the measured and predicted traffic noise levels at 31 sites did not exceed 3 dB(A). The maximum deviation was an overestimation of 2.96 dB(A). Deviations of less than 1 dB(A) were found at 21 out of the 31 sites. The CRTN model

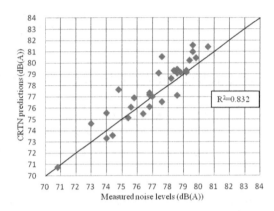

Figure 2. Predicted traffic noise levels by the CRTN model against measured values.

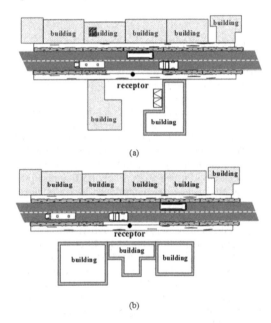

Figure 3. (a) Without building façade right behind the reception point; (b) With building façade right behind the reception point.

overestimated traffic noise level by 0.52 dB(A) on the average. Figure 2 shows the scatter plot drawn between the measured and predicted values of traffic noise. Using the regression analysis, it was found that the predicted traffic noise levels by the CRTN model correlated well with the measured values with an R^2 of 0.832. The results suggest that the predicted traffic noise levels by the CRTN model correlate closely with the measured values in this study.

Figure 3 shows two typical cases in a high-density city. In one case there is no building façade right behind the reception point, see Figure 3(a), while in another one there is building façade right behind the reception point, see Figure 3(b). In the CRTN calculation, a correction of +2.5 dB(A) is required to take into account the reflection of noise from the building façade behind the reception point. Even though the reception point is away from the building façade, the same addition of the 2.5 dB(A) is required. This might lead to an overestimation of traffic noise level. Some studies in the literature have reported that the CRTN model overestimated the road traffic noise (Samuels and Saunders 1982; Chew and Lim 1994). Therefore, the calibration of CRTN model is suggested by considering whether or not there is building façade right behind the reception point.

4 CONCLUSIONS

Rapid economic growth in Macao in recent years has been accompanied by serious environmental pollution problems, among which road traffic noise is a major concern. This study evaluated the validity of the CRTN model by comparing the predicted traffic noise levels using the CRTN model and the measurements. It was found that the predictions using the CRTN model correlated well with the measured results with an R^2 of 0.832 and a mean difference of +0.52 dB(A). This indicates that the CRTN model is a reliable and suitable traffic noise prediction model for Macao. However, the calibration of CRTN model should be performed in the future by considering local environment and conditions.

ACKNOWLEDGEMENT

The research work is supported by Macau Science and Technology Development Fund (FDCT) under Grant No. 120/2012/A. It is also supported by Faculty Research Grant (Nov.2014) at Macau University of Science and Technology.

REFERENCES

Anderson G.S., Menge C.W., Rossano C.F., Armstrong R.E., Ronning S.A., Fleming G.G. & Lee C.S.Y. 1996. FHWA traffic noise model, version 1.0: introduction to its capacities and screen components. *The Wall Journal* 22: 14–17.

Bendtsen H. 1999. The Nordic prediction method for road traffic noise. *Science of the Total Environment* 8: 235–331.

Chew C.H. & Lim K.B. 1994. Façade effects on the traffic noise from the expressway. *Applied Acoustics* 41(1): 47–62.

CRTN. 1988. *Calculation of Road Traffic Noise*. Department of Transport, Welsh Office. London: HMSO.

Delany M.E., Harland D.G., Hood R.A. & Scholes W.E. 1976. The prediction of noise levels L10 due to road traffic. *Journal of Sound and Vibration* 48(3): 305–325.

DSEC. 2013. *Yearbook of Statistics 2013*. Statistics and Census Service, Macao SAR Government.

EMPA. 1987. *Modéle de calcul de bruit du trafic routier pour ordinateur, lére partie: manuel d'utilisation du logiciel STL-86, version 1.0*. L'Office fédéral de la protection de l'environnement: Berne, Switzerland.

Lam W.H.K. & Tam M.L. 1998. Reliability analysis of traffic noise estimates in Hong Kong. *Transportation Research Part D* 3(4): 239–248.

Leung B.K.H. & Mak C.M. 2008. Is the CRTN method reliable and accurate for traffic noise prediction in Hong Kong? *The Hong Kong Institution of Engineers Transactions* 15, 17–23.

Li B., Tao S. & Dawson R.W. 2002. Evaluation and analysis of traffic noise from urban roads in Beijing. *Applied Acoustics* 63: 1137–1142.

Mak C.M., Leung W.K. & Jiang G.S. 2010. Measurement and prediction of road traffic noise at different building floor levels in Hong Kong. *Building Services Engineering Research & Technology* 31: 131–139.

Mak C.M. & Leung W.S. 2013. Traffic noise measurement and prediction of the barrier effect on traffic noise at different building levels. *Environmental Engineering and Management Journal* 12: 449–456.

Murphy E., King E.A., Rice H.J. 2009. Estimating human exposure to transport noise in central Dublin, Ireland. *Environment International* 35: 298–302.

Petraitis E., Pranskevicius M., Idzelis R.L., Vaitiekunas P. 2011. Predictive modelling of environmental noise levels in Lithuanian urban areas. *Environmental Engineering and Management Journal* 10: 1935–1941.

RLS-90. 1990. *Richtlinien für den Lärmschutz an Straßen*. Der Bundesminister für Verkehr: Bonn, Germany.

Samuels S.E. & Saunders R.E. 1982. The Australian performance of the UK DoE traffic noise prediction method. In: *Proceedings of the Eleventh Australian Road Research Board Conference*: 30–44. ARRB Group Limited.

Steele C.M. 2001. A critical review of some traffic noise prediction models. *Applied Acoustics* 62: 271–287.

Suksaard T., Sukasem P., Tabucanon S.M., Aoi I., Shirai K. & Tanaka H. 1999. Road traffic noise prediction model in Thailand. *Applied Acoustics* 5: 123–130.

Tachibana H. & Sasaki M. 1994. ASJ prediction methods of road traffic noise. *Inter-noise 94*: 283–288.

Tang U.W. & Wang Z.S. 2007. Influences of urban forms on traffic-induced noise and air pollution: results from a modelling system. *Environmental Modelling & Software* 22: 1750–1764.

To W.M., Ip C.W., Lam C.K. & Yau T.H. 2002. A multiple regression model for urban traffic noise in Hong Kong. *Journal of the Acoustical Society of America* 112: 551–556.

Biomedical Engineering and Environmental Engineering – Chan (Ed.)
© 2015 Taylor & Francis Group, London, ISBN: 978-1-138-02805-0

Microalgae treatment of wastewater from the Henan oilfield

Feng He
Beijing Institution of Fashion and Technology, China

Hua Zhou
State Intellectual Property Office of the People's Republic of China

ABSTRACT: Microalgae treatment of heavy crude oil production wastewater is a well-established method for remediation of these wastes. We have developed effective biological treatments by: (1) utilizing microbes with high oil-degrading abilities, (2) allowing greater organic loads while increasing both the process stability and the resistance to shock loading, (3) minimizing the production of waste sludge by-products, and (4) adopting anaerobic and aerobic biological processes to improve the biodegradation of the wastewater. Fixed-film bioreactors with 15 h hydraulic retention times have decreased chemical oxygen demand by 74.8%, total suspended solids by 90.9%, oil by 80.6%, and phenols and sulphides by 100%. The results with an in situ pilot system show that the bioreactor's hydrolytic acidulation and contact oxidation tanks are suitable for treating oilfield wastewater, and that water quality after treatment fully meets national drainage standards.

1 INTRODUCTION

The majority of oil fields in China have now entered into the third oil recovery stage (after primary and secondary recovery procedures have been completed), which is carried out by injecting large quantities of water to flush out light crude oil, or injecting steam to flush out heavy crude oil (Deng and Zhu, 2008) This approach is considered to be the most efficient technique for tertiary oil recovery during the development of the oil fields (Richard, G. 2007)

A unique feature of fixed-film bioreactors is the tremendous surface area of the media available for the attachment and growth of the microorganisms. The aim of the present study was to develop an effective, less expensive biological treatment process that produces less waste sludge from the wastewater generated by the production of heavy crude oil.

2 EXPERIMENT

2.1 *The characteristics of the wastewater*

Oil and sulphides are the main contaminants in the wastewater generated by the production of heavy crude oil at the Henan oilfield, but Chemical Oxygen Demand (COD) is also high in this water. The characteristics of statistical wastewater data in the present study are listed in Table 1.

The wastewater sample described in Table 1 was taken from the settling and filtration components of the production process for heavy crude oil.

Table 1. The characteristics of the wastewater generated by the production of heavy crude oil.

Parameter	Value
pH	7.06 ± 0.02
Temperature (°C)	55 ± 2
Oil content (mg/L)	29.8 ± 28.9
Sulphides (mg/L)	2.3 ± 0.6
CODcr (mg/L)	537 ± 166
Volatile phenols	0.45 ± 0.3
Total salinity (mg/L)	4466.3

2.2 *The pilot wastewater treatment system*

The pilot wastewater treatment plant (Figure 1) is composed of six concrete tanks.

These bacteria were chosen based on a preliminary screening (data not shown) that demonstrated their ability to grow well at a range of higher temperatures and acidity levels, and salinity levels. Their concentrations reached $3–5 \times 10^8$ cells/ml in the liquid culture medium of salt concentrations 3% (w/w)), pH = 6.0–9.0, at 55°C.

3 RESULTS AND DISCUSSION

To select a suitable treatment process, it is important to be able to treat realistic volumes of wastewater from heavy crude oil production, which is characterized by large fluctuations in the quantity and quality of the wastewater. The fluctuation in wastewater quality from

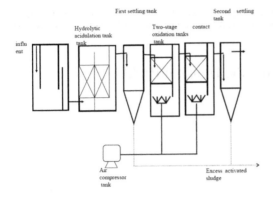

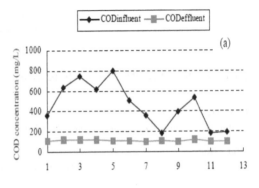

Figure 1.

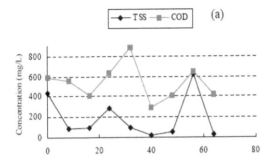

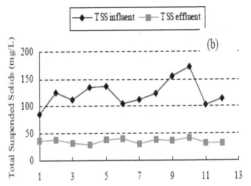

Figure 3. COD, TSS, and oil changes for the pilot system at a 15–h HRT.

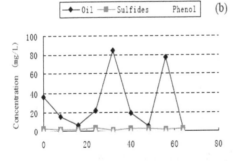

Figure 2. Changes in the main contaminant concentrations in wastewater produced from the production of heavy crude oil: (a) Maximum, minimum, and average TSS = 620, 21, and 190 mg/L; Maximum, minimum, and average COD = 880, 292, and 537 mg/L. (b) Maximum, minimum, and average oil = 84.7, 3.1, and 29.8 mg/L; Maximum, minimum, and average sulphites = 84.7, 3.1, and 29.8 mg/L; Maximum, minimum, and average Phenol = 1.12, 0.19, and 0.45 mg/L.

26 to 29 March 2012 during the pilot experiment is depicted in Figure 2.

3.1 Biological treatment system using anaerobic and aerobic bioreactors

3.1.1 Oil separation and flux regulation tank
The treated influent that enters the first tank is unstable, and varies greatly in both its quantity and its quality after a batch treatment by settling and filtration.

Table 2. Results of the treatment of the wastewater generated by the production of heavy crude oil.

parameter	Before treatment	After treatment	Reduction (%)
Oil content	25.9	5.0	80.6
TSS	110	10	90.9
Sulphides	2.6	0	100.0
CODcr	362	91	74.8
Volatile	0.37	0	100.0

The results are listed in Table 2.

After the anaerobic and aerobic biological treatments, the fixed-film bioreactors with a 15-h hydraulic retention time (HRT) reduced COD by 74.8%, TSS by 90.9%, oil by 80.6%, and phenols and sulphides by 100%.

4 CONCLUSION

We developed an effective biological treatment process, which was suitable for the treatment of wastewater with lower costs and less production of waste sludge. It is summarized as follows:

The biological treatment of the wastewater was made possible through an effective combination of anaerobic hydrolytic acidulation and aerobic contact oxidation. Despite large fluctuations in the quantity and quality of the influent wastewater, the pilot system performed well, producing stable and high-quality effluent, while demonstrating good resistance to shock loads. Several main characteristics of the effluent (COD, oil content, TSS, sulphides, and phenols) fully met the national standards for drainage water.

Four strains of bacteria (8-B, 7-D, 5-B, and I) were identified that exhibited excellent abilities to degrade heavy crude oil at temperatures of around ± 55°C, and played an important role in the success of the anaerobic and aerobic degradation processes.

5 THE FOUNDATION ASSISTANT

The project was encouraged by AJ2014-06 and PTTBIFT_YC_008.

REFERENCES

Defilippi, L.J. & Lewandowski, G.A. Biological *Treatment of Hazardous Wastes*, John Wiley and Sons. ISBN no. 0471-0486-5. (1998).

Deng, B. & Zhu, W. Biochemical process of treatment wastewater from the exploitation of heavy crude oil. Chinese water and wastewater engineering, Vol.76–78(2008), p: 19

Dong, X.Z. & M. Y. Manual of Determinative Bacteriology, ISBN no.7-03-008460-8/Q·968, 2001.Science Publication Press, Beijing.

Elmaleh, S., Ghafforn. Cross-flow UF of hydrocarbon and biological solid mixed suspension. NJ of Membrane Sci, 1996(1): 111–120.

Huy, N.Q., Jin, S., Amada, K., Haruki, M. et al. Characterization of petroleum-degrading bacteria from oil-contaminated Sites in Vietnam. J. Biosci. and Bioeng. Vol. 100–102(2009), p. 88.

Ismail and Serageldin. Water Resources Management. Vol 2015–2021 (2007).

Jou, C.G. & Huang, G.C. A pilot study for oil refinery wastewater treatment using afixed-film bioreactor. Adv. in Environ. Res, Vol: 463–469(2007) p. 7.

Leahy, J.G. & Colwell, R.R. Microbial degradation of hydrocarbons in the environment. Microbiol. Rev. Vol: 305–312. (2009), p. 54.

Pujol, R., Hamon, M. & Kandel, X. Biofilters: flexible, reliable biological reactor [J]. Wat. Sci. Tech. Vol: 33–38. (1999), p. 29.

Radwan, S.S., A1-awadi, H.A. & Majeda Khanafer, Effects of lipids on n-alkane attenuation in media supporting oil-utilizing microorganisms from the oily Arabian Gulf coasts. FEMS microbial. Lett. 2007, 198, 99–103.

Biomedical Engineering and Environmental Engineering – Chan (Ed.)
© *2015 Taylor & Francis Group, London, ISBN: 978-1-138-02805-0*

Determination of 16 Phthalic Acid Esters in water by solid phase extraction gas chromatography-mass spectrometry

Wen-hua Zhao
Zhongshan Polytechnic, ZhongShan, Guang Dong, China

Feng Zeng
Sun Yat-sen University, GuangZhou, Guang Dong, China

ABSTRACT: A method was developed for the determination of 16 phthalic acid esters in water by solid phase extraction gas chromatography mass spectrometry (SPE-GC-MS). Phthalic acid esters in water can be enriched with a solid-phase extraction by the AmberliteXAD-2 resin. A column was loaded with 1 g resin soaked in methanol, water loading follow rate was about $1.0 \, mL \cdot min^{-1}$, 25 mL acetone. N-hexane (V:V) 5:5 was the best eluting solvent; the eluting rate was $0.20–0.25 \, mL \cdot min^{-1}$. Using the method to detect PAEs in water of the ZhuJiang River in order to find that there were 4 types of PAEs. The concentration was $0.0046 \sim 0.2100 \, ug \cdot L^{-1}$. The results showed that the linear response of the method was 0.200–1000 (or 400) $ug \cdot L^{-1}$, and the linear correlation $R^2 > 0.99$. The limit of detection (LOD) of all the PAEs were $0.0533–1.24 \, ng \cdot L^{-1}$, the average recoveries of 16 phthalic acid esters in water were 60.1%–120%, and the relative standard deviations were less than 19.9%. The method is sensitive, accurate and cheap, so it can be applied to the detection of trace phthalic acid esters in water.

1 INTRODUCTION

1.1 *Phthalic Acid Esters (PAEs)*

Phthalic Acid Esters (PAEs) are also called phthalates, which are types of synthetic organic compounds. PAEs have widespread production and can be extensively applied all over the world (Chandola et al. 2008). PAEs can be used as pesticide carriers, cosmetics, fragrances, lubricants and decontaminants in the production of many types of products. Especially as a plasticizer, the consumption has increased sharply. Because PAEs and polyolefin were combined by hydrogen bonds or Vander Wals forces, so PAEs easily escape from the manufacturing process or products. Now we can find PAEs in the atmosphere, water, soil, even the organisms (Liu et al., 2012; Liu et al. 2007; Wang et al. 2014). PAEs are poisonous, especially in reproductive organs. There are many researches about this project focusing on rodents and humans (Kondo et al. 2006; Kleinsasser et al. 2004; Reddy et al. 2006). The conclusion shows that PAEs can lead to chromosome damage.

PAEs can migrate throughout the environment all over the earth, including water. People have already realized that PAEs pollution in water are generally existent. The concentration of PAEs in the water is very low, so we must enrich them before detection. Enrichment methods usually have a solid phase extraction (SPE) (Lin al. 2011; Zhang et al. 2014; Li et al. 2011)), a dispersive solid phase extraction (D-SPE) (Li et al.

2012; Wang et al. 2007), and Liquid-Liquid Extraction (LLE) (Wu et al. 2014), etc. Detection methods of the PAEs usually have gas chromatography-mass spectrometry (GC-MS) (Li et al., 2012; Wu et al. 2014), gas chromatography (GC) (Liu et al. 2013), liquid chromatography – mass spectrometry (LC-MS) (Wang et al. 2007; Yang et al. 2012) and high performance liquid chromatography (HPLC) (Shi et al. 2011), etc. In this paper, the macroporous resin XAD-2 was used to pre-concentrate the PAEs of the water and then determine the content by gas chromatography – mass spectrometry.

2 THE EXPERIMENTAL PART

2.1 *Chemicals and instruments*

Sixteen PAEs standard mixture, containing dimethyl phthalate (DMP), diethyl phthalate (DEP), diisobutyl phthalate (DiBP), di-n-butyl phthalate (DnBP), dimethylglycol phthalate (DMGP), di(4-methyl-2-pentyl) phthalate (DMPP), di(2-ethoxyethyl) phthalate (DEEP), di-n-amyl phthalate (DnAP), di-n-hexyl phthalate (DnHP), butylbenzyl phthalate (BBP), di(hexyl-2-ethylhexyl) phthalate (HEHP), di(2-n-butoxyethyl) phthalate (DBEP), dicyclohexyl phtha-late (DCHP), di(2-ethylhexyl) phthalate (DEHP), di-n-nonyl phthalate (DnNP), di-n-octyl phtha-late (DnOP) at $100 \, \mu g \, mL$ (they came from Dr. Ehrenstorfer GmbH, Germany). Internal standard

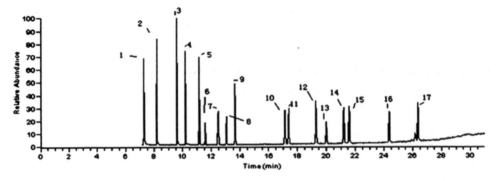

Figure 1. Total ion chromatograms of 16 phthalic acid esters (Nos. 3 peak is Benzyl benzoate, other16 peaks are PAEs, which are appearing in order in Table 1.

material: benzyl benzoate (which came from SIGMA, the USA). Methyl alcohol (chromatographic pure, which came from Fisher, the USA)

AMBERLITE XAD-2 resin (which came from Trademark of The Dow Chemical Company, the USA): the resin was wrapped in filter paper, chromatographic pure dichloromethane Soxhlet extraction solvent to extract 48 h, fresh 24 hours, dried into glass bottles, standby in the dryer.

Instrments: Agilent GC 4890D with FID analyser, Thermo-finnigan GC-MS analyser, Rotary evaporator RE-52AA (which come from ShangHai YaRong biochemical instrument factory), NP-8-38-300 ultrasonic extraction instrument (which come from Xin Dong Li Ultrasonic Electronic Equipment Co Ltd), 1810B automatic double water distiller (which come from Shanghai ShenLi Glass Instrument Co Ltd.).

Water was filtered and double distilled. All organic solvents used were of analytical grade, and redistilled using glass system. Laboratory glassware was soaked overnight in K_2CrO_7/H_2SO_4, then heated to 230°C for 12 h, and the use of plastic products was thus avoided.

2.2 Standard solution

Standard solution mixture: 0.100 g of 16 PAEs standard material were dissolved by 50 mL acetone in flask, and kept under seal in a dark and dry place. A single solution: 0.100 g of every type of the PAEs standard material was dissolved by 50 ml acetone flask, and kept it under seal in a dark and dry place.

2.3 Instrumental analysis

Standard solution mixture: 0.100 g of 16 PAEs standard material was dissolved by 50 mL acetone in a flask, and kept under seal in a dark and dry place. A single solution: 0.100 g of every type of PAEs standard material was dissolved by 50 ml acetone in a flask, and kept under seal in a dark and dry place.

Chromatographic column was DB-5MS capillary column (30 m × 250 μm i.d.; 0.25 μm film thickness). Mass analyser parameters were follows: ionization mode was EI; ionization energy was 70 eV, monitoring mode was selective ion monitoring (SIM), scan quality number was 50–650 amu. The transfer line and the ion source temperature were maintained at 230°C and 280°C respectively. The column temperature program was initiated at 80°C for 1.0 min, increased to 220°C at a rate of 20°C·min^{-1}, then ramped at 2°C·min^{-1} to 250°C, and finally ramped at 5°C·min^{-1} to 280°C. The flow rate of the carrier gas helium was kept constant at 0.6 mL·min^{-1}. The solution (1.0 μL) was injected onto GC-MS in a splitless mode with an inlet temperature of 280°C. Quantitation was performed using the internal calibration method based on a five-point calibration curve for an individual PAE. Benzyl benzoate was used as internal standards for the quantification of PAEs; 16 types of phthalic acid ester's retention time and ions are shown in Table 1. The total ions chromatograms of 16 phthalic acid esters are shown in Figure 1.

3 THE EXTRACTION PARAMETERS

3.1 Optimization of SPE conditions

SPE column: SPE column was made by hand, loaded with 1 g XAD-2 resin which was soaked in MeOH. The column was the U sharp, the tube inner diameter was 5 mm, water was loaded form one side and flowing out from another end. Extraction column was packed under the condition of solution existence. Two solutions were tested: acetone and methanol. The result shows that the average loss rates of DMP were 2.43% and 0.540% respectively. Therefore we chose a pure chromatographic methanol solvent to load the SPE column.

Sample loading conditions: Three loading Rate were tested: 0.40–0.50, 1.0–1.2, and 1.6–1.8 mL·min-1, the average loss rate of 16 PAEs were 1:1.85:2.46. The results showed that lower loading rate had lower loss rate but needed more much time. Comprehensive the average loss rate of 16 PAEs and the time of absorption, loading rate about 1.0 mL·min^{-1} was selected.

Elution conditions:3 elution solutions were tested respectively:acetone, dichloromethane and n-hexane.

226

Table 1. Retention time and characteristic ions of phthalic acid esters.

phthalic acid esters (PAEs)	Mass spectrometric parameters		
	retention time (min)	Qualitative ions (m/z)	Quantitative ions (m/z)
1 Dimethyl phthalate	7.27	163	77, 163
2 Diethyl phthalate	8.16	149	149, 150, 176
3 Diisobutyl phthalate	10.18	149	57, 76, 105, 149
4 Di-n-butyl phthalate	11.14	149	149, 205, 223
5 Dimethylglycol phthalate	11.54	149	59, 76, 149, 209
6 Di(4-methyl-2-pentyl) phthalate	12.44	149	107, 149, 150
7 Di(2-ethoxyethyl) phthalate	12.99	149	72, 73, 149,
8 Diamyl phthalate	13.60	149	149, 150
9 Di-n-hexyl phthalate	17.07	251	84, 85, 251
10 Butylbenzyl phthalate	17.31	206	206
11 Di(hexyl-2-ethylhexyl) phthalate	19.21	149	149, 251
12 Di-n-butoxyethyl) phthalate	19.89	149	55, 149, 193, 293
13 Dicyclohexyl phthalate	21.08	149	55, 83, 149, 168
14 Di(2-ethylhexyl) phthalate	21.50	149	83, 166, 168, 279
15 Dinonyl phthalate	24.56	149	149, 293
16 Di-n-octyl phthalate	26.25	149	149, 150

Table 2. Linearity range, linear equations and detection limits of 16 phthalic acid esters in water.

PAEs	Linearity range ($\mu g \cdot L^{-1}$)	Linear equations	Correlation coefficient	LOD ($ng \cdot L^{-1}$)
1 Dimethyl phthalate	0.200–1000	$y = 3279x - 82543$	0.9986	0.191
2 Diethyl phthalate	0.200–1000	$y = 3548x - 73127$	0.9971	0.104
3 Diisobutyl phthalate	0.200–1000	$y = 6637x + 194582$	0.9979	1.12
4 Di-n-butyl phthalate	0.200–1000	$y = 5584x + 126486$	0.9982	0.580
5 Dimethylglycol phthalate	0.200–1000	$y = 4025x - 4698$	0.9984	0.158
6 Di(4-methyl-2-pentyl) phthalate	0.200–1000	$y = 2152x - 69241$	0.9992	0.0847
7 Di(2-ethoxyethyl) phthalate	0.200–1000	$y = 3094x + 3578$	0.9982	0.147
8 Diamyl phthalate	0.200–1000	$y = 2812x - 1835$	0.9987	0.0553
9 Di-n-hexyl phthalate	0.200–1000	$y = 5569x - 12654$	0.9988	0.477
10 Butylbenzyl phthalate	0.200–1000	$y = 4005x + 27984$	0.9992	0.292
11 Di(hexyl-2-ethylhexyl) phthalate	0.200–400	$y = 4158x + 156741$	0.9907	0.101
12 Di(2-n-butoxyethyl) phthalate	0.200–400	$y = 3924x + 18946$	0.9976	0.328
13 Dicyclohexyl phthalate	0.200–400	$y = 2067x - 28951$	0.9981	0.153
14 Di(2-ethylhexyl) phthalate	0.200–400	$y = 5671x + 62302$	0.9966	1.24
15 Dinonyl phthalate	0.200–400	$y = 5372x - 60205$	0.9984	0.153
16 Di-n-octyl phthalate	0.200–400	$y = 5072x - 55377$	0.9991	0.0865

Each elution solution's volume was 25 mL. The result show that the small molecule PAEs were easy eluted by acetone, the elution rate was 84%, larger molecule PAEs were easily eluted by n-hexane, the acetone and n-hexane were mixed with two ratios of 9:1,7:3,5:5,3:7,1:9 (V/V) to further test, the results demonstrated that 5:5 (V:V) ratio was better, and the recovery rates of PAEs were 71–102%. The elution rate of 0.25 mL·min^{-1} was selected. Ultrasound as an assistant can increase 6% elution of the DnOP.

3.2 The linear range and detection limit

Quantitation was based on a five-point calibration curve with 0.200, 1.00, 100, 200, 400 and 1000 ug·L^{-1}

PAEs in the blank sample which was purified. Samples were detected by Thermo-finnigan GC-MS analyser The results show that: linear range of the method were 0.200–1000 (or 400 ug·L^{-1}) and the correlation coefficient was better than 0.99. The method detection limits for PAEs were 0.0553–1.24 ng·L^{-1}. All the results are shown in Table 2.

3.3 Results and discussion of determination of PAEs in water from the Zhu Jiang River

30 L water samples were collected from the river of Guangzhou region, then filtered with 0.75 um glass fibre membrane to clean; acetone and n-hexane were mixed as eluents. The PAEs in the water samples were

Table 3. Concentration of PAEs of the water samples in the Zhujiang River (Guangzhou).

phthalic acid esters	LLC-GC-MS Cone ($\mu g \cdot L^{-1}$) (n = 5)	SPE-GC-MS Cone ($\mu g \cdot L^{-1}$) (n = 5)
1 Diethyl phthalate	0.00786	0.00464
2 Diisobutyl phthalate	0.135	0.210
3 Di-*n*-butyl phthalate	0.0211	0.0195
4 Di(2-ethylhexyl) phthalate	0.0442	0.0573

Table 4. Recoveries and RSDs of 16 phthalic acid esters in water of the Zhujiang River.

phthalic acid esters (PAEs)	Spiked 1.00 $\mu g \cdot L^{-1}$ Recovery%	RSD%, n = 5	Spiked 2.00 $\mu g \cdot L^{-1}$ Recovery%	RSD%, n = 5	Spiked 12.00 $\mu g \cdot L^{-1}$ Recovery%	RSD%, n = 5
1 Dimethyl phthalate	64.1	7.99	62.3	12.8	60.8	12.3
2 Diethyl phthalate	62.4	1.2	71.4	12.9	79.5	13.4
3 Diisobutyl phthalate	78.5	9.96	105	7.11	100	19.0
4 Di-*n*-butyl phthalate	75.5	13.1	104	5.17	90.8	18.4
5 Dimethylglycol phthalate	118	8.71	116	5.45	69.2	13.5
6 Di(4-methyl-2-pentyl) phthalate	101	3.65	80.3	10.9	83.1	16.4
7 Di(2-ethoxyethyl) phthalate	120	10.4	115	5.67	75.5	13.7
8 Diamyl phthalate	64.4	16.4	88.8	8.98	86.9	16.3
9 Di-*n*-hexyl phthalate	60.1	12.5	46.7	8.55	76.7	15.0
10 Butylbenzyl phthalate	70.1	16.2	112	6.04	88.0	16.4
11 Di(hexyl-2-ethylhexyl) phthalate	67.9	9.20	62.2	10.7	75.0	14.8
12 Di(2-*n*-butoxyethyl) phthalate	116	7.78	118	3.52	78.4	13.1
13 Dicyclohexyl phthalate	102	7.74	114	3.91	79.1	14.5
14 Di(2-ethylhexyl) phthalate	83.0	5.72	78.9	19.9	60.2	15.4
15 Dinonyl phthalate	81.4	5.16	74.3	19.7	70.5	14.4
16 Di-*n*-octyl phthalate	60.3	3.97	63.4	11.8	66.5	14.9

deceted by SPE-GC-MS and LLE-GC-MS methods respectively.

Comparing the result, finding that there were 4 types of PAEs in the water of the Zhujiang River, other 12 kinds of PAEs were not detected, the determination results of 4 types of PAEs by SPE-GC-MS and LLC-GC-MS are shown in Table 3.

The determination results in Table 3 shows that the results of two detection methods are extremely similar, the two methods are reliable. But if great amount of water been detected, method of SPE-GC-MS will save a large amount of solvent. Values of recovery found by the standard addition method were shown in Table 4.

From Table 4, we can find that values of recovery found by the standard addition method in water samples were 60.1%–120%. RSDs were less than 20%, which is illustrated using XAD-2 resin to enrich the trace PAEs in the water, and the quantity detection of PAEs then taking by GC-MS are to detect the quantity have good. The method is simple to operate, and can save the extraction solvent in order to protect the environment. Therefore, the determination of 16 PAEs of water by SPE (XAD-2)-GC-MS has a good precision and accuracy, which shows that the method is reliable.

REFERENCES

Chandola M, Marathe S. A. QSPR for the plasticization efficiency of polyvinyl- chloride plasticizers. [J] Molecular Graphics and Modelling, 2008, 26(5): 824–828.

Kondo T, Shono T, Suita S. Age-specific effect of phthalate ester on testicular development in rats. Journal of Pediatric Surgery, 2006, 41 (7): 1290–1293.

Kleinsasser NH, Harréus UA, Kastenbauer ER, etc. Mono (2-ethylhexyl) phtha- late exhibits genotoxic effects in human lymphocytes and mucosal cells of the upper aerodigestive tract in the comet assay. Toxicology Letters, 2004, 148(1–2): 83–90.

Li Ting, Tang Zhi, Hong Wu-Xing, Determination of 17 Phthalic Acid Esters in Fatty Food by Dispersive Solid Phase Extraction-Gas Chromatography-Mass Spectrometry. Chinese J. Anal. Chem. 2012. 40(3) 391:396.

Lin Xing-Tao, Wang Xiao-Yi, Xia Ding-Guo. Determination of Phthalic Acid Esters and Their Metabolites in Human Urine by Solid Phase Extraction-High Performance Liquid Chromatography, Chinese J. Anal. Chem. 2011. 39(6): 877–881.

Liu Qing, Yang Hong-Jun, Shi Yan-Xi. Research progress on phthalate esters (PAEs) organic pollutants in the environment Chinese Journal of Eco-Agriculture, Aug. 2012, 20(8): 968–975.

Liu M, Lin Y. J., Zeng F., et al. 2007. The distribution and composition of phthalate esters in the sediment of urban lakes in Guangzhou [J]. Acta Scientiae Circumstant iae, 27 (8): 1377–1383.

Liu Yu-bo, Chen Li-li, Shen Ye-bing. Gc determination of phthalates as plasticizer in disposable plastic cups by liquid-liquid extraction. PTCA (PART B: CHEM. ANAL.) 2013. 49(4): 451–454

Reddy BS, Rozati R., Reddy S, etc. High plasma concentrations of polychlorinated biphenyls and phthalate esters in women with endometriosis: a prospective case control study. Fertility and Sterility, 2006, 85(3): 775–779.

Shi Ya-mei, Xu Dun-ming, Zhou Yu. Determination of 17 Phthalate Esters in Food by QuEChERS/High Performance Liquid Chromatography. Journal of Instrumental Analysis. 2011. 30(12): 1372–1376.

Wang Shuai, Dong Shu-ying, Sun Kuo, Organic Matters Identification and Phthalates Levels in Whole Blood of Pregnant Women in Harbin. 2014. 31(3): 201–212.

Wang Li-Xia, Kou Li-Juan, Pan Feng-Yun. Determination of Phthalates in Vegetables by Liquid Chromatography-Electrospray Ionization Mass Spectrometry with Matrix Solid Phase Dispersion. Chinese Journal of Analytical Chemistry, 2007. 35(11): 1559–1564.

Wu-Peng, Yu Xiang-xiang, Miao Jian-jun. Determination of phthalate esters in waters by small volume liquis-liquid exteaction comenined with gas chermatography-mass spectrometry. J. Anal. Chemical Science. 2014. 29(1): 139–142.

Yang Rong-jing, Wei Bi-wen Gao Huan. Determination of 17 phthalate esters in packing of food with High liquid chermatography-Tandem mass spectrometry. Environment Chemistry. [J]. 2012. 31(6): 925–929

Zhagng Fan, Li Zhong-HaI, Zhang Ying. Determination of six phalate acid esters in camelliaoil by gas chromatography-mass spectrometry coupled with solid-phase extraction using single-walled carbon nanotubes as adsorbent. Chinese J Chromatography. 2014. 32(7): 735–740.

Environmental sustainability

Biomedical Engineering and Environmental Engineering – Chan (Ed.)
© *2015 Taylor & Francis Group, London, ISBN: 978-1-138-02805-0*

Study on the industrial structure evolution and its optimizing countermeasures in Xianning City of Hubei province based on the perspective of low carbon economy

Jun-wu Fang

College of Economy and Management, Hubei University of Science and Technology, Xianning, P.R. China

Zhi Chen

College of Resources, Environmental Science and Engineering, Hubei University of Science and Technology, Xianning, P.R. China

ABSTRACT: This article took Xianning City of Hubei province as the research region and drew lessons from relevant theoretical and empirical research results, on the basis of the basic determination and analysis on Xianning City's current state of industrialization and its industrial structure evolution and energy consumption since China's reform and opening up, to analyzed the energy consumption structure and the basic feature of CO_2 emission of the city's industrial sectors. By introducing two basic discrimination models of structure evolution – energy consumption and structure evolution – unit energy consumption, and two analytical models of structure evolution – carbon emission and structure evolution – unit carbon emission, this article analyzed the energy consumption and carbon emission effect of Xianning City's industrial structure evolution and offered a few targeted countermeasures and suggestions on the city's industrial structure optimizing based on its prospective of low carbon economy. This article intends to provide a basis of decision making for Xianning City's low carbon economic development and its industrial restructuring and optimizing.

Keywords: Analysis on effects, Energy consumption, Industrial carbon emissions, Industrial energy consumption, Industrial structure evolution, Low carbon economy, Optimizing countermeasures, Xianning City of Hubei province.

1 INTRODUCTION

Since the end of 20th century, the issue of global climate change characterized with climate warming has increasingly become one of the main focuses of attention of the international community. The mainstream view suggests that climate warming is a result of the increasing accumulation of the greenhouse gases released by human activity, the major part of which comes from the increasing accumulation of CO_2 from human's huge consumption of fossil fuels[1]. It has become a general consensus of the international community to reduce human activity, high carbon energy consumption in particular, curb global warming, and protect human living environment. How to ease greenhouse effect and cut CO_2 emission is being hotly debated around the academic circles. Governments and research institutions are studying and discussing the issue of CO_2 emission reduction from different perspectives. Reducing CO_2 emission through industrial restructuring has attracted a growing attention from scholars, which puts forward new requirement to industrial restructuring and optimizing and also brings new conditions and opportunities for industrial structure optimizing. The model of low

carbon economy has been embraced by an increasing number of countries and regions against a backdrop of daunting challenges forced by global warming[2]. Industrial structure evolution not only affects the amount and intensity of CO_2 emission, but is related more directly with the sustainable and steady economic growth. From the view of industrial restructuring and optimizing, seeking a balance between economic and social development and CO_2 emission reduction has attracted a growing attention from academics home and abroad.

As the world's biggest developing country, China ranks second in energy producing, energy consumption, and greenhouse gases emission all around the world[3,4]. China now is at a critical juncture of building a well-off society in an all-round way and an important stage of accelerated industrialization and urbanization. With some industries of high energy consumption and high carbon emission still dominating the development of national economy, energy demand will keep growing. As the gateway to southern Hubei province, Xianning City is a growing medium-sized city. With a rapid social and economic development and speedy industrialization and urbanization in recent years, Xianning City's energy consumption and CO_2

emission are growing, exerting quite a pressure on its resources and environment. As a core city in the pilot region for comprehensive reforms of "two types" social construction of Wuhan megalopolis, Xianning City is facing formidable tasks of developing low carbon economy, saving energy and reducing emission, especially so when the city was listed as one of the first two Hubei province's pilot cities for low carbon economy after the province became a pilot province for low carbon economy. The boosting of industrial structure optimizing and upgrading is not only an effective means to realize energy saving and emission reduction, but also an effective approach to strengthen the competitiveness of the whole industry. In order to achieve a win-win situation of social and economic growth together with resource and environment protection, Xianning City has to speed up the transformation of economic development pattern by starting with industrial restructuring and optimizing.

In recent years, as the ecological environment problem is becoming more and more serious, the impact of industrial structure changing on resource utilization and ecological environment changing has attracted a growing attention. In the research of industrial structure optimizing, promoting economic growth has no longer been the sole target of optimizing. Multiple sustainable targets like environment protection and resource reservation have also been included. Researches on the correlation between industrial structure and energy and resources started early in foreign countries[5-10].

The research on industrial structure and the issues of energy and environment in China has lagged far behind. Currently most of these researches have aimed at the empirical analysis on specific regions, focusing mainly on industrial restructuring and resource utilization[11-18].

But currently, relevant researches are weak, especially the empirical researches on specific regions, let alone those of theoretical and systematic. In view of this, this article intends to take Xianning City of Hubei province as the research region and draw lessons from relevant theoretical and empirical research results, on the basis of the basic determination and analysis on Xianning City's current state of industrialization and its industrial structure evolution and energy consumption since China's reform and opening up, to analyze the energy consumption structure and the basic feature of CO_2 emission of the city's industrial sectors. By introducing two basic discrimination models of structure evolution – energy consumption and structure evolution – unit energy consumption, and two analytical models of structure evolution – carbon emission and structure evolution – unit carbon emission, this article analyzes the energy consumption and carbon emission effect of Xianning City's industrial structure evolution and offers a few targeted countermeasures and suggestions on the city's industrial structure optimizing based on its prospective of low carbon economy. Through the empirical research on energy consumption and carbon emission of Xianning

City's industrial structure, this article intends to provide a basis of decision making for Xianning City's low carbon economic development and its industrial restructuring and optimizing.

2 THE INDUSTRIAL STRUCTURE EVOLUTION AND ENERGY CONSUMPTION IN XIANNING CITY

2.1 Judgment on the industrialization development stage

Since China's reform and opening up, Xianning City has enjoyed a rapid economic development, with a growing proportion in Hubei province's economy. In 2012, with an increase of 12.2% than the previous year, Xianning City's gross regional production reached 76.099 billion yuan, among which, the primary industry increased by 4.7% to 14.53 billion yuan, the secondary industry grew by 16.3% to 35.973 billion yuan, and the tertiary sector rose by 10.2% to 25.596 billion yuan. The ratio of these three sectors was 19.1:47.3:33.6. According to the general model of industrial structure evolution (Salguiyin – Chenery Model), when the proportion of non-agricultural industries surpasses 80% and that of agricultural industries goes down to below 20%, the industrialization reaches the second stage of the medium level. And when the proportion of non-agricultural industries continues to exceed 90% and that of agricultural industries is less than 10%, the industrialization is in its late stage or basically achieved. In light of this, Xianning City's current economic development is still in the second stage of the medium level of industrialization.

2.2 Analysis on the industrial structure evolution

Xianning City's industrial structure evolution since China's reform and opening up can be divided into two stages (see Figure 1). In the first stage (1978–1995), its industrial structure had been gradually improved. During this period, Xianning City had sped up the development of the tertiary sector while striving to develop the agricultural industry and light industry. The ratio between the three sectors had been improved, which basically reversed the serious situation of long-standing imbalance in the industrial structure. As the agricultural structure gradually improved, the proportions of forestry, animal husbandry, side-line production, and fishery in the total value of agricultural output had increased greatly. The internal structure of industry had been given a preliminary adjustment and the ratio between light and heavy industries started to improve. In the second stage (since 1996), the tertiary sector has enjoyed a rapid growth. During this period, the proportion of the primary industry has continued to decline and the proportion of secondary industry has increased steadily, while that of the tertiary sector enjoying a tremendous growth. In this time, the proportion of the primary industry has declined by 14.6%, while that of the secondary industry and the

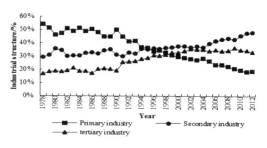

Figure 1. The evolution of industrial structure in Xianning City (1978–2012).

tertiary sector has respectively rose by 12.3% and 2.2%. Meanwhile, Xianning City's industrial structure has transformed from "primary – secondary – tertiary" type to "secondary – primary – tertiary" type, which indicates that Xianning City's industrial structure evolution has entered a new stage.

2.3 The general situation about energy consumption

As Xianning City is enjoying a rapid economic growth, some underlying, structural, and accumulated problems have risen to prominence. Considering Xianning City's energy consumption structure on the one hand, coal is its major energy consumption resource. In 2012, 5.193 million tons of coal was consumed, accounting for 77.3% of the total energy consumption. Xianning City's economic development still relies heavily on coal consumption. On the other hand, the city's unit GDP energy consumption has been declining since 2005. In 2012, Xianning City's clean energy consumption accounted for less than 4% of the total energy consumption.

The unit GDP energy consumption from 2005 to 2010 was stable, which indicates a heavy reliance of Xianning City's economic growth on energy during this period. Only from 2010, had the unit GDP energy consumption begun to decline, which, in addition to the economic development and industrial restructuring during the Eleventh Five-Year Plan period, mainly benefited from the improvement of opening up level and the change of economic growth pattern of this time. Moreover, the unit GDP energy consumption of Xianning City's secondary industry, mainly industry, still remains high, which has become one bottleneck for the city's development of low carbon economy.

3 INDUSTRIAL ENERGY CONSUMPTION AND CARBON EMISSIONS IN XIANNING CITY

3.1 Industrial energy consumption and its structure

Industrial structure determines the structure of industrial energy consumption. With the development of industrial restructuring, industrial energy consumption structure changes correspondingly. The energy consumption proportion of Xianning City's primary industry declined from 2% in 2009 to 1.5% in 2010,

while that of the secondary industry remained almost unchanged from 2009 to 2010, at 68%. It shows that the secondary industry is Xianning City's major energy consumer. Therefore, in order to realize low carbon economy, the rate of energy utilization must be raised and the energy consumption of the secondary industry must be gradually reduced. The energy consumption of the tertiary sector generally is on the rise, with a 60,000-ton of increase in 2010 than in 2009. So it is necessary not to ignore the development of the tertiary sector while reducing the energy consumption of the secondary industry. Residential energy consumption is the second largest sector after the secondary industry, rising from 459,500 tons in 2009 to 573,700 tons in 2010. Its proportion has declined from 17% in 2009 to 13.6% in 2009. Generally speaking, with the development of economy, the proportion of residential energy consumption has been declining. Historically, the secondary industry was the major energy consumer, with industry being its main consumption sector. In 2010, Xianning City's industry sector consumed 1.8362 million tons of standard coal, accounting for 97% of the secondary industry's 1.8903 million tons, which means that industry is the city's biggest energy consumer.

3.2 Industrial carbon emissions

Global warming is caused mainly by the rising density of CO_2 in the atmosphere while energy activity is the source of CO_2 emission. Carbon emission can be divided into two categories – natural emission and human emission, with the latter being caused by human activity. Carbon emission mainly includes fossil fuel consumption and biomass burning, of which the carbon emission of fossil fuel consumption accounts for 80%. Taking the carbon emission decomposition model created and improved by Xu Guo-quan et al.[19] for reference, according to research findings of DOE/EIA, Institute of Energy Economics Japan, The State Science and Technology Commission climate change project and NDRC Energy Research Institute etc organization, calculated the mean of all kinds of energy carbon emission coefficient, energy carbon emission coefficient of coal, oil, natural gas, and hydroelectricity and nuclear power are respectively 0.7329, 0.5574, 0.4226, and 0.0000. This article calculated Xianning City's carbon emission using carbon emission coefficient (see Table 1). The calculation formula is:

$$E=\alpha E_c+\beta E_0+\gamma E_t \qquad (1)$$

In the formula (1), E stands for carbon emission; E_c stands for coal consumption; α stands for carbon emission conversion coefficient of coal consumption; E_0 stands for oil consumption; β stands for carbon emission conversion coefficient of oil consumption; E_t stands for natural gas consumption; and γ stands for carbon emission conversion coefficient of natural gas consumption.

Table 1. The industrial carbon emissions and its structure in Xianning City (2010, 2012).

Year	Index	Primary industry	Secondary industry	Of which: Industrial	Tertiary industry	Total of three industries
2010	Carbon emissions (104 t CO_2)	9.50	209.84	207.76	67.01	286.35
	Ratio (%)	3.32%	73.28%	72.56%	23.40%	100.00%
2012	Carbon emissions (104 t CO_2)	9.45	306.17	305.13	75.50	391.13
	Ratio (%)	2.42%	78.28%	78.01%	19.30%	100.00%
The gain or loss	Carbon emissions (104 t CO_2)	−0.05	96.34	97.37	8.49	104.78
	Ratio (%)	−0.01	0.05	0.05	−0.04	0.00

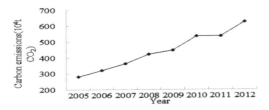

Figure 2. The changes of carbon emissions in Xianning City (2005–2012).

According to the whole energy consumption of Xianning City during 2005 and 2012, the city's carbon emission during this period can be calculated by using formula (1) (see Figure 2).

Figure 2 shows that since 2005, Xianning City's carbon emission has been increasing every year, with an annual average increasing rate of about 3.7%, indicating that although the city has enjoyed a rapid economic growth since 2005, its energy consumption has been increasing significantly.

In order to discuss the carbon emission of Xianning City's various industries, this article calculated the carbon emission of the city's industries during 2005 and 2012 (see Table 1), according to their energy consumption of this time.

Table 1 shows that the carbon emission of the secondary industry ranks first among the three sectors. In 2012, the carbon emission of the secondary sector was 32 times that of the primary industry and 4 times that of the tertiary sector. While in the secondary industry, carbon emission of industry sector accounted for the most. In 2012, carbon emission of industry sector accounted for 99% of the secondary industry's carbon emission and 78% of the total carbon emission. This is mainly because Xianning City's energy-intensive industries have constituted a considerable proportion in the city's industrial structure.

4 ANALYSIS ON EFFECTS OF INDUSTRIAL STRUCTURE EVOLUTION ON ENERGY CONSUMPTION AND CARBON EMISSIONS IN XIANNING CITY

4.1 Constructing the model

In today's society, the added value of the three sectors' output constructs the accumulation of social wealth. From the point of view of energy consumption, among the three GDP sectors, the energy consumption of primary industry and secondary industry is of social production energy consumption, while the energy consumption of the tertiary sector is of social living energy consumption. In studying the energy consumption effect of industrial structure evolution, this article introduces two basic discrimination models[15,20] of structure evolution – energy consumption and structure evolution – unit energy consumption, based on which, two analytical models of structure evolution – carbon emission consumption and structure evolution – unit carbon emission are constructed.

(1) Structure evolution – energy consumption correlation model: this model aims to explore the relation between regional industrial structure evolution and the change of one time energy consumption. The formula of this model is:

$$EEI=EU/ESD \tag{2}$$

In the formula (2), EEI stands for structure energy correlation coefficient; EU stands for regional one time energy consumption; ESD stands for regional industrial structure diversification coefficient, indicating the evolution degree of regional industrial structure. The calculation formula of ESD is:

$$ESD=\sum(P/P, \ S/P, \ T/P) \tag{3}$$

In the formula (3), P, S, and T stand for the primary industry, the secondary industry, and the tertiary sector respectively. The range of industrial structure diversification can be $(1, \infty)$.

(2) Structure evolution – carbon emission correlation model: in order to research the relation between regional industrial structure evolution and the total carbon emission visually and thoroughly, this article constructs structure evolution – carbon emission correlation model, on the basis of structure evolution – energy consumption correlation model. The formula of this model is:

$$ECI=CU/ESD \tag{4}$$

In the formula (4), ECI stands for structure carbon emission correlation coefficient; CU stands for the total regional carbon emission; and ESD stands for regional industrial structure diversification coefficient.

(3) Structure evolution – unit energy consumption correlation model: this model basically reveals the relation between regional industrial structure evolution and unit *GDP* energy consumption change, aiming to explore the energy-saving effect and the changing trend of regional industrial structure evolution. The formula of this model is:

$$EEE = EE/ESD \qquad (5)$$

In the formula (5), *EEE* stands for structure unit energy consumption correlation coefficient; *EE* stands for regional unit one time energy consumption coefficient; and *ESD* stands for regional *EC* industrial structure diversification coefficient. In $EE = EC/GDP$, *EC* stands for one time total energy consumption and *GDP* stands for regional gross domestic product.

(4) Structure evolution – unit carbon emission correlation model: in order to research the relation between regional industrial structure evolution and unit *GDP* carbon emission visually and thoroughly, this article constructs structure evolution – unit carbon emission correlation model, on the basis of structure evolution – unit energy consumption correlation model. The formula of this model is:

$$ECE = CE/ESD \qquad (6)$$

In the formula (6), *ECE* stands for structure unit carbon emission correlation coefficient; *CE* stands for unit *GDP* carbon emission coefficient; *ESD* stands for regional industrial structure diversification coefficient. In $CE = CU/GDP$, *CU* stands for regional total carbon emission and *GDP* stands for gross regional production.

4.2 The calculation results and analysis

According to the data of Statistical Yearbook of Xianning City from 2005 to 2010, we calculated using the above-mentioned formulae, and the results are shown in Table 2.

(1) From 2005 to 2010, with the development of industrial structure evolution year by year, *ESD* continued to rise, and then declined slightly in 2012. *EEI* and *ECI* increased as a whole, with some dropping fluctuations in some isolated years. *EEE* and *ECE* were generally falling year by year.

(2) From the aspect of correlation between these indexes, the correlation coefficients between *EEI*, *ECI*, *EEE*, and *ECE* and *ESD* are 0.865, 0.865, −0.990, and −0.989 respectively, indicating that during 2005 and 2010, the effect of industrial restructuring and upgrading on the change of energy consumption was not so obvious and the energy-saving effect of industrial structure evolution was unsound, while industrial restructuring and upgrading was essential to reducing unit *GDP* energy consumption and carbon emission, and restructuring had a considerable potential for the realization of energy saving and emission reduction.

Table 2. The effects of industrial structure evo-lution on energy consumption and carbon emissions in Xianning City (2005–2012).

Year	2005	2006	2007	2008
ESD	3.79	4.25	4.27	4.52
EEI	9.70	9.60	11.35	12.69
ECI	7.11	7.04	8.32	9.30
EEE	4.75	4.07	3.89	3.43
ECE	3.48	2.98	2.85	2.52

Year	2009	2010	2011	2012
ESD	4.81	5.15	5.49	5.32
EEI	12.70	14.14	12.95	15.01
ECI	9.31	10.36	9.49	11.00
EEE	3.04	2.72	1.99	1.97
ECE	2.23	1.99	1.46	1.42

Note: the units of EEI, ECI, EEE, and ECE are respectively $10^5_{t\,ce}$, $10^5_t\,CO_2$, $10^3_{t\,ce}/10^8$ yuan, and $10^5_t\,CO_2/10^8$ yuan.

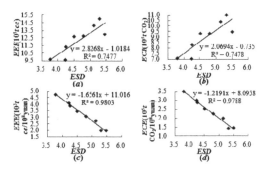

Figure 3. Correlation analysis and regression analysis between the various variables..

(3) The further correlation and regression analyses on *EEI*, *ECI*, and *ESD* (see Figure 3(*a*) and (*b*)) suggest that from 2005 to 2010, Xianning City demonstrated an obvious acceleration effect of structure evolution energy consumption and carbon emission. The city's economic growth and rapidly growing *GDP* still relied heavily on energy consumption and thus led to huge amount of greenhouse gases emission.

Some developed provinces, such as Jiangsu and Guangdong, have already entered the deceleration stage of structure evolution energy consumption, according to the research of some experts. With regard to total growth of energy consumption, the acceleration effect of structure evolution energy consumption and carbon emission in Xianning City, who is at the stage of accelerating development of industrialization, is mainly caused by the development of industry (manufacturing industry). On the other hand, the correlation and regression analyses on *EEE*, *ECE* and *ESD* (see Figure 3(*c*) and (*d*)) suggest that from 2005 to

2010, Xianning City demonstrated an obvious deceleration effect of unit *GDP* energy consumption of structure evolution and carbon emission, indicating that Xianning City has made remarkable achievements on adhering to the concept of low carbon development, shifting development pattern, optimizing industrial structure, transforming conventional industries, eliminating outdated production capacity, and developing clean energy.

5 COUNTERMEASURES AND SUGGESTIONS TO OPTIMIZE THE INDUSTRIAL STRUCTURE IN XIANNING CITY

Xianning City is the core city in the pilot region for comprehensive reforms of "two types" social construction of Wuhan megalopolis. Especially after the province became a pilot province for low carbon economy, the city has become one of the first two Hubei province's pilot cities for low carbon economy. During the Twelfth Five-Year Plan period, Xianning City will continue to adhere to the concept of low carbon development and promote scientific and leaping development, aiming to become a national new energy base, national outstanding tourist destination, important industrial transfer base in central region, comprehensive reform and innovation model for "two types" society of Wuhan megalopolis, and Wuhan province's first demonstration area for low carbon economy, and striving to create a well-off, ecological, livable, healthy, and harmonious Xianning City.

Industrial restructuring is instrumental in raising energy consumption rate and reducing carbon emission. It is also a challenge to Xianning City, who is at the stage of accelerating development of industrialization. As a pilot city for low carbon, Xianning has to grasp the opportunity of pilot construction, fully upgrade the economic and industrial structure, and promote its low carbon competitiveness. To help achieve these goals, here are our countermeasures and suggestions.

First, the city should optimize its industrial structure, using low carbon high technology to transform the conventional agriculture. It is extremely important for Xianning City to adjust its industrial structure, especially the industry sector's structure optimizing. In order to promote industrial upgrading, it is necessary for the city to adhere to the path of new industrialization, actively employing high technologies and low carbon technology to transform the conventional industry, so as to maximize resource productive efficiency and energy utilization rate. The government should adjust its industrial policies according to the national policies, strictly enforcing market admittance and striving to develop new industries of low energy consumption and low pollutant. Second, the city should eliminate backward production capacity and make great efforts to develop and exploit clean energy. The government should perfect relevant policies concerning the encouragement and restriction of eliminating outdated production capacity, establishing a mechanism to compensate backward production capacity. Meanwhile, the government should encourage and guide enterprises to speed up the elimination of outdated production technologies, strive to research and develop energy-saving technologies, and innovate energy-saving patterns, turning high energy consumption and low output into low energy consumption and high output. On the other side, the city should make full use of its advantageous clean energy reserve, further strengthening the exploitation of wind power, solar power, and hydropower. Third, the city should spare no efforts to develop its service industry, strengthening the deceleration effect of energy consumption and carbon emission of structure evolution. With its remarkable characteristic industries such as tourism, eco-agriculture, and green industry, Xianning City's low carbon industries and service industry featuring tourism showcases quite an industrial advantage. Take tourism for example, Xianning City boasts abundant vegetation and numerous tourist destinations, like Mount Jiugong, Sanjiang Hot-spring, Taiyi Cave, Star Bamboo Sea, as well as Underground Project 131. It should make the most of its current tourism advantages and actively develop new scenic spots, to blend the concepts of environment protection and low carbon and publicize them during travelling.

ACKNOWLEDGMENT

Foundation: The Key Science Study Program of Hubei Provincial Department of Education, No.D20112802; The Key Humanities and Social Science Study Program of Hubei Provincial Department of Education, No.2012D124.

Author: Jun-wu FANG (1964–), Master, Associate professor, specialized in applied economics. Email: fjw1964@163.com

Corresponding author: Zhi Chen (1967–), Professor, specialized in regional economy and city development. Email: chzh1967@163.com

REFERENCES

[1] IPCC. Climate change 2007: the physical science basic [M]. Cambridge: Cambridge University Press, 2007.

[2] Xie Li-Jian, Zhou Su-Hong, Yan Xiao-Pei. A review and outlook of the study on low-carbon development in china and overseas [J]. Human Geography, 2011, 26(1): 19–23, 70. (in Chinese)

[3] Chinese government net. Chinese energy conditions and policies [EB/OL]. http://www.gov.cn/zwgk/2007-12/26/content_844159.htm.

[4] CDIAC. National CO_2 emissions from fossil-fuel burning, cement manufacture, and gas flaring: 1751-2005 [R]. Carbon Dioxide Information Analysis Center, 2006.

[5] Forester. World dynamics [C]. China Environmental Management, Economy and Law Association, 1984.

[6] Herman Daley. Beyond growth: the economics of sustainable development [M]. Shanghai: Shanghai Translation Press, 2001.

[7] Miller R.E., Blair P.D. Input-output analysis: foundations and extensions [J]. Englewood Cliffs, 1985, 200–227.

[8] Grossman G., Kreuger A. Economic growth and the environment [J]. Quarterly Journal of Economics, 1995, 110(2): 353–377.

[9] David Reed. The economic structure, environment and sustainable development [M]. Beijing: China Environment Science Press, 1998.

[10] Chihiro Watanabe. Systems option for sustainable development-effect and limit of the ministry of international trade and industry's efforts to substitute technology for energy [J]. Research Policy, 1999(28): 719–749.

[11] Chen Guo-Jie. Resources utilization and adjustment of industrial structure in china. [J]. China Population, Resources and Environment, 1994, 4(1): 26–30. (in Chinese)

[12] Lu Zheng-Nan. Empirical analysis on the effects of industrial structure adjustment on energy consumption in China [J]. The Journal of Quantitative & Technical Economics, 1999, (12): 53–55.

[13] Shi Dan, Zhang Jin-Long. On the impact of industrial structure changes on energy consumption Economic Theory and Business Management, 2003, (8): 30–32. (in Chinese)

[14] He Jian-Kun, Zhang Xi-Liang. Analysis on the impact of the structural changes in the manufacturing industry on the rising of intensity of GDP resources and its trend [J]. Environmental Protection, 2005, (12): 43–47. (in Chinese)

[15] Zhu Shou-Xian, Zhang Lei. Energy-saving potentials in Beijing based on industrial structure [J]. Resources Science, 2007, 29(6): 194–198. (in Chinese)

[16] Dai Yan-De, Zhou Fu-Qiu, Zhu Yue-Zhong, Xiong Hua-Wen. Approaches and measures to achieve the anticipated goal of reducing China's energy intensity of GDP by 20% to 2010 [J]. China Industrial Economy, 2007, 22(4): 29–37. (in Chinese)

[17] Guo Guang-Tao, Guo Ju-E, Xi You-Min, Meng Le. Calculation of energy-saving effect and research of implementation strategy on industrial structure adjustment in Western China [J]. China Population, Resources and Environment, 2008, 18(4): 44–49. (in Chinese)

[18] Li Ming-Yu, Li Kai, Yu Pei-Li, Chen Hao. An empirical study for industrial structure in Liaoning Province in the perspective of energy saving [J]. Journal of Northeastern University (Natural Science), 2009, 30(1): 145–148. (in Chinese)

[19] Xu Guo-Quan, Liu Ze-Yuan, Jiang Zhao-Hua. Decomposition model and empirical study of carbon emissions for China, 1995–2004 [J]. China Population, Resources and Environment, 2006, 16(6): 158–161. (in Chinese)

[20] Zhang Lei, Huang Yuan-Xi. Potential analysis on structural energy-saving in China [J]. China Soft Science, 2008, (5): 27–34, 51. (in Chinese)

Biomedical Engineering and Environmental Engineering – Chan (Ed.)
© 2015 Taylor & Francis Group, London, ISBN: 978-1-138-02805-0

Study of the spatial differences of coordinative development between urbanization and human settlement in Wuhan City Circle

Ping-fan Liao & Zhi Chen
School of Resources Environmental Science and Engineering, Hubei University of Science & Technology, Xianning, P.R. China

ABSTRACT: Human settlement and urbanization mutually influence each other, the coordinated development of which is an important goal for sustainable urban development. This paper took the Wuhan City Circle as the area of study, utilized the data from the statistics yearbook and annual statistics report for Hubei Province and each of the nine cities within the Circle. Based on the comprehensive assessment index system for urbanization and human settlement, the factor analysis method was adopted to calculate the urbanization composite index and the human settlement composite index. Then, by establishing the model of coordination degree for urbanization and human settlement, the coordination degree between urbanization and human settlement of the nine cities within the Wuhan City Circle were worked out and their spatial differences were analysed. Through discussing the situation of coordinated development between urbanization and human settlement in this region and revealing the spatial differences in the coordinated development, the paper could offer reference for management and decision-making regarding the promotion of the development of regional human settlement science and urban sustainable development.

Keywords: Coordinative degree, factor analysis method, human settlement, spatial differences, urbanization, Wuhan City Circle

1 INTRODUCTION

Since the 20th century, urbanization has been developing at an accelerated pace, driving the development of industrialization and informatization and promoting the rapid growth of social and economic development. Meanwhile, the intense increase of urban populations and the expansion of city scales have brought a series of problems such as tight housing space, traffic jam, severe air pollution, ecological environment degradation, and overloading of the operation of urban infrastructure, causing serious damage to the inhabitant environment of urban residents. The issue of urban human settlement has increasingly attracted wide attention from the public[1,2]. Therefore, to improve the quality of urban human settlement and realize sustainable development of human settlement are the goals of urbanization. The question of how the relationship between urbanization and the human settlement, on which the urban residents all rely for survival, could be well handled during the urbanization process has become an unavoidable realistic problem in the development process of mankind.. Moreover, it has become a great challenge for sustainable development how the coordinated development of human settlement and urbanization could be realized, which is of significant theoretical and practical meaning.

Foreign studies of human settlement began in the late 1990s and the early 2000s, the ideas of which have always been incorporated in the realm of urban planning science[3–8]. Until the 1950s when C. A. Doxiadis initiated the *Ekistics* magazine, systematic study of human inhabitation commenced[9–12]. Later on, experts and scholars from different disciplines joined the study one after another, continuing to enrich the contents of this discipline. By far, various schools have been formed in the foreign study of human settlement, such as the urban planning school, the group inhabitation school, the geological school, and the ecological school[4–16]. Currently, domestic and foreign studies are mostly focused on environmental influence assessment, establishment of RS and GIS database, and national-level studies and practices[17]. Domestic study of human settlement started late in the 1980s. Since the beginning of reform and opening up, with the improved livelihood and the housing system reform pressing ahead, China has made remarkable achievements in its human settlement practices. Meanwhile, relevant problems concerning human settlement began to appear, the study of which has drawn attention from governments, experts, and scholars, who began to explore these issues within an academic framework[18]. From 1993, the systematic study of human settlement in the true and strict sense has been developing momentously. Wu Liang-yong, among others, established human settlement science under the inspiration of Western theories on human inhabitation science. From then on, China's systematic study of human settlement theories began to gradually develop. In 2001, Mr. Wu Liang-yong published his works *"Introduction to Human Settlement Science"*, which basically laid an academic framework for human settlement science and is a milestone of significance for the development of residential[19] science in China. Later on, with the endeavours of many experts and scholars, human settlement science has achieved substantial progress and fruitful achievements have been made in various fields. Now, urban human settlement has become an important issue for many disciplines such as planning, geology, architecture, and environment, which all studied it in depth t from different perspectives. In this regard, there are mostly studies of ecological residential quarter by Zhu Xi-jin[20], studies of urban ecological health system by Guo Xiu-rui and Hu Ting-lan et al.[21,22], studies of urban human settlement ecology by Wang Ru-song et al.[23], studies of metropolitan and township human settlement by Ning Yue-min et al.[24,25], and the urban human settlement assessment by Li Xue-ming[26], et al. In recent years, many scholars have carried out empirical studies in certain fields and proposed an index system for the sustainable development of an ecological city, human settlement, and the urban environment. A few scholars also conducted explorative studies of coordinated development between urban human settlement, urbanization and economic growth. These studies have yielded few results and most of them are qualitative.

On 14 December, 2007, the Wuhan City Circle was officially approved as a comprehensive supporting reform and pilot zone for the building of a "dual-type society". With Wuhan as the centre, the Circle accommodates eight cities of Huangshi, E'zhou, Huanggang, Xiaogan, Xianning, Xiantao, Qianjiang, and Tianmen to form a "one plus eight" regional economic coalition. It is the core area for economic development in Hubei Province and an important strategic pivot for the rise of the central region, and now it has become a rapid growth polar for the social and economic development of Hubei Province. With the improved urbanization and rapid social and economic growth of the Wuhan City Circle, comes the increasingly intense contradiction between urbanization level and human settlement, the coordination between which needs further improvement. This paper intends to take the Wuhan City Circle as the area of study, utilizing the data from the statistics yearbook and annual statistics report for Hubei Province and each of the nine cities within the Circle. Based on the comprehensive assessment index system for urbanization and human settlement, the factor analysis method will be adopted to calculate the urbanization composite index and human settlement composite index. Then, by establishing the model of coordination degree for urbanization and human settlement, the coordination degree between urbanization and human settlement of the nine cities within the Wuhan City Circle will be worked out and their spatial differences will be analysed. Through discussing the situation of coordinated development between urbanization and human settlement in this region and revealing the spatial differences in the coordinated development, the paper could provide a reference for management and decision-making regarding the promotion of the development of regional human settlement science and urban sustainable development.

2 BASIC THEORIES AND APPROACHES

2.1 *Theoretical basis*

Urbanization refers to the gradual transformation of population, industrial structure, and lifestyle from agricultural to non-agricultural. It is a comprehensive manifestation of the economic development and social progress of a city. Urban human settlement refers to the organic combination of the various material and non-material components needed for maintaining human activities, including the tangible space for living and activities of urban residents and various related aspects such as population, resources, environment, social policies, and economic development[17,26]. The development of urban human settlement has a close relationship with the urbanization process and sustainable urban development. The urbanization process is a process of development and construction of both a city and its human settlement. A human settlement situation is the specific embodiment of urbanization, and also the basis for further development of urbanization. Sustainable, healthy and orderly urbanization and sustainable development of urban human settlement can be each other's precondition and result. Human settlement and urbanization mutually influence each other, the coordinated development of which is an important goal for sustainable urban development.

While the urbanization system and human settlement system respectively belong to different systems, they both belong to the huge socioeconomic system, and are mutually interactive and coupling to each other. This makes the various factors of the two systems mutually influential and penetrating.

2.2 *Establishment of index system*

In the light of the complexity of the abovementioned two systems, in order to reveal their coordinative nature, this paper, by following the principles of scientific selection and operability to pick up representative, multilevel and dynamic data, and by referring to existing research results[17,26], obtains a comprehensive assessment index system for urbanization and human settlement. Comprehensive assessment indexes for urbanization include nine ones in the four aspects of economic structure, population, lifestyle, and environment. Comprehensive assessment indexes for human settlement include eleven ones in the four aspects of human settlement, ecological environment, public service facilities, and spiritual and cultural life (See Table 1).

2.3 Data source and index standardization

The original data in this paper comes from Hubei Provincial Statistical Yearbook (1999–2010), and the statistical yearbook and annual statistical report (1999–2010) of the nine cities within the Wuhan City Circle, including Wuhan, Huangshi, E'zhou, Huanggang, Xiaogan, Xianning, Xiantao, Qianjiang and Tianmen.

In order to eliminate the screening effect among the data caused by different measuring standards of the original indexes, there is a need for standardized treatment of the original data. Considering the characteristics of the following study approaches, the paper adopts the approach of standard deviation standardization, with the formula being:

$$x'_{ij} = \frac{x_{ij} - X_j}{S_j} ; \quad y'_{ij} = \frac{y_{ij} - Y_j}{S_j} \tag{1}$$

In the formula, $i = 1, 2, \ldots, n$; $j = 1, 2, \ldots, m$. i refers to i year index, j refers to j item index, x'_{ij} and y'_{ij} respectively refers to the standardization value of the index of urbanization and human settlement, S_j refers to the standard deviation of the j item original index of urbanization and human settlement, X_j and Y_j are respectively the average value of the j item original index of urbanization and human settlement.

2.4 Calculation of a composite index of urbanization and human settlement

In order to overcome the information overlapping between multiple index variables and artificially determine the subjectivity of the index weight, the paper adopts the approach of factor analysis to calculate the regional composite index of urbanization and human settlement. Factor analysis is a type of multivariate statistical approach. It identifies the potential factors hidden in the measurable variables, which cannot be, or cannot easily be, detected but can influence or control the measurable variables, and then estimate the degree of impact of the potential factor on the measurable variables and the correlation between the potential factors. Its working principle is to find main factor variables, estimate the factor model, calculate the value of the main factor variable, and provide a reasonable explanation for the main factor variables. The paper employs statistical analysis software SPSS19.0 to conduct the factor analysis for the urbanization and human settlement indexes of the nine cities within the Wuhan City Circle respectively. The specific steps are as follows: (1) Standardizing the original index data of urbanization and human settlement by using the above formula (1) for calculation; (2) Working out the relevant coefficient matrix R and R's characteristic root and characteristic vector, and establishing a factor model to determine the factor contribution rate and

accumulative contribution rate. Through calculation and analysis, the characteristic root and contribution rate of the urbanization and human settlement factors are obtained. According to the principle of accumulative variance contribution rate ≥85%, the main factors of urbanization and human settlement were extracted respectively. Among the urbanization indexes, 3 main factors U_1, U_2 and U_3 were selected, which have an accumulative contribution rate of 85.31% and a variance contribution rate of 41.33%, 30.77% and 13.22% respectively. Among the human settlement indexes, 4 main factors E_1, E_2, E_3 and E_4 were chosen, which have an accumulative contribution rate of 92.41% and a variance contribution rate of 42.29%, 21.15%, 14.59%, and 14.38% respectively. The abovementioned main factors can all fully explain and provide enough information that can be expressed by the original data. (3) Varying the factor loading by rotating and calculating to obtain the ideal main factor loading matrix and the main factor score coefficient matrix. Then establishing the urbanization and human settlement factor loading matrix for the main factors of urbanization and human settlement. In order to make it convenient for a reasonable explanation of the various factor loadings, the variance maximum method was adopted for the orthogonal rotation in order to obtain a clearer factor loading matrix of urbanization and human settlement. Using calculations, the main factor score coefficient matrix was obtained for the urbanization and human settlement. (4) Calculating the comprehensive score of the main factors of urbanization and human settlement, i.e. to work out the weighted sum of the various main factors, with the weight being the variance contribution rate of various main factors. Hence the comprehensive scores of the various main factors, namely the composite indexes of urbanization and human settlement are obtained. The formula is:

$$U = \sum_{i=1}^{p} \lambda_i u_i ; \quad E = \sum_{j=1}^{q} \lambda_j e_j \tag{2}$$

In the formula, $i = 1, 2, \ldots, p$; $j = 1, 2, \ldots, q$. i is the i main factor of urbanization, j is the j main factor of human settlement, U and E are respectively the comprehensive score of the main factors of urbanization and human settlement, namely, the composite index of urbanization and human settlement, λ_i and λ_j are respectively the weight of the various main factors of urbanization and human settlement, i.e. the variance contribution rate of the main factors, u_i and e_j are respectively the score of the various main factors of urbanization and human settlement.

2.5 A coordination degree model and type discrimination

Coordination refers to the harmony and consistency between the system component factors during the process of evolution. Development refers to the process

243

Table 1. Index system of comprehensive evaluation for urbanization and human settlement.

Target layer	Criterion layer	Index layer	Unit
Urbanization level/X	Economic structure/X_1	GDP per capita/x_{11}	Yuan/person
		Second industries accounted for the proportion of GDP/x_{12}	%
		Third industries accounted for the proportion of GDP/x_{13}	%
	Population/X_2	Non-agriculture population proportion/x_{21}	%
	Lifestyle/X_3	Average wages of staff and workers/x_{31}	Yuan
		Per capita of post and telecommunications services/x_{32}	Yuan/person
		Per capita total retail sales of social consumer goods/x_{33}	Yuan/person
		Per capita public library/x_{34}	Book/person
	Environment/X_4	Target area of environmental noise/x_{41}	hm^2
Human settlement/Y	Living environment/Y_1	Per capita living space/y_{11}	m^2/person
		Population density of built-up area/y_{12}	People/hm^2
		Per capita public green area/y_{13}	hm^2/person
	Ecological environment/Y_2	City green coverage/y_{21}	hm^2
		Target area of environmental noise/y_{22}	hm^2
		Sewage treatment rate/y_{23}	%
	Public service establishment/Y_3	Per thousand people internet users/y_{31}	doors/10^3 people
		Per capita Road area/y_{32}	m^2/person
		Per ten thousand people public transport vehicles/y_{33}	Cars/10^4 people
	Spiritual and cultural life/Y_4	Per hundred people public library/y_{41}	Book/10^2 people
		Per capita educational expenditure/y_{42}	Yuan/person

of the system component factors growing from bigger, more complicated, more advanced and more orderly. Hence we can know that coordination means better correlation between system factors, while development is the evolution process of the whole system. The coordinated development between urbanization and human settlement refers to the harmony between people's human settlement, sanitation, education, and cultural environment. As the weight of the non-farming population increases, the ratio of the secondary and tertiary industries is enhanced, and the people's livelihood keeps rising. The coordination degree between urbanization and human settlement refers to the degree of coupling between urbanization development and human settlement quality. It is a quantitative index measuring the coordination between systems and within each system, as specifically embodied in the interface characteristics featuring a mutual influence between the urbanization with time and space as the reference system.

The coordination degree is a quantitative model that measures the coupling and coordination between systems, which can describe the coordination between urbanization and human settlement in a quantitative way. The formula is:

$$M = \frac{U + E}{(U^2 + E^2)^{\frac{1}{2}}} \quad (3)$$

In the formula, M is the coordination degree between urbanization and human settlement, U is the composite index for urbanization, and E is the composite index for human settlement. M value is jointly determined by U and E. When $U > 0$, $E > 0$ and $U = E$, M reaches its maximum, 1.414. Otherwise, when $U < 0$, $E < 0$ and $U = E$, M reaches

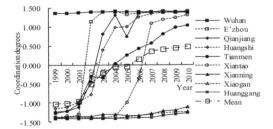

Figure 1. The coordination degrees between urbanization and human settlement of the nine cities within Wuhan City Circle (1999–2010).

its minimum, −1.414. The results of any other circumstances are between the former two, hence $-1.414 \geq M \geq 1.414$. According to the values of M, U and E, the coordination degree can be divided into the following types (see Table 2).

3 RESULTS AND ANALYSIS

By using the statistical data of the nine cities within the Wuhan City Circle during 1999–2010 and applying the abovementioned algorithm, the composite indexes for urbanization and human settlement of the nine cities were worked out, and further, the coordination degrees of urbanization and human settlement of the nine cities within the Wuhan City Circle during 1999–2010 were obtained (see Table 3), and the coordination degree change trend chart can be drawn (see Figure 1).

We can know from Table 3 and Figure 1, combining the composite indexes of urbanization and human settlement, that: (1) In terms of the whole Wuhan City

Table 2. Type and characteristics of coordination degree between urbanization and human settlement.

M	U, E	Type	Characteristics
$0 \leq M \leq 1.414$	$U \geq 0, E \geq 0$	I Coordinated	Close to equilibrium, the whole system is optimal.
$M \geq 0$	$U \geq 0, E \leq 0$	II Basically coordinated	Urbanization level is higher than the human settlement, the coordination degree is positive, system tends to be optimized.
$M \geq 0$	$U \leq 0, E \geq 0$	III Weakly coordinated	Urbanization level is lower than the human settlement, the coordination degree is positive, but in a certain range of threshold values.
$M < 0$	$U > 0, E < 0$	IV Less coordinated	Urbanization level is higher than the human settlement, the coordination degree is negative, system tends to be optimized, the whole system performance is poor.
$M < 0$	$U < 0, E > 0$	V Uncoordinated	Urbanization level is lower than the human settlement, the coordination degree is negative. It is not conducive to the whole society progress.
$-1.414 \leq M \leq 0$	$U \leq 0, E \leq 0$	VI Maladjusted	Urbanization level and human settlement are low, system recession, society development slow.

Table 3. Its type and coordination degrees between urbanization and human settlement of the nine cities within Wuhan City Circle (1999–2010).

City	1999	2000	2001	2002	2003	2004	2005	2006	2007	2008	2009	2010
Wuhan	1.368	1.367	1.392	1.406	1.407	1.405	1.402	1.405	1.397	1.391	1.385	1.392
	I	I	I	I	I	I	I	I	I	I	I	I
E'zhou	−1.385	−1.405	−1.309	1.134	1.401	1.412	1.338	1.362	1.407	1.414	1.411	1.413
	VI	VI	VI	I	I	I	I	I	I	I	I	I
Qianjiang	−1.245	−1.223	−1.131	−0.449	0.802	1.319	0.747	1.373	1.384	1.414	1.399	1.408
	VI	VI	VI	V	III	I	III	I	I	I	I	I
Huangshi	−1.159	−1.123	−0.978	−0.795	0.383	0.989	1.003	1.264	1.342	1.381	1.401	1.409
	VI	VI	IV	IV	II	II	I	I	I	I	I	I
Tianmen	−1.282	−1.231	−1.031	−0.311	−0.333	0.024	0.252	0.421	0.573	0.814	0.987	1.034
	VI	VI	VI	V	V	III	III	III	III	III	I	I
Xiantao	−1.395	−1.344	−1.378	−1.410	−1.409	−1.360	−0.999	−0.361	1.093	1.198	1.253	1.321
	VI	VI	VI	VI	VI	VI	V	V	I	I	I	I
Xianning	−1.394	−1.379	−1.368	−1.361	−1.322	−1.319	−1.357	−1.344	−1.310	−1.221	−1.208	−1.125
	VI	VI	VI	VI	VI	VI	VI	VI	VI	VI	IV	IV
Xiaogan	−1.410	−1.414	−1.414	−1.414	−1.414	−1.414	−1.412	−1.408	−1.406	−1.358	−1.317	−1.283
	VI	VI	VI	VI	VI	VI	VI	VI	VI	VI	IV	IV
Huanggang	−1.414	−1.414	−1.414	−1.412	−1.413	−1.412	−1.383	−1.349	−1.304	−1.282	−1.261	−1.243
	VI	VI	VI	VI	VI	VI	VI	VI	VI	VI	V	V

Circle, the coordinated development between urbanization and human settlement can be divided into two phases. One is the less-coordinated period of 1999–2005, when the coordination degrees of eight cities (Wuhan non-included) were all negative and the average coordination degree of the nine cities was also negative. The urbanization and human settlement of the Wuhan City Circle as a whole are at a low level, and the coordination degrees are mainly VI, V and IV types. The other one is the basically-coordinated period of 2006–2010, when the coordination degrees of five cities (Xiaogan, Huanggang and Xianning non-included) were positive, and the average coordination degree of the nine cities was also positive. The urbanization level of the Wuhan City Circle as a whole has kept rising, especially driven by the rapid urbanization of the four cities of E'zhou, Qiangjiang, Huangshi and Tianmen. This has further stimulated economic growth and improved the human settlement to a certain extent. The overall development of the Wuhan City Circle is now being optimized, and the coordination degrees are mainly I, II and III types.

(2) There are enormous regional differences in the coordinated development of urbanization and human settlement of the nine cities within the Wuhan City Circle. Judging by the trend of change of the coordination degree, the nine cities can be divided into three types. The first type includes only Wuhan City, which is a stable and coordinated type. During the 12 years of study period, the annual coordination degree was all above 1.36, and the approximated average was 1.4. The coordination degree of each year all belonged to I type, with a small margin of change, the maximum being 1.407 and the minimum being 1.367, with a margin of only

245

0.04. This indicates that the urbanization and human settlement of Wuhan City during the 12 years were quite coordinated, showing that Wuhan, as a central city in the Wuhan City Circle, had played an exemplary role in the coordinated development between urbanization and human settlement. The second type includes the five cities of E'zhou, Qianjiang, Huangshi, Tianmen, and Xiantao, which is a fast-growth, coordinated and optimized type. During the 12 years, the coordination degree increased rapidly year by year, from less than -1.1 in 1999 to a positive value 3–8 years later, and reached above 1.0 in 2010. In particular, the coordination degrees of E'zhou, Qianjiang and Huangshi rose to more than 1.4, with a large margin of growth. The coordination degree growths of the five cities were all at above 2.3, with those of E'zhou, Qianjiang, Huangshi, and Xiantao reaching over 2.5. The coordination degrees, from the coordination-losing VI type, non-coordinated IV type and less-coordinated IV type, were gradually optimized to the weakly-coordinated III type, basically-coordinated II type or coordinated I type in the later phase of the study period, indicating significant improvement in the coordinated relationship between urbanization and human settlement in the five cities during the study period. The third type includes the three cities of Xianning, Xiaogan and Huanggang, which is a slow-growth, non-coordinated type. During the 12 years of study, the coordination degree each year was below -1.1, with the average being under -1.3. During 1999–2008, the coordination degree each year was all non-coordinated VI type. During 1999–2010, despite a slow growth, the growth margin was small, with the maximum being less than 0.27. As for 2009 and 2010, the coordination degree was still a non-coordinated V type or less-coordinated IV type, indicating that the relationship between urbanization and human settlement of Xianning, Huanggang, and Xiaogan had always been uncoordinated during the study period. To coordinate the relationship is an arduous task, and the coordination is yet to be improved.

(3) Judging from the coordination degrees in 2010, the nine cities can be divided into two types. First, the six cities of Wuhan, E'zhou, Huangshi, Qianjing, Tianmen, and Xiantao had positive coordination degrees of above 1.0, with high-level urbanization and superior human settlement, belonging to the coordinated I type. Second, the three cities of Xianning, Xiaogan, and Huanggang have coordination degrees below -1.1, with low-level urbanization and inferior human settlement, belonging to the non-coordinated V type or less-coordinated IV type, with a poor coordination. In particular, the urbanization of Xianning and Xiaogan was higher than their human settlement levels. The education input and expenditure and public service facilities in the two cities are yet to be improved. Particularly, the urbanization level of Huanggang lagged behind its human settlement level. Therefore, Huanggang shall strengthen its construction of an urban infrastructure, improve the investment and financing climate, expand the radiation effect of the city, and

at the same time, deepen a reform in the household registration system, employment system, and welfare and social security system to form a normal development mechanism for the population urbanization, and endeavour to enhance the urbanization level.

4 CONCLUSION AND DISCUSSION

(1) Judging from the overall Wuhan City Circle, the coordinated development between urbanization and human settlement can be divided into two phases, one is the less-coordinated period of 1999–2005, and the other is the basically-coordinated period of 2006–2010. In this process, the urbanization level of the Wuhan City Circle as a whole has kept enhancing, the human settlement has been improved to a certain extent, and the overall development of the Circle tends to be optimized.

(2) There are enormous regional differences in the coordinated development of urbanization and human settlement of the nine cities within the Wuhan City Circle. Judging from the trend of a change of the coordination degree, the nine cities can be divided into three types. Wuhan belongs to the stable and coordinated type, E'zhou, Qianjiang, Huangshi, Tianmen, and Xiantao belong to the fast-growth, coordination-optimizing type, while Xianning, Xiaogan, and Huanggang belong to the slow-growth, non-coordinated type.

(3) Judging by the coordination degree of 2010, the nine cities can be divided into two types: The six cities of Wuhan, E'zhou, Huangshi, Qianjing, Tianmen, and Xiantao belong to the coordinated I type, while the three cities of Xianning, Xiaogan, and Huanggang belong to the non-coordinated V type, or the less-coordinated IV type.

(4) Given the characteristics of the coordinated development of urbanization and human settlement of the nine cities of the Wuhan City Circle, the relevant measures shall be taken. Wuhan has high-level urbanization and human settlement, and its urbanization has driven development in people's life, education, and culture. While maintaining the existing development scale, the city shall better realize sustainable development between urbanization and human settlement, and its development experience shall be drawn on by other cities within the Wuhan City Circle. E'zhou, Qianjing, Huangshi, Tianmen, and Xiantao have high-level urbanization and human settlement, and Xianning, Xiaogan, and Huanggang have low-level urbanization and human settlement. For cities such as Huangshi, Xianning, and Xiaogan, in which the construction of human settlement lags behind urbanization, measures shall be taken to strengthen education, healthcare, and public facilities, in order to adapt to the urbanization process. For cities such as E'zhou, Qianjiang, Tianmen, Xiantao, and Huanggang, measures shall be taken, in order to adjust the urban and rural layout and to

reform the household registration system, so as to improve their urbanization level.

ACKNOWLEDGMENT

Foundation: The Key Science Study Program of Hubei Provincial Department of Education, No.D20112802; The Key Humanities and Social Science Study Program of Hubei Provincial Department of Education, No. 2012D124, No. 13g393.

Author: Ping-fan Liao (1977–), Master, Prelector, specialized in land resources and realty management. Email: liaopfboy@sina.com

Corresponding author: Zhi Chen (1967–), Professor, specializing in regional economy and city development. Email: chzh1967@163.com

REFERENCES

Wu Ji-Lin, Mu Ming. Re-examine the city problem in twenty-first century [J]. City and Disaster Reduction, 2001, (6): 18–19. (in Chinese)
Deng Qing-Hua, Ma Xue-Lian. Urban residential ideal and urban problems [J]. Journal of South China Normal University (Natural Science Edition), 2002, (1): 129–135. (in Chinese)
Qi Xin-Hua, Cheng Yu, Chen Lie, Chen Jun. Review of literatures on human settlements [J]. World Regional Studies, 2007, 16(2): 17–24. (in Chinese)
E. Howard. Garden cities of tomorrow [M]. Faber and Fabel, London, 1946.
P. Geddes. Cities in evolution: an introduction to the town planning movement and the study of civism [M]. New York: Howard Ferug, 1915.
L. Mumford. The city in history: its origin, its transformation, and its prospects [M]. Haccourt, Brace & World, Inc. 1961.
L. Donald. Miller: Lewis Mumford—a life [M]. Weidenfeld & Nicloson, 1989.
P. Hall. Urban and regional planning [M]. London and New York: Routledge, 1992.
C.A. Doxiadis. Ekistics: an introduction to the science of human settlements [M]. Athens Publishing Center, 1968.
C.A. Doxiadis. Action for human settlements [M]. Athens Publishing Center, 1975.
C.A. Doxiadis. Athroplpolis: City for human development [M]. Athens Publishing Center, 1975.
C.A. Doxiadis. Ecology and ekistics (edited by Gerald Dix) [M]. Elek Boods Ltd., 1977.
I.L. Mcharg. Design with nature [M]. Natural History Press, New York, 1969.
O. Yanitsky. Social problems of man's environment. The city and ecology [M]. Moscow: Nauka, 174, 198.
R. Register. Ecocity Berkeley. Berkeley [M]. North Atlantic Books, 1987.
R.E. Pach, E.W. Burgis. City sociology [M]. Beijing: Huaxia press, 1987.
Wang Qing. The evolution research of coordination of urbanization and human settlement: taking Dalian for a case [D]. Liaoning Normal University, 2013. (in Chinese)
Qi Xin-Hua, Mao Jiang-Xing, Cheng Yu, Fan Jian-Hong. Development of theoretical study on human settlement in China since reform and opening-up [J]. Planners, 2006, 22(8): 14–16. (in Chinese)
Wu Liang-Yong. An introduction to the science of human settlement environment [M]. Beijing: Chinese Architecture Industry Press, 2001. (in Chinese)
Zhu Xi-Jin. Overview of human ecological residential district planning in the twenty-first century [J]. Urban Planning Forum, 1994, (5): 1–6. (in Chinese)
Guo Xiu-Rui, Yang Ju-Rong, Mao Xian-Qiang. Primary studies on urban ecosystem health assessment [J]. China Environmental Science, 2002, 22(6): 525–529. (in Chinese)
Hu Ting-Lan, Yang Zhi-Feng, He Meng-Chang, Zhao Yan-Wei. An urban ecosystem health assessment method and its application [J]. acta scientiae circumstantiae, 2005, 25(2): 269–274. (in Chinese)
Wang Ru-Song. A transition of urban human settlement planning towards eco-integrity, naturalizing, low cost and humanity orientation [J]. Urban Environment & Urban Ecology, 2001, 14(3): 1–5. (in Chinese)
Ning Yue-Min, Cha Zhi-Qiang. Study on evaluation and optimization of the metropolis human settlements—a case study of Shanghai City [J]. City Planning Review, 1999, 23(6): 15–20. (in Chinese)
Ning Yue-Min, Xiang Ding, Wei Lan. Study on the human settlement environment with three small towns in the Shanghai suburban areas as the case [J]. City Planning Review, 2002, 26(10): 31–35. (in Chinese)
Li Xue-Ming. Study on city human settlements in Dalian [M]. Changchun: Jilin People's Press, 2001. (in Chinese)

Biomedical Engineering and Environmental Engineering – Chan (Ed.)
© 2015 Taylor & Francis Group, London, ISBN: 978-1-138-02805-0

An air pollution monitoring system based on a Vehicular Ad-hoc Network (VANET)

L. Zhao & Y. Ma
College of Information Science and Engineering, Wuhan University of Science and Technology, Wuhan, Hubei, China

J. Gao
Shanghai Inforstack Information Technology Ltd., Shanghai, China

M. Cheng & F. Xiao
College of Information Science and Engineering, Wuhan University of Science and Technology, Wuhan, Hubei, China

ABSTRACT: The limitations of existing air pollution monitoring systems are discussed. A novel air pollution monitoring system based on VANET is designed. By means of the VANET technology, sensors are mounted in vehicles which can collect the density values of pollutants in the air. The data are sent to the database for storage, query, and further analysis for different user applications. The infrastructure of the system is introduced in the paper. An experimental study of the air pollution analysis is presented by using the Inforstack data analysis platform.

1 INTRODUCTION

1.1 *Motivation*

In recent years, with the rapid development of civilization and industrialization, air pollution in the cities in China has become a severe problem. According to *2013 Environment Report of China*, only three cities (Haikou, Zhoushan, and Lasa) in seventy four cities have qualified air quality, which means that the air pollution in 95.9% cities cannot reach the standard[1]. Hence, air pollution monitoring and control are very important to China.

Much effort has been made in this area in China. An interesting method, which makes use of plants to monitor air pollution is proposed in [2]. In [3], the system uses the technique of differential optical absorption spectra to monitor the density of NO2 in the air. A limestone/lime-gypsum desulfurization method is introduced in [4] which is applied to SO2 discharged from thermal power stations in Xinjiang province. However, as these systems sample air pollutants by using fixed located sensors, the monitoring scope and sampling density cannot meet the requirements of the air pollution monitoring in large cities, which have very complex geographic environments. Furthermore, as most of the existing systems only provide ordinary information publishing, users can only make a simple query with coarse granularities of position and time. As a result the QoS cannot be satisfied.

Therefore, in this paper we designed an air pollution monitoring system based on a Vehicular Ad-hoc Network (VANET). A VANET uses cars as mobile nodes in a MANET to create a mobile network. A VANET turns every participating car into a wireless router or node, allowing cars approximately 100 to 300 meters far from each other to connect and, in turn, create a network with a wide range[5]. The feature of our system is to mount air pollution sensors on the vehicles, so that the sensors can sample and transmit data in the city in a mobile way. The advantages of such a design are: first, it is unnecessary to choose the locations to install the sensors. The sensors can move within a city area in arbitrary styles; second, the sensors can synchronize with the cars to be switched on or off. Hence the power supply can be provided by the car and the energy consumption will never be a problem. Furthermore, our system uses a powerful platform – Inforstack for data analysis. The analysis results can either be stored in a database or presented to users on different user application platforms, so that users can understand the air pollution conditions in real time. When the density of an air pollutant is too high, the system will send SMSs to users as alerts.

1.2 *Paper outline*

The rest of the paper is organized as follows: related work of air pollution monitoring systems is introduced in Section 2. The overall design of our system is proposed in Section 3. A case study of the air pollution data analysis in Inforstack platform is shown in Section 4, and we conclude our research in Section 5.

2   RELATED WORK

Much attention has been paid to air pollution monitoring in recent years. However, most of the approaches have disadvantages in one or more aspects. For example, the research in [6] only focuses on metallurgical dusts and it cannot be applied to different air pollutants. In [7] a quantitative remote sensing monitoring method is used, which can precisely sense the pollution data. However, because of the high cost, this method is very difficult to be widely used. The approach in [8] is effective in an emergency but ineffective in routine monitoring. Other similar research, such as [9–11], can monitor one or several air pollutants, but they all have the following limitations.

Limited monitoring range: using stationary sensors to collect the environmental data requires a large amount of sensors to cover a wide range of geographical area. As a result, to build such a system will be fairly expensive. To save the cost, a smaller number of sensors will be installed which results in a smaller monitoring range or a lower monitoring density.

Low sampling frequency: for a sensor network without a power supply, the energy is always a big problem. To save energy and prolong the lifetime, a common method used is to reduce the sampling frequency.

Deficient data volume: in order to prolong the lifetime of a sensor network, power-efficient routing protocols were designed. These protocols usually decrease the number of working sensors or the working time of a sensor. This inevitably cuts down the data volume being collected. Hence, it is unsuitable for some special applications [12, 13] for which a large amount of data is essential.

Inadequate data usage: the air pollution publishing systems in most countries, including China, only show the reading of $PM_{2.5}$ accompanied with the weather forecast. Other readings of air pollutants are sent to related departments for research only [14, 15]. Thus, these data are not fully used by normal residents.

The design of our system can overcome the above limitations and we will describe the infrastructure in detail in the following section.

3   SYSTEM DESIGN

3.1   *System infrastructure*

Our air pollution monitoring system based on VANET is shown in Figure 1. Figure 1a is the system infrastructure and 1b is the system logical structure. From Figure 1a we can see that this is a hierarchical system with three layers: a Data Collection Layer (DCL), a Data Process Layer (DPL), and a Data Application Layer (DAL). In a DCL, four air pollution (NO, NO_2, SO_2 and O_3) sensors are installed on vehicles to sample the density of pollutants. Hence sensors can move to any location with the vehicles. A DPL can receive and

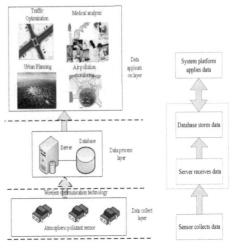

a. Infrastructure   b. Logical structure

Figure 1.   System design.

store sampled raw data. As the volume of sampled data is very high and the data types are different, to handle this problem, a sensor middleware was designed for traffic control and a data type conversion. A DAL can be divided into two sub-layers, one is for data analysis that is based on the Inforstack platform, and another one is for users' online queries and is based on the Web or Android clients.

3.2   *Data collection layer*

The system can collect density readings of NO, NO_2, SO_2, and O_3 in the air. As sensors are installed in the vehicles and have unlimited power supply, the working time is not restricted. The sampling frequency can be one sample per minute. The sampled analogue data is converted into digital data and then sent to the DPL through a network interface.

With a very high sampling frequency and a very large number of vehicles, sensors can collect data up to hundreds of thousands of items per minute. If the system sends these raw data directly to a DPL, a heavy traffic load could cause data collision and data lost. For this reason, we adopt the two-layer network topology as shown in Figure 2: the large number of sensors mounted on vehicles constructs the mobile sensors network and the small number of sensors installed on the roadside constructs the stationary sensor network [12]. A mobile sensor network transmits collected data to a stationary sensor network nearby. The stationary sensor can store data and then send them to the sink when necessary. Such a design cannot only avoid a data loss, but it can also maintain connectivity of the network when vehicle traffic is low.

3.3   *Data process layer*

Because the raw data collected by a DCL has the features of noise, incompleteness, incompatible data

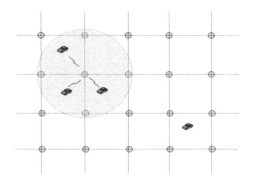

Figure 2. A two-layer network topology.

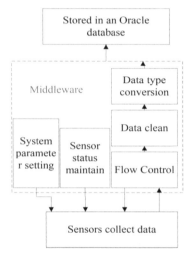

Figure 3. The structure of the middleware.

types, and a complex logical relationship, they cannot be used by user applications directly. For this reason, a middleware is designed to handle these problems.

The main tasks of the middleware include a data clean, a data type conversion, a system parameter setting, a sensor status maintenance, a flow control, etc. It can set the parameters of the whole sensor network according to the status of the sensors and the environment. It also converts the data type from raw data to the types that the database can accept. In the meantime, it controls the data flow sent from sensors to the sink to avoid data collision and loss. The middleware integrates data collection and data storage effectively, in order to guarantee the communication and data exchange among different modules of the system. The structure of the middleware is shown in Figure 3.

3.4 Data application layer

There are two application platforms in this layer: Air pollution real-time Query Platform (AQP) and Air pollution data Analysis Platform (AAP). AQP mainly provides services to normal users for pollutant density queries and emergency alerts. A Google map is embedded into the platform as illustrated. A user can input time, location, and pollutant type to query the pollution conditions online. The results will be shown in the Google map. Clicking different option tabs will show different information, such as timestamp, location, and pollution value.

AAP provides a powerful data analysis capability which is very useful for research organizations. We will introduce AAP in Section 4 in detail.

4 AIR POLLUTION DATA ANALYSIS PLATFORM AND CASE STUDIES

4.1 Inforstack data analysis platform

Inforstack is a comprehensive integrated analysis and data mining platform which was developed by our research group. It can be applied to bioinformatics, medicine and health, business intelligence, research organizations, etc. Inforstack has the advantage of a working flow to utilize different data. All the analysing components, including data sources, executions, and results, are presented to users by means of icons. Thus, it is very convenient for users to set up a data analysis working flow using the dragging mode. Therefore, Inforstack has very powerful data analysis and prediction capabilities and it can provide interactive data and model visualization.

The data collected by sensors are stored in the database after a pre-process. They are maintained by the database using different formats. The database builds relations between tables according to their physical relationships. Consequently, when Inforsense analyses data, it has to identify the data, which is imported from a database to the Inforstack platform, according to the user requirements. It then utilizes different existing, or user self-defined, components to process the data, in order to meet the user requirements.

4.2 Case study of data analysis

In this section, we will import the collected NO, NO_2, SO_2, and O_3 data to the Inforstack platform for analysis, in order to understand their density features. 140 sensors are distributed in an area of east London as shown in Figure 4. The sensors sample the density of NO, NO_2, SO_2, and O_3 with a data unit PPM, once per minute, from 8 am to 6 pm. Hence the total data items are 336000. In this case study, we chose the samples at the locations of schools, factories, and hospitals as analysis objects. As we have known that, the pollution distribution will not change distinctly within 15 minutes, and in order to reduce the data volume being analysed, we only chose the data in 5 minutes intervals from the data set.

Now we analyse the sampled data at the location around a school in the Inforstack platform. The workflow is shown in Figure 5. In the figure, 4 scatter plots correspond to 4 pollutants of NO, NO_2, SO_2,

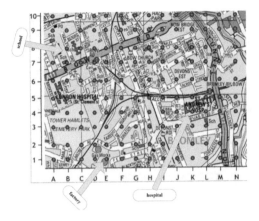

Figure 4.   140 sensors distributed in an area in east London.

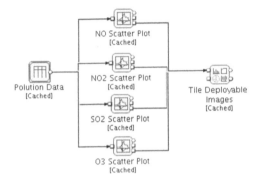

Figure 5.   Workflow for data analysis around a school.

and O_3 respectively. The analytical results are shown in Figure 6.

From Figure 6 we can see that the density of O_3 always increases from morning to afternoon, whilst the densities of NO, NO_2, and SO_2 have similar peaks during the day. The peaks all appear at about 8:30 am and 15:30 pm. These are the times that pupils go to school or leave school. At these two time periods, the number of cars in the school area increases rapidly. If we ignore the wind and the mechanical reaction, the comparison between the density of pollutants and number of cars will reveal that the vehicle emission contributes much NO, NO_2, and SO_2 to the air. Besides, the maximum values of the density of the pollutants are at least 5 times, even tens of times of the minimum values. Therefore, we can conclude that the nitrides and sulphide are mainly caused by vehicle emissions and people who are sensitive to them should avoid getting exposed to them at these time periods.

Next, we chose to sample the pollutants around a factory. Generating and running workflow similar to that in Figure 5, we can get the results as shown in Figure 7.

In the figure we can see that the density of O_3 also increases as time elapses. The change of NO_X is more complex. In the morning when work at the factory begins, its emissions can increase the density of NO and NO_2 in the air. As NO will transform to NO_2 after

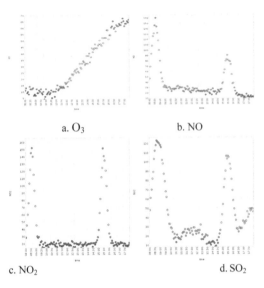

a. O_3 b. NO

c. NO_2 d. SO_2

Figure 6.   The density of 4 pollutants around a school vary during the day.

a period of time, after the density of NO reaches the peak, it will decrease afterwards. For NO_2, on the one hand, it can react with H_2O in the air to produce HNO_3 and H_2NO_2, which makes the density of NO_2 decrease. However, when HNO_3 and H_2NO_2 saturate in the H_2O in the air, NO_2 will not be absorbed and the density will then increase again. On the other hand, as NO_2 is vulnerable to the influence of temperature and pressure, the density is comparatively high at dawn and dusk, while low at midday. For these reasons, the density of NO_2 is increasing at the beginning then decreasing later, and increasing again afterwards. For SO_2, its density does not change much in the morning. It is because SO_2 can be absorbed by the water in the air to produce H_2SO_3. When H_2SO_3 is saturated, the volume of SO_2 in the air starts to increase rapidly. After 4 pm, the factory gradually stops manufacturing and the densities of NO_2 and SO_2 start to decrease. According to the discussion above we can see that water can reduce the volume of nitrides and sulphides in the air.

Form above figures we can see that the density of O_3 always increases no matter where, when, and how the data was sampled. To better understand the features of O_3, we compare the volumes of O_3 during the day at the locations of the factory, hospital and school in Figure 8.

We have known that the pollution around the factory mainly comes from the burning of coal and gas. The pollution around the school mainly comes from vehicle emissions and the pollution around the hospital includes medical treatments and vehicle emissions. At the same time, there is a brook near the hospital (from 1N to 10L in Figure 4). In Figure 8, the variation tendencies of O_3 in 3 locations are similar exponential increasing. However, the density around the school is much higher than that around the factory and hospital. As the data is collected in the same day, thus

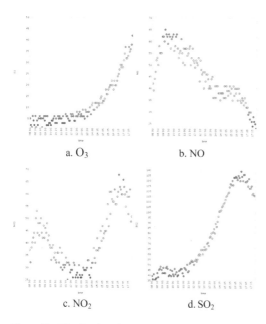

a. O₃   b. NO

c. NO₂   d. SO₂

Figure 7. The density of 4 pollutants around factory vary during the day.

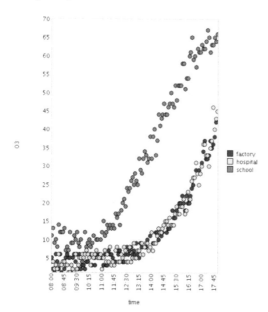

Figure 8. The density of O₃ at 3 locations during a day.

the weather conditions at 3 locations are nearly the same. Such a phenomenon is due to the geographical differences that there are no high buildings in the school area. Hence this area can accept more sunlight. It can prove the conclusion that the volume of O_3 in the air depends partly on the illumination intensity and time.

From the discussion above we can know that the volumes of pollutants in the air depend not only on the emission of pollutants but also on the geographical

environments and weather conditions. Therefore, the air pollution management and research have to take all the objective conditions into account. Using the Inforstack data analysis platform, our VANET-based air pollution monitoring system can perform different experiments and present valuable results according to different user requirements.

5 CONCLUSION

In this paper, we introduced an air pollution monitoring system based on VANET. We also conducted an experimental study of the air pollution distribution by using an Inforstack data analysis platform. The study showed the relationship among the air pollution, environments, and human activities. Furthermore the results also verified the powerful data analysis capability of Inforstack.

With the rapid popularization of cars and the development of VANET, such a system will find its applications in many fields including everyday life, health, traffic scheduling, and city planning. At that moment, the collected data will explosively increase, which may bring about the challenges of data storage, data query, and data analysis. How to address these problems will be our future research.

ACKNOWLEDGEMENTS

This work is supported by the National Science Foundation of China project Grant No. 61104215, and the Scientific Research Foundation for the Returned Overseas Scholars, the State Education Ministry.

REFERENCES

[1] Xinsheng Liu, Nahuan Liu. The status quo, the consequences and countermeasures of urban air pollution. *2014 Beijing-Tianjin-Hebei steel industry cleaner production, and environmental shifting exchange Proceedings*. 2014(4).
[2] Wei Li. Plant monitoring in the application of atmospheric pollution monitoring. *Science and Technology Communication*. 2014(3).
[3] Zhi Cao, Yong Yang, Zicong Chen, etc. The study of detection NO2 concentration based on the visible laser difference. *Chinese Optical Society 2011 Academic Conference Abstracts*. 2011(9).
[4] Yucheng Zhao. In the application of the lime gypsum flue gas desulfurization technology in air pollution. *Xinjiang Chemical*. 2011(6).
[5] Vehicular ad hoc network, http://en.wikipedia.org/wiki/VANET.
[6] Lihua Tian. Metallurgical dust pollution detection method based on wireless sensor networks. *Electrical Applications*. 2012(12).
[7] Chenglin Li. The study of the quantitative remote sensing detection method of urban air pollution. *Lanzhou University*. 2012.
[8] Jianwei Ren, Kaixu Han, Fengcai Huo. The atmospheric pollution emergency response system based on the rules

of the actuating and dynamic path. *Chemical Automatic Meter*. 2014(6).

[9] Jianhong Chen. The detection technology of heavy metals in atmospheric particulate matter $PM_{2.5}$. *Chemical Management*. 2014(5).

[10] Guihua Lei, Ruqiang Min. The evaluation of determination of sintering flue gas SO_2. *Heilongjiang Environmental Bulletin*. 2013(3).

[11] Jianling Fan, Zhengyi Hu, Jing Zhou, etc. The study of woodland atmospheric nitrogen deposition fluxes comparative. *China Environmental Science*. 2013(5).

[12] Yajie Ma, Mark Richards, Moustafa Ghanem, etc. Air Pollution Monitoring and Mining Based on Sensor Grid in London. *Sensors*. 2008(8).

[13] Rahman A., Smith D.V. and Timms G. A Novel Machine Learning Approach Toward Quality Assessment of Sensor Data. *IEEE Sensors Journal*. Vol. 14, No. 4, 2014.

[14] Peng Xu, Haiyan Xie, Yuanyuan Sun.The research of Urumqi Atmospheric Pollution and Control effectiveness. *Environmental Science Survey*. 2014(8).

[15] Qinqian Zhou, ShichunZhang, Weiwei Chen, et al. The changes characteristics and sources of Changchun SO_2, O_3 and NO_X in atmospheric. *Environmental Sciences Study*. 2014(7).

[16] Yiwen Hong. Why do people feel the air pollution detection lopsided. *Southern People Weekly*. 2011(7).

[17] Xiaocun Xiao, Xuena Wang. The prevention and pollution of atmospheric nitrides. *Coal Technology*. 2005(8).

[18] Xuesong Wang, Jinlong Li, Yuanhang Zhang, etc. The analysis of sources of ozone pollution in Beijing. *Chinese Science*. 2009 (2):548–559.

Biomedical Engineering and Environmental Engineering – Chan (Ed.)
© 2015 Taylor & Francis Group, London, ISBN: 978-1-138-02805-0

Degradation of anode material in ozone generation via PEM water electrolysis

G.B. Jung, C.W. Chen, C.C. Yeh, C.C. Ma & C.W. Hsieh
Department of Mechanical Engineering, Yuan Ze University Taoyuan, Taiwan

ABSTRACT: Membrane electrode assemblies (MEAs) using commercial PbO_2 powder as the anode catalyst to generate ozone via water electrolysis were investigated. We found that commercial MEAs evinced the typical degradation phenomenon after current interruption during operation, where the performance degraded but gradually recovered after the resumption of current. Then, homemade MEAs using PbO_2 powder and additives were developed, which ameliorated the degradation phenomenon. SEM and XRD analyses were used to compare the anode structure of the homemade and commercial MEAs after short- and long-term operation post-resumption of current after an interruption.

1 INTRODUCTION

Using membrane electrode assemblies (MEA) to generate ozone via water electrolysis has been proposed for the degradation of increasingly problematic pollutants. MEAs currently used to produce high concentrations of ozone have a Pt/Nafion/PbO_2 (cathode/electrolyte/anode) system. To enhance the efficiency of ozone production and the durability, researchers have added various substances to the PbO_2 anode.

2 EXPERIMENTAL METHODS AND PROCESS

2.1 Structure of PEM water electrolysis cell

Figure 1 shows the structure of the PEM water electrolysis cell, which included a proton exchange membrane, gasket, flow-field plate, and current collector. The active area was 9 cm^2.

2.2 Experimental equipment

2.2.1 PEM water electrolysis system
The hardware and testing system are shown in Fig. 2. The water tank was connected to the anode and cathode.

2.2.2 Ozone concentration meter
The concentration of ozone generated was determined by measuring the oxidation-reduction potential of the ozone bubbled into water using an ORP-15 digital ozone concentration meter.

2.2.3 X-ray diffraction
Crystals are composed of atoms arranged with a certain periodicity, and the lattice planes are differentiated by their Miller indices. The Miller indices h, k, l of a

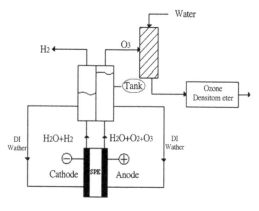

Figure 2. The experimental system configuration for measuring ozone generation.

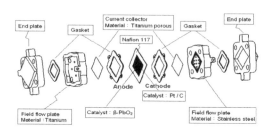

Figure 1. The structure of the water electrolysis cell.

lattice plane are related to the intercept of the lattice plane with the three crystallographic axes and these indices also designate the direction <*hkl*> orthogonal to the lattice plane (*hkl*). When a crystal powder or thin film sample is irradiated by monochromatic X-rays there will be diffraction spots or rings when the incident angle satisfies Bragg's law:

$$n\lambda = 2d\sin\theta, \qquad (1)$$

where d is the vertical distance between two adjacent lattice planes, λ is the wavelength of the X-rays, θ is the incident angle, and n is the number or class of diffraction. The diffraction pattern can be used to analyze the atomic structure, which can help determine composition

2.2.4 SEM analysis

In scanning electron microscopy (SEM), an electron beam is focused on a sample surface, and the surface is scanned point by point. When the sample is a bulk material or particles, the electron beam may generate secondary electrons, backscattered electrons or electron donating which is the most important secondary electron imaging signal. The electron beam energy of 5–35 keV is emitted by an electron gun which is then focused after passing through the second condenser lens and the objective lens to form a fine electron beam with a certain energy, strength and width This beam scans the sample surface in a grid pattern with a certain time and space sequence The focused electron beam interacts with the sample, resulting in secondary electron emission (and other emissions). The secondary electron emission amount changes with the sample surface topography, and it is converted into an electrical signal by the detector The signal is amplified and displayed on a CRT or computer screen. The resulting secondary electron image reflects the surface topography of the sample.

2.3 MEA preparation

The PbO_2/Nafion solutions were prepared by mixing commercially available PbO_2 powder with Nafion solution. The commercially available Nafion 117 membranes (DuPont) were combined with the as-prepared anode and a gas diffusion cathode (5 g m^{-2} Pt catalyst loaded on a carbon structure) to fabricate the MEA.

2.4 Experiment process

The system used to evaluate the MEA/electrolyzer performance was composed of a pair of porous titanium plates and two stainless steel end plates holding the titanium plates in place. The assembled MEA was protected in between two pieces of rubber gasket, prior to fixing it in the center of the test hardware (Fig. 1).

Table 1. Specifications of homemade and commercial MEA.

MEA	Thickness mm	Active area mm^2	Anode loading mg/cm^2
Commercial	0.85	30 × 30	127
Homemade	0.52	30 × 30	16

Table 2. Proportion of catalyst.

Sample	PbO$_2$	Nafion	Acid	PTFE
A	9	1	5	1
B	9	1	5	3
C	9	1	5	5
D	9	1	10	2
E	9	1	15	3

Table 3. Current interruption test conditions.

Sample	Voltage	Active time V	Current interruption time h	Current recovery time h
Test 1	4.5	12	1 min/	1
Test 2	4.5	12	30 min/1 h	1
Test 3	4.5	12	1 min-6 cycle/ 10 min/1 h 2.5 V-1 min/ 2 V-1 min/ 1.8 V-1 min	1
Test 4	4.5	3	3 h	3

3 RESULTS AND DISCUSSION

3.1 Properties of homemade and commercial MEA

The specifications of the homemade and commercial MEAs are shown in Table 1. The anode loading in the commercial MEA was four times the amount in the homemade MEA, so the thicknesses were significantly different. The catalyst proportions and test conditions are shown in Tables 2 and 3, respectively.

3.2 Performance of homemade and commercial MEA

Under an operating voltage of 4.5 V, the performance of the homemade and commercial MEA are compared in Fig. 3 The commercial MEA shows a higher current than the homemade MEA; however, the concentration of ozone generated was the same, as shown in Fig. 4. One possible reason is the generated ozone failed to fully mix with the water in the gas–liquid mixing tower, and the high current produced more heat, so the ozone was broken down due to high temperature.

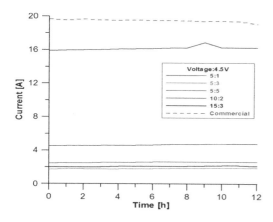

Figure 3. The current output of the homemade and commercial MEAs.

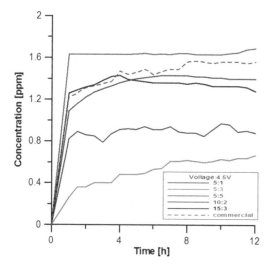

Figure 4. Concentration of ozone generated with the homemade and commercial MEAs.

3.3 Current interruption/restart test

When power is interrupted during ozone generation via an MEA the performance drops significantly after the resumption of power [3] Power interruption comparisons of the homemade and commercial MEAs are shown in Figs. 5 and 6. When the commercial MEA restarted, current slightly decreased after power interruptions of 1 min and 10 min, but after a power interruption of 1 h there was a significant decrease in current In comparison, the homemade MEA showed only slight decreases in current after power interruptions of 1 min 10 min and 1 h The difference between the homemade and commercial MEA were particularly significant after the 1 h power interruption. All the homemade MEAs showed good resistance to degradation after power interruption. Beaufils et al have noted that any interference will change the battery performance in an electrochemical reaction. After a

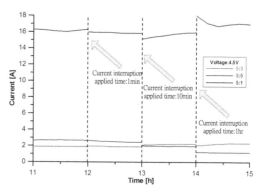

Figure 5. The performance of the homemade MEA restarted after power interruptions.

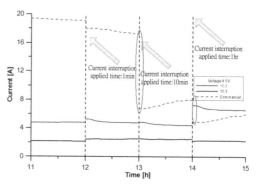

Figure 6. Performance comparison of the homemade and commercial MEAs restarted after power interruptions.

power interruption, the reaction needs time to restabilize; hence, the performance will decrease. The effect of a power interruption must be related to the chemical or electrochemical reaction on the surface of the electrode and membrane. Ozone concentrations stayed at 1–2 ppm for both the commercial and homemade MEAs One reason for this result is that ozone for an ambient temperature of around 25°C and the commercial MEA produced more current, heating the system, which caused thermal decomposition of the ozone.

3.4 One-min current interruption

We also operated with several 1-min power interruptions to observe the effect of repetitive short-term power interruptions. Figs. 7 and 8 show that the commercial MEA suffered no significant drop in effectiveness after six 1 min power interruptions, alternating with 1 h periods of operation. When the current was interrupted for 10 min in the 7th cycle, performance significantly decreased. When the power interruption was 1 h in the next cycle, the performance was even more attenuated. One can see that the performance attenuation was independent of the total operating time and only depended on the interruption time. One possible reason is that the commercial MEA has similar colloidal substances that can protect the catalyst of

257

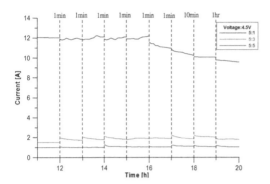

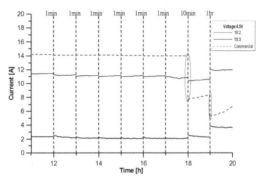

Figure 7. The performance analysis of homemade MEAs for a series of short-term power interruptions followed by longer interruptions.

Figure 8. The performance comparison of the homemade and commercial MEAs after a series of short-term power interruptions.

PbO_2 for short power interruptions. So we used different proportions of catalyst in our homemade MEA to try to improve performance after current interruption. The current of the homemade MEA gradually increased after each power interruption This is because the structure was more porous allowing easy access to the catalyst layer, which encourages the electrochemical reaction and increases the current. The new homemade of currents are rising slightly after each time of power down to investigate this particular case, especially for an extended outage test, while ozone concentrations are maintained at 1–2 ppm. According to Takahiro Ohba and Yann Beaufils et al. cannot return to the initial performance after current interruption Our previous experimental results show that current performance of some of the homemade MEAs slightly improve after current interruption. After testing long breaks in power and return, power supply and power outages in 3 h for observed changes of the performance in a cycle. The result was the same as that before the power is interrupted several times with a short experiment, performance of the homemade MEA improved significantly after a power outage and then restart. Since the parameters of the experiment were fixed, the cause of the improvement was that element B was lost from the catalyst, leaving more space for the electrochemical reaction. The reason for gradual decline in performance over time was that the MEA structure was eventually destroyed by electrolysis.

3.5 XRD analysis

The XRD analyses of scraped PbO_2 from the anode after operation are shown in Figs. 9 and 10. The intensity of lattice planes (110) and (101) grew significantly after eight straight hours of electrolysis for the homemade MEA, while power interruption and restarts decreased the lattice plane intensity. According to literature, lattice planes will develop toward a state for ozone formation during the electrolysis of water. However lattice stability will tend toward another new direction due to the electrochemical or

chemical reactions that change after power interruptions, which leads to a reduction in ozone generation of the lattice planes. When the current is restarted, the lattice will slowly increase toward the direction of the ozone; hence, performance will respond slowly. The intensity of the lattice planes grew substantially in the homemade MEA after operation. The signal intensity of the first peak (110) grew relative to the second peak (101) after electrolysis by about a factor of 2. The intensity from lattice planes (110) and (101) were basically the same before the experiment. This suggests that power interruption affected lattice plane (110), but not lattice plane (101) or (200). The performance of the commercial MEA declined slowly, as seen in Figs. 9 and 10 A possible explanation is that commercial MEA contains some unknown additives and the catalyst amount was much higher than the homemade MEA. Changes in intensity are not as obvious for the commercial MEA, but both the (110) and (101) lattice plane peaks intensify after operation. After restarting after the power interruption, the intensity ratio was approximately 1:1 for the (110) and (101) peaks. These results suggest that ozone generation via PEM water electrolysis is related to the intensity of lattice plane (110).

3.6 SEM analysis

Images of the commercial and homemade anode surfaces before and after electrolysis are shown in Figs. 11 and 12, respectively. Fig. 11 shows white streaks around the commercial membrane after water electrolysis. When electrolysis later, the surface particle of homemade will be attached to the current collector slightly due to the structure of homemade is relatively loose and close contact with current collector.

SEM images of the homemade MEAs made with elements A and B in ratios of 5:3 and 15:3 are shown in Figs. 13 and 14. The PbO_2 looked like stacked particles before electrolysis, and presented fine crushing block and dense gatherings after electrolysis relatively, which might make it more difficult for the electrolyte

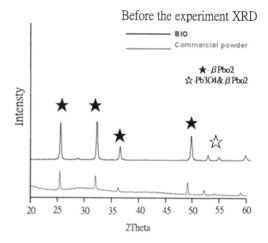

Figure 9. The XRD analysis of the homemade and commercial MEAs before the experiment.

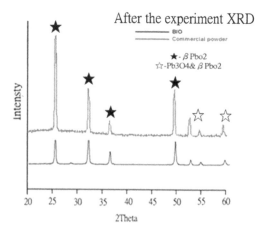

Figure 10. The XRD analysis of the homemade and commercial MEAs after the experiment.

Figure 11. The commercial anode surface morphology before and after electrolysis.

to penetrate the interior and react; PbO_2 will produce $(-OH)$ during ozone generation via PEM water electrolysis, and the reaction is as follows [9]:

$$PbO_2 + H_2O \rightarrow PbO_2 \ (\cdot OH) \ + H^+ + e^- \qquad (2)$$

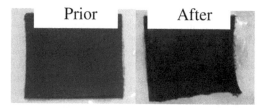

Figure 12. The homemade anode surface morphology before and after electrolysis.

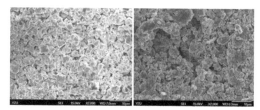

Figure 13. SEM images of the MEA made with elements A and B in a ratio of 5:3. The left picture is the anode before the experiment, and the right is after.

Figure 14. SEM images of the MEA made with elements A and B in a ratio of 15:3. The left picture is the anode before the experiment, and the right is after.

The PbO_2 is easily absorbed (–OH) and reacts to generate H_2O_2 [10] [11] [18]. The power is strong that MEA contact with current collector around and will cause the higher voltage locally, the reaction of H_2O_2 and thus more likely to cause corrosion on PbO_2, and loss of function and easier to produce H_2O_2 causing further corrosion of PbO_2 and loss of function finally.

4 CONCLUSION

In this study, an anode with proportions of PbO_2/Nafion of 9:1 and catalyst loading of 30 mg/cm^2 showed the best performance for ozone generation. When the ratio of elements A and B was 5:3, it had a better ability of resist current degradation. Although the commercial MEA had a higher current value in the beginning, it also produced more heat, causing a temperature increase, and the ozone decomposes with heat. The homemade MEA overcame this problem with a different catalyst ratio, and the ozone concentration was more stable. The structural difference between the homemade and commercial MEAs was that the porosity of the commercial MEA was larger. Large porosity lets water diffuse in the catalyst layer, and this increases the reaction efficiency. The homemade MEA had a better ability to resist decay than commercial MEA in current interruption tests. Although the commercial MEA showed good recovery after short-term current interruptions, performance could not recover to the original levels after long-term current interruption. The homemade MEA performed better than the commercial MEA for both short- and long-term current interruptions, perhaps because the loss of element B in the catalyst layer led to more holes that let water diffuse into the catalyst layer easily, which raised performance.

REFERENCES

Abaci S., Pekmez K., Hokelek T. & Yildiz A. 2000. Investigation of some parameters influencing electrocrystallisation of PbO_2. *Journal of Power sources* 88: 232–236.

Aoki M., Uchida H. & Watanabe M. 2006. Decomposition mechanism of perfluorosulfonic acid electrolyte in polymer electrolyte fuel cells. *Electrochemistry Communications* 8: 1509–1513.

Arihara K., Terashima C. & Fujishima A. 2007. Electrochemical Production of High-Concentration Ozone-Water Using Perforated Diamond Electrodes. *Journal of Electrochemical Society* 15: 71–75.

Awad M.I., Sata S., Kaneda K., Ikematsu M., Okajima T. & Ohsaka T. 2006. Ozone electrogeneration at a high current efficiency using a tantalum oxide-platinum composite electrode. *Electrochemistry Communications* 8: 1263–1269.

Beaufils Y., Cornninellis C. & Bowen P. 1999. Preparation and Characterisation of Ti/Iro₂/Pb Electrodes for Ozone Prouction in a Spe Electrochemical Cell. *Institution of Chemical Engineers Symposium Series* 145: 191–200.

Bender G., Zawodzinski T.A. & Saab A.P. 2003. Fabrication of high precision PEFC membrane electrode assemblies. *Journal of Power Sources* 124: 114–117.

Cao M., Hu C., Peng G., Qi Y. & Wang E. 2003. Selected-Control Synthesis of PbO_2 and Pb_3O_4 Single-Crystalline Nanorods. *Journal of The American Chemical Society* 125: 4982–4983.

Da Silva L.M., Franco D.V., Forti J.C., Jardim W.F. & Boodts J.F.C. 2006. Characterisation of a laboratory electrochemical ozonation system and its application in advanced oxidation processes. *J. Applied Electrochemistry* 36: 523–530.

Da Silva L.M., De Faria L.A. & Boodts J.F.C. 2003. Electrochemical ozone production: influence of the supporting ctrolyte on kinetics and current efficiency. *Electrochimica Acta* 48: 699–709.

Fitas R., Zerroual L., Chelali N. & Djellouli B. 1996. Heat treatment of α- and β-battery lead dioxide and its relationship to capacity loss. *Journal of Power sources* 58: 225–229.

Kraft A., Stadelmann M., Wunsche M. & Blaschke M. 2006. Electrochemical ozone production using diamond anodes and a solid polymer electrolyte. *Electrochemistry Communications* 8: 883–886.

Morales J., PetKova G., Gruz M. & Caballero A. 2006. Synthesis and characterization of lead dioxide active material for lead-acid batteries. *Journal of Power sources* 158, 831–836.

Ohba T., Kusunoki H., Sunakawa D., Arakis T. & Onda K. 2005. Improvement Characteristics of Ozone Water Production with Multilayer Electrodes and Operating Conditions in a Polymer Electrolyte Water Electrolysis Cell. *Journal of Electrochemical Society* 152 (10): 177–183.

Sires I., Brillas E., Cerisola G. & Panizza M. 2008. Comparative depollution of mecoprop aqueous solutions by electrochemical incineration BDD and PbO_2 as high oxidation power anodes. *Journal of Electroanalytical Chemistry* 613: 151–159.

Stucki S., Theis G., Kotz R., Devantay H. & Christen H.J. 1985. In Situ production of Ozone in Water Using a Membrel Electrolyzer. *The Electrochemical Society* 132: 367–371.

Tong S.P., Ma C.A. & H. Feng. 2008. A novel PbO_2 electrode preparation and its application in organic degradation. *Electrochemica Acta* 53: 3002–3006.

Wen T.C. & Chang C.C. 1993. The Structural Changes of PbO_2 Anodes during Ozone Evolution. *Journal of Electrochemical Society* 140: 2764–2770.

Zhou D. & Gao L. 2007. Effect of electrochemical preparation methods on structure and properties of PbO_2 anodic layer. *Electrochimica Acta* 53: 2060–2064.

Biomedical Engineering and Environmental Engineering – Chan (Ed.)
© 2015 Taylor & Francis Group, London, ISBN: 978-1-138-02805-0

Research on a highway maintenance low-carbon framework and key technology

Jing Zhang
Henan Highway Tolling Centre, Henan Zhengzhou, China

ABSTRACT: For reducing greenhouse gas emissions and to address climate warming, we build a highway low-carbon framework and key technologies. Highway maintenance, including low-carbon, are the way to increase energy conservation and carbon sequestration. This is done mainly to reduce greenhouse gas emissions, which is to absorb greenhouse gases, mainly by car exhausts. Key technologies include energy conservation, resource recycling, and green includes three dimensions, the key technologies involved: preventive maintenance techniques, coordinating control technology, rubber asphalt materials, pavement recycling technology, composting and pest ecological control technology. Achieving the goal of low-carbon and highway maintenance provides decision makers with a range of relevant countermeasures.

Keywords: Highway maintenance; low-carbon; preventive maintenance; green conservation

1 INTRODUCTION

The IPCC's (Intergovernmental Panel on Climate Change) Fifth Assessment Report has further enhanced the human activity on global warming credibility, for it to reach 95%[1]. The main factors that contribute to climate warming, greenhouse gas emissions are the third largest transportation sector greenhouse gas emissions, accounting for 13.1%, 25.9%, second only to the energy, industry 19.4%. As an important branch of the highway traffic, since the end of the last century, that is for over 80 years, rapid development has occurred in our country.

As of the end of 2013, the total mileage of high-speed has reached 104,500 km, while the total planned mileage of 165,000 kilometres. With such a large road network, low carbon highway research has important practical significance.

Xiao-Li Chen[2], Xin-Jin[3] studied the key technical framework of a low carbon highway, including: planning technology, design technology, construction technology, operations technology, and energy management technology in five sections, taking the preventive maintenance technology as an important element of operational techniques. Bing fei Chen[4] has researched the highway system from four aspects: service areas, tunnels, highways maintenance study of low carbon operations: carbon conservation modes include: maintenance management, road maintenance, green conservation, supporting conservation facilities. Xin Han[5], Wen-Yan Qi[6] analysed and discussed the freeway low carbon conservation. Mainly based on ecological green conservation, four new technologies (new materials, new equipment, new technology, new technologies) of carbon maintenance, preventive maintenance the full life cycle, and based on four aspects of fine saves jobs, etc. are described. Highway construction work for more than a low-carbon research laid the foundation and framework. This paper focuses on highway maintenance low-carbon research. The main contents include a highway maintenance system for low-carbon construction. We then proceed to present our key technical analysis and conclude by recommending measures.

2 A LOW-CARBON CONSTRUCTION HIGHWAY MAINTENANCE SYSTEM

2.1 Highway maintenance carbon connotations

After the highway was opened to traffic, with the passage of time for operators, traffic growth, an increase in the frequency of highway facilities and ancillary facilities, will cause varying degrees of damage. Highway Maintenance refers to the timely detection and repair of these damages in order to maintain normal highway use and fully functioning ancillary facilities, and conduct regular, periodic, preventive maintenance and repair work to maintain highways and ancillary facilities in good condition. This, in turn, will ensure good road conditions and the use of good quality materials. Highway maintenance is an important part of highway operations.

Highway maintenance is a complex system of engineering. From the conservation point of view objects, it includes road (road embankment, slope), bridge, tunnel, green, disaster prevention (snow, sand), and traffic

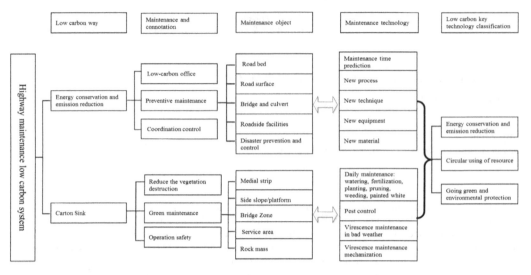

Figure 1. Highway maintenance connotation and low-carbon system.

engineering, control, communications, lighting, signage, fence, along facilities, etc. From the technical point of view, conservation it includes the conservation of timing decisions, new materials, new processes, new technology and new equipment with integrated use. Content from the conservation point of view, includes routine maintenance to keep the roads and facilities intact carried out for reinforcement to improve roads and facilities operations carried out special projects for the restoration and improvement of the original design, and carry out the repair works, in response to disasters and repair and emergency response conducted harsh climatic conditions, as well as along the landscape, green landscaping and environmental protection.

2.2 Highway maintenance low-carbon systems

2.2.1 A low-carbon pathway highway maintenance
Given the complexity of highway maintenance, this is a long-term endeavour, involving a lot of resources, capital, energy, human consumption, knowledge of highway maintenance, how to implement low-carbon, energy saving ideas, as well as new technologies, equipment, technology, materials, and increased use of waste resources. Reducing the impact on the environment, as well as strengthening the green while increasing the carbon sink, absorbing vehicle emissions, is an important and necessary part of ways to achieve a low-carbon transport.

2.2.2 Highway maintenance classification of low-carbon technologies
Highway maintenance of low-carbon technologies can be classified as energy conservation, resource recycling, and environmental protection categories. Energy conservation, highway maintenance and construction mainly refer to the process of reducing carbon dioxide emissions, reducing construction and

high energy consumption, reducing waste and other waste emissions, as well as coordinated control mechanisms to increase vehicle capacity, smooth traffic conditions conducted in order to reduce vehicle during the energy consumption and exhaust emissions. They also include reducing the highway maintenance management process such as paper waste, energy consumption, and other green low-carbon office and so on. Resource recycling is about the use of other waste resources in highway maintenance processes, such as rubber asphalt, road material regeneration of waste, and green waste ecological utilisation. Environmental protection includes the reduction of vegetation damage, strengthens green conservation, and security operations. Greening beautifies the landscape. At the same time it can preserve a lot of wood to absorb large amounts of automobile emissions. Highway maintenance connotations and low-carbon systems are shown in Figure 1.

3 TECHNICAL ANALYSIS OF KEY LOW-CARBON HIGHWAY MAINTENANCE

According to the conservation of the content and low carbon highway system, highway maintenance key low-carbon technologies, including energy conservation, resource recycling, green three categories, the following key technologies involved in each type of analysis.

3.1 Energy conservation

3.1.1 Preventive Maintenance
Preventive maintenance of road yet, which is before the damage occurred or producing mild disease has not been broken before, take proactive, predictable and effective means of conservation measures, and the

262

highway diseases and diseases caused by factors found earlier, former Treatment to prevent the development of an organizational disease management behaviour. It mainly aims to improve the quality of road maintenance, and actively explores the development of highway maintenance in line with technical standards and specifications under the local geography and climate conditions, and improves the durability of the project, reducing road maintenance cycle costs and improving the efficiency of road use. This paper is conscientiously summing up the successful experience of highway maintenance engineering model demonstration projects, and demonstrating an eco-friendly way of road maintenance and actively carrying out new environmental engineering design, construction, and management. The following main areas are included:

First, the new technologies are the vibration layer type method, the grouting compound, the glass fibre packaging and testing, and the synchronous pavement surface.

Second, the new processes are thermal regeneration technology, micro-surfacing technology, and a flexible base.

Third, the new devices are radar vehicles, rubber synchronous chip sealer machines, self-propelled pavement grooving potting machines, bridge inspection vehicles, road marking trucks, and hand held sweepers.

Fourth, the new materials are modified bitumen, CFRP, asphalt pavement protection agents, and the new de-icing salt.

3.1.2 Coordinated control

First, daily cleaning and maintenance work should be carried out outside of rush hour, especially before the start of the holidays maintenance and cleaning should make use of mechanized tools, integrated enhanced operating efficiency so that it does not affect the normal traffic.

Second, on the road, along small engineering maintenance facilities etc., try to avoid rush hour and put in place safety warnings and precautions., Also use advance notification prompt cards in order to remind passing vehicles to avoid or bypass.

Third, when dealing with roadbeds, bridges and culverts, when planning road maintenance and/or other large projects, do make alternate arrangements for the road, alongside the safety warnings and precautions Inform passengers on appropriate websites accordingly.

3.2 Resource recycling

3.2.1 Rubber asphalt

Rubber asphalt is first processed from scrap tire rubber powder of original quality, and then a certain thickness grading scale combination, while adding a variety of high polymer modifiers, and under high temperature conditions sufficient mixing of the above (180°), the full swell with asphalt to form a modified asphalt

cements. Rubber asphalt with high temperature stability, low temperature flexibility, ageing resistance, fatigue resistance, resistance to water damage and other properties, is an ideal environmentally friendly pavement material, mainly used in road construction in the stress-absorbing layer and the surface layer. Rubber asphalt pavements, while improving highway pavements, consume large amounts of scrap tires and are therefore rather wasteful in using resources for recycling.

3.2.2 Waste recycling road materials

Roads uses waste material recycling technology renovation and expansion of existing roads, and the use of old asphalt pavement materials and aggregates, new asphalt mixing appropriate ratio to become a good way to use recycled asphalt mixture properties, can be used for each grade highway big Here in the repair and the grass-roots level and the renovation and expansion project, to achieve the recycling of resources.

3.2.3 Eco-use and handling green waste

Reasonable disposal and utilisation continued is to increase litter and other green waste, from the perspective of ecological resources freeway green waste, according to the ecological functions of litter, the use of ecological methods take advantage of green waste, such as: green waste composting technology, with better application performance and market prospects. Based on this, we believe that composting is an effective way to deal with green waste which can provide a practical and workable model for the highway greening industry to develop the recycling economy[7].

3.3 Environment protection

3.3.1 Low carbon green office

First, water saving. Non-life production water run, run, drip, leak, prevent long-running water.

Second, reduce the production of official vehicles and vehicles consuming. Establish a sound management system using official vehicles, official requirements and overall maintenance costs of oil production car fixed lump sum. When the vehicle was equipped with updated facilities, choose low-emission, economical environmentally friendly cars. Promote activities in the collective public official car ride together. Winter snow removal equipment to support reasonable use is to reduce fuel consumption and wear and tear.

Third, cut down the cost of running an office. Office phones and office supplies are used to implement a fixed contract system. A network version of otherwise printed newspapers and magazines should be subscribed to in principle.

Fourth, accelerate the process of the paperless office. Minimise paper documents, presentations, and other information issued. We should make full use of the website portal for releasing information which the company publishes for the public. Between units within the department people should try to transfer files using a mailbox-based network, under the premise of confidentiality, to take full advantage of the network to transfer files and publish notifications. To accelerate the "cooperative office platform", there should be expanded use of applications in the shortest possible time, in order to achieve inter flow within each unit.

Fifth, control conferences and official hospitality expenses incurred. Meet strict examination and approval system control the number and size of compressed time, under normal circumstances, shall be in the form of video. People do need to focus on the organisation's meetings, events and should adhere to the principle of efficiency, thrift council and strictly controlling official reception standards. Strictly control the number of meals to accompany advocate eating meals.

Sixth, office electricity saving. Lights in offices, meeting rooms and other places should be switched off when not in use. Office appliances when not in use should be promptly shut down, turning off the power switch after work. A reasonable set of air temperature, not the elevator advocate holidays. Some public places could be changed to start using energy-saving lamps. All lighting should be gradually replaced with LED lights.

Seventh, cleaning, inspection of vehicles using green energy, such as with electric patrol vehicles.

3.3.2 Increase the green area

Strengthen the greening isolation belt, green slope, highway shelter green belts, three-dimensional traffic area, green highway toll station landscaping, parking areas, and service areas greenery.

3.3.3 To reduce damage to vegetation

Green conservation includes timely watering, pruning, pest control, reducing vegetation mortality; using new materials, the snow piled reasonable, reduce damage to vegetation; weeds, leaves disposal, combustion cannot be used the way.

3.3.4 Green conservation

First, the main biological pest control to affect the ecological environment is (A) physical control, such as: cleaning up litter, scraping bark, soil plowing and other killing methods; light trapping, trapping food, booby traps and other trapping potential of the method; trunks whitewashed, glue and other isolation methods. (B) biological control, such as: microbial agents, natural enemies, beneficial birds and so on. (C) Integrated Control: worm harmless, natural control; biological diversity, mutual restraint; human intervention, mainly to biological factors to affect the ecological environment, free from pest damage.

Second, try to use mechanised maintenance operations, improve the efficiency of maintenance operations, in order to reduce the impact of road traffic. These are: sprinklers, a drug vehicle, brush saws, hedge trimmers, chain saws, lawn mowers etc.

4 HIGHWAY MAINTENANCE CARBON COUNTERMEASURES PROPOSED MEASURES

Since China is still in a massive highway construction period, people have started to think about building conservation. Therefore, low-carbon highway maintenance has to take place. Therefore, it is necessary to introduce relevant countermeasures to accelerate the implementation of conservation and implement low-carbon concepts.

4.1 Implement low carbon concepts, construction laws and regulations

For the individual who is aware of the current situation, the greenhouse effect has brought great disaster to human beings, especially less per ca pita resources, energy relative shortage. Therefore, to enhance a low-carbon office, low carbon travel, the concept of low-carbon life needs to be spread and further popularised. On national, provincial and other highway construction, operation and managers, should strengthen the construction of highway maintenance low carbon relevant laws and regulations. These are: technical standardisation, management processes, scientific decision-making, strengthening financial support, tax incentives and goals of building low-carbon highway maintenance.

4.2 R&D and introduction of new technologies

New technologies, processes, materials, equipment, highway maintenance is to achieve support for a low-carbon foundation. This should be based around geography, meteorology, penology and other conditions, according to local conditions related technology research and development, introduction, experiment and do a pilot project, to be technical, the promotion of a comprehensive process maturity. Strengthen collaboration management units, construction units, research units, in order to establish long-term mechanisms which will enable innovation to become the norm. This includes: preventive maintenance, precuring involving timing, optimaszing capital allocation, conservation technology selection and so on. Interim related to construction management, construction supervision, etc., the latter related to the effect evaluation, traffic detection. These jobs tend to have multiple units and departments which need comprehensive and coordinated action; therefore cooperation needs to be strengthened.

4.3 Sound evaluation mechanism

Highway maintenance which is meant to achieve the goal of low-carbon and landing, must establish a sound evaluation mechanism. First, we should establish working groups and expert advisory committee leadership, responsible for the overall goal of decomposition, implementation and evaluation. Second, we envisage the establishment of who carries ultimate responsibility, where are the chains of accountability linked to carbon targets and how is work performance being measured. Again, the establishment of supervision and management system, such as: carbon accounting association and so on. Finally, streamline maintenance management system, the reform of the conservation of the market, the introduction of third-party rating agencies and various conservation assessment objectives and specific work projects combined need to be achieved.

5 CONCLUSIONS

This article focuses on constructing a highway maintenance system including a low-carbon path, meaning, object and technology from energy conservation, resource recycling, green three-dimensional analysis of key low-carbon technologies. Aspects of energy conservation include: comprehensive pavement maintenance, preventive maintenance techniques, increase of vehicle traffic coordination and control technology. Resource recycling aspects include: rubber asphalt technology, asphalt pavement recycling technology and green waste composting. Green areas include: low-carbon office, increasing green space, reduce damage to vegetation and green pest control techniques. This paper also introduces countermeasures to achieve low-carbon objectives for management and decision makers. With the concept of depth, implement sound, preferential policies, conservation and technological innovation, as well as evaluation mechanism laws and regulations, highway maintenance is bound to sell low-carbon solid pace in reducing emissions, conserve resources and increase the green, carbon sequestration and other aspects of the industry to make its due contribution to the highway, in response to climate warming slightly into the humble.

REFERENCES

Bing Fei Chen, Highway carbon operations research [D]. Chang'an University Master Thesis, 2011.5.

Jing Bai, Hong Yan Shen, Shi-kui Dong, ecological use of highways and green waste treatment [J]. Anhui Agricultural Sciences, 2010, 38 (34): 19488–19490.

Shao-Wu Wang, Yong Luo, Zong-Ci Zhao, et al. IPCC 5th assessment report came [J]. Advances in Climate Change, 2013, 9(6): 436–39.

Wen-Yan Qi. Discussion on the freeway carbon conservation [J]. Changsha Railway Institute (Social Science Edition), 2011, 12 (4): 212–213.

Xiao-Li Chen, Study of low-carbon technologies critical highway construction framework [J]. Highway and Transport, 2012, 12 (6): 139–142.

Xin Jin, Carbon frame analysis of key technologies in highway construction [J]. Communications Science and Technology, 2012, 5: 196.

Xin Han. Reflections on the highway carbon conservation [J]. Shanxi Architecture, 2011, 8 (23): 158–159.

Biomedical Engineering and Environmental Engineering – Chan (Ed.)
© 2015 Taylor & Francis Group, London, ISBN: 978-1-138-02805-0

Field experiment on nitrogen and phosphorus removal of SBR process of a WWTP

C.M. Li & H.P. Chen
School of Environmental Science and Engineering, Taiyuan University of Technology, Taiyuan, Shanxi, China

Z.H. Yang & H.Z. Zhao
Zhengyang Wastewater Treatment Plant, Shanxi International Electricity Group Limited Company, Jinzhong, Shanxi, China

ABSTRACT: A field experiment was carried out to provide substantial data for the upgrading of a SBR WWTP to improve the nitrogen and phosphorus removal effect. For the chemical method of the phosphorus removal, when the PAC dosage was greater than 120 mg/L, the effluent TP could achieve the discharge standard. The addition of the PAC decreased the COD_{Cr} removal efficiency but the effluent COD_{Cr} still meet the standard. The PAC had less effect on the other properties of active sludge. Among the several tested biological nitrogen removal processes, the effluent TN of filling-aeration-stirring (adding glucose)-aeration-settling-decanting-idling procedure and the step-feed procedure could achieve discharge standard. The effluent TN and TP of latter procedure could both achieve discharge standard yet without the supplement of glucose as carbon source.

1 INTRODUCTION

A traditional SBR WWTP has been designed to meet the class II standard of "Wastewater and Sludge Disposal Standard for Municipal Wastewater Treatment Plants" (CJ3025-93). Now it was required to meet the more stringent class A standard of "Discharge standard of pollutants for municipal wastewater treatment plant (GB18918-2002)" (hereinafter referred to as the class A standard). The effluent of the existing SBR process could meet all the requirements of the new standard except the TN and TP. Therefore, a field experiment was carried out to investigate the feasibility of upgrading project aimed at improving the nitrogen and phosphorous removal effect of existing SBR process.

2 SELECTION OF TREATMENT METHODS

2.1 Phosphorous removal

2.1.1 Selection of phosphorous removal method
Biological method for the phosphorous removal is affected by influent quality and operation therefore the treatment effects are unstable and the effluent is difficult to achieve the Class A standard. While chemical method for the phosphorous removal is simple and can guarantee the compliance with the Class A standard, it was increasingly used in WWTP. Therefore this experiment used chemical method for the phosphorous removal. The commonly used flocculating agents are aluminium potassium sulfate, aluminium sulfate, aluminium sulphate ammonia, crystalline aluminium chloride, aluminium sulfate, aluminum ammonium sulfate, polyaluminium chloride (PAC) etc. Among them, PAC is the best for the chemical phosphorous removal.

2.1.2 Experimental design
PAC is used as the flocculating agent in the field experiment. The procedure of the pilot scale SBR was set as: instant filling, 4 h aeration, 1 h stirring, 1 h settling, 1h decanting (decanting ratio 1/3) and sludge waste, 0–0.5 h idling. The DO at the end of the aeration was controlled to be 2–4.5 mg/L. The flocculating agent (10% solution) is added in the last 5 minutes of each aeration step. The dosage is calculated according to the influent volume of each batch. They were 60, 90, 120, 130, 140 and 150 mg/L respectively in the experiment and the operating time of each dosage was one week.

The TP, COD_{Cr}, NH_4^+-N, MLSS, MLVSS and SVI of the influent and effluent of each dosage were measured and compared with a parallel test without addition of the flocculating agent to study influence of the PAC on the treatment effect and the characteristic of the sludge.

2.2 Nitrogen removal

2.2.1 Selection of nitrogen removal method
The physicochemical method for the nitrogen removal consumes large quantity of chemical agents and produces huge amount of sludge therefore the biological method is preferred for the nitrogen removal in the treatment of municipal wastewater.

2.2.2 Experimental design

The four tested procedures of the pilot scale SBR were: instant filling-stirring-aeration-settling-decanting-idling; instant filling, pre-stirring, aeration, post-stirring, settling, decanting, idling; instant filling-preaeration-stirring (adding carbon source)-post aeration-settling-decanting-idling; step feed procedure. The TN, NH_4^+-N, COD_{Cr} and TP etc. of influent and effluent were measured to study the treatment effects of different procedures.

3 MATERIALS AND METHODS

3.1 Installations

Two sets of pilot scale SBRs.

(1) SBR: 700 mm × 350 mm × 650 mm, effective volume 120 L.
(2) Aeration system and flow control: one set of ACQ-008, P = 120 W, Q = 110 L/min air compressor; one 0–1600 L/h glass rotameter, porous diffuser.
(3) Stirring: one set of JJ-1/1A, P = 100 W mixer.
(4) Decanting: inlet of the drainage rubber pipe is hung under a floating board and is just below the water surface. Turn on/off the valve at the end of the drainage pipe outside bottom of the SBR to control the decanting.
(5) Sludge: The sludge was taken from one of the SBR of this WWTP and was capable of nitrification. After mixing with raw water in the reactor, the mixture was aerated for 1 d before started the full cycle. The MLSS was adjusted to around 3500 mg/L and the SRT was between 16–20 d.
(6) Raw water: the water was siphoned from the outlet of the grit chamber of this WWTP. The temperature of raw water was between 15–23°C, the pH was between 7.50–7.85 during the experiment.
(7) Carbon source: crude dextrose.
(8) Automatic control: automatic time switch to fulfill the on/off of the air compressor and mixer.

3.2 Analysis items and methods

Water quality was analyzed according to Monitoring and Analysis Method of Water & Wastewater (the fourth edition) (see Table 1).

4 RESULTS AND DISCUSSIONS

4.1 Phosphorus removal

The mixture was sampled after the completion of filling step and before the aeration step. After 0.5 h settling, the supernatant was analyzed. The effluent was sampled after the end of settling step and was directly analyzed. The removal effects (the average of one-week) of TP, COD_{Cr} and NH_4^+-N under different dosage of PAC was listed in Table 2, Table 3 and Table 4.

The removal rate of TP increased along with the increasing of PAC dosage. When the PAC dosage increased to 120 mg/L, the TP could meet the Class A Standard with an average value of 0.48 mg/L and removal rate of 92%. Then the removal rate could increase no more along with the further increasing of PAC dosage. But without the adding of PAC the average removal rate of TP was only 46%–49% as demonstrated in the parallel experiment (data not

Table 1. Methods of water quality analysis.

Item	Unit	Method
COD_{Cr}	mg/L	potassium dichromate
NH_4^+-N	mg/L	nesster's reagent spectrophotometry
NO_3-N	mg/L	thymol spectrophotometry
TN	mg/L	potassium persulfate-ultraviolet spectrophotometry
TP	mg/L	molybdate salt spectrophotometry
DO	mg/L	dissolved oxygen meter (JPB-607)
MLSS	mg/L	gravimetric method
MLVSS	mg/L	gravimetric method
SV_{30min}	%	volume method

Table 2. Removal results of TP.

Dosage of PAC	Mixture mg/L	Effluent mg/L	Removal rate %
60	5.28	1.18	78.00
90	5.01	1.06	79.00
120	5.73	0.48	92.00
130	5.54	0.48	91.00
140	5.39	0.47	91.00
150	5.19	0.47	91.00

Table 3. Removal results of COD_{Cr}.

Dosage of PAC	Mixture mg/L	Effluent mg/L	Removal rate %
60	125.47	28.34	77.41
90	117.93	24.69	79.00
120	135.54	26.07	80.77
130	116.82	25.13	78.49
140	117.22	23.44	80.00
150	106.64	23.17	78.27

Table 4. Removal results of NH_4^+-N.

Dosage of PAC	Mixture mg/L	Effluent mg/L	Removal rate %
60	17.69	0.63	96.45
90	16.21	0.65	95.99
120	12.47	0.32	97.43
130	14.65	0.43	97.06
140	12.34	0.42	96.60
150	14.76	0.47	96.82

listed). Therefore the addition of PAC improved the TP removal rate greatly.

The average effluent COD_{Cr} was between 23.17 mg/L–28.34 mg/L in the range of the adopted PAC dosage in this experiment, which all meet the Class A standard. The removal rate was between 77.41%–80.77% which did not change apparently with the variation of the PAC dosage. The average COD_{Cr} removal rate was 93.25%–95.72% without PAC as demonstrated in the parallel experiment. Therefore the addition of PAC decreased the COD_{Cr} removal rate. Since the PAC was added in the end of the aeration in each cycle, it did not directly influence the organic degradation process during aeration period of this cycle. Theoretically the COD_{Cr} removal rate should be improved by the flocculation of PAC. So it can be considered that the addition of the PAC influenced the characteristic of activated sludge and consequently decreased the COD_{Cr} removal rate at the aeration step of next cycle. Even though the effluent COD_{Cr} could still meet the Class A standard in the range of the tested PAC dosage.

The average effluent NH_3-N was between 0.32 mg/L–0.63 mg/L in the range of the tested PAC dosage, which all meet the Class A standard. The removal rate was between 96.45%–97.43%. For the parallel experiment without the addition of PAC, the average NH_3-N removal rate was 94.13%–97.46%. There was no apparent difference. It suggested that the addition of PAC affected the organic degradation ability of the heterotrophic bacteria but did not affect the nitrification ability of autotrophic nitrobacteria. The influence of PAC on the denitrification and TN was not studied for the unsatisfactory denitrification effect during the experiment.

After the addition of PAC, the floc size was bigger than before and the color was changed from brown to dark brown. There was no other significant sensory property variation. The MLVSS/MLSS was 0.09–0.13 when added with PAC and was 0.15–0.16 in the parallel experiment. The decrease of MLVSS/MLSS proved that PAC reduced the active parts of the activated sludge.

The results were already put into practice in this WWTP after a further full scale experiment. PAC was dissolved and mixed and added by a dosing pump. PAC was added in the last 0.5 h of aeration. The dosage was 120 mg/L.

The chemical method for the phosphorous removal can ensure the effluent TP to achieve the Class A standard.

4.2 Nitrogen removal

The influent COD_{Cr} was between 312–420 mg/L, NH_4^+-N 42.2–57.94 mg/L and TN 58–70 mg/L in the experiment.

4.2.1 Filling-stirring-aeration-settling-decanting-idling (procedure 1)

This procedure was set as: instant filling, 1.5 h stirring, 4 h aeration, 1 h settling, 1 h decanting (decanting ratio

Table 5. Effluent quality of procedure 1.

Item	NO_3^--N mg/L	TN mg/L	NH_4^+-N mg/L	COD_{Cr} mg/L
Data	18–20	20–22	<1.2	<30

Table 6. Effluent quality of procedure 2.

Item	NO_3^--N mg/L	TN mg/L	NH_4^+-N mg/L	COD_{Cr} mg/L
Data	18–20	20–22	<0.8	<50

1/3) and sludge waste, 0–0.5 h idling. The results of 30 d stable running were listed in Table 5.

The results (not listed in Table 5) showed that the NO_3^--N was already below 2 mg/L in the end of the stirring. After the beginning of aeration, the NH_4^+-N in the influent was continuously transformed into NO_3^--N and the NO_3^--N increased to 18–20 mg/L in the end of aeration. The effluent TN was between 18–22 mg/L and could not meet the Class A standard.

4.2.2 Filling-pre stirring-aeration-post stirring-settling-decanting-idling (procedure 2)

In procedure 1, the NH_4^+-N was nitrified to NO_3^--N after aeration but without denitrification it could not be removed from the mixture. Therefore the TN exceeded the limit of Class A standard. The residual NO_3^--N after decanting have to be denitrified in the stirring step of next cycle.

A post stirring step was added following the aeration step to provide an anoxic condition for the denitrification to remove the NO_3^--N produced in the aeration step. The procedure 2 was set as: instant filling, 0.5 h pre stirring, 3.5 h aeration, 1.5 h post stirring, 1 h settling, 1 h decanting (decanting ratio 1/3) and sludge waste, 0–0.5 h idling. The DO at the end of the aeration was controlled to be 2–4.5 mg/L. The results of 30 d stable running were listed in Table 6.

The result (not listed in Table 6) showed that the NO_3^--N was denitrified completely in the pre stirring step and the DO was below 0.45 mg/L after 0.75 h post stirring which satisfied the anoxic condition needed for the denitrification. But the dinitrification process did not proceed because of the shortage of carbon source. The effluent TN couldn't achieve the standard while COD_{Cr} and NH_4^+-N could.

4.2.3 Filling-preaeration-stirring (adding carbon source)-post aeration-settling-decanting-idling (procedure 3)

Compared with the procedure 2, procedure 3 canceled the pre stirring step and added carbon source during the post stirring step. A post aeration step was also added following the post stirring to remove the residual carbon.

Table 7. Effluent quality of procedure 3.

Item	NO_3^--N mg/L	TN mg/L	NH_4^+-N mg/L	COD_{Cr} mg/L
Data	5.55	11.07	2.49	29.45

Table 8. Effluent quality of procedure 4.

Item	NO_3^--N mg/L	TN mg/L	NH_4^+-N mg/L	COD_{Cr} mg/L
40 L:0 L	20.56	4.65	48.12	0.27
32 L:8 L	19.06	2.77	23.71	0.31
28 L:12 L	13.46	0.22	44.90	0.16
25 L:15 L	10.19	0.45	24.84	0.26
20 L:20 L	14.02	0.36	32.24	0.21

The procedure 3 was set as: instant filling, 3 h pre aeration, 2 h stirring (adding carbon source), 0.75 h post aeration, 1 h settling, 1 h decanting (decanting ratio 1/3) and sludge waste, 0-0.5h idling. The DO at the end of the aeration was controlled to be 2–4.5 mg/L. The results of 10 d stable running were listed in Table 7.

The effluent TN, COD_{Cr} and NH_4^+-N could all meet the standard after the addition of carbon source.

4.2.4 Step-feed procedure (procedure 4)

The procedure 3 could guarantee the effluent meet the standard but the operation cost was high because of the addition of glucose as carbon source. To reduce the cost, only part of the influent of each batch was added at the beginning of each cycle and the rest was substituted for the glucose as carbon source and was added at the beginning of the stirring step (procedure 4). The total influent of each batch was 40 L. The results of 10d stable running of each assignment (filling-step influent/stirring-step influent) were listed in Table 8.

As the portion of the influent added at the stirring step increased, the effluent TN, NH_4^+-N decreased in general in spite there were slight fluctuations. When the influent assignment fell between 28 L:12 L and 20 L:20 L, the effluent TN and NH_4^+-N decreased to the lowest level. The effluent of procedure 4 could meet the standard. Although procedure 4 did not use chemical method to remove the phosphorus, the TP was also incompliance with the standard.

The results of the nitrogen removal experiment showed that the traditional SBR procedure (procedure 1) could effectively remove the organic matter. If guaranteed with sufficient aeration, it had satisfactory nitrification effect but denitrification effect yield in vain. Procedure 2 added a stirring step before the aeration step therefore provided the anoxic condition for the denitrification. In this step, the residual NO_3^--N of last cycle was denitrified to N_2 with organic matter in

the raw water as carbon source. However, the influent NH_4^+-N of this cycle was transfer to NO_3^--N during the aeration step and part of the NO_3^--N was discharged in the following decanting step without denitrification. The residual NO_3^--N was denitrified in the stirring step of the next cycle. Procedure 3 added a post stirring step following the aeration step to provide an anoxic condition for the denitrification. But the NO_3^--N produced in this cycle still could not be denitrified in the followed stirring step for the deficiency of carbon source inhibited the denitrification. So the effluent TN of procedure 1 to 3 could not achieve the Class A standard.

In the procedure 4, the influent NH_4^+-N was transferred to NO_3^--N and then denitrified in the following anoxic stirring step with sufficient added glucose as the carbon source. The residual glucose was removed in the post aeration step. The effluent TN, NH_4^+-N and COD_{Cr} all achieved the Class A standard. Nevertheless, the need for extra carbon source increased the operation cost and was unsuitable for the full scale operation.

Procedure 4 fed the SBR in step and enhanced the utilization of carbon source in the raw water by anaerobic and facultative bacteria. The nitrogen and phosphorus remove effect was better than the other procedure when in the similar influent quality. The field experiment shows that all the effluent indexes of Procedure 4 including TP could achieve the Class A standard. This procedure neither needs adding glucose as carbon source nor the need of chemical method for the phosphorus removal, thus significantly decreased the operation cost.

5 CONCLUSION

The addition of PAC into the SBR significantly improved the TP removal efficiency. With sufficient amount of PAC, the effluent TP could be guaranteed to achieve the Class A standard.

Within the range of tested dosage, the added PAC could decrease the COD_{Cr} removal effect but did not have apparent influence on the nitrification. That means PAC affected the characteristic of the heterotrophic bacteria in the biological treatment system but hardly affected the autotrophic nitrifying bacteria. Unfortunately, this experiment did not study the effect of PAC on the denitrification and nitrogen removal.

After the addition of PAC, the floc size was bigger than before and the color was changed from brown to dark brown. The decrease of MLVSS/MLSS rate shows that the addition of PAC can reduce the active parts in the activated sludge.

After the experiment on the several SBR procedures capable of nitrification and denitrification, it was concluded that effluent TN of procedure 3 could achieve the Class A standard. But this procedure had to add carbon source to the system therefore increased the operation cost. Meanwhile, the effluent TN, NH_4^+-N, COD_{Cr} and TP of procedure 4 could all achieve the Class A standard without the addition of carbon source.

The step-feed influent procedure has many advantages such as high efficiency, low capital and operation cost etc. This procedure is suitable for the SBR upgrading of existing WWTP and the design of new WWTP.

The results of this experiment are influenced by the influent quality, treatment process and the operation configuration. It may differ greatly with results of other studies or the operation of other WWTP. But some of the results in this experiment are already been put into practice in this WWTP.

REFERENCES

Cui Yu-chuan et al. 2004. *Design and Calculation of Municipal Wastewater Treatment Plant Facilities.* Beijing: Chemical Industry Press.

Editorial Board on Monitoring and Analysis Method of Water & Wastewater of Ministry of Environmental Protection of the People's Republic of China. 2002. *Monitoring and Analysis Method of Water & Wastewater (the fourth edition).* Beijing: China Environmental Science Press.

Ge Shi-jin. & Peng Yong-zhen. 2009. Analysis and optimization control of the continuous step feed biological nitrogen and phosphate removal process. *Acta Scientiae Circumstantiae* 29(12): 2465–2470.

Jin Xue-biao et al. 2002. Comparison between Biological and Chemicobiological Phosphorus Removal. *Journal of Shanghai Normal University(Natural Sciences)*31(1): 78–82.

Liu Zhi-gang et al. 2011. Study on the Morphological Variation of Phosphorus of Municipal Wastewater Treatment Plant. *Water & Wastewater Engineering* 37(2): 50–53.

Luo Bin et al. 2010. Study Progress of Domestic Sewage Dephosphorization Technics. *Journal of Anhui Agricultural Sciences* 38(9): 4769–4771.

Wang Lin-na et al. 2009. Experimental Study on Feeding $FeCl_3$ into SBR Reactor for Accessorial Phosphorus Removal. *Environmental Science & Technology* 32(8): 80–87.

Xie Li-guo. 2010. Research on the Deep Phosphorus Removal in North Suburban Sewage Treatment Plant of Taiyuan. *Sci-Tech Information Development & Economy* 19(20): 150–166.

Xu Wei-yong et al. 2009. Study on the Comparative Experiment of PAC & PFS against Phosphorus Removal of Municipal Wastewater Treatment Plant's Tail-water. *Guangxi Journal of Light Industry* 2009(4): 115–116.

Biomedical Engineering and Environmental Engineering – Chan (Ed.)
© 2015 Taylor & Francis Group, London, ISBN: 978-1-138-02805-0

A new fuzzy dynamic evaluation model for the energy consumption of public buildings

Min Jiang, Lin Sun, Qiang Wang, Rang Zhao, Xin Zhang, Danshi Yu & Zhiping Zhou
Key Laboratory of Advanced Process Control for Light Industry (Ministry of Education), Jiangnan University, China
Logistics Management office, Jiangnan University, China

ABSTRACT: We propose a fuzzy dynamic evaluation model for energy consumption of large-scale public buildings based on an Improved Adaptive Genetic Algorithm (IAGA) and fuzzy theories. We extract membership functions from quantitative data, which adopt real-number coding. The fitness of each set of membership functions is evaluated by the fuzzy support of the item sets and their suitability consisting of the overlap factor and the coverage factor of the derived membership functions. It confirms the fuzzy interval by genetic evolution through improving the rate of cross operators and mutation operators automatically. We apply our model to assess the rank of offices in an on-campus building in terms of its energy consumption. Experiment results show that our method has higher robustness and the membership function achieves higher fitness. The proposed model provides a dynamic quantitative evaluation standard according to the concrete situation, so as to assisting managers in making more reasonable decisions.

Keywords: Energy Consumption Evaluation Method, Fuzzy membership functions, IAGA, Dynamic

1 INTRODUCTION

Nowadays, energy management systems provide an opportunity to collect vast amounts of building of building-related data. The collected data contain abundant knowledge about all of a building's energy consumption and influencing factors. In order to improve building energy performance, it is highly desirable that hidden knowledge can be extracted from building-related data (Yu et al., 2013). Thus, enhancing the energy management of a building will be significant for the development of the economy and society in terms of reducing energy consumption and improving energy efficiency (Tong and Zhao, 2013) (Zhang et al., 2010).

Therefore we must enhance the energy management of campus buildings for the purpose of promoting the saving of energy and developing a green campus (Liu et al., 2013). However, the energy consumption of a campus building can be affected by weather, building orientations, usage and other uncertain factors (Balta et al., 2013). For example, the energy consumption of a building used for teaching is generally lower than that of scientific laboratories. Thus, it is unreasonable to adopt the same static evaluation criterion. It is therefore essential to develop more effective data analysis techniques to deal with complex building-related data. For this reason, it can be more realistic to use membership functions and fuzzy intervals in order to comprehensively analyse and evaluate the energy consumption of campus buildings.

Recently, fuzzy theory has been used more and more frequently in intelligent systems because of its simplicity and similarity to human reasoning. Fuzzy theory has been applied in various fields, such as economics, engineering and manufacturing (Bede, 2013) (Garibaldi, 2005). Several fuzzy learning algorithms for inducing membership functions from quantitative data have been designed and applied to some specific domains. As to fuzzy data mining, Di Nola et al. proposed an algorithm to mining fuzzy intervals. They apply the Fuzzy Clustering Method (FCM) to deal with the problem of a sharp boundary (Di Nola et al., 2002). Hong, Chen el al. proposed an algorithm for mining membership functions based on the divide-and-conquer strategy (Chen et al., 2011). They transformed quantitative data into a fuzzy set and used a genetic operation to find and obtain a suitable membership function, but the membership function models were assumed to be known as isosceles-triangle functions in advance which had certain restrictions on the fuzzy interval (Tzung-Pei et al., 2008) (Hong et al., 2012). Although many approaches for learning membership functions were proposed, most of them were usually used for controlling problems (Cordon et al., 2001) (Yu and Kuang, 2010) (Long et al., 2006).

In this paper, we thus propose a fuzzy dynamic evaluation method based on the IAGA model. The proposed method adopts a multiple membership function model for fuzzy intervals, and is applied to the comprehensive analysis and evaluation of the energy consumption of on-campus building.

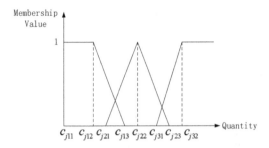

Membership
Value

Quantity

$c_{j11}\ c_{j12}\ c_{j21}\ c_{j13}\ c_{j22}\ c_{j31}\ c_{j23}\ c_{j32}$

Figure 1.   Membership function.

2   FUZZY DYNAMIC EVALUATION BASED ON GENETIC MODEL

2.1   *Membership function selection*

In general, the selection of the membership function is the key to fuzzy interval. Generally there are three mainmembership models: Gaussian, Triangular and Trapezoid (Garibaldi, 2005) (Bede, 2013). In this paper we construct a mixed model containing Triangular and Trapezoid.

As shown in Figure 1, the left and the right membership function model adopt the Trapezoid, and the middle one adopts the Triangular model. Each of them represent three linguistic terms: Low, Middle and High (Bilgiç, 2013). The membership values of these three linguistic terms are thus defined as:

$$f_{Low} = \begin{cases} 1 & if\ x < c_{j12} \\ \dfrac{c_{j13} - x}{c_{j13} - c_{j12}} & if\ c_{j12} < x < c_{j13} \\ 0 & if\ otherwise \end{cases} \quad (1)$$

$$f_{Mid} = \begin{cases} \dfrac{x - c_{jk1}}{c_{jk2} - c_{jk1}} & if\ x < c_{j12} \\ \dfrac{c_{jk3} - x}{c_{jk3} - c_{jk2}} & if\ c_{j12} < x < c_{j13} \\ 0 & if\ otherwise \end{cases} \quad (2)$$

$$f_{High} = \begin{cases} 1 & if\ x > c_{j12} \\ \dfrac{c_{j12} - x}{c_{j12} - c_{j11}} & if\ c_{j12} < x < c_{j13} \\ 0 & if\ otherwise \end{cases} \quad (3)$$

2.2   *Chromosome representation*

It is important to encode membership function as a string representation for the Genetic Algorithm (GA) to be applied. In this paper, each set of membership functions is encoded as a chromosome and handled as an individual with real-number strings (Tzung-Pei et al., 2008) (Klein, 2005). According to the proposed

Memership Function Coding

$\underbrace{c_{j11}\ c_{j12}\ c_{j13}}_{\substack{R_{j1}\\ Low}} \quad \underbrace{c_{j21}\ c_{j22}\ c_{j23}}_{\substack{R_{j2}\\ Middle}} \quad \underbrace{c_{j31}\ c_{j32}\ c_{j33}}_{\substack{R_{j3}\\ High}}$

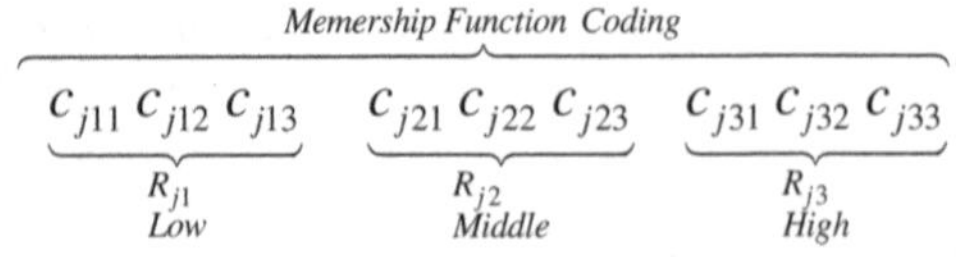

Figure 2.   Membership function coding in Figure 1.

membership functions in Figure 1, for item I_j, where the region R_{jk} denotes the membership function of the k_{th} (linguistic term for I_j), each membership function is encoded as a real-number string $(c_{jk1}c_{jk2}c_{jk3})$ and the inequality condition of the three parameters is $c_{jk1} < c_{jk2} < c_{jk3}$. The inequality condition of the three centre values of the membership functions is $c_{j12} < c_{j22} < c_{j32}$. The chromosome for the fuzzy set of membership functions in Figure 1 is encoded as shown in Figure 2.

2.3   *Genetic progress*

According to the above-mentioned proposed real-number encoding scheme, the initial population is randomly generated within some constraints as shown in the section Chromosome Representation for forming membership functions with high fitness values.

2.3.1   *Fitness and selection*

In order to get a series excellent sets of membership functions from an initial population and select parent sets of membership functions with high fitness values, an evaluation function is designed to qualify the derived sets of membership functions (Hong et al., 2012). The fitness value of a chromosome is defined as:

$$f(C_q) = \frac{\sum\limits_{X \in L1} fuzzy\_support(X)}{suitability(C_q)} \quad (4)$$

The evaluation function is determined by two main factors: the suitability of each membership function and the fuzzy support of large 1-item sets. The suitability function contains two main parts: overlap factor and coverage factor, both of which are defined as below. The suitability of each membership function of the chromosome C_q is defined as:

$$suitablity(C_q) = Cov(C_q) + Lap(C_q) \quad (5)$$

The overlap factor of the membership function for item I_j in chromosome C_q is defined as:

$$Lap(C_q) = \sum_{k < i} \left[\max\left(\left(\frac{lap(R_{jk}, R_{ji})}{\min(c_{jk3} - c_{jk2}, c_{ji2} - c_{ji1})} \right), 1 \right) - 1 \right] \quad (6)$$

The design of the overlap factor is to avoid the fuzzy interval of membership functions overlapping too much.

The coverage factor of the membership function for item I_j in chromosome C_q is defined as:

$$Cov(C_q) = \frac{1}{\frac{range(R_{j1},...,R_{jl})}{\max(I_j)}} \tag{7}$$

The design of the coverage factor is to avoid the fuzzy interval of membership functions separating too much.

2.4 Improved genetic model

2.4.1 Adaptive genetic algorithm

In the genetic algorithm, the convergence speed of the generating process is directly affected by the crossover as well as mutation rate. The generating rate of new individuals in a population is mainly determined by the crossover rate. The genetic algorithm with a high crossover rate may result in the destruction of excellent individuals; with a low crossover rate, it will lead to premature individuals. The mutation rate is the key factor to avoid trapping in the local optimal solution (Klein, 2005) (Herrera et al., 1997).

In order to solve the problems effectively as low convergence and prematurity in a Simple Genetic Algorithm (SGA), Srinivas et al (Srinivas and Patnaik, 1994) proposed a self-adaptive genetic algorithm in which the values of cross over and mutation rate could be adjusted automatically according to the fitness of the individual. In the traditional AGA, the crossover rate and mutation rate are defined as:

$$P_c = \begin{cases} K_1 * \dfrac{f_{max} - f'}{f_{max} - f_{avg}} & f' \geq f_{avg} \\ K_2 & f' \leq f_{avg} \end{cases} \tag{8}$$

$$P_m = \begin{cases} K_3 * \dfrac{f_{max} - f'}{f_{max} - f_{avg}} & f' \geq f_{avg} \\ K_4 & f' \leq f_{avg} \end{cases} \tag{9}$$

where f_{max} is the highest fitness of the individual in a population, f_{avg} is the average fitness of all individuals in a population, and f' is the larger fitness of the two parent chromosomes. The four parameters value $K_1 K_2 K_3 K_4$ are all ranging in (0, 1). With genetic progress, the crossover rate P_c and mutation rate P_c will keep changing automatically until the algorithm converges.

2.4.2 Improved adaptive genetic algorithm

We can find clearly that the adjustment strategy of the AGA is only suitable for the early evolution of generation process because the best individuals will remain in the early evolution, which will likely result in local optimal solution rather than global optimum (Venugopal et al., 2009).

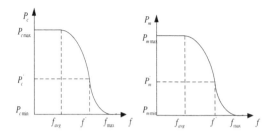

Figure 3. Relation curves between fitness and genetic operators.

Now we relate this to logistic functions (Holmgren, 1994) (Im and Lee, 2008) (Hou et al., 1997). If we apply the new strategy into the crossover rate and mutation rate, the Logistic function can keep a better balance between linear and nonlinear, as is defined below:

$$\varphi(x) = \frac{1}{1 + \exp(-ax)} \tag{10}$$

According to the definition of the Logistic function, when $ax \geq 9.9034$, $\psi(x) = 1$, when $ax \leq -9.9034$, then $\psi(x) = 0$.

The new definition of crossover rate and mutation rate is demonstrated below:

$$P_c = \begin{cases} \dfrac{P_{c\,max} - P_{c\,min}}{1 + \exp(A * (\frac{2(f - f_{avg})}{f_{max} - f_{avg}} - 1))} + P_{c\,min} & f \geq f_{avg} \\ P_{c\,min} & f \geq f_{avg} \end{cases} \tag{11}$$

$$P_m = \begin{cases} \dfrac{P_{m\,max} - P_{m\,min}}{1 + \exp(A * (\frac{2(f - f_{avg})}{f_{max} - f_{avg}} - 1))} + P_{m\,min} & f \geq f_{avg} \\ P_{m\,min} & f \geq f_{avg} \end{cases} \tag{12}$$

We set $A = 9.9034$ and we can find that crossover rate and mutation rate will be adjusted smoothly between f_{avg} and f_{max}, which can promote the progress of evolution and be significant in terms of avoiding a local optimal solution and preventing the algorithm from remaining stagnant.

3 THE PROPOSED DYNAMIC APPROACH

According to the above description of the design of the membership function and the IAGA, the proposed approach for mining the membership functions is described below.
INPUT:

1. A body of quantitative data D.
2. An initial population P.
3. A number of Predefined terms.
4. The max and min of crossover and mutation rate $P_{cmax}, P_{cmin}, P_{mmax}, P_{mmin}$.
5. A support threshold α.

OUTPUT: A set of high fitness membership functions.

Step 1: Randomly generate an initial population. Each individual in the initial population represent a possible set of membership functions.

Step 2: Encode each set of membership functions into a real-number string as above.

Step 3: Calculate the fitness value of each chromosome in the population by the definition.

Step 3.1: For the energy data, transfer each quantitative value in D into a fuzzy set.

Step 3.2: For each region R_{jk} in the fuzzy set, calculate the sum $count_{jk}$ of membership values on the transaction.

Step 3.3: For each $count_{jk}$, check whether its value is larger than or equal to the minimum support threshold. If R_{jk} can satisfy the following condition, put it (the membership function) in the set of large 1-item sets ($L1$), as follows:

$$L_1 = \{R_{jk} | count_{jk} \geq \alpha, 1 \leq j \leq m \ \& \ 1 \leq k \leq |I|\}$$

Step 3.4: For each individual in population P, according to the definition of fitness, calculate the fitness of the chromosome.

Step 4: Execute the crossover operations on the population.

Step 5: Execute the mutation operations on the population.

Step 6: Check whether the individuals satisfy the criterion, if not then go to step 3, otherwise execute the next step.

Step 7: Gather the final sets of membership functions with the highest fitness value in the final population.

Step 8: Analyse results and make decisions.

4 EXPERIMENTS

4.1 Description of the initial parameters and experimenta lDataSets

In this section, experiments were carried out in MATLAB2010b on an Intel Pentium(R) Dual-Core personal computer with 3.06 GHz and 2 GB RAM. The size of the initial population P was set at 50, the parameter d of the crossover operator was set at 0.35, the parameter ε of the mutation operator was set at 3, according to (Herrera et al., 1997), and the minimum support α is set at 0.25. According to (Im and Lee, 2008) (Venugopal et al., 2009), we set $P_c = 0.8$, $P_m = 0.02$, $P_{cmax} = 0.8$, $P_{cmin} = 0.6$, $P_{mmax} = 0.02$, and $P_{mmin} = 0.005$, for comparing the experiments of the GA, the SGA, and the IAGA.

The experiment data sets were collected from the electricity energy consumption of the Administration Department at *Jiangnan* University, China from 2009 to 2012. The statistics of Energy Consumption are shown in Table 1. The experiments were conducted by applying the proposed approach on the analysis and evaluation of energy consumption of an on-campus office building.

Table 1. The characteristics of energy.

Energy Consumption	Year			
Kw*h/m^2	2009	2010	2011	2012
Average	30.1	27.18	32.6	37.7
Maximum	103.7	83.58	110.7	141.6
Minimum	0.25	0.46	0.82	1.14
Standard	22.64	20.18	25.12	28.71

Table 2. The best fuzzy intervals with different GAs.

Year	Genetic	Best Fuzzy Interval (Centre)			Fitness
		Low	Middle	High	Value
2009	IAGA	18.12	35.12	69.35	0.5565
	AGA	19.81	38.40	69.01	0.5465
	SGA	27.17	56.01	74.43	0.5165
2010	IAGA	17.18	33.36	67.52	0.5719
	AGA	12.24	21.17	53.10	0.5032
	SGA	21.08	51.05	62.53	0.4754
2011	IAGA	13.07	39.76	84.53	0.5872
	AGA	15.81	42.08	76.98	0.5667
	SGA	14.26	39.71	85.05	0.5626
2012	IAGA	23.01	55.43	87.17	0.5984
	AGA	25.83	78.04	89.19	0.5326
	SGA	25.06	78.01	91.11	0.5139

Table 3. Energy use index of public buildings.

Applications	EUI (Kw*h/m^2)		
	Low	Middle	High
Administration	50	65	80
Entertainment	60	80–100	120–150
Medical	50	65	80
Education	45	65	80

4.2 Comprehensive analysis on campus building energy consumption

After 200 generations, the final set of membership functions were generated with high fitness which were apparently much better than the initial ones. To evaluate the improved method, experiments were made comparing the SGA with the AGA, it can be easily deduced from Table 2. For the same experiment environment and data, the best fuzzy intervals with the proposed membership functions using the IAGA are higher than the approaches using the SGA and the AGA.

EUI (Energy Use Intensity) is a unit of measurement that describes energy consumption of buildings. It is mainly utilised to survey building energy use patterns and analyse the energy consumption, which was shown in Table 3 (Yu et al., 2013).

Figure 4 gives the comparisons between the evaluated sample energy consumption on campus and EUI.

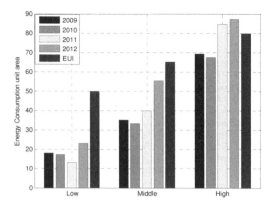

Figure 4.   Comparisons between year data and EUI.

Take the data in 2012 for example: the low consumption fuzzy region can be represented as (0, 23.01), the Middle consumption fuzzy region can be represented as (23.01, 55.43), and the High consumption fuzzy region can be represented as (55.43, 87.17). From Figure 4, we compared our final results with EUI. By doing so, we can learn that Low consumption and Middle consumption are all less than the Energy Use of Index, but the High consumption in 2012 is higher than EUI.

According to these experiment results, we can clearly find that the overall level of energy consumption of the samples are significantly higher than EUI. Moreover, historical data clearly reveal the rising trend of energy consumption which will directly lead to the aggravation of the energy cost burden for the university. Penalties could be considered to constrain the exorbitant energy consumption. Meanwhile, the assessment standard of low consumption is relatively low indicating that some offices may be not fully used as effectively as they might have been.

It is clear that the average fitness values of the chromosomes along with the different numbers of generations in the population P are shown in Figure 5. As expected, the curve went upward gradually, finally reaching a fixed value. To evaluate the improved genetic algorithm (IAGA), experiments were made comparing the SGA with the AGA. As can easily be seen from Figure 5, the fitness of the membership functions using the IAGA are higher than the approach using the SGA and the approach using the AGA.

5   CONCLUSION AND FUTURE WORK

In this paper, we introduced a new dynamic fuzzy evaluation method for analysing and interpreting the energy consumption in public buildings. We improved a GA-based fuzzy data mining algorithm for extracting membership functions from quantitative data. Since the fitness of each set of membership functions is evaluated by fuzzy-support values and the suitability of the derived membership functions, the membership function finally achieves higher fitness. The

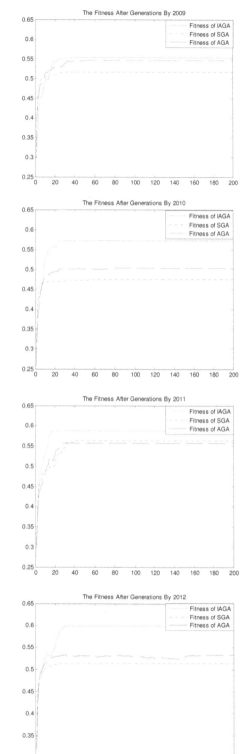

Figure 5.   The fitness each year after 200 generations.

improved approach has been applied to the analysis of the energy consumption of an on-campus building. The experiment was conducted in order to classify the quantitative data into three fuzzy intervals low, middle and high, based on the concrete situation for each year. Our dynamic fuzzy evaluation model is particularly useful for helping managers make instructive and reasonable decisions about the management of energy and eventually to build a sustainable, resource-saving green campus of the University.

ACKNOWLEDGEMENT

This project was supported by the Jiangsu Cooperative Innovation Fund of Production, Education & Research (BY2013015-33) and the National Natural Science Foundation of China (61362030, 61201429).

REFERENCES

Balta, M.T., Dincer, I. & Hepbasli, A. 2013. Environmental Impact Assessment of Building Energy Systems. *In:* Dincer, I., Colpan, C.O. & Kadioglu, F. (eds.) *Causes, Impacts and Solutions to Global Warming.* Springer New York.

Bede, B. 2013. Fuzzy Sets. *Mathematics of Fuzzy Sets and Fuzzy Logic.* Springer Berlin Heidelberg.

Bilgi, T. 2013. The Membership Function and Its Measurement. *In:* Seising, R., Trillas, E., Moraga, C. & Termini, S. (eds.) *On Fuzziness.* Springer Berlin Heidelberg.

Chen, C.-H., Hong, T.-P. & Tseng, V. 2011. Genetic-fuzzy mining with multiple minimum supports based on fuzzy clustering. *Soft Computing,* 15, 2319–2333.

Cordon, O., Herrera, F. & Villar, P. 2001. Generating the knowledge base of a fuzzy rule-based system by the genetic learning of the data base. *Fuzzy Systems, IEEE Transactions on,* 9, 667–674.

DI Nola, A., Loia, V. & Staiano, A. 2002. An Evolutionary Approach to Spatial Fuzzy c-Means Clustering. *Fuzzy Optimization and Decision Making,* 1, 195–219.

Garibaldi, J.M. 2005. Fuzzy Expert Systems. *In:* Gabrys, B., Leivisk, K. & Strackeljan, J. (eds.) *Do Smart Adaptive Systems Exist?:* Springer Berlin Heidelberg.

Herrera, F., Lozano, M. & Verdegay, J.L. 1997. Fuzzy connectives based crossover operators to model genetic algorithms population diversity. *Fuzzy Sets and Systems,* 92, 21–30.

Holmgren, R. 1994. The Logistic Function, Part I. *A First Course in Discrete Dynamical Systems.* Springer US.

Hong, T.-P., Chen, C.-H. & Tseng, V. 2012. Genetic-Fuzzy Data Mining Techniques. *In:* Meyers, R. A. (ed.) *Computational Complexity.* Springer New York.

Hou, R.-H., Hong, T.-P., Tseng, S.-S. & Kuo, S.-Y. 1997. A New Probabilistic Induction Method. *Journal of Automated Reasoning,* 18, 5–24.

Im, S.-M. & Lee, J.-J. 2008. Adaptive crossover, mutation and selection using fuzzy system for genetic algorithms. *Artificial Life and Robotics,* 13, 129–133.

Klein, R. 2005. Genetic Algorithms. *In:* Stadtler, H. & Kilger, C. (eds.) *Supply Chain Management and Advanced Planning.* Springer Berlin Heidelberg.

Liu, Z.-L., Li, S.-Q., Li, M.-Q. & Feng, L.-X. 2013. Building Energy Consumption in the Universities of China: Situation and Countermeasures. *In:* Qi, E., Shen, J. & Dou, R. (eds.) *The 19th International Conference on Industrial Engineering and Engineering Management.* Springer Berlin Heidelberg.

Long, Z., Dawei, M., Xiaoli, W. & Guigao, L. Strategy of Adaptive Fuzzy Control of Crab-Like Robot.Intelligent Control and Automation, 2006. WCICA 2006. The Sixth World Congress on, 0-0 0 2006. 3787–3790.

Srinivas, M. & Patnaik, L.M. 1994. Adaptive probabilities of crossover and mutation in genetic algorithms. *Systems, Man and Cybernetics, IEEE Transactions on,* 24, 656–667.

Tong, D. & Zhao, J. 2013. Analysis of energy saving optimization of campus buildings based on energy simulation. *Frontiers in Energy,* 7, 388–398.

Tzung-Pei, H., Chun-Hao, C., Yeong-Chyi, L. & Yu-Lung, W. 2008. Genetic-Fuzzy Data Mining With Divide-and-Conquer Strategy. *Evolutionary Computation, IEEE Transactions on,* 12, 252–265.

Venugopal, K. R., Srinivasa, K. G. & Patnaik, L. M. 2009. Self Adaptive Genetic Algorithms. *Soft Computing for Data Mining Applications.* Springer Berlin Heidelberg.

Yu, S.-Y. & Kuang, S.-Q. 2010. Fuzzy adaptive genetic algorithm based on auto-regulating fuzzy rules. *Journal of Central South University of Technology,* 17, 123–128.

Yu, Z., Fung, B. M. & Haghighat, F. 2013. Extracting knowledge from building-related data — A data mining framework. *Building Simulation,* 6, 207–222.

Zhang, S., Yang, X., Jiang, Y. & Wei, Q. 2010. Comparative analysis of energy use in China building sector: current status, existing problems and solutions. *Frontiers of Energy and Power Engineering in China,* 4, 2–21.

Biomedical Engineering and Environmental Engineering – Chan (Ed.)
© *2015 Taylor & Francis Group, London, ISBN: 978-1-138-02805-0*

Nitrogen removal and transformation in constructed wetlands inoculated with the activated sludge from sewage treatment

S.J. Hu
College of Environmental Science and Engineering, Beijing Forestry University, Beijing, China

L.L. Zheng
College of Forestry, Beijing Forestry University, Beijing, China

X.H. Wang, F.Z. Li, Y. Zhao, L.J. Liu & W.Y. Liang
College of Environmental Science and Engineering, Beijing Forestry University, Beijing, China

ABSTRACT: The paper studied the nitrogen removal and transformation in a simulated vertical-flow wetland inoculated with the activated sludge. The concentrations of total nitrogen (TN), NO_3^--N, NO_2^--N, and NH_4^+-N were detected along the depth of the reactor to the influent with different concentrations of NO_3^--N, NH_4^+-N, and organic N (org-N). The results showed that the average removal efficiency of TN was as high as 85%. And the TN removal was fulfilled in the top of the wetlands (-0.25 m). In the processes of nitrogen transformation, the NO_2^--N of 0.61–1.25 mg/L occurred at the depth of -0.05 and then reduced sharply. Both NO_3^--N and NO_2^--N were completely converted at the depth of -0.45 m and were undetectable in the effluent. The wetland reactor could not remove the NH_4^+-N produced by org-N ammoniation and dissimilatory reduction of NO_3^--N.

1 INTRODUCTION

During rainfall events, the nitrogen pollutants on the impervious surface such as pavement, roof, parking lot, and highway are swept, mobilized, and transported by stormwater into the receiving water bodies, causing eutrophication and polluting aquatic environment. (Han et al., 2014). Constructed wetland, one of the most popular measures, has been employed widely because of its capacity to improve the water quality and beautify the ecological environment. It is effective at reducing a series of pollutants, including organics, solids, phosphorus, heavy metals, and microorganism. (Bulc and Slak, 2003).

Constructed wetlands have proven potential for removing nitrogen, but in stormwater wetlands, the proportion of nitrogen removed by wetlands is considered to be noticeably lower (Lee et al., 2009). This could be due to the lower concentration of total nitrogen (TN) in the influent, the wetland design-limited space, and the weather humidity (Birch et al., 2004, Yi et al., 2010). Commonly, microbial degradation is regarded as the most effective way to remove nitrogen in the wetlands (Lee et al., 2009). Especially, nitrification-denitrification makes great contributions to reducing nitrogen (Lee et al., 2009). The deficiency of microbial community and microbial activity influence the pollutants removal and lead to the poor performance of nitrogen removal (Tonderski, 2009).

Nitrogen in stormwater is present in many chemical forms, including ammonia (NH_3-N), ammonium (NH_4^+-N), nitrate (NO_3^--N), nitrite (NO_2^--N), dissolved organic nitrogen (org-N), and particulate org-N (Lee et al., 2009). The composition varies with hydrologic conditions and land utilization (Li and Davis, 2014). But the nitrogen in stormwater is predominantly dissolved ($\sim$80%), with ammonia the least-abundant form ($\sim$11%) (Taylor et al., 2005). The nitrogen behavior in constructed wetland systems is complicated on account of the complexity of the nitrogen species and the transformation and removal mechanisms.

In the present study, the activated sludge from the sewage treatment plant was inoculated into a simulated vertical-flow wetland reactor to improve the abundance of microbial groups. The concentrations of different nitrogen forms, namely NO_3^--N, NO_2^--N, NH_4^+-N, and TN, were detected along the depth of wetland reactor to investigate the nitrogen removal effects and nitrogen transformation processes.

2 MATERIALS AND METHODS

2.1 Experimental setup

Fig. 1 shows the schematic diagram of a simulated vertical-flow wetland reactor. It was fabricated from polymethyl methacrylate with a diameter of 220 mm, a height of 1200 mm, and a working volume of 12 L.

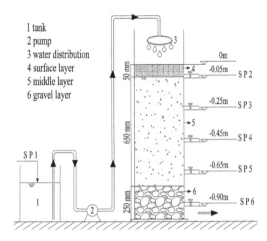

1 tank
2 pump
3 water distribution
4 surface layer
5 middle layer
6 gravel layer

SP1

0m
-0.05m — SP2
-0.25m — SP3
-0.45m — SP4
-0.65m — SP5
-0.90m — SP6

50 mm
650 mm
250 mm

Figure 1. Schematic of the simulated wetland reactor. SP = Sampling Point.

Table 1. The nitrogen composition of the influent in the experiments.

Exp. No.	TN (mgN/L)	Nitrogen composition (mgN/L)		
		NH_4^+-N	NO_3^--N	NO_2^--N
1	16.0	3.2 (20%)	12.8 (80%)	0 (–)
2	16.0	0 (–)	16.0 (100%)	0 (–)
3	16.0	0 (–)	12.8 (80%)	3.2 (20%)
4	16.0	3.2 (20%)	9.6 (60%)	3.2 (20%)

Note: Percentage in brackets: concentration ratio of nitrogen composition to TN. Exp = Experiment.

The gravel layer (Φ30–50 mm) was placed in the bottom of the reactor. The middle layer was composed of the mixture of zeolites (Φ30–50 mm), volcanic rocks (Φ10–60 mm), and ceramic rings (Φ9–17 mm). The surface layer was filled with the smaller gravels of 5 mm in size.

2.2 Start-up and operating strategies

Approximately 10 L of activated sludge with the mixed liquor suspended solid of 6000–7000 mg/L was inoculated into the wetland reactor, which was obtained from Beixiaohe Wastewater Treatment Plant, Beijing, China. After inoculation, the reactor was kept stationary for 24 h. Then, the synthetic wastewater was fed into the reactor at the flow rate of 15 ml/min. The wastewater was prepared by dissolving the glucose, KH_2PO_4, $(NH_4)_2SO_4$, KNO_3, $CO(NH_2)_2$, and trace elements ($FeSO_4$ 7 H_2O, $MnCl_2$ 4 H_2O, $ZnSO_4$, $CaCl_2$, $CuSO_4$ 5 H_2O, H_3BO_3, and $NiCl_2$ H_2O) into the distilled water. The concentrations of chemical oxygen demand (COD), total phosphorus (TP), and TN were 100 mg/L, 1.0 mg/L, and 16.0 mg/L, respectively. When the biofilm that adhered on the media was visible to the naked eyes and the removal efficiency of COD and TN kept stable for 10 days, the start-up period was finished.

After the start-up period, the composition of TN in the experiments was changed by varying the ratio of nitrogen in different forms as shown in Table 1. The hydraulic retention time was fixed on 13.3 h at the flow rate of 15 ml/min. When the experiments were changed from one to another, it took 8 days for the reactor to make itself work stable. During the subsequent running of 6 days, the water samples were taken from the sampling points (SP) 1 to 6 every day. The concentrations of COD, TN, NH_4^+-N, NO_3^--N, NO_2^--N, dissolved oxygen (DO), and pH were analyzed immediately after sampling.

2.3 Analytical methods

The COD was measured with a COD analyzer (CTL-12, Chengde Huatong Environmental Protection Apparatus Co., China) (Yin et al., 2011). The concentrations of TN, NH_4^+-N, NO_3^--N and NO_2^--N were examined according to National Standard Methods (APHA, 2005). DO and pH were measured using a DO detector (Multi 3410, WTW Co., Germany) and a pH meter (PB-10, Sartorious Co., LTD., Germany), respectively. The presented data are the average values of 6 days.

3 RESULTS AND DISCUSSION

3.1 TN removal

The activated sludge comes from the biological treatment of wastewater, which is abundant in microbial groups such as bacteria, fungi, protozoa, and rotifers. Based on the requirement of oxygen for growth, the activated sludge is often divided into aerobic and anaerobic sludge. (Von Sperling and Oliveira, 2009). In this study, the aerobic activated sludge was chosen as the inoculated one because of its strong acclimation and abundance of microbial community on nitrification-denitrification (Lee et al., 2009). Fig. 2 gives the variation of TN and COD along the depth. Although the nitrogen compositions in the influent of the four experiments were different, the profiles of TN concentrations along the depth were similar. When the wastewater reached a depth of -0.25 m, approximately 85% of TN was removed. Thereafter, the TN concentrations kept constant until the discharge. The results indicated that the activated sludge inoculated in the reactor promoted the nitrogen removal effectively.

The nitrogen removal mechanisms in constructed wetlands are divided into two aspects: physicochemical processes such as volatilization, filtration, adsorption, sedimentation, and mineralization; biological transformation including ammonification, nitrification-denitrification, anammox, and plant uptake (Lee et al., 2009). Because there were no plants in the reactor, the biological action depended mainly on the microorganisms. Ammonification converts org-N into NH_4^+-N, and DO in the wastewater can accelerate its rates greatly (Lee et al., 2009). In the nitrification process, two sequential oxidative stages – ammonia to nitrite and nitrite to nitrate – are performed under strict aerobic conditions (Ruiz et al., 2003). Nitrate is reduced to nitrite and then to nitrous oxide and nitric oxide and finally to nitrogen gas by denitrification (Ruiz et al., 2003).

In the denitrification process, denitrifying microbes require a very low oxygen concentration of less than 10%, as well as organic carbons for energy (Lee et al., 2009). As shown in Fig. 2, both the TN and COD concentrations decreased rapidly from the depth of 0 to -0.25 m. The high removal efficiency of nitrogen indicated that the COD provided an adequate carbon source in the denitrification process. The simultaneous nitrification and denitrification are often accomplished under the DO concentration of 0.5–1.0 mg/L (Zeng et al., 2003). Although the DO concentration of the influent reached 8.56–10.56 mg/L, it was only 0.51–1.08 mg/L at the depth of -0.05 m, which was in favor of the simultaneous nitrification and denitrification. Therefore, even if the influence contained a certain amount of org-N and NH_4^+-N, the TN could also be removed effectively at the depth of -0.25 m. The results reflect that the nitrification and denitrification are realized successfully in the top layer of the wetland (-0.25 m), if the adequate carbon is provided.

Below the depth of -0.25 m, both the TN and COD concentrations kept constant till the discharge. It is known that the C/N ratio of 6.0–7.0 in the wastewater is considered to be optimal for denitrification (Lu et al., 2009). However, the kinds of carbon sources also influence the denitrification effects tremendously (Sirivedhin and Gray, 2006). The carbon sources are often divided into three categories in the nitrification-denitrification, i.e., easy degradation carbon source such as carbohydrate, slow degradation carbon source such as protein, and cellular components (Sirivedhin and Gray, 2006). Although the C/N ratio was still 5.0–6.0 below the depth of -0.25 m, the organic carbons (glucose) in the influent have been transformed to the unavailable carbon sources such as cellular components during the COD and TN degradation processes. So, there was not any more removal of TN below the -0.25 m in the reactor.

3.2 Transformation of nitrogen

The variations of NO_3^--N, NO_2^--N, and NH_4^+-N along the depth were shown in Fig. 3. Although the NO_3^--N concentrations in the influent changed from 9.6 to 16.0 mg/L, the profiles along the depth in the four experiments were very similar to each other. From the depth of 0 m to -0.25 m, the NO_3^--N concentrations decreased fast, and they were completely used up below the depth of -0.45 m. Nearly 100% of NO_3^--N was removed or converted to other forms of nitrogen in the effluent. The removal of NO_3^--N was attributed to the denitrification of microorganism and the dissimilatory nitrate reduction to NH_4^+-N (Giblin et al., 2013).

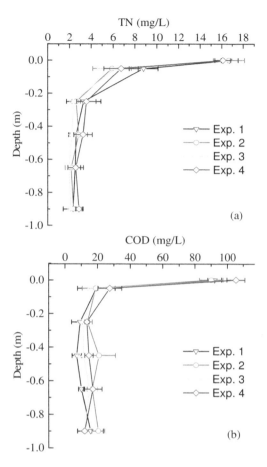

Figure 2. The variation of TN and COD concentrations along the depth of the simulated wetland reactor. Data: average ± standard deviation.

Unlike NO_3^--N, there was no NO_2^--N in the synthetic influent. However, the NO_2^--N concentration rose to a peak at the depth of -40.05 m. Because NO_3^--N can be converted into NO_2^--N in the denitrification, the occurrence of NO_2^--N at -0.05 m were resulted from this process. It was obvious that the NO_2^--N concentrations of Exp. 3 and 4 were higher than those of Exp. 1 and 2. This could be explained by that the org-N in the influent produced more amount of NO_2^--N by the ammoniation and nitrification (Lee et al., 2009). The phenomenon proved that org-N influenced the production of NO_2^--N more intensively than other forms. Under the depth of -0.05 m, the NO_2^--N concentrations were reduced sharply and became undetectable in the effluent.

Although there was no NH_4^+-N in the influent in Exp. 2 and Exp. 3, the NH_4^+-N occurred at -0.05 m and its concentration increased with the depth of the reactor, reaching a peak at the depth of -0.45 m. The org-N assimilated by microorganism was released afresh and transformed to NH_4^+-N by ammoniation (Lee et al., 2009). In addition, nitrate also produced the NH_4^+-N by dissimilatory reduction (Giblin et al.,

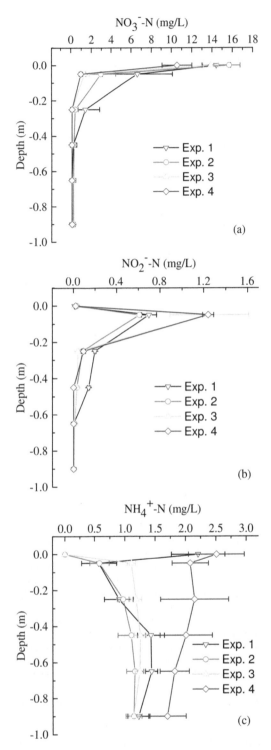

Figure 3. The variation of $NO_3^- $-N, NO_2^--N and NH_4^+-N concentrations along the depth of the simulated wetland reactor. Data: average ± standard deviation.

2013). These processes resulted in the production of NH_4^+-N in the wetland reactor. The DO concentrations (0.7–1.7 mg/L) were very lower below the depth of 0.25 m, which was not appropriate for nitrification to transform the NH_4^+-N to NO_3^--N or NO_2^--N. Thus, the NH_4^+-N in the influent and the excess NH_4^+-N produced from the ammoniation and dissimilatory reduction were hard to remove. Besides, the adsorbed NH_4^+-N on the media in constructed wetlands might also released into the effluent again and brought about the increase of NH_4^+-N in the effluent (Zhu et al., 2011). Therefore, the nitrogen forms in the effluent were composed mainly by NH_4^+-N with the concentrations of 1.1–1.7 mg/L. The rest nitrogen chemicals in the effluent were the refractory org-N produced by the microbial metabolization, approximately 0.9–1.2 mg/L.

4 CONCLUSIONS

The inoculation of the aerobic activated sludge into the constructed wetlands improved the nitrogen removal effectively. When the TN in the influent was 16.1–16.7 mg/L, approximately 85% of TN could be removed in the top part of the wetland (−0.25 m). The NO_3^--N and NO_2^--N were all undetectable in the effluent, though the NO_2^--N reached a peak of 0.51–1.21 mg/L at the depth of −0.05 m. The NH_4^+-N produced by org-N ammoniation and dissimilatory reduction of NO_3^--N was very hard to remove. The effluent was composed by NH_4^+-N (1.1–1.7 mg/L) and org-N (0.9–1.2 mg/L).

ACKNOWLEDGMENTS

The authors acknowledge the financial assistance generously provided by Major Science and Technology Program for Water Pollution Control and Treatment of China (2012ZX07307-001-006 and 2013ZX07209001-003).

REFERENCES

Apha 2005. *Standard Methods for the Examination of Water and Wastewater*, American Public Health Association: Washington, DC.
Birch, G.F., Matthai, C., Fazeli, M.S. & Suh, J. 2004. Efficiency of a constructed wetland in removing contaminants from stormwater. *Wetlands*, 24, 459–466.
Bulc, T. & Slak, A.S. 2003. Performance of constructed wetland for highway runoff treatment. *Water Science and Technology*, 48, 315–322.
Giblin, A.E., Tobias, C.R., Song, B., Weston, N., Banta, G.T. & Rivera-monroy, V. H. 2013. The Importance of Dissimilatory Nitrate Reduction to Ammonium (DNRA) in the Nitrogen Cycle of Coastal Ecosystems. *Oceanography*, 26, 124–131.
Han, J.C., Gao, X.L., Liu, Y., Wang, H.W. & Chen, Y. 2014. Distributions and transport of typical contaminants in different urban stormwater runoff under the effect of

drainage systems. *Desalination and Water Treatment,* 52, 1455–1461.

Lee, C.G., Fletcher, T.D. & Sun, G.Z. 2009. Nitrogen removal in constructed wetland systems. *Engineering in Life Sciences,* 9, 11–22.

Li, L.Q. & Davis, A. P. 2014. Urban Stormwater Runoff Nitrogen Composition and Fate in Bioretention Systems. *Environmental Science & Technology,* 48, 3403–3410.

Lu, S.L., Hu, H.Y., Sun, Y.X. & Yang, J. 2009. Effect of carbon source on the denitrification in constructed wetlands. *Journal of Environmental Sciences-China,* 21, 1036–1043.

Ruiz, G., Jeison, D. & Chamy, R. 2003. Nitrification with high nitrite accumulation for the treatment of wastewater with high ammonia concentration. *Water Research,* 37, 1371–1377.

Sirivedhin, T. & Gray, K.A. 2006. Factors affecting denitrification rates in experimental wetlands: Field and laboratory studies. *Ecological Engineering,* 26, 167–181.

Taylor, G.D., Fletcher, T.D., Wong, T.H.F., Breen, P.F. & Duncan, H.P. 2005. Nitrogen composition in urban runoff – implications for stormwater management. *Water Research,* 39, 1982–1989.

Tonderski, K.S. 2009. Molecular and microbial advances in wetland science. *Ecological Engineering,* 35, 959–960.

Von sperling, M. & Oliveira, S.C. 2009. Comparative performance evaluation of full-scale anaerobic and aerobic wastewater treatment processes in Brazil. *Water Science and Technology,* 59, 15–22.

Yi, Q.T., Yu, J. & Kim, Y. 2010. Removal patterns of particulate and dissolved forms of pollutants in a stormwater wetland. *Water Science and Technology,* 61, 2083–2096.

Yin, H.B., Yun, Y., Zhang, Y.L. & Fan, C.X. 2011. Phosphate removal from wastewaters by a naturally occurring, calcium-rich sepiolite. *Journal of Hazardous Materials,* 198, 362–369.

Zeng, R.J., Lemaire, R., Yuan, Z. & Keller, J. 2003. Simultaneous nitrification, denitrification, and phosphorus removal in a lab-scale sequencing batch reactor. *Biotechnology and Bioengineering,* 84, 170–178.

Zhu, W.L., Cui, L.H., Ouyang, Y., Long, C.F. & Tang, X.D. 2011. Kinetic Adsorption of Ammonium Nitrogen by Substrate Materials for Constructed Wetlands. *Pedosphere,* 21, 454–463.

Biomedical Engineering and Environmental Engineering – Chan (Ed.)
© *2015 Taylor & Francis Group, London, ISBN: 978-1-138-02805-0*

Effect of air pollution on physical function and air quality of competitive sports

Guangfa Jiao, Haiying Wang, Hui Liu, Yongtao Xie & Jintian Yang
HeBei Institute of Physical Education, Shijiazhuang, China

ABSTRACT: Air pollution can produce short-term acute and long-term chronic health hazard. The athletes may be more dangerous at training under poor air because of special exercise physiological condition. This paper was to review the study of air pollution and exercise, and to discuss the interacting between motor function of air pollution and exercise, and to enhance the awareness and the control of air pollution for athlete training. The air pollution had a greater influence on the ability of aerobic exercise, and smaller effect on anaerobic exercise capacity based on analyzing the study of motor function test. Athlete training under air pollution induced airway hyperresponsiveness and asthma, and reduced aerobic exercise capacity through increasing myocardial ischemia and reducing the blood oxygen content. Air pollution was mainly impact on long time endurance sports, but the impact on performance is not clear. There is also no air quality standard for environment of sports training and competition. The further research should be strengthened the low levels of air pollution impacted on the motor function, and protect athletes against air pollution.

Keywords: Air pollution; physical function; physiological mechanism; air quality

1 INTRODUCTION

Air pollution is a problem which affects the human health, public health and environment; it has drawn more and more attention and is explored by scholars in recent years. The main pollutants in the air include carbon monoxide (CO), nitrogen oxide (NOX), ozone (O_3), sulfur dioxide (SO_2) and particulate matter (PM) and others. PM is divided into PM10, PM5, and PM2.5 according to its aerodynamic diameter in micron, the so- called superfine particle PM0.1 is also beginning to draw the attention of scholars. The air pollutants of high levels will bring short-term acute and long-term chronic hazards to the health, and it will also have a long-reaching impact on the national physique, but the impact of air pollutants of low levels on motor function remains unclear. The body is in a special physical state in exercise training, so the impact of air pollution is more serious. This paper integrates the research data of air pollution and motion at home and abroad, explores the impact of the interaction of air pollution and motion on motor function, so as to enhance the understanding and prevention and control of the exercise training on the environment and air pollution.

2 AIR POLLUTION AND HEALTH

It can be determined by integrating domestic and foreign research that air pollution is closely related to the increase in the morbidity and mortality of cardiopulmonary disease, including pulmonary and systemic inflammatory response, atherosclerosis, vascular dysfunction, coronary heart disease and other diseases (Dong, et al., 2012). Air pollution caused by the automobile exhausts can cause lung complex reaction, which includes oxidative stress, redox sensitive transcription factor, inflammatory cell infiltration and proinflammatory component activation and secretion. These substances can come into the bloodstream by the lungs, causing systemic inflammatory response, resulting in the cardiovascular system diseases (Poursafa and Kelishadi, 2010). Survey of adults conducted in Canada found that (Cakmak, et al., 2011), air pollution is related to blood pressure when ventilation declines and in a quiet condition. It is found in America in 1980 that the content of carboxyhemoglobin (COHb) of the city residents is 2 times than that of rural residents (Lee, et al., 1994), which indicated that in the environment of air pollution, the body of the city residents is in a long-term mild hypoxia state, it may also be the pathogenetic basis of heart and lung disease. O_3 is a strong oxidant, has a strong stimulating effect on the respiratory mucosa, can also cause neurological poisoning. SO_2 and NO can form acid on the respiratory tract mucosa after entering the respiratory tract, induce or aggravate asthma. While NO is a normal biological active substance in vivo, and an atypical neurotransmitter of the central and peripheral nervous system as well; low- concentration NO can enhance myocardial contractility, whereas high-concentration

NO decreases myocardial contractility, the hazard of NO on health still lacks in-depth research.

The hazard of PM on the health is related to particle aerodynamics diameter. Because of aerodynamic characteristics, the hazard of PM10 on health mainly deposits in the upper respiratory tract; and PM2.5 is more easily to get into the alveolar for diffusion. With the decrease in the volume of PM, the number of intrapulmonary deposition increases with no gender differences. PM can be acute and chronic hazard caused by the adsorption of components, and the mechanisms inducing inflammation of the airway are free radicals and oxidative damage, transcription factors and inflammatory factors activation, cell calcium steady destruction, resulting in fibrosis effect and mutagenic effect. PM2.5 and PM10 are to increase the risk of myocardial ischemic attack three times , and the toxicity of PM on cardiovascular is mainly reflected in the increase of leukocytes and endothelial adhesion molecules of the blood, the strengthening of coagulation activity, and it can change the normal electrophysiological of the heart as well. After PM exposure, cardiac function changes, at the same time, it reflects the increase in the C reactive protein of myocardial injury. PM can also change the normal activities of the neural pathway, and has an adverse impact on the muscle microcirculation, but these impacts have yet to be further confirmed.

3 IMPACT ON OF AIR POLLUTION ON MOTOR FUNCTION

3.1 *Air pollution and motor function test*

Current research suggests that the decrease in exercise capacity is related to air pollution. It is analyzed through the study of motion function test that the air pollution has a greater impact on adult aerobic capacity, and has a smaller impact on adult anaerobic exercise capacity. In two regions with high and low air pollution index of Isfahan, an Iran industrial city, a Cooper test was conducted on the college student volunteers (Kargarfard, et al., 2011), the maximal oxygen uptake, and the number of red blood cells and hemoglobin of the subjects decreased, while the lactic acid increased. The healthy young people who breathed and the air mixed with CO which was comparable to the city air pollution level, carried on motor test of Bruce program, it was found in the study that young people's maximal aerobic capacity decreased when they were in short-term acute CO exposure (Adir, et al., 1999). The study also found lactate and pyruvate levels in the blood showed no difference when breathing CO air and breathing normal air; and it also showed no difference in lactate and pyruvate ratio representing the level of anaerobic metabolism after exercise under the two kinds of conditions, and the maximum heart rate and blood pressure also showed no difference. The young men carried out the maximal aerobic capacity test in 50 ppm CO air (Drinkwater, et al., 1974), and

the results of the study showed that the air with CO had no impact on the maximal aerobic capacity of the subjects, but the motion time of the subjects reduced.

Healthy male college students carried out a 6-minute power cycling test, and inhaled the air with low or high content of PM1, during the motion, heart rate showed no difference, but in the high-concentration PM1 condition, maximal motion work capacity decreased (Rundell and Caviston, 2008), which indicated that the short-term inhalation of large amount of PM1 will affect the strength endurance. The results of Rome traffic police power bicycle incremental exercise test showed that (Volpino, et al., 2004), the maximal voluntary ventilation and blood oxygen partial pressure in a quiet state was lower than the general population, the maximum aerobic capacity and anaerobic capacity decreased, at the same time, after 3 minutes of exercise, cardio pulmonary function was not complete restored to normal condition. Healthy subjects did exercises in 0.37, 0.50 or 0.75 ppm concentration O_3, for 15 minutes and took a break for 15 minutes in 2 hours, under different O_3 levels, the maximal oxygen uptake and per-minute ventilation of the subjects did not change, but with the increase in O_3 concentration, the respiratory frequency increased and the tidal volume decreased (Folinsbee, et al., 1975). The above study suggests that the decrease in aerobic exercise capacity is mainly caused by air pollution.

3.2 *Basic physiological mechanisms of the impact air pollution on motor function*

Integrating the related research, it can be inferred the air pollutants O_3, SO_2, CO and particulates affect the nerves, heart and lung function mainly by acting on the respiratory system, thus affecting motor capacity; in which, it is found that in the low-concentration O_3, the high-level endurance athletes endurance sports simulation game showed the psychology and subjective symptoms reaction; and that athletic endurance and decreased lung function , and intermittently exposed to the ambient O_3 reaction lung function obviously decreased. While the lung function and subjective symptoms reaction of the athletes who were intermittently exposed to the O_3 environment decreased significantly. In the polluted air, it is not entirely confirmed that exercise increases the mechanism of myocardial ischemia, one possible explanation may be a reduction of myocardial oxygen supply, which may be related to vasoconstriction and a temporary thrombosis; Another possibility is to reduce the oxygen transport capacity, which may be related to the increase of COHb and oxygen demand, as well as changes in myocardial energy metabolism (Mittleman, 2007).

The hazards of motion on the athletes in polluted air harm have three main factors (Mirkin, 1982): the first factor is the increase in per- minute ventilation volume causes the increase in the quantity of inhaled pollutants in motion time; the second factor is the taken air gets into the lungs through oral in motion time and without the effective filtration of nasal cavity and humidification; the third factor is the speed of the gas increases, pollutants can be taken deeply into the respiratory tract in motion time. The mass rate of pulmonary gas increased in motion time, so the air pollutants in the blood will increase. Strenuous exercises increase the oronasal breathing, affecting athlete's sports ability (Flouris, 2006). The nasal mucociliary clearance of a long-distance runner after consecutive days' training is impaired (Muns, et al., 1995), and this will increase the health hazards of air pollution during continuous movement training and competition.

In the polluted air, compared with the quiet state, athletes' bronchial stenosis and obstruction of respiratory tract symptom increased in the motion state (Pierson, 1989). For airway hyperresponsiveness (AHR) and asthma symptoms when athletes appeared in training and in the matches, the current basic ideas are as follows (Kippelen, et al., 2012): the first one is the athletes' common medical condition in the Summer and Winter Olympic Games, the average incidence rate is about 7-8%; the second one is the athletes' respiratory epithelium at the highest levels of competition is prone to injure, this is caused by the necessity of athletes' maintaining high volume in the cold and dry air and polluted air; The third one is for the athletes with high airway sensitivity, the repeated epithelial injury and repair process of the airway leads to the change of epithelial structure and function, this may be the underlying mechanism of recrudesce. For the highly-trained athletes, air pollutants can aggravate airway obstruction, airway high reaction state and asthma symptoms, but some studies suggested that the change in the pulmonary function of athletes can at least partially recovered after stopping the training (McKenzie and Boulet, 2008).

4 AIR QUALITY OF EXERCISE ENVIRONMENT

Traffic conditions and other environmental factors can all affect the level of air pollutants. In recent years, World Health Organization (WHO) and countries have developed or revised the corresponding air quality standards, air quality standard guidelines in the newly revised WHO in 2005, O_3 100 $\mu g/m^3$ (8 h), PM10 50 $\mu g/m^3$ (24 h), PM2.5 25 $\mu g/m^3$ (24 h), NO_2 200 $\mu g/m^3$ (1 h), SO_2 20 $\mu g/m^3$ (24 h), SO_2 500 $\mu g/m^3$ (10 min). Currently, no matter it is environmental science or sports science, no air quality standard for environmental movement authority is formulated; but some studies have put forward a preliminary standard of the air quality according to the concentration of air pollutants on the sports ability in sports training and competition under the condition of the motion. The movement ability of the residents in Los Angeles is obviously damaged, and the conditions of exposure to air pollution were O_3 concentration greater than 0.18 ppm and doing exercise 1 hour every day, 180 days a year (Adams, 1987).

According to the level of pollutants of air pollution affecting lung function in the motion time, the basic standard of the air quality of the marathon is as follows: O_3 is 0.05 ppm, NO_2 is 1 ppm, SO_2 is 0.2 ppm. The exercise intensity of the above standard data sources in the experiment is less than 65%VO2max, the time is about 1 hour, but the CO and PM standards are not put forward. Hesterberg et al (Hesterberg, et al., 2009) integrated a great number of studies of NO_2 exposure time from 30 minutes to 6 hours, the concentration from 0.1 to 3.5 ppm, put forward the concentration of NO_2 with no human health hazards in motion time was 0.2–0.6 ppm. Some of the above air quality indexes are higher than the WHO standards, and some are lower than the WHO standards. Therefore, the air quality standards of sports training and competition needs further study.

ACKNOWLEDGEMENTS

Foundation item:
2013 Nonprofit industry research projects of Hebei Province Environmental Protection Department.
2014 Sports scientific research projects of Hebei Province Sports Bureau (NO.20141002)

REFERENCES

Adir Y, Merdler A, Ben HS, Front A, Harduf R, Bitterman H. 1999. Effects of exposure to low concentrations of carbon monoxide on exercise performance and myocardial perfusion in young healthy men. Occup Environ Med. 56(8): 535–538.

Cakmak S, Dales R, Leech J, Liu L. 2011. The influence of air pollution on cardiovascular and pulmonary function and exercise capacity: Canadian Health Measures Survey (CHMS). Environ Res. 111(8): 1309–1312.

Dong GH, Zhang P, Sun B, et al. 2012. Long-term exposure to ambient air pollution and respiratory disease mortality in Shenyang, China: a 12-year population-based retrospective cohort study. Respiration. 84(5): 360–368.

Drinkwater BL, Raven PB, Horvath SM, et al. 1974. Air pollution, exercise, and heat stress. Arch Environ Health. 28(4): 177–181.

Flouris AD. 2006. Modelling atmospheric pollution during the games of the XXVIII Olympiad: effects on elite competitors. Int J Sports Med. 27(2): 137–142.

Folinsbee LJ, Silverman F, Shephard RJ. 1975. Exercise responses following ozone exposure. J Appl Physiol. 38(6): 996–1001.

Hesterberg TW, Bunn WB, McClellan RO, Hamade AK, Long CM, Valberg PA. 2009. Critical review of the human data on short-term nitrogen dioxide (NO2) exposures: evidence for NO2 no-effect levels. Crit Rev Toxicol. 39(9): 743–781.

Kargarfard M, Poursafa P, Rezanejad S, Mousavinasab F. 2011. Effects of exercise in polluted air on the aerobic power, serum lactate level and cell blood count of active individuals. Int J Prev Med. 2(3): 145–150.

Kippelen P, Fitch KD, Anderson SD, et al. 2012. Respiratory health of elite athletes – preventing airway injury: a critical review. Br J Sports Med. 46(7): 471–476.

Lee K, Yanagisawa Y, Spengler JD, Nakai S. 1994. Carbon monoxide and nitrogen dioxide exposures in indoor ice skating rinks. J Sports Sci. 12(3): 279–283.

McKenzie DC, Boulet LP. 2008. Asthma, outdoor air quality and the Olympic Games. CMAJ. 179(6): 543–548.

Mirkin G. 1982. Air Pollution and Athletic Performance. JAMA: The Journal of the American Medical Association. 248(12): 1511.

Mittleman MA. 2007. Air pollution, exercise, and cardiovascular risk. N Engl J Med. 357(11): 1147–1149.

Muns G, Singer P, Wolf F, Rubinstein I. 1995. Impaired nasal mucociliary clearance in long-distance runners. Int J Sports Med. 16(4): 209–213.

Pierson WE. 1989. Impact of air pollutants on athletic performance. Allergy Proc. 10(3): 209–214.

Poursafa P, Kelishadi R. 2010. Air pollution, platelet activation and atherosclerosis. Inflamm Allergy Drug Targets. 9(5): 387–392.

Rundell KW, Caviston R. 2008. Ultrafine and fine particulate matter inhalation decreases exercise performance in healthy subjects. J Strength Cond Res. 22(1): 2–5.

Volpino P, Tomei F, La Valle C, et al. 2004. Respiratory and cardiovascular function at rest and during exercise testing in a healthy working population: effects of outdoor traffic air pollution. Occup Med (Lond). 54(7): 475–482.

Biomedical Engineering and Environmental Engineering – Chan (Ed.)
© 2015 Taylor & Francis Group, London, ISBN: 978-1-138-02805-0

Recovery of rare earth elements from waste optical glass

Jinxiu Yuan & Jianhua Wang
Weifang University of Science & Technology, Weifang, China

Zhijuan Bao
Linxi Jinding Industrial Park Management Committee, Chifeng, China

Tao Xu
Inner Mongolla Baotou Steel Rare-Earth Hi-Tech co. Ltd, Baotou, China

Xiqing Dong & Xin Cui
Weifang University of Science & Technology, Weifang, China

ABSTRACT: A hydrometallurgical process was performed in order to recover rare earth elements, from waste optical glass. In this process the hydrochloric acid leaching method was used to extract the rare earth elements. The leaching rate of the rare earth was affected by hydrochloric acid concentration, temperature, leaching time, liquid-solid ratio, and stirring rate. The optimal lixiviated condition was obtained in the experiment. Under the optimal combination, the leaching rate of Y, La, and Gd was 100.00%, 99.98%, and 99.96%, respectively.

Keywords: Rare earth, waste optical glass, hydrochloric acid

1 INTRODUCTION

In the late 1930s, the United States Kodak Company developed the first optical glass which contained lanthanum, thorium, tantalum, and other elements, had a high refractive index, and was of a low-dispersion type [1]. At present, the world annual demand for rare earth optical glass is about 20,000 t [2]. RE oxides content in the glass was 10% to 60%. It is estimated that the amount of the rare earth glass waste is approximately 4,000 t annually. As a precious and limited resource, rare earth should be recovered from rare earth glass waste. We must attach great importance to rare optical glass recycling. However, there has been very little in the literature to recover rare earth elements from optical glass waste [3–5]. In this paper, we attempted to recover the rare earth from the optical glass waste using acid dissolution.

2 EXPERIMENT

2.1 Materials and instrument

The waste optical glass, supplied by a china company, was washed with distilled water, dried at in an oven at 105°, crushed, and milled with a grinding machine, and sieve residue 150 mesh was obtained. The prepared powder was used as an experimental sample. The chemical composition obtained is shown in Table 1. Water used for the experiment was deionized water,

Table 1. Chemical compositions of rare earth optical glass waste.

Component	La_2O_3	Y_2O_3	Gd_2O_3	ZrO_2
Content/(wt.%)	36.21	5.84	15.25	6.87
Component	Nb_2O_5	B_2O_3	SiO_2	ZnO
Content/(wt.%)	8.18	19.85	3.57	0.80

and all reagents were of analytical grade. ICP-AES (US Varian, Inc) and ICP-MS were used to detect.

2.2 Experimental methods

Hydrochloric acid was used to extract the rare earth in this study. L_{16} (4^5) orthogonal experiment was designed with four factors at five levels. In order to investigate the effects including hydrochloric acid concentration, extraction temperature, leaching time, liquid-solid ratio, and stirring rate on the leaching rate of the rare earth elements. The results of the experimental influencing factors and levels are shown in Table 2.

2.3 Results and analysis

2.3.1 Range analysis

Table 3 shows that the leaching rates of La Y and Gd are at their highest in the 16th experiment, and that

Table 2. $L_{16}(4^5)$ orthogonal experiment design.

	C_{HCl} mol/	Temperature	Leaching time	Liquid-solid ratio	Stirring rate
Level	L (A)	°C (B)	min (C)	mL/g (D)	r/min (E)
1	1	60	20	6	150
2	2	70	40	8	200
3	3	80	60	10	300
4	4	90	80	12	400

Table 3. Results of the orthogonal experiment.

Entry	A	B	C	D	E	η Y/%	η La/%	η Gd/%
1	1	2	3	3	2	59.24	59.10	57.45
2	2	4	1	2	2	61.96	76.75	76.15
3	3	4	3	4	3	100.00	98.28	97.88
4	4	2	1	1	3	75.00	75.62	75.59
5	1	3	1	4	4	58.70	58.53	58.58
6	2	1	3	1	4	65.22	65.17	65.76
7	3	1	1	3	1	45.11	45.82	45.35
8	4	3	3	2	1	90.22	90.67	92.03
9	1	1	4	2	3	55.98	56.35	56.88
10	2	3	2	3	3	90.22	90.59	93.16
11	3	3	4	1	2	88.04	88.33	91.46
12	4	1	2	4	2	77.17	76.75	79.74
13	1	4	2	1	1	46.20	46.39	47.24
14	2	2	4	4	1	65.22	65.74	67.46
15	3	2	2	2	4	85.33	84.52	90.70
16	4	4	4	3	4	100.00	99.90	99.58

Y

k_1	55.03	60.87	60.19	68.61	61.68
k_2	70.65	71.20	74.73	73.37	71.60
k_3	79.62	81.79	78.67	73.64	80.30
k_4	85.60	77.04	77.31	75.27	77.31
R_j	30.57	20.92	18.48	6.66	18.61

La

k_1	55.09	61.02	64.18	68.88	62.16
k_2	74.56	71.24	74.56	77.07	75.23
k_3	79.24	82.03	78.31	73.85	80.21
k_4	85.74	80.33	77.58	74.83	77.03
R_j	30.64	21.01	14.13	8.20	18.05

Gd

k1	55.04	61.93	63.92	70.01	63.02
k2	75.63	72.80	77.71	78.94	76.20
k3	81.35	83.81	78.28	73.89	80.88
k4	86.73	80.22	78.85	75.92	78.66
Rj	31.70	21.87	14.93	8.93	17.86

the leaching rates are 100.00%, 99.90%, and 99.58%, respectively.

The correlation values in Table 3 are calculated by the following formulas:

$$\eta = \frac{c \times V}{m \times w}\%$$ (1)

where, $\eta(\%)$ is leaching rate, c(g/L) is the concentration of corresponding element in the filtrate, V(L) is lixivium volume, m(g) is the quality of the sample, and w(%) is the content of corresponding element.

$$k_n = \frac{K_n}{4}\%$$ (2)

where, kn is the average experimental indicator with the corresponding factors of n levels, and kn is the sum of experimental indicators with the corresponding factors of n levels. Rj is the range:

$$R_j = k_{n(max)} - k_{n(min)}$$ (3)

R_j of the La, Y, and Gd show a decreasing order of $R_A > R_B > R_E > R_C > R_D$, which indicates that the relationship between primary and secondary factors affecting experimental indicators is A > B > E > C > D. Therefore, hydrochloric acid concentration shows the maximum effect, while the liquid-solid ratio shows the minimum effect. The optimal extraction conditions of Y, La, and Gd are $A_4B_3C_3D_4E_3$, $A_4B_3C_3D_2E_3$, and $A_4B_3C_4D_2E_3$, respectively. When the leaching time is 60 min, Y and La reach the maximum leaching rate, while the leaching time of Gd is 80 min. However from 60 min to 80 min, the leaching rate of Gd changes very little. Considering the efficiency and energy consumption, a leaching time should be chosen as 60 min. When the liquid-solid ratio is 8, La and Gd reach the maximum leaching rate, and the liquid-solid ratio of Y is 12. When the liquid-solid ratio difference is from 8 to 12, the variation on the leaching rate of Y is less than 2%; the Y content in the waste glass is much smaller than that of La and Gd. Therefore the ratio of 8 is more conducive to the total rare earth leaching rate. In summary, under the combination $A_4B_3C_3D_2E_3$, the total leaching rate of rare earth is optimal.

2.3.2 Factor analysis of the leaching rate of rare earth elements

The effect of HCl concentration on the extraction of La, Y, and Gd is shown in the Figure 1. The results indicate that the leaching rate of La, Y and Gd have a strong relationship with HCl concentration. Increases in HCl concentration sharply induce the rare earth elements extraction, and the leaching rate reaches a maximum in 4 M. Therefore, 4 M HCl could be seen as an acceptable level in the experiment.

Figure 2 shows the effects of temperature on La, Y and Gd extraction in the orthogonal experiment. It can be seen that the leaching rate of La, Y and Gd gradually increases in the range of 60~80°; the highest leaching rate could be obtained under the 80°. Due to the influence of other factors in the orthogonal experiment, at the temperatures of 80~90° all leaching rates decline slightly. The optimal temperature of 80° which is identified in the experiment falls into the appropriate ranges.

It can be clear from Figure 3 that leaching time shows a significant effect on the leaching rate of La, Y,

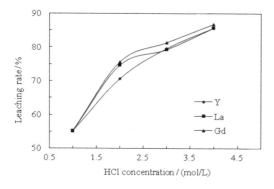

Figure 1. Effect of HCl concentration on the leaching of rare reath elements.

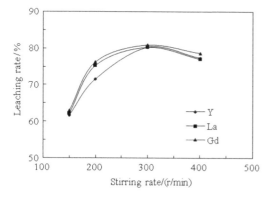

Figure 2. Effect of stirring rate on the leaching of rare reath elements.

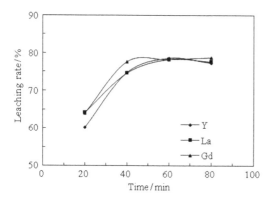

Figure 3. Effect of time on the leaching of rare reath elements.

and Gd. All the leaching rates of La, Y and Gd increase rapidly in the range of 20~40 min, increase slowly from 40 to 60 min, and remain constant after 60 min. Therefore, for an optimal performance, the leaching time should be 60 min.

Figure 4 reveals that the leaching rate of La, Y, and Gd rises along with the liquid-solid ratio increasing from 6 to 8, and achieves a maximum in L/S of 8. Due to the effects of other factors in the orthogonal experimental, the leaching rates decline in L/S from 8

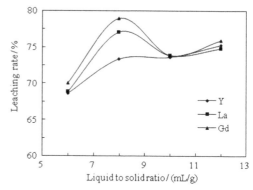

Figure 4. Effect of liquid to solid on the leaching of rare reath elements.

Table 4. Chemical compositions of HCl filtrate and the leaching rate of elements.

Elements	Y	La	Gd	B	Si	Zn
Concentration (g/L)	1.84	12.35	5.29	2.49	0.019	0.23
Leaching rate/%	100.00	99.98	99.96	99.76	3.10	89.84

Elements	Nb	Zr
Concentration (mg/L)	<0.01	0.045
Leaching rate/%	0	0

to 10, and increase slightly in L/S from 10 to 12. The results of the fluctuation in leaching rates also indicate that the influence degree of the liquid-solid ratio on the leaching rate is lower than other factors. The optimal L/S of 8 should be selected in the experiment.

The dependence of the leaching rate of La, Y, and Gd on stirring rate is shown in Figure5. As can be seen, leaching rates rise rapidly with the increase of stirring rates in the range 150~200 r/min, and increase slowly from 200 to 300 r/min. Because of the influence by other factors in the orthogonal experimental, leaching rates have a slight reduction in the range 300~400 r/min. Therefore, the 300 r/min is the most appropriate as the optimal stirring rate. Chemical compositions of HCl filtrate and the leaching rate of elements are shown in Table 4.

The above results reveal that the optimal conditions for leaching of the rare earth elements are 4 M HCl, a temperature of 80°, a leaching time of 60 min, an L/S of 8, and a stirring rate of 300 r/min. Under the optimal conditions of leaching of rare earth, lixivium of rare earth was obtained.

3 CONCLUSIONS

The hydrometallurgical process was designed to recover the rare earth elements from the optical glass wastes. The optimal conditions for leaching the rare earth were identified as 4 M HCl and a liquid-solid ratio of 8 at 80°, with stirring of 300 r/min for 60 min. Under the above conditions, the leaching rates of La, Y, and Gd were 100.00%, 99.98%, and 99.96%, respectively.

REFERENCES

[1] Liu Guanghua. Rare earth materials and application technology [M], Beijing: Chemical Industry Press, 2005. (in Chinese)

[2] Li Weimin. The present situation and the development of rare earth optical glass [J]. Rare Earth Information, 2012, 338: 19

[3] Zhao Guoyan, Huang Dongfeng, Guo Jinfu, et al. Research on extraction cerium dioxide from glass refuse [J]. Inorganic Chemicals Industry. 2012, 44: 37.

[4] Jiang Yuren, Atsushi shibayama, Kejun liu, et al. A hydrometalurgical process for extraction of lanthannum yttrium and gadolinium from spent optical glass [J]. Hydrometallurgy, 2005, 76: 1.

[5] Mitsuaki MATSUDA, Atsushi SHIBAYAMA, Keiei MATSUSHIMA, et al. Recovery of rare earth from waste optical glass by precipition and solvent extraction [J]. The mining and materials processing institute of Japan, 2003, 119: 668.

Biomedical Engineering and Environmental Engineering – Chan (Ed.)
© 2015 Taylor & Francis Group, London, ISBN: 978-1-138-02805-0

Approaches for the promotion of enterprise technology management capability, based on the sci-tech innovation platform cluster

Lijun Deng & Qiuming Wu
Management School of Fuzhou University, Fuzhou, Fujian, P.R. China

ABSTRACT: In the era of a knowledge-based economy, technology management capability becomes key for enterprises to achieve competitiveness. The approaches for the promotion of technology management capacity are studied, based on the analysis of the pattern of a sci-tech innovation platform cluster and its construction path, in order to impact science and technology resource management, technical management, and technology quality management.

Keywords: Technology management, sci-tech innovation platform, cluster

1 INTRODUCTION

With the rapid development of world science and technology, science and technology have become important aspects of reflecting the comprehensive national strength, where the capability of independent innovation is its core content. To further enhance the capability of independent innovation, policies have been drafted to promote the development of a science and technology innovation platform. Studies on how to improve the technology management capability of enterprises, based on a sci-tech innovation platform cluster are important in both theory and practice.

2 RESEARCH STATUS

Although the enterprise managers and scholars recognize that "the technology itself is not competitive, management is the competitiveness", there are few studies of enterprise technology management capacity. Beatriz (2001) proposed that the keys to enhancing the technical capacity were evaluation techniques, the optimization of organization, rich sources of technology, the protection of intellectual property rights, enhanced absorptive capacity, and surveillance technologies. SPRU (2004) argued that upgrading technological management capacity was performed based on scan, focus, resources, and network of learning these five activities sequentially or simultaneously. Chinese scholar, Li Yongming (2002), proposed that in order to improve the capability of enterprise technology management was to pay attention to research and development, using "project management" to achieve the goals. Wu Guisheng (2002) believed that the

keys to raising the level of technical management of the enterprise were advanced design, production, the application of technical and organizational implementation, the introduction of secondary innovation on the basis of digestion and absorption, and independent technology development. Most of the researches on technology innovation platforms are for individual platforms, only some of the scholars from the relations between platforms in terms of the framework and network. Domestic scholar Wu Shouhui (2009) proposed the "top design + backbone nodes + regional integration" model, the building of the science and technology platform system structure. Sun Qing (2012) built the system framework of the regional science and technology innovation platform, proposing a path for the government-led research cooperation and business network development. Literature has not seen similar researches abroad.

To sum up, some scholars have recognized the impact of science and technology innovation platforms on the promotion of enterprise technology management, but no scholars at home and abroad have conducted researches on the promotion of enterprise technology management based on an innovation platform cluster. Therefore, this text has extended and deepened the research on business technology management.

3 A CONSTRUCTION PATH FOR A SCI-TECH INNOVATION PLATFORM CLUSTER

Since the proposal of an innovative State target, the pace of the construction of science and technology innovation platforms is speeding up in all parts. Take

Fujian province for example, where regulations were continuously enacted. Such medium-and long-term plans in Fujian province (2006-2020) science and technology platform development plan, Science and technology platform development plan for the 11th five-year in Fujian province, Administrative measures on construction plan of science and technology innovation platform in Fujian province. The platform and operation mechanism has initially been set up. It has promoted the construction of regional innovation system, and it has also provided important support on industrial, economic, and social development. In 2011, there were 123 key laboratories, 183 engineering centres, 35 industrial research and development centres, 297 technological centres, 14 science and technology resources sharing platforms, 112 sci-tech intermediate service platforms, and 8 science and technology achievement transformation platforms in Fujian province. In order to further play the role of science and technology innovation platform, we propose to construct a science and technology innovation cluster.

3.1 Construction of a science and technology innovation platform cluster mode

We built science and technology innovation platform cluster mode based on the present situation and characteristics of the cluster. Firstly, the science and technology innovation platforms in accordance with the same or similar industries, and according to different research and innovation capacity, form unions with core science and technology innovation platforms as the main lines in the leaves. Secondly, a coalition of science and technology innovation platforms in accordance with the same or similar industries is formed by issues. Finally, a science and technology innovation platform in fields of Unions will be integrated with the project, forming a tree of technology innovation platform cluster.

The cluster of innovation platforms in Fujian province, straightened out the relationships between the various platforms, can keep abreast of all areas, developments in the various industries and various platforms, and can complete major projects by the appropriate virtual teams.

3.2 Project-oriented innovation platform cluster operation mode

An innovation platform cluster, an organizational model of the mutual collaboration of the innovation platform, has features such as relevance, complementarity, networking, virtuality, self-organization, and concertedness. An innovation platform cluster is project-oriented. First, identify the core platform according to the project, then break the project into subprojects, next find the right cooperation platforms, and form a virtual organization to finally finish the major projects.

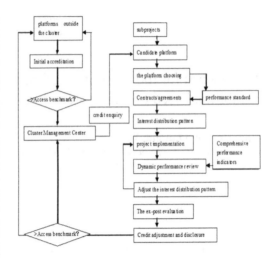

Figure 1. Innovation platform cluster management process.

3.3 Operation and management mechanism of innovation platform cluster design

Innovation platform cluster is a quasi-market mechanism, which is between market and enterprises mechanism. It is inapplicable of using the long-term contracts enterprise governance model, or the short-term contract market governance model. The technology innovation platform cluster adopts the dynamic multi-continuous contracts governance model, which is determined by the coupling degree of cooperation platforms. Information asymmetry exists between the platforms. Therefore, the key to the operation and management of the science technology innovation platform cluster is to establish an administrative centre with the functions of competency assessment, credit transfer, credit services, and credit security. The management of innovation platform cluster including prior appraisals, monitoring and ex-post evaluations, will be based on the characteristics of the project as well as the relationship between subprojects. (As shown in Figure 1).

Science and technology are the primary productive forces. Therefore the knowledge infrastructure comprising those who promote the development of science and technology of public research institutions, universities, standards organizations and the protection of intellectual property rights plays an important role in the national sci-tech innovation capacity development. The science and technology innovation platform as the important contents of knowledge infrastructure plays a crucial role in upgrading the enterprise's technology innovation capability.

A case study of Fujian province, according to the 2010 comprehensive evaluation index of the technology development level and the science and technology innovation platform in all districts and cities of Fujian province (Table 1), compared to Fuzhou city (capital cities, infrastructure, science and technology talent pool), and Xiamen city (one of the earliest special economic zones with high level of economic development,

294

Table 1. The comprehensive evaluation of development levels of science and technology and science and technology innovation platforms in districts and cities of Fujian province.

Cities	R&D personnel (man-year)	R&D expenditure (Billion)	Regional GDP (Billion)	High-tech industry output value (Billion)	Total number of technology platforms
Fuzhou	27044	54.55	3123.41	1697.25	286
Xiamen	22648	50.98	2060.07	2591	138
Putian	1978	4.82	850.33	346.08	40
Sanming	3531	7.02	975.10	327.86	40
Quanzhou	9698	24.53	3564.97	1165.63	123
Zhangzhou	5317	12.16	1430.71	449.56	45
Nanping	2407	6.07	728.65	152.74	34
Longyan	2893	8.58	990.90	264.46	39
Ningde	1221	2.19	738.61	79.03	24

and advantage funds for science and technology inputs and outputs), the R&D funding in Quanzhou city was not the most, but the total GDP was the largest. Learned through further analysis, most of the science and technology innovation platforms in Fuzhou and Xiamen are in colleges, universities, and research institutes while more than 84% of science and technology innovation platforms belong to the Enterprise Technology Centre in Quanzhou City.

4 THE MEANS OF IMPROVING THE TECHNOLOGY MANAGEMENT CAPACITY OF ENTERPRISES IN INNOVATION PLATFORM CLUSTER

4.1 The upgrade of enterprises' technical resource management capability in the innovation platform cluster

The way to upgrade the enterprise resource management skills in the innovation platform cluster is embodied in five main areas:

(1) Funds management. Because cluster technology innovation platform is a project-oriented, each enterprise is responsible for enterprise's essential part of the project, which is more suitable for enterprises to understand exactly the amount of resources required in a project, and to estimate the accuracy of different funding requirements, which is more likely to come to the technical activities of the usage of the funds.
(2) Device management. Since the innovation platform cluster is a dynamic multi-continuous contract governance model, companies can learn the enterprise's equipment management experience from cooperative groups in the process, as well as sharing related equipment according to the needs of technology projects, thus saving the cost to enterprise.
(3) Human resource management. The innovation platform clustering is a virtual organization, therefore businesses can build virtual teams to address the enterprise's technical personnel

required for the project, reducing the cost of the enterprise's technology mining and cultivation of talents.
(4) Information management. The key to operating and managing the innovation platform cluster is the management centre, whose specific features include the assessment of quality and ability of scientific and technical personnel, platform scale, management, technology, reliability, compatibility and ability of cooperation as well as providing information, disclosure inquiries, credit information, and credit monitoring services. The innovation platform cluster establishes an effective technical information delivery channel to promote the information management level of enterprises.
(5) Results management. The innovation platform cluster includes technical research and development collaboration platforms, sci-tech intermediate service platforms, technological resource sharing platforms and technology platforms. Therefore, enterprises in cluster can quickly promote their scientific and technological achievements to achieve the market.

4.2 The upgrade of enterprises' technical management ability in the innovation platform cluster

The path to upgrading the enterprise's technical management ability in the innovation platform cluster lies in two main areas:

(1) Cultural management. The Science Technology Park as a transformation innovation platform, has been hailed as the "plots" of science and technology environment innovation in China. Many policies were extensively promoted after being successfully tested in the Science Technology Park, building conducive innovation environment in a larger context. Therefore, the enterprises within the cluster are more easily focused on innovation environment changes, timely adjusting the enterprise strategies, eliminating hindering factors of enterprise technical innovation culture

and corporate culture, which has always been consistent with the innovation environment.

(2) Organization management. The innovation platform cluster is a flat and networked virtual organization. Companies need to adjust the technology strategy to adapt the new requirements.

4.3 The means of improving the technology management capacity of enterprises in the innovation platform cluster

The method of improving the enterprise technology management ability in the innovation platform cluster is found in three main areas:

(1) Quality management. The innovation platform cluster management centre implements the total quality management of the project by prior appraisals, monitoring, and ex post evaluations, prompting companies to establish a quality management system in order to ensure the quality of the technical completion of the project.

(2) Standardization management. The innovation platform cluster is an integrated system, while the cluster access benchmark is a standardized interface design. Therefore, the technology innovation cluster promoted the establishment of a standardization system of enterprises within the group, including technology and management standards, and facilitated resource sharing within the group.

(3) Risk management. Since enterprises within the cluster were only responsible for a subproject, by greatly reducing the technical risks, they improved the level of enterprise risk management.

5 SUMMARY

This article firstly reviewed the studies on technical and management capacity enhancement of enterprises within an innovation platform cluster. It found that while scholars have begun to focus on the impact of science and technology innovation platforms for enterprise technology management, there was no scholarly research on approaches to the promotion of the technology management capabilities of enterprises based on a sci-tech innovation platform cluster. Then, the innovation platform cluster building mode and path

were put forward, after the research into the science and technology innovation platform in Fujian province. Finally, taking Fujian province as an example, the article explained the role of the science and technology innovation platform for enterprise promotion, and proposed the path of upgrading the enterprise technology management capabilities from the enterprise technology resource management, the enterprise organization and management, and the enterprise technology quality management.

This is an exploratory research, and there are limitations. This article only discussed the matter in principle, and further research needs to be extended and deepened.

6 SUPPORT

This research is financially supported by the Humanities and Social Sciences project of the Ministry of Education (No. 11YJA630153).

REFERENCES

Beatriz Cristina Brito Vinas, John Bessant, Gilberto Hernandez Perez, Arnaldo Alvarez Gonzalez. A conceptual model for the development of technological management process in manufacturing companies in developing countries[J]. Technovation, 2001, 21: 345–352.

Li Yongming. Advanced manufacturing technology and the independent development ability of Enterprise [M]. Beijing: China machine press, 1998: 28–42.

Sun Qing, Wang Hong. System and operational mechanism of local science and technology innovation platform [J]. China Science and technology forum 2010 (3): 16–19.

Tik Tak, Benshante, Keith pavitt. Innovation management: integration of technological change, market changes and organizational changes [M]. Beijing: Tsinghua University Press, 2004: 54–55.

Wu Gui, Wang, Xie Wei. Technology growth and management of enterprises in China [J]. Management of research and development, 2002 (10): 34–40.

Wu Shouhui. System construction and countermeasures of science and technology basic conditions platform [J]. China Science and technology forum, 2009 (10): 3–8.

Xinyue Hu. Task-oriented virtual enterprise construction and operation management [M]. Beijing: Science Press, 2011: 124.

Biomedical Engineering and Environmental Engineering – Chan (Ed.)
© *2015 Taylor & Francis Group, London, ISBN: 978-1-138-02805-0*

Interest demands analysis of the core stakeholders in the Chinese rural residential electrovalent policy

Bin Luo & Hui Li
Beijing Institute of Technology, Beijing, China

ABSTRACT: The core stakeholders of Chinese rural residential electrovalent policy are power grid enterprises, government price management departments, and the rural residents. This paper analyses the interest demands of the core stakeholders and the key variables of interests, and establishes the utility function of interests through the historical data and data analysis methods.

1 INTRODUCTION

Although there are many stakeholders in Chinese rural residential electrovalent policy, including power grid enterprises, rural residential users, government price control departments (hereinafter referred to as the Government), power grid enterprises shareholders, internal employees in power grid enterprises, tax authorities, banks, power generation companies, other energy industries, the media, the power industry associations and agricultural management department, the core stakeholders are mainly power grid enterprises, government price management departments and the rural residential users (Luo Xiyang, 2012). This article will analyze interest demands of core stakeholders respectively and discuss key variables and utility functions.

1.1 *Literature review*

There are many applications of utility functions, such as pricing (Chen He, 2006), investment (Koichi Matsumoto, 2006), and decision-making (Bugera, V. 2002). The representation of utility function methods is widely used, such as S-type value function (Xiong Jixia, 2013), Epstein-Zin utility function (Tan Peng, 2006), E-V utility function (Qiang Shao, 2011), mean variance utility functions (Jiang Tiejun, 2009), power utility function (Yao Shengbao, 2013), recursive utility function (Zhu Weili, 2004), double exponential utility function (Zhou Qingjian, 2011), Von Neumann utility function(Fishburn,1979), and so on. This paper will analyze the interest demands of Chinese rural residential electrovalence, and try to use a new method to analyze the utility function.

2 INTEREST DEMANDS AND UTILITY FUNCTION ANALYSIS OF POWER GRID ENTERPRISES

2.1 *Proposal of key variables and a quantitative index*

Chen Qi (2008) found the utility function of hospitals mainly focused on cost, revenue, and influence through the analysis of the doctor-patient relationship and its own management, its income mostly includes government subsidies, the income of selling medicine and labor service fees. Wang Bin (2014) pointed out that the two major factors influencing the profits of power grid enterprises are sales of electricity and the electricity purchasing cost based on the analysis of the double auction game model when the grid corporation purchases electricity. These analyses have reference values for the research of this paper.

For grid corporations, spending is the main power supply cost for rural residents, and the income has two sources: income from selling electricity and government subsidies. The government gives subsidies for power grid enterprises (using the way of cross subsidy), because the cost of Chinese rural residential electricity power is much higher than electrovalence. Therefore, this paper argues that the grid corporations in rural residential electrovalent policy focus on three variables: government subsidies, the income of selling electricity and the cost of power supply.

2.2 *The analysis of utility function*

Mao Sheng (2013) argued that the government subsidies should be greater than the difference of average price and average cost, and then he put forward the method of government subsidies for quasi-public

Table 1. Key variables and index of rural power users.

key variables	fairness	endurance capacity to tariff	satisfaction with the demand for electricity	power quality	safety
index	rural per capita net income/urban per capita income	relative tariff level	reliability of power supply	voltage qualified rate	number of accidents

goods. Lu Xue (2013) holds that public transport companies always sought to maximize their own profit; their profitability can be expressed as $\pi = pq - c + d$, where p is the ticket price, q is the number of the transportable people, c is the cost, and b is the government subsidies. Electricity is quasi-public goods. At present, the Chinese government pricing under the sales price has seven major categories, and the income of the power grid enterprises in each category is different. Therefore, in order to seek the benefit maximization, grid corporations should make rural residential power proceedings not lower than that of industry average.

Based on the above analysis, the utility function of power grid enterprises can be expressed as:

$$u_1 = (P - C + M)/\pi, \quad \pi = \overline{P} - \overline{C}, \quad \pi \neq 0,$$

where P = rural residential electrovalence; C = unit electricity costs of rural residents; M = government subsidies for unit power of grid enterprises; $\overline{P}$ = average electrovalence of grid industry; $\overline{C}$ = average cost of unit power of grid industry; $\overline{\pi}$ = average profit of unit power.

3 INTEREST DEMANDS AND UTILITY FUNCTION ANALYSIS OF RURAL RESIDENTS

3.1 Proposal of the key variables and a quantitative index

Wang Bo (2013) has put forward the idea that consumers are supposed to take their own endurance capacity and public service satisfaction into consideration in the usage and consumption of public goods, and correspondingly it is user's endurance capacity of electricity price and satisfaction of the power supply service in the electricity price policy. In addition, due to the universality of electrovalent policy and the uniqueness of power, fairness and security are two factors that rural electricity users also focus on. In conclusion, this paper argues that Chinese rural residential electricity users focus on five key variables including fairness, endurance capacity of tariffs, satisfaction with the demands for electricity, power quality, and safety index.

Huang Chao (2014) pointed out that reliability of the power supply is an important index of the power supply quality, while the significant index of the power quality is measured by a voltage qualified rate (Zou

Jian, 2013). The ratio of electricity expenses per capita of farmers and the rural net income per capita can replace the endurance capacity of the tariff index. For rural users, safety of the power index mostly refers to the number of accidents, including personal accidents and accidents related to the power grid. In addition, we use A to represent a fairness index, which is the ratio of B and C. B is the ratio of the electricity price level and the urban per capita income, and C is the ratio of the electricity price level and the rural per capita net income amongst them. Then, the fairness index can be replaced by the ratio of the rural per capita net income and the urban per capita income. This is shown in Table 1.

3.2 The analysis of utility function

Yang Xianju (2009) has studied a regulation model about sales of electrovalence, and has put forward the consumers' utility function:

$$u(q) = \int p(q)dq - pq - \gamma mp,$$

γmp is the government transfer payments and the sum of the shadow of the transfer payment, $\int p(q)dq$ is consumer utility, and pq is the cost that the consumers paid. This can calculate the consumer utility more accurately in the current ladder electrovalent policy, but it cannot grasp the interest demands of consumers because it just focuses on the electricity number. Ding Bo (2005) analysed doctor-patient relationships and proposed the utility function:

$$u = \alpha_h \Delta h - \alpha_{cc} c_c + \alpha_a v_a (\alpha_h - \alpha_{cc} + \alpha_a = 1),$$

Consumers pay attention to the improvement of health conditions Δh, cost c_c, and additional values v_a; α represents the consumer's health preferences, the cost, and the additional value, respectively. It is easy to express or explain all the variables effectively by using weighting function to represent the consumers' utility function. The disadvantage is that some variables cannot be quantified, such as improvement of health.

This paper will seek quantifiable indicators in order to achieve quantified results on the basis of the entropy weight method and the existing data. The main steps that entropy weight method determines weights of each variable are shown as below.

Table 2. Original data of index affecting the interests of the rural users.

Year	voltage qualified rate %	reliability of power supply %	accidents	electricity sales (billion kwh)	rural net income per capita (yuan)	urban income per capita (yuan)	Net income of the farmers in Hubei province (yuan)	Farmers' electricity bills in Hubei province (yuan)
2003	96.670	99.278	60	2883	2622.2	8472	2566.76	21.840
2004	97.180	99.308	48	2972.9	2936.4	9422	2890.01	24.780
2005	95.800	99.382	54	2680.72	3254.9	10493	3099.2	30.440
2006	96.200	99.490	16	3550.88	3587	11759.5	3419.4	38.170
2007	96.768	99.540	13	4365.6	4140.4	13785.8	3997.41	46.430
2008	97.050	99.589	9	4665.86	4760.6	15780.8	4656	58.770
2009	97.250	99.615	10	5006.98	5153.2	17174.7	5035	70.080
2010	97.477	99.636	7	6275.57	5919	19109.4	5832	80.500
2011	97.688	99.665	4	7378.65	6977.3	21809.8	6898	85.040
2012	98.074	99.735	12	8492.82	7917	24565	7852	112.57

*The data source: The data of power supply reliability, voltage qualification rate, and accidents is from the Chinese electric power yearbook; Rural net per capita income and urban per capita income data is from the Chinese statistical yearbook; Data of net income and electricity bills of the farmers in Hubei province is from a statistical yearbook of Hubei province (because there is no national per capita electricity bill data, and the level of net per capita income of farmers in Hubei province is mostly close to the national average, so this paper adopts the related data of Hubei province)

Table 3. Data of index.

Year	voltage qualified rate	reliability of power supply	the accident rate	safety	relative tariff level
2003	96.670	99.278	0	0.310	0.991
2004	97.180	99.308	0.224	0.312	0.991
2005	95.800	99.382	0.0321	0.310	0.990
2006	96.200	99.49	0.783	0.305	0.989
2007	96.768	99.54	0.857	0.300	0.988
2008	97.050	99.589	0.907	0.302	0.987
2009	97.250	99.615	0.904	0.300	0.986
2010	97.477	99.636	0.946	0.309	0.986
2011	97.688	99.665	0.974	0.320	0.987
2012	98.074	99.735	0.932	0.322	0.986

Standardize the raw data matrix, and get the standardization of the matrix for each index. Suppose the standardization of the matrix is r_{ij}, then the standardized formula is:

$$r_{ij} = \left(a_{ij} - \min_j |a_{ij}|\right) \Big/ \left(\max_j |a_{ij}| - \min_j |a_{ij}|\right)$$

In an evaluation problem of m evaluation index and n evaluation objects, the entropy of index for 'i' is:

$$h_i = -k \sum_{j=1}^{n} f_{ij} \ln f_{ij}.$$

In the formula,

$$f_{ij} = r_{ij} \Big/ \sum_{j=1}^{n} r_{ij}; k = 1/\ln n; f_{ij} \ln f_{ij} = 0 \, (f_{ij} = 0)$$

calculate the deviation degree after defining the entropy of the index of the ith, defined as $g_i = 1 - h_i$. Finally, calculate the weight, of which the weight calculation formula for factor 'i' is:

$$w_i = g_i \Big/ \sum_{i=1}^{m} g_i.$$

In order to express the utility function more accurately, this paper processes the original data to some extent, making numerical values reflect the farmers' satisfaction as far as possible to facilitate the writing of interest expression. For instance, we adopt accidents numbers of per unit of electricity in the handling of accidents, it can be more conducive to compare the change of accident numbers. And calculate the ratio of each year with the year of maximum accident number as a base. But such results cannot well reflect the satisfaction of rural residential users, so we need 1 minus the ratio of each year to get the final value. The index of relative tariff level also makes the similar treatment. As shown in Table 2.

To deal with the raw data, and get the data of each quantitative index, as shown in Table 3, calculate the entropy value of each index: $e = (0.918, 0.893, 0.889, 0.833, 0.853)$.

Then, calculate the deviation of each index and carry out the normalization processing, so as to get the weight of each factor; the result is: $w = (0.135, 0.174, 0.18, 0.272, 0.239)$.

299

The utility function of rural residential power users can be expressed as:

$$\dot{u}_2 = 0.272e + 0.239n + 0.174r + 0.135q + 0.18s$$

where e = fairness; b = relative tariff level; r = reliability of power supply; q = voltage qualified rate; s = safety.

Because of the same residential electrovalence between urban and rural areas, the final result of the fairness variable is equal to the ratio of rural residential income and urban residential income, so we can regard it as a constant. Variable of bearing capacity of tariffs is related to price and subsidy; this article will see the sum of price P and the subsidy M as one variable of the negotiation between power grid enterprises and rural residents, considering the relationship between the electrovalence and subsidy, assuming that $W_i = P_i + M_i$. The electric power quality such as the voltage qualified rate, the reliability of the power supply, and the safety index are relevant to cost (the cost also refers to the cost of per unit of electricity). Based on this, we do the corresponding regression analysis using the data from 2004 to 2012 by SPSS software; if the fitting degree is satisfactory, then get the following results:

$$u_2(W, C) = 0.272e + 0.239 - 0.00853W + 0.868C$$

where e = fairness; W = sum of electrovalence and subsidy; C = cost of per unit of electricity.

4 INTEREST DEMANDS AND UTILITY FUNCTION OF GOVERNMENT

The government is seeking to maximize the social welfare, which is the sum of consumer surplus and the profits of grid corporations (Yang Xianju, 2009); simply speaking, the government pursues the interests of the power grid enterprises and rural users. That is to say, the government's utility function is the sum of utility between the power grid enterprises and rural residents.

In terms of the actual situation of the Chinese rural residential power supply, the grid corporations form a monopoly while farmers remain a typically vulnerable group. Therefore, we recommend that the government should pay attention to protecting the interests of farmers in this game of interests.

5 CONCLUSIONS AND RECOMMENDATIONS

(1) Research on interest demands and utility function of core stakeholders in rural residential electrovalent policy is the foundation for building game models.

(2) In the game of rural residential electricity, rural users focus on security issues, voltage qualified rates, the reliability of the power supply, the endurance capacity of the tariff and fairness, and the weight of each index increase, in turn.;he grid corporation pays attention to residential electricity prices, government subsidies, and the average profit of power industry; the government focuses on maximizing the interests of both.

(3) Due to the monopoly of the power industry, we suggest that the government pay more attention to protecting the interests of the farmers in the game, because farmers are a vulnerable group.

REFERENCES

Bugera, V., Konno, H. and Uryasev, S., Credit cards scoring with quadratic utility functions [J]. Multi-Crit. Decis. Anal., 2002 (11): 197–211.

Ching-Ter Chang, Zheng-Yun Zhuang.The Different Ways of Using Utility Function with Multi-choice Goal Programming [J]. 2014(275): 407–417.

Fishburn, P. C. Kochenberger, G. A. Two-piece Von Neumann-Morgenstern Utility Functions [J]. Decision Sciences, 1979 (10): 503–518.

Huang, J., Demand Functions in Decision Modeling: A Comprehensive Survey and Research Directions [J]. Decision Sciences, 2013(44): 557–609.

Koichi Matsumoto. Optimal portfolio of low liquid assets with a log-utility function [J]. Finance Stochast. 2006 (10): 121–145.

Lu Xue, Sun yingjun. Analysis of Urban Preference Public Transportation under Game Theory [J]. Wuhan University of Technology (Social Science Edition), 2013 (1).

Luo Xiyang, Luo Bin. Analysis on Rural Power Price Policy Stakeholders and Mutual Benefit [J]. Issues in Agricultural Economy, 2012 (5).

Mao Sheng. How to Make up Bus Enterprise's Losses [J]. Hubei Today, 2013 (2).

Qiang Shao, Zhongbing Wu, Feng Zhou. E-V Utility Function and Its Application in Shanghai Securities Market [M]. Innovative Computing and Information.2011. 109–114.

Wang Bin, Jiang Wen, Wang Jinyu, Li Hua, Zhang Yuelong. Research on Double Auction Game of Grid Purchasing Electricity[J]. Science Technology and Industry, 2014,(1).

Yang Xianju. Sales Price of Electricity Regulation Model and Policy Analysis [D]: Hu Nan University, 2009.

Biomedical Engineering and Environmental Engineering – Chan (Ed.)
© 2015 Taylor & Francis Group, London, ISBN: 978-1-138-02805-0

Daily variation characteristics and refined prediction of ozone pollution in Beijing during summer

L.S. Wang, L.N. Zhang, J. Li, Y.P. Chen, Y. Liu & Y.H. Chen
CECEP LiuHe Talroad Environmental Technology Co. Ltd., Beijing, China

ABSTRACT: Based on hourly O_3 concentrations from a representative air quality monitoring station, and the hourly temperature, relative humidity, atmospheric pressure, wind direction, wind speed, and weather event data from a nearby meteorological station during June, July, and August 2014, the daily variation characteristics of O_3 were analysed, and the correlation coefficients of O_3 with the six meteorological parameters were calculated using Spearman correlation analysis. The results showed that O_3 was significantly positively correlated with temperature, and significantly negatively correlated with relative humidity and wind speed. There was also significant correlation between O_3 concentration and wind direction. The hourly O_3 concentrations were predicted by using a Radial Basis Function (RBF) neural network. The curve trends of the predicted values are similar to the curve trends of the monitored values, and the predicted values have a significant linear relationship with the monitored values, which demonstrates the possibility of refined prediction for O_3 pollution.

1 INTRODUCTION

In recent years, the air pollution problem in Beijing has become a focus for public concern. Normally, fine particulate matter is the primary pollutant, but during summer the ground ozone (O_3) can replace fine particulate matter and become the primary pollutant. Due to the strong corrosivity of O_3, it can cause respiratory infection of human beings, and the high concentration of ground ozone can also do harm to the ecological environment and the material cycle (Lippman 1989, Ashmore & Bell 1991). Since the O_3 pollution could change a lot in a day, and the research showed that the residential daily mortality was more relevant to the short-time highest concentration of O_3 than to the average daily concentration of O_3 (Yang 2012), it is necessary to analyse the daily variation characteristics of O_3 pollution and to make more refined prediction of O_3 concentration.

In this paper, the hourly O_3 concentrations from a representative air quality monitoring station, and the hourly temperature, relative humidity, atmospheric pressure, wind direction, wind speed and weather event data from a nearby meteorological station during June, July, and August 2014 were collected. The daily variation characteristics of O_3 were analysed and the possible reasons were discussed. The correlation of O_3 concentrations with the meteorological parameters were calculated by using Spearman correlation analysis, and the hourly O_3 concentrations were predicted by using a RBF neural network method, all of which could provide some theoretical support for refined air quality prediction and air pollution prevention work in the future.

2 DATA AND METHODS

2.1 Data sources and processing

In this paper, the Yizhuang Development Zone air quality monitoring station (39.795°N, 116.506°E) in Daxing District was chosen for data collection because it is surrounded by residential areas thus can represent the air quality in a residential area and it is very close to the No. 54511 international meteorological station (39.80°N, 116.47°E), from which the meteorological data could be easily obtained. The hourly O_3 concentration data from the Yizhuang Development Zone air quality monitoring station during June, July, and August 2014 were collected from the official website of Beijing Municipal Environmental Monitoring Centre (http://zx.bjmemc.com.cn). The hourly temperature, relative humidity, atmosphere pressure, wind direction, wind speed and weather event data from No. 54511 station during June, July, and August 2014 were collected from the Weather Underground Website (http://www.wunderground.com). The non-numerical data: wind direction and weather events were processed to numerical data in order to facilitate data analysis (see Tables 1 and 2).

2.2 Correlation analysis

As the distribution of O_3 concentration data were not known, the Spearman non-parameter analysis method was used. This method is more flexible than the parameter analysis method, and especially effective when the overall distribution of samples is not clear (Li et al. 2009).

Table 1. Change of non-numerical to numerical data for wind direction.

Original data	Processed data
North	1
NNE	2
NE	3
ENE	4
East	5
ESE	6
SE	7
SSE	8
South	9
SSW	10
SW	11
WSW	12
West	13
WNW	14
NW	15
NNW	16
Variable	17
Calm	18

Table 2. Change of non-numerical to numerical data for weather events.

Original data	Processed data
No event	0
Rain or Thunderstorm	1

2.3 O_3 concentration prediction

Because air pollution is affected by meteorological conditions, emission sources, complex underlying surface and physical, chemical, and biological processes, it has a strong non-linear characteristics (Raga & Moyne 1996). An artificial neural network is therefore widely used to predict air pollution since it is an efficient tool for non-linear analysis (Jiang et al. 2004). The RBF neural network method which was approved to be fast at training and better at function approximation ability (Wang et al. 2003), was used to predict the hourly O_3 concentrations.

3 RESULTS AND DISCUSSIONS

3.1 Daily variation characteristics of O_3 concentration

According to the hourly O_3 concentration data through June, July, and August 2014, the average daily variation of O_3 concentrations in June, July, and August is shown in Figure 1. It can be seen that the three curves are similar, with one peak and one valley. The O_3 concentrations increase to a peak among 15.00 to 17.00 hours and decrease to a valley among 05.00 to 07.00 hours, which demonstrates that O_3 concentration is highly affected by temperature, and always increases to a peak in the afternoon when the temperature is high.

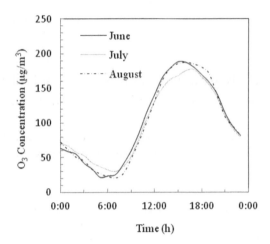

Figure 1. The average daily variation of O_3 concentrations in June, July, and August.

The daily variation curves with obvious peaks in the afternoon during three typical days are shown in Figure 2. The corresponding weather condition in the three days are as followings: In the afternoon of 19 June, the temperature increased to a peak of 27°C, the relative humidity was about 60%, and there was a southeast wind with 3–5 m/s. In the afternoon of 3 July, the temperature increased to a peak of 30°C, the relative humidity was between 60% and 70%, and there was a gentle wind with a speed of less than 2 m/s. The temperature in the afternoon of 14 July increased a high of 36°C, the relative humidity was only about 30%, there was a southeast wind of 2–4 m/s, and the O_3 peak concentration in this day was the highest. Therefore, higher temperature and lower relative humidity with a southeast wind or a gentle wind is considered to increase the O_3 peak concentration in the afternoon.

The daily variation curves with no obvious peak in the afternoon during three typical days are shown in Figure 3. The corresponding weather conditions for the three days are as follows: In the afternoon of 6 June, it rained with a highest temperature of only 19°C, the relative humidity was about 90%, and there was a 3–11 m/s north wind. In the afternoon of 23 July, the highest temperature was 32°C, the relative humidity was less than 40%, and there was a 2–4 m/s northwest wind. In the afternoon of 4 August, it rained with a highest temperature of only 23°C, the relative humidity was about 80%, and there was a 3–6 m/s north wind. Therefore, a lower temperature and a higher relative humidity with a north or northwest wind is considered to obviously decrease the O_3 peak concentration in the afternoon.

3.2 Correlation of O_3 concentration with meteorological factors

The Spearman correlation coefficients of O_3 concentration with temperature, relative humidity, atmospheric pressure, wind direction, wind speed and

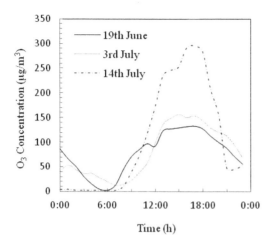

Figure 2. The typical daily variation curves of O₃ concentration with a peak in the afternoon.

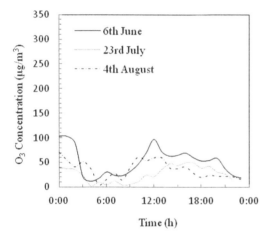

Figure 3. The typical daily variation curves of O₃ concentration without a peak in the afternoon.

weather event factors were derived by using the Spearman correlation analysis based on the hourly O₃ concentration from Yizhuang Development Zone monitoring station and the meteorological data from No.54511 meteorological station during June, July, and August 2014. The results are listed in Table 3. It can be seen that O₃ concentration is significantly correlated with temperature, relative humidity, wind speed, and wind direction, which is compliant with the result of 3.1.

O₃ concentration is significantly positively related to temperature. When the temperature rises, the photochemical reaction will be accelerated and thus the production of O₃ will be faster (Sillman & Samson 1995, Yan et al. 2013). O₃ concentration is significantly negatively related to the relative humidity. When the relative humidity is lower, there are always fewer clouds and the solar radiation is higher, the photochemical reaction will be accelerated and thus the

Table 3. The Spearman correlation coefficients of O₃ concentration with meteorological factors.

Meteorological factors	Correlation coefficient with O₃ concentration
Temperature	0.634*
Relative humidity	−0.550*
Atmosphere pressure	−0.124*
Wind direction	0.206*
Wind speed	−0.336*
Weather event	−0.018

*Correlation is significant at the 0.01 level.

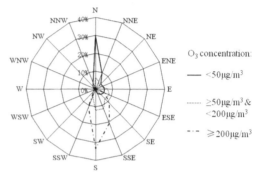

Figure 4. The wind direction frequency distribution with different ranges of O₃ concentration.

production of O₃ will be faster (Tan et al. 2007). O₃ concentration is significantly negatively correlated with the wind speed. When the wind speed is low, the atmospheric layer near the ground is stagnant and the accumulation of O₃ is increased (Cheung & Wang 2001).

O₃ concentration is significantly correlated with wind direction. In order to analyse their relationship more specifically, the wind direction frequency distributions under different O₃ concentration ranges were drawn as Figure 4.

It can be seen that when O₃ was in a lower concentration range, the wind direction was more likely to be from the north, and when O₃ was in a higher concentration range, the wind direction was more likely to be from the south. By summarizing the result of 3.1, it can be concluded that the wind from the north or the northwest can decrease the O₃ peak value in the afternoon, while the wind from the south or the southeast has no obvious effect on the O₃ peak value in the afternoon. The reason is that the north or northwest wind in summer is always cooler and brings in air with lower pollutants, while the wind from the south or the southeast is always warmer and brings in air from the North China Plain with higher pollutants (Wang et al. 2008).

3.3 Refined O₃ concentration prediction

The above result shows that O₃ concentration has a significant correlation with temperature, relative

Table 4. The data used for training and for predicting in the four prediction periods.

Prediction period No.	Data used for training	Data used for predicting
1	1 Jun to 3 Aug (448 sets of data)	4 Aug to 10 Aug (168 sets of data)
2	1 Jun to 10 Aug (616 sets of data)	11 Aug to 17 Aug (168 sets of data)
3	1 Jun to 17 Aug (784 sets of data)	18 Aug to 24 Aug (168 sets of data)
4	1 Jun to 24 Aug (952 sets of data)	25 Aug to 31 Aug (168 sets of data)

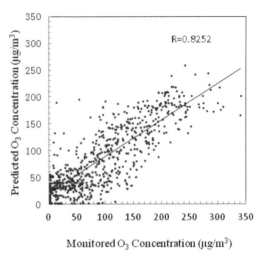

Figure 6. The correlation between predicted values and monitored values of O_3 concentrations.

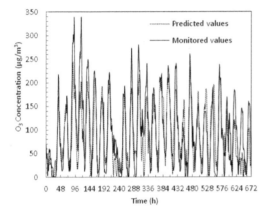

Figure 5. The comparison between predicted values and monitored values of O_3 concentrations.

humidity, wind speed, and wind direction. Therefore, in the prediction model, the temperature, relative humidity, wind speed, and wind direction at time T and the O_3 concentration at time (T-1) were set as the model input, and the O_3 concentration at time T was set as the model output. The RBF neural network method was used for data training and predicting. The hourly O_3 concentration was predicted up to 168 hours ahead in four prediction periods. The data used for training and for predicting in the four prediction periods are listed in Table 4. During predicting, the O_3 concentrations predicted for time T was used as the input value to predict the O_3 concentration for time (T+1), and so on.

The predicting results in the four prediction periods are summarized in Figure 5. The curve trends of predicted values are similar to the curve trends of monitored values.

A linear correlation analysis was done between the predicted values and the monitored values of O_3 concentrations during the whole four prediction periods (Figure 6). It shows that the linear correlation coefficient is 0.8252 and the correlation is significant at the 0.001 level, which proves the good predicting ability of the RBF neural network.

4 CONCLUSIONS

According to the hourly O_3 concentration data from a representative air quality monitoring station through June, July, and August 2014, the monthly average daily variation curves of O_3 concentrations in June, July, and August were studied. The curves are similar to one peak and one valley. The O_3 concentrations increase to a peak between 15.00 to 17.00 hours, and decrease to a valley between 05.00 to 07.00 hours, which demonstrates that O_3 concentration is highly affected by temperature, and always increases to a peak in the afternoon when the temperature is high.

The correlation coefficients of O_3 with the six meteorological parameters were calculated by using Spearman correlation analysis. The results showed that O_3 is significantly positively correlated with the temperature, the concentration of O_3 always increases to a peak in the afternoon when the temperature is high. O_3 is significantly negatively correlated with the relative humidity and the wind speed. There is also a significant correlation between the O_3 concentration and the wind direction; wind from the north or northwest can decrease the O_3 peak concentration, while wind from the south or southeast has no obvious effect on the O_3 peak concentration.

The hourly O_3 concentrations were predicted by using the RBF neural network method. The prediction period is as long as one week, i.e. 168 hours. The curve trends of the predicted values are similar to the curve trends of the monitored values, and the predicted values have a significant linear relationship with the monitored values, which demonstrates the possibility of refined prediction for O_3 pollution.

REFERENCES

Ashmore, M.R. & Bell, J.N.B. 1991. The role of ozone in global change. *Annals of Botany* 67: 39–48.

Cheung, V.T.F. & Wang, T. 2001. Observational study of ozone pollution at a rural site in the Yangtze Delta of China. *Atmospheric Environment* 35(29): 4947–4958.

Jiang, D. et al. 2004. Progress in developing an ANN model for an air pollution index prediction. *Atmospheric Environment* 38: 7055–7064.

Li, J. et al. 2009. Non-parameter statistical analysis of impacts of meteorological conditions on PM concentration in Beijing. *Research of Environmental Sciences* 22 (6): 663–669. (in Chinese)

Lippman, M. 1989. Health effects of ozone: A critical review. *Journal of the Air & Waste Management Association* 39(5): 672–695.

Raga, G.B. & Moyne L.L.E. 1996. On the nature of air pollution dynamics in Mexico City – I. Nonlinear analysis. *Atmospheric Environment* 30(23): 3987–3993.

Sillman, S. & Samson P.J. 1995. Impact of temperature on oxidant photochemistry in urban, polluted rural and remote environments. *Journal of Geophysical Research* 100:11497–11508.

Tan, J.G. et al. 2007. Analysis and prediction of surface O3 concentration and related meteorological factors in summertime in urban area of Shanghai. *Journal of Tropical Meteorology* 23(5): 515–520. (in Chinese)

Wang, W. et al. 2003. Three improved neural network models for air quality predictioning. *Engineering Computations* 20(2): 192–210.

Wang, Z.F. et al. 2008. Simulation of the impacts of regional transport on summer ozone levels over Beijing. *Chinese Journal of Nature* 30(4): 194–198. (in Chinese)

Yan, R.S. et al. 2013. Characteristics of typical ozone pollution distribution and impact factors in Beijing in summer. *Research of Environmental Sciences* 26(1): 43–49. (in Chinese)

Yang, C.X. 2012. *Acute effects of fine particulate matter and ozone on daily mortality in China*. Shanghai: Fudan University. (in Chinese)

Biomedical Engineering and Environmental Engineering – Chan (Ed.)
© 2015 Taylor & Francis Group, London, ISBN: 978-1-138-02805-0

A study on compound corrosion inhibitor for environmentally friendly snow melting agents

Xiqing Dong
Weifang University of Science and Technology, Shouguang, Shandong, China

Lina Zhao
Beijing Juneng Pharmaceutical Co. Ltd., Beijing, China

Shanshan Zhang & Jinxiu Yuan
Weifang University of Science and Technology, Shouguang, Shandong, China

ABSTRACT: A simulation is conducted to study the corrosion of carbon steel by a snow melting agent by using the alternate wetting and drying test in a real environment. A weight loss experiment and electrochemical test were adopted to investigate the corrosion of carbon steel with added single and compound corrosion inhibitors under the condition of alternate wetting and drying. The corrosion morphologies and corrosion products of the rust samples were characterized by scanning electron microscopy. The results indicated that the corrosion rate dropped significantly and inhibitor efficiency increased by 31.9% when a compound corrosion inhibitor was added, compared to a single corrosion inhibitor. There were almost no corrosion characteristics on the surface of carbon steel when the compound corrosion inhibitor was added.

1 INTRODUCTION

During the snowing periods in winter in the northern areas of China, snow melting agents were often used on the road to promote snow melting and ensure the safety of traffic. The traditional snow melting agent was chlorine salt. It could cause serious harm to the steel concrete pavement, cars, and even plants by the side of the road[1,2]. A snow melting agent with no chlorine salt was used only in some important areas such as airports and so on because of its high price and ordinary effect. As to the problem of the corrosivity of a traditional snow melting agent, adding a corrosion inhibitor to it to delay the corrosion process of carbon steel, was an important direction of development of efficient and environment friendly snow melting agents in recent years[3].

Polyphosphate is currently one of the most widely used and most economical corrosion inhibitors. As a kind of cathodic type of corrosion inhibitor, sodium hexametaphosphate is used in many fields such as anti-corrosion, water treatment, food, etc. because of its advantages of being non-toxic, environment friendly etc.[4].

The compound corrosion inhibitor would have a great improvement in corrosion inhibition efficiency than the single corrosion inhibition[5]. Research involving a compounding experiment among different kinds of corrosion inhibition has been paid much attention by the corrosion science researcher with the development and progress of corrosion inhibitor science and technology. The principles are generally compounding between cathodic type and anodic type corrosion inhibitor. By comparing and analysing single and compound corrosion inhibitors in this paper, we select a kind of high efficient compound corrosion inhibitor, which is added to the snow melting agent. The snow melting agent was both economically and environmentally friendly and could be launched into the market.

In the actual situation, it often presents alternation of wetting and drying on the surface of carbon steel while in the process of melting. The corrosion process of carbon steel cannot be reflected only by an immersion test under such condition. So we can evaluate the corrosion inhibition effect of a compound corrosion inhibitor in the close real environment by using an alternation of a wetting and drying test[6].

2 EXPERIMENT

2.1 Materials and specimen

The material in this study was 20# carbon steel. The samples were processed into size of $50.0\,mm \times 25.0\,mm \times 2.0\,mm$ for the corrosion weight loss test under the condition of alternation of wetting and drying, and size of $10.0\,mm \times 10.0\,mm \times 2.0\,mm$ for electrochemical testing after alternation of wetting and drying.

The test solution used in the experiment was made up of 20% (wt) snow melting agent of chlorine salt. Furthermore, a single corrosion inhibitor (sodium hexametaphosphate) and a compound corrosion inhibitor (sodium hexametaphosphate and sodium silicate) were added into the solution. The concentration of the single corrosion inhibitor was $4 \, g \cdot L^{-1}$ while the compound corrosion inhibitor contained sodium hexametaphosphate $4 \, g \cdot L^{-1}$ and sodium silicate $2 \, g \cdot L^{-1}$.

2.2 *Alternation of wetting and drying test*

The same volume of 20% snow melting agent with different kinds of corrosion inhibitor added was drawn into the beaker at room temperature before the alternation of the wetting and drying test. The specimens were immersed into the solution for 10 minutes and taken out in the air for 50 minutes. This process lasted from six o'clock to eighteen o'clock in the daytime. The above process repeated every 6 hours in the night. The whole day constituted one cycle. The alternation of wetting and drying test lasted for 30 days; distilled water needed to be replenished regularly to maintain the constant concentration.

2.3 *Electrochemical test*

The electrochemical test was measured with CS300 potentiostat and performed in a typical three-electrode cell. The working electrode was the rusted carbon steel after alternation of wetting and drying. A Pt counter electrode and a saturated calomel reference electrode were used. The scan rate of polarization curves was 0.5 mV/s and the range was −500 to +1000 mV (vs. OCP).

2.4 *Characterization of rust samples*

The specimens of carbon steel were taken out after 30 cycles of alternation of wetting and drying. After drying in the air, the images of rust layers were characterized with Scanning Electron Microscopy (SEM).

3 RESULTS AND DISCUSSION

3.1 *Corrosion rate and inhibitor efficiency after alternation of wetting and drying test*

The carbon steel specimens were taken out from the snow melting agent solution after 30 cycles of alternation of wetting and drying. The corrosion product was washed off from the surface of the specimen. The corrosion rate and corrosion inhibition effect would be calculated according to weight loss before and after the test.

Table 1 shows the corrosion rate and inhibitor efficiency of carbon steel in a simulated solution with different kinds of inhibitor added. It can be seen from the table that inhibitor efficiency in the snow melting

Table 1. The corrosion rate and inhibitor efficiency of carbon steel in melt solution with different kinds of inhibitor added.

Category	Corrosion rate $v/mm \cdot a^{-1}$	Inhibitor efficiency $\eta/\%$
–	0.0695	0
sodium hexametaphosphate	0.0230	66.9
sodium hexametaphosphate + sodium silicate	0.0008	98.8

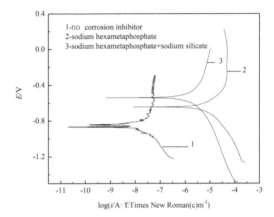

Figure 1. The polarization curve of carbon steel after alternation of the wetting and drying test.

agent solution with the compound corrosion inhibitor added increases by 31.9%, to 98.8%, than when a single corrosion inhibitor was added. The corrosion rate compound corrosion inhibitor plays an important inhibiting role in the corrosion process on the carbon steel.

3.2 *The polarization curve of carbon steel after alternation of the wetting and drying test*

We could find out that the compound corrosion inhibitor showed better corrosion inhibition effects from the corrosion weight loss test under the condition of alternation of wetting and drying. The polarization curve of carbon steel after 30 cycles of alternation of wetting and drying was measured to investigate if the corrosion inhibitor possesses a long-term corrosion inhibition effect in such an environment. The result are shown in Figure 1 and the parameters fitting data of polarization curve are shown in Table 2.

We could observe from Figure 1 and Table 2 above that the corrosion potential of both No. 2 and No. 3 curves was higher than that of No. 1, which was with no added corrosion inhibitor. It indicated that the corrosion resistance of carbon steel increased significantly

Table 2. The parameters fitting data of polarization curve.

Category	E_0/V	I_0/A·cm^{-2}	Ba/mV	Bc/mV
–	−0.84	4.4E−9	98.1	133.8
$(NaPO_3)_6$	−0.65	6.7E−6	254.4	270.3
$(NaPO_3)_6 +$ Na$_2$SiO$_3$	−0.54	2.5E−6	541.4	377.5

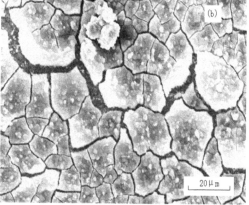

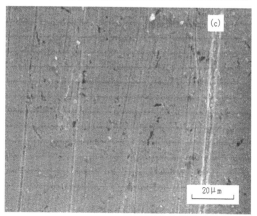

Figure 2. The microphotograph of carbon steel after 30 cycles of alternation of wetting and drying in a solution with different corrosion inhibitors added.

after the addition of the corrosion inhibitor. The electrochemical parameters current density of No. 1 was very small because of the thick rust layer generated on the surface of carbon steel after 30 cycles of alternation of wetting and drying. β_a, β_c represented the polarizability of the anode curve and cathdic curve. The β_a, β_c of No. 2 and No. 3 curves became much larger than that of No. 1's curve according to the fitting data in Table 2. That is, the corrosion resistance film formed on the surface of carbon steel remained intact after alternation of wetting and drying. The corrosion resistance of carbon steel improved greatly.

3.3 The microphotograph of carbon steel after alternation of the wetting and drying test

Figure 2 shows the microphotograph of carbon steel after 30 cycles of alternation of wetting and drying in snow melting agent added different corrosion inhibitor. Its resulting corrosion product with no corrosion inhibitor was in clutter, as shown in Figure 2(a). That is, a lot of product was generated on the surface of carbon steel and was in disorder.

Figure 2(b) is its microphotograph with a single corrosion inhibitor sodium hexametaphosphate. There is a layer of protective film on the surface of the carbon steel. However, the film is not smooth. Several large cracks divide the film.

The microphotograph with a compound corrosion inhibitor is shown in Figure 2(c). There is little corrosion product on the surface. Only a little scratch can be seen from the microphotograph, that is, the mark left when the specimen was treated before the test. It indicates that corrosion behaviour on the carbon steel was restrained to a high degree.

Sodium hexametaphosphate, which is a common kind of cathode type corrosion inhibitor, can form chelation with Fe^{2+} and deposit on the surface of carbon steel. The formed protective film resists the diffusion of the dissolved oxygen towards the cathodic electrode and thus reduces the corrosion reaction of carbon steel.

Sodium silicate can form colloidal negative particles in the solution. The hydrogen in the colloidal negative particles reacts with the oxygen adsorbed on the surface of steel and forms hydrogen bonds and then constitutes a dense ferrosilicon film. Both sodium hexametaphosphate and sodium silicate play their role respectively in controlling the corrosion process, and manifest a nice synergistic effect.

4 CONCLUSIONS

(1) The corrosion rate dropped significantly and inhibitor efficiency increased by 31.9% when a compound corrosion inhibitor was added, compared to a single corrosion inhibitor.

(2) Sodium hexametaphosphate can restrain the cathodic reaction and sodium silicate can restrain the anodic reaction, both of which manifest a nice synergistic effect.

ACKNOWLEDGEMENTS

This work was supported by Startup Foundation for Doctors of Weifang University of Science and Technology (NO. W13K019).

REFERENCES

[1] Li Yuming, Liu Jingmin, Ma Guangchao. 2004. Inhibitors of Molybdate Combined With Phosphate or Silicate. Corrosion & Protection 25(6): 248–251.

[2] Luo Hong, Luo Libin, Zhang Jing. 2004. Affecting and Strategies for Environment of the Solvent of Snow-melted. Environmental Monitoring in China. 20(1): 55–57.

[3] Li Han, Gong Yunlan. 1994. The Study on the Inhibiting Mechanism of Polyphosphate Corrosion Inhibitor and the Composition of the Membrane of Corrosion Inhibitor. Journal of Tianjin University of Commerce 3: 6–10.

[4] Tao Peng, Xu Chunchun. 2007. Testing of Corrosion Behaviors of Steel Bar in Deicing Salt by Using Alternate Immersion. Corrosion & Protection 28(9): 452–454.

[5] Wu Qingyu. 1996. The Synergistic Effect of Inhibitor. Materials Protection 29(10): 16–18.

[6] Zhao Yingying, Huang Mingyue, Xiao Guang. 2005. The Effect of Snow Melting Agent on the Environment. Journal of Jilin Institute of Chemical Technology 22(4): 25–28.

Biomedical Engineering and Environmental Engineering – Chan (Ed.)
© 2015 Taylor & Francis Group, London, ISBN: 978-1-138-02805-0

Effect of desulfurization Ash-FeCl$_3$ on sludge dewatering performance

W. Chen, Y. Chen & C.Y. Song
Civil and Environmental Engineering Institute, University of Science and Technology Beijing, Beijing, China

ABSTRACT: Combining desulfurization ash with FeCl$_3$ as a conditioning agent, its effect on the performance of sludge dewatering was investigated by Capillary Sop Time (CST) and the water content of the sludge cake (W_C). The mechanism was also explored by Extracellular Polymeric Substance (EPS). The results indicated that when the desulfurization ash combined with FeCl$_3$, the treatment effect was significantly better. It is shown that a large number of Tightly Bound EPS (TB-EPS) cracked. One portion turned into Soluble EPS (S-EPS) and Loosely Bound EPS (LB-EPS), the other portion was removed by a hydroxo complex. The ideal condition would include a desulfurization ash dosage of 300 mg·(g DS)$^{-1}$ and a FeCl$_3$ dosage of 60 mg·(g DS)$^{-1}$. CST and W_C were reduced to 14.3s and 70.22%, falling by 98.48% and 16.10% respectively.

Keywords: Sludge; desulfurization ash; FeCl$_3$; dewaterability; extracellular polymeric substance.

1 INTRODUCTION

In recent years, China's sewage treatment scale has gradually expanded, which has led to a sharp increase in the total amount of sludge (Liu et al. 2013). A survey showed that, the sludge treatment cost accounted for about 25%–40% of the total operational cost in a city sewage treatment plant, sometimes even up to 60% (Low & Chase 1999). Because of the high hydrophilic (Li & He 2004), chemical agents (such as flocculants) were added to improve the dewatering performance of sludge (Lu et al. 2008). Flocculants can be divided into two categories: inorganic and organic polymers. The most widely used in sludge treatment was Polymeric Aluminum Chloride (PAC) and Polyacrylamide (PAM) (Ye et al. 2009). Although they can effectively improve the dehydration property, the flocculants were not easy to be degraded and had the potential problem of secondary pollution (Yuan et al. 2007).

Desulfurization ash was the reaction product of the sintering flue gas and desulfurizing agent in the semi dry flue gas desulfurization process (Wang et al. 2004a, b, Liu et al. 2004). It was mainly composed of calcium oxide (CaO) and fly ash. Desulfurization ash, which has no biological toxicity, may improve the sludge dewatering performance by using the main component of CaO. However, the dosage was very large when the desulfurization ash added alone. The flocculants were therefore added at the same time. FeCl$_3$ was chosen as the traditional ferric flocculant for its low cost, high efficiency, and wide application range.

This study used desulfurization ash combined with FeCl$_3$ for sludge conditioning, providing a new way for the resource utilization of desulfurization ash and the

Table 1. The sludge properties.

	Parameter	Value
	pH	7.03 ± 0.17
	Rate of water content/%	97.01 ± 0.15
	CST/s	937.9 ± 6.4
S-EPS	S-protein/(mg/L)	2077.8 ± 83.2
	S-polysaccharide/(mg/L)	185.4 ± 13.6
LB-EPS	LB-protein/(mg/L)	427.6 ± 25.4
	LB-polysaccharide/(mg/L)	26.3 ± 1.7
TB-EPS	TB-protein/(mg/L)	2895.7 ± 124.3
	TB-polysaccharide/(mg/L)	607.0 ± 36.8

final disposal of sludge. Its effect on the performance of sludge dewatering was investigated by measuring the CST and W_C. The mechanism was also explored by focusing on the EPS.

2 MATERIALS AND METHODS

2.1 Experimental materials

Sludge samples were obtained from concentrated pool at Qinghe sewage treatment plant in Beijing. The rate of water content was concentrated further to about 97%. Following an analysis of the basic properties, the sludge was stored in 4°C. The sludge properties are shown in Table 1. All experiments were completed within 72 hours.

The desulfurization ash samples were obtained from the desulfurization workshop at a steel plant. Its main chemical composition is shown in Table 2. We can see that the content of Ca (count in CaO) was 50.34%.

Table 2. The chemical composition of desulfurization ash (%).

Component	Content
CaO	50.34
SiO$_2$	6.49
Al$_2$O$_3$	4.75
Fe$_2$O$_3$	9.41
MgO	5.52
TiO$_2$	1.24
SO$_3$	19.58
Cl	0.44
Σ	97.77

2.2 Conditioning methods

We put 300 ml sludge into a 500 ml beaker, add a certain amount of desulfurization ash, then quickly mix it for 30s under the stirring intensity of $350r \cdot min^{-1}$. We let it stand for 10 minutes, then add a certain amount of FeCl$_3$, before quickly mixing it for 30s under the stirring intensity of $350r \cdot min^{-1}$ and then slowly mixing it for 15 min under the stirring intensity of $50r \cdot min^{-1}$. The dosage of desulfurization ash was 0, 50, 100, 200, 300, 400, 500 mg·g DS)$^{-1}$. In each dosage, the dosage of FeCl$_3$ was 0, 10, 20, 40, 60, 80, 100 mg·(g DS)$^{-1}$.

3 RESULTS AND DISCUSSION

3.1 Effect on dewatering performance

The CST of raw sludge was up to 937.9s, and its W_C was up to 83.69%. As Figure 1 shows, when the amount of desulfurization ash increased separately, the value of CST and W_C decreased to the lowest point at first (244.9s and 75.77% respectively), then it went up. At the lowest point, the dosage of desulfurization ash was 300 mg·(g DS)$^{-1}$. When the amount of FeCl$_3$ increased separately, the value of CST and W_C continued to decrease until it reached the lowest point (62.2s and 76.85% respectively). At the lowest point, the dosage of FeCl$_3$ was 100 mg·(g DS)$^{-1}$. When the desulfurization ash combined with FeCl$_3$, the consequence was noticeably better. As the amount of the two increased together, the CST went down to around 20s (Area 1) and WC went down to around 70% (Area 2).

In Area 1, the combination dosage of desulfurized ash and FeCl$_3$ included 300/60 and 300/80 mg·(g DS)$^{-1}$. In Area 2, the combination dosage of desulfurized ash and FeCl$_3$ included 300/60, 300/80 and 300/100 mg· (g DS)$^{-1}$. As the amount of conditioning agent was the least and the dewatering performance was the best (the value of CST and W_C were 14.3s and 70.22% respectively), 300/60 mg·(g DS)$^{-1}$ was the best combination dosage.

3.2 Mechanism analysis

EPS mainly consisted of protein and polysaccharide, which accounted for 75%~89% (Zhang et al. 2012).

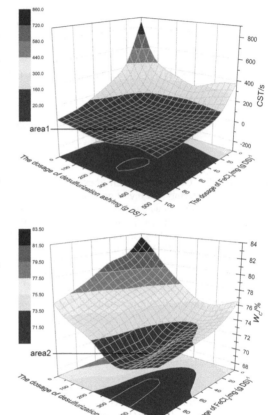

Figure 1. The conditioning effect on CST and W_C.

Analysing the variations of protein and polysaccharide in different layers could contribute to clarification of the mechanism of sludge dewatering.

The total extraction amounts of protein and polysaccharide in raw sludge were 3759.9 mg/L and 671.5 mg/L respectively. The amounts of protein and polysaccharide in the TB-EPS layer were 2895.7 mg/L and 607.0 mg/L, which occupied 77.02% and 90.39% of the total amount respectively. The amounts of protein and polysaccharide in LB-EPS layer were 200.4 mg/L and 15.7 mg/L, which occupied 5.33% and 2.34% of the total amount respectively. Amount of protein and polysaccharide in S-EPS layer were 663.8 mg/L and 48.8 mg/L, occupied 17.65% and 7.27% of the total amount respectively. In conclusion, the amount of protein and polysaccharide varied in different sludge EPS layers, with the most in the TB-EPS layer while the least in the LB-EPS.

In Figure 2, with the dosage of desulfurization ash increased, the total amounts of protein and polysaccharide first decreased and then increased. When the dosage of desulfurization ash was 300 mg·(g DS)$^{-1}$, the contents of protein and polysaccharide reached the lowest point which were 2805.0 mg/L and 349.8 mg/L,

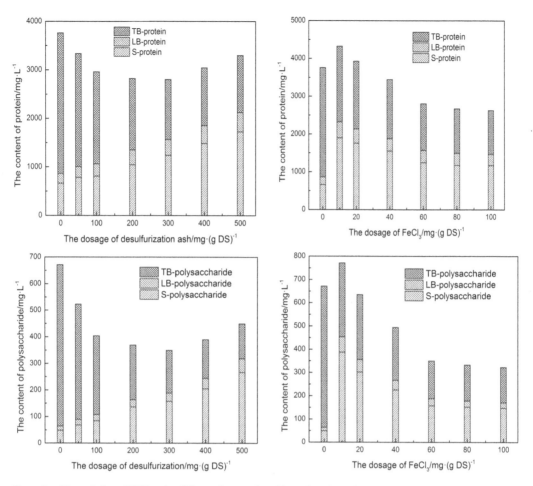

Figure 2. The variation of EPS under different dosage of desulfurization ash.

Figure 3. The variation of EPS under different dosage of FeCl₃.

falling by 25.40% and 47.91% respectively. When the dosage of desulfurization ash was less than $300\,\mathrm{mg}\cdot(\mathrm{g\,DS})^{-1}$, the TB-EPS decreased while the S-EPS and LB-EPS increased. Due to the decrement was more than the increase, the total amounts of extractions decreased. This may result from the exfoliation of the TB-EPS and it turning into S-EPS and LB-EPS. Then most of it was adsorbed by $Fe(OH)_3$ and removed from the supernatant. Research has shown that the desulphurization ash could raise pH by a wide margin so that the microbial cell is crushed and releases organic matter (Foster 1985). Fe^{3+} dissolved in the water and produced hydroxo complexes which accumulated organic matters through electrostatic adsorption (Li et al. 2006). By continuing to increase the dosage of desulfurization ash, more microbial cells were destroyed. But the adsorption amount of hydroxo complexes gradually saturated. As the increasing amplification of S-EPS and LB-EPS increased, the total amounts of protein and polysaccharide began to rise.

In Figure 3, with the dosage of FeCl₃ increased, the total amounts of protein and polysaccharide first increased and then decreased. Specifically, the TB-EPS continued to decrease while the S-EPS and LB-EPS first increased and then decreased. When the dosage of FeCl₃ was less than $10\,\mathrm{mg}\cdot(\mathrm{g\,DS})^{-1}$, electrostatic adsorption was insufficient to make the EPS gather and precipitate. Therefore, the total amounts of protein and polysaccharide increased by 15.08% and 14.83% respectively compared to the original sludge. When the dosage of FeCl₃ increased to $60\,\mathrm{mg}\cdot(\mathrm{g\,DS})^{-1}$, the supernatant already had lots of hydroxyl complexes and enhanced electrostatic adsorption. Therefore the total amounts of protein and polysaccharide decreased greatly. By continuing to increase the dosage of FeCl₃, the trend of decline slowed down, indicating that the EPS in the supernatant had been largely adsorbed.

4 CONCLUSIONS

When the desulfurization ash combined with FeCl₃, the treatment effect was noticeably better. The best combination dosage was $300/60\,\mathrm{mg}\cdot(\mathrm{g\,DS})^{-1}$. In this circumstance, the value of CST and W_C were 14.3s and 70.22% respectively, reduced by 98.48% and 16.10% compared to the raw sludge. It is shown that a large number of TB-EPS cracked. One portion turned into S-EPS and LB-EPS, the other portion was removed by the hydroxo complex.

REFERENCES

Forster, C. F. 1985. Factors involved in the settlement of activated sludge. *Water research* 19(10): 1259–1264.

Li, D. Z. & He, J. J. 2004. The sludge properties, floc structure and disposal. *Science and Technology Review*, 9: 26–30.

Li, J. X. & Sun, S. Y. & Yuan, X. H. 2006. Comparison of the application of chemical phosphorus removal reagent of city life sewage. *Guangdong Trace Elements Science* 13(1): 19–22.

Liu, L.Z. & Zhang C.Z. & Huang, X.M. 2004. Contrast test research on three kinds of calcium based desulfurizer in CFBA sintering flue gas desulfurization. *Environmental Pollution and Control* 26(6): 418–420.

Liu, P. & Liu, H. & Yao, H. 2013. Effect of fenton and skeleton construction on sludge dewatering performance. *Environmental Science and Technology* 36(10): 146–151.

Low, E.W. & Chase, H. A. 1999. Reducing production of excess biomass during wastewater treatment. *Water Research* 33(5): 1119–1132.

Lu, W. & Zhang, D.F. & Hu, K. L. 2008. The effect and mechanism of cationic surface active agent on sludge dewatering. *Environmental chemistry* 27(4): 444–448.

Wang, F. & Zhang, F. & Wang, H. M. 2004. Industrial application of semi dry flue gas desulfurization technology. *Environmental Pollution and Control* 26(3): 209–211.

Ye, H.L. & Ye, J.S. & Zhong, Z.J. 2009. Research on sludge dewatering performance of microbial flocculant. *Environmental chemistry* 28(3): 414–417.

Yuan, L.J. & Tang, B. & Xue, J.Y. 2007. Research progress in the biochemical sludge conditioning technology. *Industrial safety and environmental protection* 33(1): 27–29.

Zhang, L.H. & Li, J. & Guo, J. B. 2012. Effect of EPS on flocculation, settleability and surface properties of the activated sludge. *Journal of Chemical Engineering* 63(6): 1865–1871.

Biomedical Engineering and Environmental Engineering – Chan (Ed.)
© 2015 Taylor & Francis Group, London, ISBN: 978-1-138-02805-0

Thermodynamics of Cu^{2+} adsorption on Amino Ion Exchange Fibre (AIEF)

Jinxin Qian, Qiang Gan, Mingyu Li & Changgen Feng
Beijing Institute of Technology, Beijing, P.R. China

ABSTRACT: The removal of Cu^{2+} from an aqueous solution by Amino Ion Exchange Fibre (AIEF) was achieved using batch adsorption experiments. The effect on contact time, initial Cu^{2+} concentration, and adsorption isotherms at various temperatures were obtained. The results showed that the maximum adsorption capacity of the AIEF was 177.6 mg/g, and a Langmuir linear equation model can well describe the adsorption equilibrium data, suggesting that the adsorption process involves both chemisorption and physisorption. The values of thermodynamic parameters, including ΔH, ΔG, and ΔS, indicate that the adsorption of Cu^{2+} is a spontaneous, entropy-driven, and endothermic process.

1 INTRODUCTION

Copper-containing wastewater is mainly produced by copper mining, smelting, and processing. It will cause serious damage to humans and animals, especially livers and digestive systems. The copper content of wastewater can reach dozens of mg per gram, while the Chinese Standard for Cu^{2+} in industrial wastewater is only 0.25 mg/l[1–2]. Traditional processing methods of Cu^{2+} include a chemical precipitation method, electrolysis method, extraction method and reverse osmosis method, but they all have their own limitations. For example, the cost of the electrolysis method is much too high and the chemical precipitation method will produce heavy metal sludge which may cause secondary pollution[3–5]. In recent years, the ion exchange method including ion exchange resin (IER) and ion exchange fibre (IEF) has experienced rapid development[6–7].

IEF is a new kind of fibrous surface adsorption and separation material after IER. It has been proved to be the most promising method for the removal of the Cu^{2+}. That is because IEF has higher absorption/elution rates, good mechanical strength, high generation ability, and much freedom in form use when compared with IER[8–9]. IEF has been applied to metal recovery, gas purification and separation, extraction and separation of rare earth and radioactive elements, and water purification etc.[10] Recently, a polyamine type ion exchange fibre (AIEF) as a new type of fibre has been widely used to adsorb heavy metal ions, such as Cr, Cd, and Cu etc.[11] In 2004, Deng [12] synthesized the styrene type strongly basic anion exchange fibre, which is applied in the adsorption heavy metal ions of Cd, Cr. In this article, amino ion exchange fibre was used for the removal of Cu^{2+} from an aqueous solution. Equilibrium isotherm models were applied to obtain the thermodynamic property and the mechanism of Cu^{2+} adsorption on AIEF.

2 EXPERIMENTAL

2.1 Apparatus and materials

AIEF has been prepared from chloramethylated styrene grafted polypropylene fibre reacted with TEPA, which was made by my own laboratory.

An aqueous solution of 250 mg/g Cu^{2+} was prepared by dissolving copper sulphate (AR, Sinopharm) in distilled water, 5-Br-PADAP (AR, Sinopharm), P-nitrophenol (AR, Sinopharm), sodium fluoride (AR, Sinopharm), ammonium hydroxide (AR, Sinopharm), hydrochloric acid (AR, Sinopharm), and sodium acetate (AR, Sinopharm). The apparatus included a UV-visible spectrophotometer (725P Apl), an elec-trothermal contant-temperature dry box (DHP-9032, Blue Pard), a water batch oscillator (SHA-B, Blue Pard), an electronic analytical balance (AR2140 Ohaus), and mechanical pipettes (20–200 μL Biohit).

2.2 Analysis

The concentration of Cu^{2+} was determined by UV-visible spectroscopy using 5-Br-PADAP as the colour agent at the wavelength of 560 nm[13–14]. The linear equation of the standard curve could be described as:

$$A = 0.14076 + 0.12965c \tag{1}$$

where A is the absorbance of the purple coloured solution, and c is the concentration of Cu^{2+}. The correlation coefficient R^2 is 0.9989.

2.3 Adsorption experiment

A 25 ml of Cu^{2+} aqueous solution with particular initial concentration and pH value was placed in a set of 50 ml glass conical flasks, respectively. A fixed quality of AIEF was added into the adsorbent solutions. In the

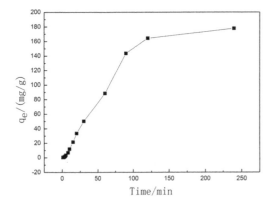

Figure 1.   Effect of time on adsorption capacity to Cu^{2+}.

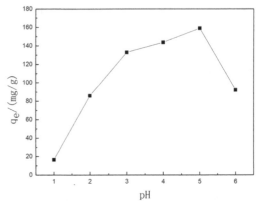

Figure 2.   Effect of pH on adsorption capacity to Cu^{2+}.

water batch oscillator, the mixture of Cu^{2+} and AIEF was agitated at a speed of 120 rpm under a constant temperature for a certain time. The aqueous solution was sampled at predetermined time intervals. Adsorption capacities at time t, q_t, and equilibrium, q_e, are expressed respectively as:

$$q_t = \frac{(c_0 - c_t)V_t}{m} \qquad (2)$$

$$q_e = \frac{(c_0 - c_e)V_e}{m} \qquad (3)$$

where c_0, c_t, and c_e are the Cu^{2+} concentrations of original, residual, and equilibrium Cu^{2+} solution, respectively; V is the volume of solution; m is the mass of AIEF.

3   RESULTS AND DISCUSSION

3.1   *Adsorption curve*

An adsorption curve, expressed in terms of adsorption capacity (q_e) of AIEF as a function of time (t), is adopted to determine the time of equilibrium. The adsorption curve of AIEF was carried out (25 ml 400 mg/l Cu^{2+}, 0.050 ± 0.0005 g AIEF, 298 K, 120 r/min). The results in Figure 1 show that the adsorption equilibrium between $CuSO_4$ solution and AIEF almost finished within 120 min, with the adsorption amount of Cu^{2+} reaching a maximum of 177.6 mg/g. The adsorption of Cu^{2+} onto AIEF was not as quick as a usual IEF, for the mechanism of Cu^{2+} adsorption to AIEF is different from IEF, it is a kind of chemical adsorption[15].

3.2   *Effect of pH*

Some experiments were carried out to examine the influence of initial pH on the adsorption of Cu^{2+} (25 ml 400.0 mg/l Cu^{2+}, 0.05 g (± 0.0005 g) AIEF, 298 K, 120 r/min, 12 h). The initial pH value of the

Cu^{2+} solution was adjusted from 1.00 to 6.00 with HCl or $NH_3 \cdot H_2O$ solution.

The adsorption capacity of Cu^{2+} increases from 15.8 mg/g to 157.9 mg/g, as the pH value increases from 1.00 to 5.00. The AIEF reached a maximum adsorption capacity of 157.9 mg/g at pH 5.00. When the pH value is higher than 5.00, the adsorption capacity sharply decreases. That is because when the pH is lower, the concentration of H^+ is rather high, the amino of AIEF is easier to be combined with H^+, resulting in the decrease of AIEF's ability to coordinate with metal ions. However, when the pH is higher than 5.00, a copper ammonia complex will be generated, affecting the Cu^{2+} adsorption ability of the AIEF[16].

3.3   *Effect of initial Cu^{2+} concentration*

The effect of initial Cu^{2+} concentration on adsorption was conducted at the Cu^{2+} concentration from 10 mg/l to 500 mg/l, 0.05 g $\pm$ 0.0005 g AIEF was added to the Cu^{2+} solution at 298 K. The mixture was agitated at 120 r/min until the adsorption reached equilibrium. Figure 3 shows that the adsorption capacity of Cu^{2+} increased with the increasing of the initial Cu^{2+} concentration and the maximum adsorption capacity of Cu^{2+} was 177.6 mg/g. This could be explained by the fact that increasing the concentration gradient was attributed to the efficiency of Cu^{2+} adsorption.[17] Besides, q_e and c_0 fitted a good linear relationship before the platform of adsorption capacity appeared.

3.4   *Isotherm studies*

Equilibrium isotherms at different temperatures for Cu^{2+} adsorption on AIEF experiments were performed at pH 5 (25 ml Cu^{2+} solution with concentration from 50 mg/l to 500 mg/g, 0.050 ± 0.0005 g AIEF), the results are given in Figure 4. It can be figured that the isotherm rises in the initial stage until the plateau appears at 400 mg/l~500 mg/l, and the Cu^{2+} adsorption capacity increases with temperature, which implies an endothermic process of Cu^{2+} adsorption on AIEF in the temperature range from 298K to 318K.

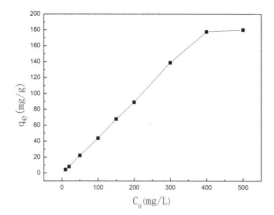

Figure 3. Effect of Cu^{2+} concentration on adsorption capacity.

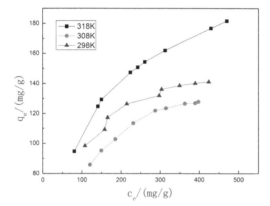

Figure 4. Adsorption isotherms of Cu^{2+}.

Several adsorption isotherm models, such as Langmuir and Freundlich have been applied to describe the adsorption process, of which the Langmuir model was used to describe this experiment.

The Langmuir model[18] could be described as:

$$q_e = \frac{q_m k_L c_e}{k_L c_e + 1} \tag{4}$$

Where c_e is the equilibrium concentration of Cu^{2+} (mg/l); q_e and q_m are the equilibrium adsorption amounts of Cu^{2+} and the saturated monolayer adsorption capacity of Cu^{2+} (mg/g), respectively; k_L is an adsorption equilibrium constant related to the affinity of binding sites. Equation (4) is transformed to the linear form of the Langmuir equation:

$$\frac{c_e}{q_e} = \frac{c_e}{q_m} + \frac{1}{k_L q_m} \tag{5}$$

Figure 5 shows the fitting lines of the Langmuir linear equation at different temperatures. Table 1 presents some adsorption parameters which are derived from the linear regression method. As shown in Table 1,

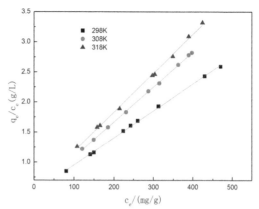

Figure 5. Langmuir plot for adsorption of Cu^{2+}.

Table 1. Adsorption equilibrium constants obtained from Langmuir isotherms.

T/K	Langmuir constant		
	q_m/(mg/g)	k_L	R^2
298	197.1	0.0150	0.9958
308	223.5	0.0119	0.9947
318	240.9	0.0122	0.9961

the high correlation coefficients ($R^2 > 0.99$) demonstrate that the Langmuir equation could be used as the thermodynamic model of the Cu^{2+} adsorption process. Therefore, it could be concluded that Cu^{2+} adsorption on AIEF is a monolayer adsorption and the chemisorption process.

3.5 Thermodynamic study

3.5.1 Enthalpy change ΔH

ΔH can be obtained by the Clausius-Claoeyron equation represented as[19]:

$$\ln c = \frac{\Delta H}{RT} + k \tag{6}$$

Where c is the Cu^{2+} equilibrium concentration, R is the mole gas constant, and k is constant, while T is the thermodynamic temperature. The values of ΔH were obtained from the slope of the curve. The values of ΔH shown in Table 2 are positive, indicating that the adsorption is endothermic.

3.5.2 Gibbs free energy change ΔG

ΔG is usually obtained by the Gibbs equation[20]:

$$\Delta G = RT \int_0^x q \frac{dx}{x} \tag{7}$$

Where x is the molar fraction of absorbate, and q is the adsorption capacity. The values of ΔG are shown

Table 2. Thermodynamic parameters for adsorption of Cu^{2+} on AIEF.

q_e/ mg/g	ΔH/ kJ/mol	ΔG/(kJ/mol)			ΔS/J(mol·K)		
		298K	308K	318K	298K	308K	318K
100	−25.89	−38.55	−30.76	34.57	42.32	15.8	27.2
110	−27.94	−38.55	−30.76	34.57	34.45	9.25	20.8
120	−29.34	−38.55	−30.76	34.57	30.90	4.61	16.4

in Table 2. The negative values of ΔG indicate that the adsorption of Cu^{2+} was spontaneous.

3.5.3 *Entropy change ΔS*

ΔS can be achieved by Gibbs-Helmholtz equation[21]:

$$\Delta S = \frac{(\Delta H - \Delta G)}{T} \tag{8}$$

Where ΔH is the enthalpy change; ΔG is the Gibbs free energy change. The values of ΔS are shown in Table 2. The values of ΔS are positive, indicating an increase in the randomness at the solid/liquid interface.

4 CONCLUSION

(1) AIEF was effective in the removal of Cu^{2+} from an aqueous solution and the maximum adsorption capacity was 177.6 mg/g under optimal conditions.

(2) The maximum adsorption capacity of Cu^{2+} could be obtained at pH = 5.00 and the effect of temperature on the adsorption is obvious.

(3) The values of all the isotherm model constants indicate that an adsorption process involved both chemisorption and physisorption. The Langmuir model is in good agreement with the experimental data with high R^2 and reasonable calculated parameters.

(4) The values of thermodynamic parameters were calculated. The values of ΔH are positive, indicating the endothermic nature of the adsorption. The negative values of ΔG reflect that the adsorption of Cu^{2+} was a spontaneous process, and the values of ΔS are positive indicating an increase in the randomness at the solid/liquid interface.

REFERENCES

[1] C. L. Song, Z. W. Chen, H. M. Fan. Reviews of copper wastewater treatment technologies. Chemical Defence on Ships, 2008, 2:22–25.

[2] World Health Organization. Guidelines for drinking-water quality [R]. 3rd ed. Geneva, 2004, 1: 334.

[3] C. L. Song, Z. W. Chen, H. M. Fan. Study on treatment of copper-containing wastewater by superfast chelating fiber with amidocyanogen. Ship Science and Technology, 2009, 1:125–128.

[4] B. Li, S. P. Liu. The technologies for treating wastewater containing copper and research progress. Multipurpose Utilization of Mineral Resources, 2008, 5:33–38.

[5] C. L. Song, Z. W. Chen, H. M. Fan. Comparison of adsorption kinetics of copper ion on chelating fiber and ion exchange resin. Chemical Defence on Ships, 2008, 6:1–7.

[6] X. L. Sun, Q. X. Zeng, C. G. Feng. Progress in research on the synthesis of chelating ion-exchange fiber. Science & Technology Review, 2008, 26(8):75–78.

[7] S. J. Zhou. Research on application of ion exchange fiber in the separation and extraction of heavy metal Ion. Xinjiang Nonferrous metals, 2012, 2:49–52.

[8] Q. Shu, H. W. Wang, Y. Chen. Application of ion-exchange fiber in sewage treatment. Chemical industry and engineering technology, 2006, 27(6): 41–43.

[9] H. Wu, J. J. Chen, G. D. Wu. Chelating fiber and its application. China Synthetic Fiber Industry, 2006, 28(6):52–54.

[10] J. Guo, Y. L. Chen, Y. Luo, et al. Current research and prospects on new ion-exchange fibers. 2005, 30(6): 35–38.

[11] Q. Zhang, H. M. Fan, X. T. Guo, et al. Chelating fiber remove trace of copper lead and cadmium from water. Ship Science and Technology, 2003, 25:38–42.

[12] Q. Deng, Q. X. Zeng, C. G. Feng. Effect of preparation conditions on property for strong alkali polypropylene anion exchange fiber. Polymer Materials Science and Engineering, 2006, 22:157–160.

[13] Y. Ni, X. H. Ai, M. G. Ran. 5-Br-PADAP spectrophotometry measuring copper in food. Journal of Sichuan Continuing Education College of Ms. 1999, 18, 3: 164–167.

[14] Z. S. Liu, Y. J. Chen, D. Y. Fu. 5-Br-PADAP spectrophotometry for measuring trace copper of microemulsion in food. Chinese Journal of Health Laboratory Technology, 2011, 11(4):391–395.

[15] M. Y. Li, X. Liu, J. J. Lv. Study on optimize conditions of preparation reaction of amino ion exchange fiber. Journal of Functional Materials 2012, 43(6):779–782.

[16] C. B. Xia, H. W. Shi. Multipurpose Utilization of Mineral Resources. 2001, 3:15–17.

[17] C. G. Feng, M. Y. Li, Y. R. Sun. FiI-IR spectroscopic study on synthesis and adsorption performance of amino phosphonic acid chelating fiber. Journal of spectroscopy and spectrum, 2012, 32(12):3188–3192.

[18] Langmuir I. The constitution and fundamental properties of solids and liquids [J]. Am Chem Soc, 1916, 38:2221–2295.

[19] X. C. Fu, W. X. Shen, T. Y. Yao, et al. Physical chemistry [M]. Beijing: Higher Education Press, 2006.

[20] Meena A. K., Kadirvelu K., Mishra G. K., Rajagopal C., Nagar P. N. Adsorptive removal of heavy metals from aqueous solution by treated sawdust [J]. Journal of Hazardous Material, 2008, 150:604–611.

[21] Ngah W. S. W., Fatinathan S. Adsorption of Cu (II) ions in aqueous solution using chitosan beads, chitosan-GLA beads and chitosan-alginate beads [J]. Chemical Engineering Journal, 2008, 143:62–72.

Biomedical Engineering and Environmental Engineering – Chan (Ed.)
© *2015 Taylor & Francis Group, London, ISBN: 978-1-138-02805-0*

The distribution characteristics of inorganic phosphorus forms in the soil of the high phosphorus region in the Xiangxi River watershed

Gangzhi Peng, Jianxia Xu, Heng Xie, Lei Li & Jianzhu Wang
College of biology & pharmacy, China Three Gorges University, Yichang, Hubei, China

ABSTRACT: Based on the phosphorus mineral region of the Xiangxi River (XXR) watershed, the content of different Inorganic Phosphorus (IP) forms in soil in a Mining Zone (MZ), Transportation Zone (TZ), and Hydro-fluctuation Zone (HFZ) were respectively measured. The results showed that occluded phosphorus (O-P) and calcium-bounded phosphorus (Ca-P) were the main existing forms of IP in the high phosphorus region of the XXR watershed. The contents of water-soluble phosphorus (WS-P) and aluminium-bounded phosphorus (Al-P) in 0–20 cm soil layer were lower than those in the 20–40 cm soil layer in HFZ, while those of iron-bounded phosphorus (Fe-P), O-P, and Ca-P were the opposite trend in all regions. The contents of WS-P, Al-P, Fe-P, and O-P all showed an increasing trend from the MZ to the HFZ, while the content of Ca-P first decreased and then increased significantly. All the contents of the IP forms in the HFZ were far higher than those in the MZ; IP in the MZ gradually transferred to the HFZ.

1 INTRODUCTION

The XXR watershed is rich in mineral resources, especially its reserve of phosphate rock which amounts to 357 million tons, making it one of the three biggest phosphorus mineral regions in China (Fang et al. 2006). The phosphorus chemical industry, including the mining and processing of phosphate rock, is a pillar of local industry, with a number of phosphate ore plants distributed in towns alongside the XXR, such as Gufu and Xiakou. The rich phosphate ore resources laid a strong foundation for local economic development. However, with the surface runoff and the emission of wastewater containing phosphorus from mine tailings and phosphate ore plants, large amounts of phosphorus flow into the XXR and bring rich nutrients, becoming non-point sources of pollution, which will lead to the eutrophication of water bodies and bring a great threat to the water quality and ecological environment of XXR (Luo & Tan 2000). Soil phosphorus is made up by organic phosphorus and inorganic phosphorus, and the latter is the main part, making up about 60%~80% of the total phosphorus.

In this paper, measurements of five common IP forms in the soil of the phosphorus mineral region in the XXR watershed were made, and the changes of them and their distribution were analysed. The relationship between them was discussed in order to provide basic data for eutrophication management and water quality recovery, as well as theory reference for preventing the pollution of phosphorus in mine exploration areas, and also improving the ecological environment of the XXR watershed.

2 GENERAL SITUATION OF SURVEY AREA

XXR, the first tributary near the Three Gorges Dam, is located at 110°25′E~111°06′E, 30°57′N~31°34′N, flowing through Xingshan in Hubei Province, with a total length of 94 km and basin area of 3099 km². The watershed is characterized by a subtropical continental monsoon climate with maximum precipitation in summer. The average annual precipitation is 900~1200 mm, and flood seasonal rainfall makes up 68% of the annual precipitation. It has complex geological conditions (Hui et al. 2000), and many different soil types (Jiang et al. 2002). Furthermore, the vertical distribution of vegetation is highly significant (Jin & Liu 1996).

3 MATERIAL AND METHOD

3.1 Site settings and soil sample collections

Based on a field survey of the XXR watershed conducted in April 2014, we chose the phosphorus mineral region along the shore of Xiakou in the middle of the XXR as the study area in accordance with the principles of representation, operability, and safety. Three typical sites were set to collect the soil samples respectively (Figure 1). Site ① represents the mining zone (MZ) near the mine mouth (180~185 m), site ② represents the transportation zone (TZ) near the mine exploration area (175~180 m), and site ③ represents the hydro-fluctuation zone (HFZ) at the upper area of the XXR (170~175 m). 5~8 sampling plots were then

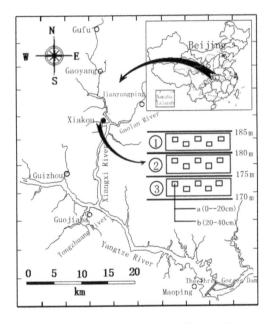

Figure 1. The location and distribution of the sampling sites in the study area.

randomly set in each site respectively. Soil samples of both 0–20 cm and 20–40 cm layers were collected with a soil drill at each sampling plot (refer with: Figure 1). The soil of each layer at each sampling plot was adequately mixed. Afterwards, visible soil fauna, roots, stones, and other debris were all removed, and 600 g soil was reserved using the quartering method.

3.2 *Soil samples measurement and data analysis*

The IP forms of soil samples were determined in the laboratory according to the sequential extraction method in the fractionation of soil phosphorus proposed by Chang & Jackson (1957). The contents of soil WS-P, Al-P, Fe-P, O-P, and Ca-P were measured with a colorimetric method at 700 nm. In addition, SPSS 20.0 software and Microsoft Excel 2003 were used for data processing and statistical analysis, as well as for drawing graphs.

4 RESULTS AND ANALYSIS

The contents of different IP forms in soil showed differences in different regions of the XXR watershed (Figure 2). The content of soil WS-P was between $0.0014\,g\cdot kg^{-1}$ and $0.0113\,g\cdot kg^{-1}$. There was an increasing trend from the MZ to the HFZ in both the 0–20 cm and the 20–40 cm soil layers. The content of soil WS-P in the HFZ was significantly higher than that in the MZ and the TZ in each layer ($p < 0.05$), while that between the MZ and TZ had no significant difference ($p > 0.05$). The content of soil WS-P in 0–20 cm soil layer was higher than that in the 20–40 cm soil layer in the MZ and TZ, while it was the opposite trend

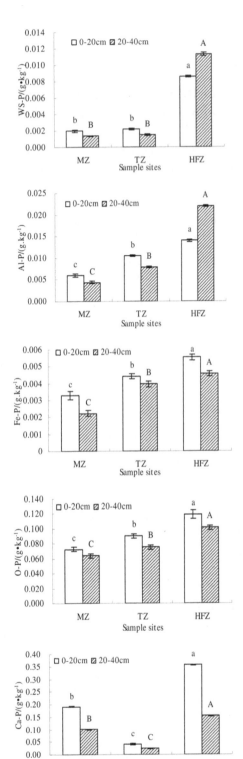

Figure 2. The content of IP forms in the 0–20 cm and 20–40 cm soil layer of different regions.

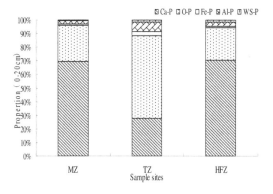

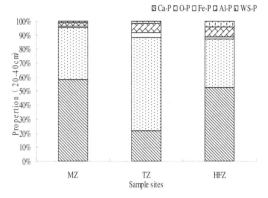

Figure 3. The proportion of IP content in the soil of different regions.

in the HFZ. The content of soil Al-P was between 0.0044 g·kg^{-1} and 0.0220 g·kg^{-1}. There was also an increasing trend from the MZ to the HFZ in both 0–20 cm and 20–40 cm soil layers, and that in the MZ, TZ and HFZ had significant difference with each other in each layer ($p < 0.05$). The content of soil Al-P in the 0–20 cm soil layer was also higher than that in the 20–40 cm soil layer in the MZ and TZ, while it was the contrary in the HFZ. The content of soil Fe-P was between 0.0014 g·kg^{-1} and 0.0113 g·kg^{-1}. There was also an increasing trend from the MZ to the HFZ in both the 0–20 cm and the 20–40 cm soil layers, and that in the MZ, TZ and the HFZ had significant difference with each other in each layer ($p < 0.05$). The content of soil Fe-P in the 0–20 cm soil layer was higher than that in the 20–40 cm soil layer in the MZ, TZ, and HFZ, while it was the contrary in the HFZ. The content of soil O-P was between 0.0014 g·kg^{-1} and 0.0113 g·kg^{-1}. There was an increasing trend from the MZ to the HFZ in both the 0–20 cm and 20–40 cm soil layers, and that in the MZ, TZ and HFZ showed significant difference with each other in each layer ($p < 0.05$). The content of soil O-P in the 0–20 cm soil layer was higher than that in the 20–40 cm soil layer in the MZ, TZ, and HFZ. The content of soil O-P was between 0.0246 g·kg^{-1} and 0.3581 g·kg^{-1}. It was in the increasing order of TZ < MZ <HFZ, and that in the MZ, TZ, and HFZ had significant difference with each other in each layer

($p < 0.05$). The content of soil Ca-P in the 0–20 cm soil layer was higher than that in the 20–40 cm soil layer in the MZ, TZ, and HFZ.

The proportion of different IP forms in soil also showed differences in different regions of the XXR watershed (Figure 3). No matter in the 0–20 cm soil layer or 20–40 cm soil layer, the content of Ca-P showed the highest proportion, followed by O-P, Al-P, Fe-P, and WS-P the lowest in MZ; the content of O-P showed the highest proportion, followed by Ca-P, Al-P, Fe-P, and WS-P the lowest in TZ; the content of Ca-P showed the highest proportion, followed by O-P, Al-P, WS-P, and Fe-P the lowest in HFZ. As a whole, the main existing IP forms were O-P and Ca-P in different regions.

5   DISCUSSION

In this study, except that the content of Ca-P was in the increasing order of TZ < HFZ < MZ, the contents of other IP forms appeared a trend of increasing from the MZ to the HFZ, with the main existing forms being O-P and Ca-P. The content of soil Fe-P, O-P, and Ca-P in the 0–20 cm soil layer was higher than that in the 20–40 cm soil layer in all the regions, while that of soil WS-P and Al-P was the contrary. All the contents of IP forms in the HFZ were far higher than those in the MZ, which means soil IP in the MZ appeared as a trend of gradually transferring to the HFZ.

The content of WS-P was quite low, only accounting for a percentage of 0.72% to 3.83% in the whole IP forms. It was possibly because of the reason that phosphides are not easy to dissolve in water. Then they are also easily reacting with ions of calcium, iron, and aluminium, generating hardly soluble sediments that were deposited on the bottom mud (Xie 2012).

Al-P and Fe-P are the part of phosphorus easily released and available for creatures (Zhu et al. 2003). Flooding leads to the conversion of iron and aluminium oxide to amorphous forms, which will increase the content of Al-P and Fe-P (Cao et al. 2006). After flooding, the soil pH tends to be neutral, which is beneficial for the formation of amorphous iron and aluminium. So the contents of Al-P and Fe-P in the HFZ experienced flooding were much higher than other regions.

We can also see from the study results that the content of O-P in different soil layers of different regions showed significant differences, but these were relatively close, which means that O-P is rather stable in soil. O-P is not easily affected by outside factors, it is always in a stable state, and is not released under natural conditions (Guo 2012). In addition, O-P is both insoluble in acid and alkali, which may become an effective source of phosphorus under waterlogging condition (Xie & Li 2011). Because O-P is a kind of invalid phosphorus, its transformation process with other IP forms is very slow, so its content is usually stable.

The content of Ca-P was lowest in the TZ but highest in the HFZ, and that in the 0-20cm soil layer was much higher than that in the 20–40 cm soil layer in all the regions. Ca-P is relatively inert phosphorus, which is considered to be the part of phosphorus not easily released and unavailable for creatures (Zhou et al. 2006). The XXR watershed is rich in phosphorus, and the soil is mostly a calcareous type. The comfortable pH and sufficient calcium as well as phosphate ions were favourable for the formation of the hard soluble Ca-P. The soil pH in the TZ was relatively low, so the content of Ca-P in the TZ was greatly reduced because of desorption. However, the soil pH in the HFZ was relatively high, which lead to the formation of precipitation due to the combination of soil phosphate and calcium ions, so the content of Ca-P in HFZ was highest.

6 CONCLUSIONS

(1) Except for the fact that the contents of WS-P and Al-P in the 0–20 cm soil layer were lower than those in the 20–40 cm soil layer in the HFZ, the contents of other IP forms in the 0–20 cm soil layer were higher than those in the 20–40 cm soil layer in all regions.

(2) The contents of WS-P, Al-P, Fe-P, and O-P showed an increasing trend from the MZ to the HFZ, while the content of Ca-P first decreased and then increased significantly, and all the contents of IP forms in the HFZ were far higher than those in the MZ; IP in the MZ gradually transferred to the HFZ.

(3) The O-P and Ca-P were the main forms of inorganic phosphorus in the high phosphorus region of the XXR watershed.

ACKNOWLEDGEMENTS

This research work was financially supported by the National Natural Science Foundation of China (No. 51179094).

REFERENCES

Cao, M. Cai, Q.H. Liu, R.Q. et al. 2006. Comparative research on physicochemical factors in the front of Three Gorges reservoir before and after the initiate impounding. *Acta Hydrobiologica Sinica*, 30(1): 12–19.

Chang, S.C. & Jackson, M.L. 1957. Fractionation of soil phosphorus [J]. *Soil Science*, 84(2): 133–144.

Fang, T. Fu, C.Y. Ao, H.Y. et al. 2006. The comparison of phosphorus and nitrogen pollution status of the Xiangxi Bay before and after the impoundment of the Three Gorges Reservoir. *Acta Hydrobiologica Sinica*, 30 (1): 26–30.

Guo, S.S. 2012. Phosphorus fractions and phosphate sorption-release characteristics of the surface soil in Water-Level-Fluctuating Zone of Three Gorges Reservoir. *Chongqing University*.

Hui, Y. Zhang, X.Y. & Chen, Z.J. 2000. Present situation and strategy about the natural environment of the Xiangxi River Basin. *Resources and Environment in the Yangtze Basin*, 9(1): 27–31.

Jiang, M.X. Deng, H.B. Tang, T. et al. 2002. On spatial pattern of species richness in plant communities along riparian zone in Xiangxi River watershed. *Acta Ecologica Sinica*, 22(5): 629–635.

Jin, T. & Liu, Y. 1996. Geographical conditions for soil erosion and water loss in Xiangxi Valley in the Three Gorges region and its renovation. *Research of Soil and Water Conservation*, 3(4): 98–102.

Luo, W.S. & Tan, G. 2000. Three Gorges Reservoir Xiangxihe bay water quality prognosis. *Hydroelectric Energy*, 18(4): 46–48.

Xie, C.S. 2012. Role of alkaline phosphatases in phosphorus transfer and transformation in point and nonpoint pollution. *Zhejiang University*.

Xie, F. & Li, Y.F. 2011. Discussion on phosphorus forms and transformation in soil. *Journal of Yang ling Vocational & Technical College*, 10(1): 4–8.

Zhou, X.N. Wang, S.R. & Jin, X.C. 2006. Influences of submerged vegetation hydrilla verticillata on the forms of inorganic and organic phosphorus and potentially exchangeable phosphate in sediments. *Environmental science*. 27(12): 2421–2125.

Zhu, G.W. Qin, B.G. Gao, G. et al. 2003. Hydrodynamics and iron: The key factors affecting resuspension of phosphorus from shallow lake sediments. *Journal of Agro-environment science*, 22(6): 762–764.

Biomedical Engineering and Environmental Engineering – Chan (Ed.)
© *2015 Taylor & Francis Group, London, ISBN: 978-1-138-02805-0*

Establishment of a natamycin detection method produced by *Streptomyces natalensis* HDMNTE-01

Shou-Feng Huang, Fang-Yi Pei, Jing-Ping Ge, Kun Liu & Wen-Xiang Ping
Key Laboratory of Microbiology, College of Life Science, Heilongjiang University, Harbin, China

ABSTRACT: Natamycin is a broad-spectrum antifungal agent which can effectively inhibit the growth of yeast and mould: it is also effective in the therapy of medical treatment and food protection. In this paper, a detection method of natamycin produced by *Streptomyces natalensis* HDMNTE-01 was established. Natamycin yield and the diameter square of the inhibitory zone took on a good linear relation after the parameters of biopotency of natamycin production were determined by the agar diffusion method. An equation of linear regression: $y = 0.0013x - 1.3369$, $R^2 = 0.9965$ was obtained. The calibration curve showed a better linear range from 0.1 g/L–1.5 g/L, providing reference and basis for further experiments and relevant research.

Keywords: Natamycin; *Streptomyces natalensis*; Agar diffusion method

1 INTRODUCTION

Natamycin is a natural efficiency antiseptic with broad-spectrum antifungal activity which has been applied extensively to food production and preservation since it was first found in 1955[1] and named by the WHO[2]. Systematic research on physicochemical properties, bioactivity, structural stability, functional mechanism[3–6] biosecurity[7–8], and toxicity of natamycin has been conducted by several researchers. Numerous advantages of natamycin were summarized and eventually its safety and nontoxicity to the human body established. It was thus authoritatively approved as a food antiseptic by the FDA of the USA[9] in 1982. Furthermore, natamycin can actively cure skin, lung, and mucosa diseases caused by fungus without drug resistance[10], thus large-scale production of natamycin was desirable. The primary path of natamycin production at present is by microorganisms' fermentation, the producers mainly are *Streptomyces natalensis*, *Streptomyces chattanoogensis*, and *Streptomyces gilvosporeus*[11–12], hence it is an important step to accurately determine the concentration of natamycin in fermentation broth in the process of natamycin production.

Analytical methods of natamycin yield are mainly bioassay, UV spectrophotometry, and High-performance Liquid Chromatography (HPLC). Natamycin yield can be determined by UV spectrophotometry for it has maximal absorption value at 303 nm. However, this method is not too precise to quantify the bioactivity of natamycin and the influence of interference factors cannot be eliminated as well. HPLC is widely applied to detect residual content of natamycin in food due to its simple operation and high accuracy, while it is costly in terms of equipment as a one-time investment[13].

The agar diffusion method is the main bioassay to detect natamycin titer. According to the method, the diffusion permeability of an antibiotic on an agar medium will inhibit the tested strain and thus lead to the formation of an inhibitory zone. The content of natamycin can be calculated by the proportional relation between the inhibitory zone and the concentration of the antibiotic. The agar diffusion method is internationally recognized because of its higher accuracy and in accordance with reality in clinical application. What is more, the method does not need special equipment[14–15]. In this paper, the agar diffusion method was applied to establish a natamycin detection method produced by Streptomyces natalensis HDMNTE-01 and desirable results were acquired.

2 MATERIALS AND METHODS

2.1 Microorganism

Streptomyces natalensis HDMNTE-01 and *Saccharomyces cerevisiae* 776502 were kept in store by Key Lab of Microbiology of Heilong jiang University in China. The culture was maintained on an agar slant medium at 4°C and subcultured every four weeks.

2.2 Medium and culture conditions

One loopful of *Streptomyces natalensis* HDMNTE-01 cells from fresh a agar slant was transferred to 250 mL Erlenmeyer flasks with 25 mL growth medium which contain (%): glucose 1.0, starch 1.0, soybean cake powder 1.0, peptone 0.6, corn steep liquor 0.6, MgSO$_4$·7H$_2$O 0.1, K$_2$HPO$_4$ 0.05, NaCl 0.2, CaCO$_3$ 0.5, and pH 7.0. The flasks were maintained at 28°C under agitation at 220 rpm for 48 h and subsequently inoculated into the fermentation medium at 5% (v/v). The compounds of the fermentation medium were (%): soybean protein extract 1.95, yeast extract powder 0.45, glucose 4.0, and the pH was controlled at 7.0. The fermentation process was carried out for 96 h with identical conditions.

Saccharomyces cerevisiae 776502 was used as sensitive strain, which grew in a Yeast Extract Peptone Dextrose (YEPD) liquid medium (2% peptone, 1% yeast extract, and 2% dextrose; w/v) or in a YEPD solid medium containing 2% agar additionally.

2.3 Draw standard curve

A range of concentration gradients of standard natamycin were injected into a sample hole on dilayer testing plates by a trace sampler with sample size of 70 μL. Cultivated at 28°C for 24 h, the size of inhibitory zone was subsequently measured. Then the standard curve was drawn which took the logarithm of natamycin concentration as Y axis and the diameter square of the inhibitory zone as X axis, therefore an equation of linear regression could be obtained.

2.4 Preparation of sensitive strain liquor and dilayer testing plates

One loopful of Saccharomyces cerevisiae 776502 cells from a fresh agar slant was transferred to 250 mL Erlenmeyer flasks with 50 mL YEPD liquid medium, cultured at 29°C under agitation at 170 rpm until the biomass was above 10^7/mL. Four Oxford cups were equaldistant inserted into the below layer containing 10 mL YEPD liquid medium, then poured 15 mL YEPD liquid medium harbouring 1.5 mL sensitive strain liquor into the below layer of the medium.

2.5 Determination of the natamycin concentration

The fermentation liquid and methanol at a ratio of 1:1 (v/v) were mixed well and centrifuged under 4000 r/min for 20 min at room temperature, the supernatant was used to detect the titer of natamycin. 70 μL of the supernatant was put into the hole of the detection medium spread with 10^7/mL Saccharomyces cerevisiae 776502, cultured at 28°C for 24 h, the inhibitory zone to indicator strain Saccharomyces cerevisiae 776502 were measured, and the titre of natamycin were calculated according to the appropriate calibration curve.

3 RESULTS AND DISCUSSION

3.1 Equation of inhibitory zone

Antibiotic permeability spread on the surface of the agar medium in the shape of a sphere and the sensitive strain began to grow. The tested strain will be inhibited

where the concentration of antibiotic is higher and thus lead to the formation of an inhibitory zone. Equation can be deduced by molecular diffusion law:

$$\log M = (1/9.21DT)r^2 + \log(C \cdot 4\pi DTH);$$

where D = diffusion coefficient; T = diffusion time of the antibiotic; M = total content of the antibiotic; r = semi-diameter of the inhibitory zone; H = thickness of medium, and C = minimal inhibitory concentration. The equation is similar to y = ax + b, illustrating that there is a linear relationship between log M and r^2, therefore, the content of the antibiotic can be calculated by the size of the inhibitory zone.

3.2 Essential parameters of the agar diffusion method

Saccharomyces cerevisiae 776502 was applied as an indicator strain in the paper because it is sensitive to natamycin. Furthermore, it lends itself to observation and measurement of the inhibitory zone because of the neat and distinct margin of the inhibitory zone formed by the strain. As shown in Figure 1, the yeast grew well and the margin of the inhibitory zone was neat and distinct when the cell density was 10^7/mL, thus it was the best choice of for the observation and measurement of the inhibitory zone. While the margin of the inhibitory zone was unclear and the yeast grew sparse when the cell density was 10^5/mL and 10^7/mL. In addition, when the cell density was 10^8/mL, the yeast grew dense and the inhibitory zone was smaller, which may lead to error in observation and measurement. Therefore, the essential parameters of the agar diffusion method were transferred one loopful of Saccharomyces cerevisiae 776502 cells from a fresh agar slant to a YEPD liquid medium, cultured at 29°C under an agitation speed of 170 rpm until the cell density was above 10^7/mL and then it was utilized to prepare the dilayer testing plates. Cultivation of the plates with the strain was at 28°C for 24 h.

3.3 Drawing the standard curve

The essential parameters of the agar diffusion method were quantified and stably controlled, the content of natamycin was the function of the square of the diameter of the inhibitory zone. The results of the measurement are described in Table 1.

According to the consequences, we drew the standard curve which considered the logarithm of natamycin concentration as the Y axis and the diameter square of the inhibitory zone as the X axis, as shown in Figure 2, Equation of linear regression: y = 0.0013x − 1.3369, the slope a = 0.0013, the intercept b = 1.3369, and R^2 = 0.9965 were obtained. The calibration curve showed a better linear range from 0.1 g/L–1.5 g/L, which meant that the minimal and maximal detection concentrations of the curve

Table 1. The potency of a natamycin standard preparation.

Concentration (g/L)	Diameter (mm)	Concentration (g/L)	Diameter (mm)	Concentration (g/L)	Diameter (mm)
0.1	16.21 ± 0.05	0.6	29.83 ± 0.04	1.1	32.59 ± 0.05
0.2	22.38 ± 0.03	0.7	30.79 ± 0.03	1.2	32.85 ± 0.04
0.3	25.18 ± 0.04	0.8	31.32 ± 0.06	1.3	33.58 ± 0.07
0.4	26.82 ± 0.04	0.9	31.86 ± 0.04	1.4	34.22 ± 0.02
0.5	28.96 ± 0.06	1.0	32.37 ± 0.03	1.5	34.67 ± 0.04

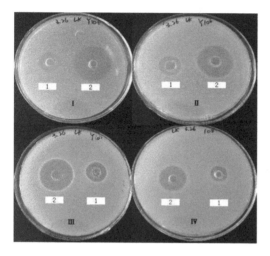

Figure 1. Bacteriostasis effect of different density of sensitive bacteria.
I: Cell density was 10^5/mL; II: Cell density was 10^6/mL; III: Cell density was 10^7/mL; IV: Cell density was 10^8/mL; 1: Inhibitory zone of 0.2 g/L natamycin; 2: Inhibitory zone of 1.2 g/L natamycin

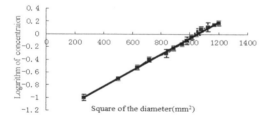

Figure 2. Standard curve of dosis agar diffusion method.

were 0.1 g/L and 1.5 g/L respectively, providing reference and a basis for further experiments and relevant research.

Whether the yield of natamycin can be quantified or not played a key role in the analysis, control, and improvement of the fermentation process. We applied the agar diffusion method to determine the content of natamycin in our work, however, many other effective ways can also be adopted such as high performance liquid chromatography (HPLC), ultraviolet spectrophotometry, and so on. Thus, a comparison based on the purpose of the results between using a

biological detection method and using HPLC will be done in our subsequent experiment. He[16] and Xun[17] applied the agar diffusion method to determine the titer of natamycin in fermentation broth. The antibiotic content could be directly described by the level of the inhibition to sensitive bacteria, which is a feature of an antibiotic. Furthermore, Xun compared the testing results with that determined by HPLC. They found that the average relative error was less than 5%, which demonstrates a high accuracy for natamycin yield determination by the agar diffusion method compared to the results achieved by HPLC as a standard.

4 CONCLUSIONS

The agar diffusion method was applied to determine natamycin produced by *Streptomyces natalensis* HDMNTE-01. Equation of linear regression: $y = 0.0013x - 1.3369$, the slope $a = 0.0013$, the intercept $b = 1.3369$, and $R^2 = 0.9965$ were obtained. The calibration curve showed a better linear range from 0.1 g/L–1.5 g/L, which meant that the minimal and the maximal detection concentrations of the curve were 0.1 g/L and 1.5 g/L, respectively.

ACKNOWLEDGEMENT

The research was supported by The National Natural Science Foundation of China (Grant No. 31270534), The National Natural Science Foundation of China (Grant No. 31270143), The National Natural Science Foundation of China (Grant No. 31470537).

REFERENCES

[1] Struyk A.P., Hoette I., Drost G. Pimaricin, a new antifungal antibiotic[J]. Antibiotics Annual, 1957–1958:878–885.

[2] Michael Davidson P., Craig H. Doan. Natamycin [J]. University of Idaho, 1993, 7:395–407.

[3] Harry Brik. Natamycin [J]. Analytical Profiles of Drug Substances. 1994, 10: 514–557.

[4] Ocstendorp J.G. Natamycin [J]. Antonie van Leeuwenhoek, 1981, 47:170–171.

[5] Pedersen J.C. Natamycin as a fungicide in agar media[J]. Applied and Environmental Microbiology, 1992, 58 (3): 1064–1066.

[6] Gill J.A, Martin J.F. Polyene antibiotics, Biotechnology of antibiotics[M]. Marcel Dekker, New York, 1997, 551–576.

[7] Elmer Y., Tu M.D. Balancing antimicrobial efficacy and toxicity of currently available topical ophthalmic preservatives [J]. Saudi Journal of Ophthalmology (2014).

[8] Hamilton-Miller J.M. Immunopharmacology of antibiotics: Direct and indirect immuno-modulation of defence mechanisms [J]. J Chemother, 2001, 13(2):107–111.

[9] Lueck E., Jager M. Antimicrobial food additives [J]. Characteristics, uses, effects, 214–218.

[10] Wen Xian, Zhong, Li Xin, Xie. Drug treatment of fungal keratitis [J]. Review of International Ophthalmology, 2006, Vol 30:332–336.

[11] Khaour S., Lehriui A., Laakel M., et al. Iuflueuce of short-chain fatty acids on the production of spirarnycin by *Streptomyces abofaciens* [J]. Appl Microbiol Biotechnol, 2002, 36:763.

[12] Yue-Yue Wang, Xin-Xin Ran, et al. Characterization of type II thioesterases involved in natamycin biosynthesis in *Streptomyces chattanoogensis* L10. FEBS Letters 588 (2014) 3259–3264.

[13] Michel G.W. Analytical Profiles of Drug substance [J]. 1997, 6:341–421.

[14] Borden G.W., Maher I.M. Process for natamycin recovery [P]. USP: 5942611, 1999.

[15] Raghoenath D., Webbers J.P. Natamycin Recovery [P]. USP: 6150143, 2000.

[16] Yan-ling He, Jian-guo Wu. Applied fed-batch fermentation to improve natamycin yield [J]. Pharmaceutical Biotechnology, 2002, 9(4):224–226.

[17] Xiao-li Xun, Xiao-bin Yu. Primary researches on fermentation process of natamycin production [J]. Journal of wuxi light industry university, 2004, 23(3):85–88.

Biomedical Engineering and Environmental Engineering – Chan (Ed.)
© 2015 Taylor & Francis Group, London, ISBN: 978-1-138-02805-0

Research on ecology sports for sustainable urban development

Ruixue Cui & Yaying Li

Hebei University of Science and Technology, Shijiazhuang, Hebei Province, China

ABSTRACT: In today's society, sport has been an important part in the urban development. Developing ecology sports is the inevitable result of influence and interaction between sports and environment. It is an inherent demand for sustainable urban development and an important symbol of social civilization. The rational development of ecology sport resources promotes urban economic development, the overall quality, construction of spiritual civilization in urban cities, health of urban residents, and the sustainable and healthy development of the city. By analyzing the role of ecology sports on sustainable urban development, this paper proposes factors that hinder the ecological development in cities and constructive conceptions that can promote development of ecology sports in cities. It puts forward constructive mode for promoting sustainable urban development with ecology sports.

Keywords: Ecology sports, City, Sustainable development, Factors

1 INTRODUCTION

Non-construction-oriented, "Ecology Sport" refers to ego behavior, environmental behavior, social behavior, as well as relevant organizations and system responses thus formed. These behaviors are conducted around sports, bodybuilding, and physical and mental experience. Facing the 21st century, the development of industrial civilization has entered a rapidly developing stage. However, it puts forward severe challenge to ecology sports. In April 2007, China Sport Science Society proposed the concept of "sunshine sports" which exhibits harmony between man and nature intuitively. In 2012, the 18th CPC National Congress presented the idea of "vigorously promoting the construction of ecological civilization", which showed urgency of constructing ecological civilization had been recognized. As the political, economic and cultural center of a country, city is the crystallization of material civilization and spiritual civilization. If a city aims to developing rapidly and sustainably, it must take saving resources, improving technology and improving the environment as the final means. Meanwhile, it should be harmonized with the external environment, resources, information and logistics. Sustainable urban development consists of three systems: sustainable social system, sustainable economic system and sustainable environmental system. Among the three systems, environmental sustainability is the basis, economic sustainability is the condition, and social sustainability is the goal. Ecology sports, as the health insurance of urban social structure, specific embodiment of urban environment system as

well as new growth point of urban economy, plays an important role in sustainable urban development. However, promoting sustainable urban development should not be achieved at the expense of the ecological environment we survive on. Ecology sport is not only the inevitable trend of future sports development, but also an important part of the eco-city construction.

2 SIGNIFICANCE OF CITY ECOLOGY SPORTS ON PROMOTING SUSTAINABLE URBAN DEVELOPMENT

In today's society, sport has become an important component of urban development. Developing ecology sports is the inevitable result of influence and interaction between sports and environment. It is an inherent demand for sustainable urban development and an important symbol of social civilization. It plays an important role in politics, economy, humanities and other aspects.

2.1 *Ecology sports promote urban planning and development*

Urban planning is the goal and plan of urban development within a certain period of time. It is the comprehensive arrangement of all kinds of urban constructions, and the basis of urban construction and management. With the development of the Olympic Games, all countries have increased construction of sporting landscapes in cities. To construct these urban buildings, the utility, uniqueness and economic value should be taken into consideration. So should the integration with the natural environment within the limited use of resources. Therefore, when planning and constructing these buildings, we are bound to utilize natural resources rationally, maintain regenerative capacity of resources, and maximize the protection of urban living environment. Construction of city sports promote the development of sports culture and ecology sports. Moreover, organizing large-scale ecological sporting events can perfect the urban infrastructure, competition facilities, traffic and public facilities, and create good city environment, thus promoting the development of urban infrastructure. This will not only greatly enhance the taste of the city, but also improve the soft environment of the city to attract more investment, promoting economic development of the city.

2.2 City ecology sports stimulate urban economic development and promote upgrading of urban management

Sport is a universal cause. It can both serve economic construction, but also be an important way of expanding domestic demand. The first is the production, manufacturing, research and development of sports products. Sports, advertising, venue management, club operations and lottery serve the society and promote the residential health. Secondly, sports consumers are mainly concentrated in cities. For example, people who buy lottery, sports equipment, clothing, beverage, food, exercise books and so on, who are involved in bodybuilding, watching games and participating in competitions are dominated by city residents. Each time the sports consumption of urban residents increases 1%, GDP grows 0.5%.

At the same time, developing the city ecology sports promotes the development of related industries. By engaging in sports and watching sports competitions personally, citizens have their community cohesion strengthened and the city's image enhanced. In addition, tertiary industries, especially industries such as tourism, business, finance, information consulting, catering, real estate and telecommunications that are related to sports, are greatly developed. They attract people's consumption, thus creating considerable income. Meanwhile, this will also put forward higher requirements for the city managers. Therefore, developing ecology sports will stimulate urban economic development and enhance the city image.

2.3 City ecology sports improve the consciousness of the masses and promote spiritual civilization

Sport is social and cultural organization that is popular and universal. It wins great concern of the residents. Wide spread of the concept "ecological construction and development of sports" promote sustainable urban development to some extent. A green, elegant and well-equipped environment is bound to actively guide

more people to get out of home, participate in exercise and train awareness of bodybuilding. During the 2008 Olympic Games, Beijing residents as well as people from all over the world who came to watch the Olympic Games, consciously fulfill responsibilities to protect the environment and nature. This shows that large-scale ecological sporting events can maximize natural, social and human nature of ecology sports; ecological development of sports can well integrate the social relations, maintain social stability and enhance ecological sustainability, etc. All these will play an important role in promoting construction of spiritual civilization and maintaining stability of social development.

3 THREAD OF DEVELOPING CTY ECOLOGY SPORTS

3.1 Positioning of the development of city ecology sports

To develop city ecology sports, the city manager should firstly change his concept ideologically, establishing ecological awareness. The city managers should strengthen the government's behavior, formulate ecological planning and construction of city and community scientifically, follow sustainable development, overall optimization, market demand and diversity principles, and reconstruct the city ecologically. The development of city ecology sports should be demand-driven. It should realize accurate and detailed market positioning based on the existing resources and environmental conditions. The design and development of city ecology sport activities should not only make people enjoy the happiness of fitness, but also make them aware that sport is humanistic, green and environmental, establish concepts of ecology sports imperceptibly and actively promote it.

3.2 Systematic integration and characteristic construction

City ecology sports should be considered and managed as a systems engineering. We should use limited city resources to arrange various types of resources and its internal components and elements, and to coordinate and optimize the structure of various development projects, so that different cities can be developed according to their own characteristics. Meanwhile, we should coordinate the macro and micro relationships. It should be coordinated with the city planning, transportation, information industry, landscape, agriculture, forestry, environmental protection, culture, commerce and other industries. In addition, the quantity, quality, characteristics, spatial layout, projects selection and other aspects should be well cohered, thus achieving overall optimization of exploitation. In the design and development process, we should take into account the carrying capacity of the natural environment. The natural environmental resources we survival on are limited. Ecological imbalance will inevitably appear if it exceeds the carrying capacity of the natural environment.

3.3 *Government guidance and multi-channel financing*

Construction of city ecology sports needs adequate capital sources. Financial support and investment of Wellbeing Projects and the government are essential. During the construction of ecology sports throughout the city, the government has always been an advocate, organizer and coordinator. Meanwhile, the government should also encourage private fund to participate in the construction through policy, coordination, grants and other methods. The capital should be raised through multiple channels. The government can offer certain preferential policies on credit guarantee, tax and other aspects, and implement policy of "Pratt & Whitney, privilege and special methods for special cases" within the law. On the basis of unified planning and management, the government should encourage economic entities to take the form of wholly-owned enterprise, joint venture and cooperation for construction, comprehensively accelerating the construction of city ecology sports, constantly perfecting the supporting facilities and improving quality and level of construction.

3.4 *Technology leads the development*

Development of city ecology sports is in harmony with the natural environment. However, at the same time, its era stigma is obvious. Modern science and technology can provide external support for the development of ecology sports. The high-tech information is applied into construction of ecology sports and protection of physical activity environment. For example, use of ecological building materials, application of eco-design, use of network resources in promoting ecological awareness, development and use of eco-products and equipment; repeated use of energy in sports events; protection of water resources in aquatic sports, postgame waste recycling and utilization, etc. To achieve these ecological environmental objectives, support of strong scientific and technological means is necessary. With the innovation and extensive application of science and technology, achievements and means of modern science and technology will be continuously applied to a variety of sporting activities.

3.5 *Rational development of multiple kinds of modes*

Planning and construction of eco-city and society emphasize the ecological background, follow the rules and laws of natural ecology and urban development, aim at sustainable development, base on ecology, and take harmony between man and nature as the core. It aims at developing healthy, efficient, civilized, comfortable and sustainable human settlements.

3.5.1 *Combination of city and community ecology sports*

Construction of large-scale city ecology sports echoes with community ecology sports. Combination of large-scale construction with small-scale one and suburban with the main city offers convenience for people to choose different ways to participate in fitness according to their individual time and characteristics.

3.5.2 *Combination of city ecology sports and eco-tourism*

According to local geographical recourses, the government can combine eco-tourism and city ecology sports. What is more, the rational use of surrounding geographical environment not only can develop city ecology sports, but also can promote the development of economic sectors around the city. For example, rich water resources can be utilized to develop boating, swimming, river trekking, fishing, rafting and other projects; forest resources can be utilized to develop hiking, horseback riding, paragliding, adventuring and other projects; plains and farms can be utilized to develop picking, kite-flying, archery, hot springs and other items; snow and ice can be utilized to develop skiing, curling, sledging and other projects; mountain can be utilized to develop camping, orienteering, mountain biking and other projects.

3.5.3 *Combination of advocacy and development*

Concept of city ecology sports needs to be gradually penetrated among humans. Heavy advocacy of media and government policies should be conducted to make it interiorize. We can apply ecological education into school sports; utilize broadcast television, Internet and other communication tools to conduct wide advocacy; use community bulletin boards and sports instructors to involve more people in this event. Meanwhile, we can develop ecological bodybuilding environment; conduct advocacy and development at the same time, hence gradually improving people's ecological awareness and promoting the upgrading of the city's overall image.

4 CONCLUSION

Ecology sport plays an important role in sustainable urban development. Meanwhile, it is the inherent demand for sustainable urban development. Although its system is imperfect in developed cities currently and the construction is just on the threshold, it is the inevitable trend of developing eco-cities. Therefore, developing city ecology sports is the request of social progress and urban development.

REFERENCES

Ran Ming. Analysis of the Value and Ecological Development Path of the Urban Community Sports [J]. Sports Science Research, 2013(4): 32–34.

Zhou Zhijun, He Liang. Philosophical contemplation of deep ecological sport view [J]. Journal of Physical Education, 2009(4): 16–19.

Tan Zhi-li. Analysis of the Current Situation and the Development of Sports Ecology Research in China [J]. Sport Science And Technology, 2013(4): 6–8.

Luo Gang. Research of Building Ecological Sports of Urban Sustainable Development [J]. Hubei Sports Science, 2011 (1): 3–4.

Biomedical Engineering and Environmental Engineering – Chan (Ed.)
© 2015 Taylor & Francis Group, London, ISBN: 978-1-138-02805-0

Comparison of capacity for 2,3-butanediol production from glucose fermentation by *K.oxytoca* 12 and *K.oxytoca* 12-1

Guang-Bin Ye, Shou-Feng Huang, Wen Sun, Shan-Shan Sun, Jing-Ping Ge & Wen-Xiang Ping
Key Laboratory of Microbiology, College of Life Science, Heilongjiang University, Harbin, China

ABSTRACT: Fermentative 2,3-butanediol (2,3-BD) production has been receiving increasing interest for its potential as a platform chemical intended for the production of synthetic rubbers, plastics, and solvents. In this study, deletion mutant of *K.oxytoca* 12-1 constructed by knocking out lactate dehydrogenase gene in laboratory previously and original strain *K.oxytoca* 12 were simultaneously used as strains in the fermentation medium utilizing glucose as sole carbon source under absolutely the same conditions, and 2,3-BD production capability were compared. The results showed that an increase of 39.03%, 39.2%, and 58.9% of 2,3-BD titers, productivity, and yield, respectively were obtained. Obviously it can be inferred that the ability of the genetic engineering mutant strains in the production of 2,3-BD was better than that of the wild-type strain, therefore it could be a promising strain for 2,3-BD production in the industrialized production.

Keywords: *Klebsiella oxytoca*, 2,3-Butanediol, Metabolic engineering, Batch fermentation

1 INTRODUCTION

2,3-Butanediol (2,3-BD), also known as 2,3-butylene glycol or dimethylene glycol, is a glycol widely used as reagent in a number of chemical synthesis[1]. In addition to its application in plastics, solvent, and anti-freeze solutions preparation, it can easily be converted to a platform chemical 1,3-butadiene (used in the manufacture of synthetic rubber), diacetyl (flavoring agent), methyl ethyl ketone (liquid fuel additive), or the precursors of polyurethane (pharmaceutical and cosmetics industries)[2]. However, conventional 2,3-BD fermentation processes suffer from low productivity, yield, and final product concentration and purity, and are not economical for commercial application[3]. While there have been some successes in improving productivity and final product concentration through various process and strain improvements[4], how to further improve 2,3-BD yield and purity in the fermentation process remains a challenge.

Glucose is used as a substrate in microbial fermentation, which not only for the synthesis of necessary proteins for cell growth, nucleic acids and other substances, but also for the formation of a series of intermediate metabolites and by-products[5]. A typical synthetic route is performed by pyruvate, α-acetolactate, acetoin, and ultimately be converted into 2,3-BD utilizing polysaccharide as substrate. The process can be summarized as the following reaction formula[6]: Glucose 2,3-BD + 2CO$_2$ + 2ATP + NADH$_2$. In the process, it is also accompanied by the formation of ethanol, acetate, lactate, succinate, 1,3-propanediol and other by-products[7]. These by-products are generated not only consuming energy, but also not conducive to the accumulation and subsequently separation and purification of 2,3-BD. Thus, 2,3-BD production can be increased by knocking metabolic pathways of by-products[8].

In this study, deletion mutant of *K.oxytoca* 12-1 constructed by knocking out lactate dehydrogenase gene in previously experiment in our laboratory and original strain *K.oxytoca* 12 were simultaneously used as strains in the fermentation medium utilizing glucose as sole carbon source under absolutely the same conditions, and 2,3-BD production capability were compared to verify whether the construction of *K.oxytoca* 12-1 was successful and whether the expression of *K.oxytoca* 12-1 was stable. The one with better ability of producing 2,3-BD will be chosen for further study on improving the yield of 2,3-BD.

2 MATERIALS AND METHODS

2.1 *Microorganisms and medium*

K.oxytoca 12 and *K.oxytoca* 12-1 were used in the work. The seed cells were prepared in 250 ml flasks containing 100 ml pre-culture medium which contained (g/L): 1.0 (NH$_4$)$_2$SO$_4$, 3.4 K$_2$HPO$_4$·3H$_2$O, 1.3 KH$_2$PO$_4$, 0.41 MgSO$_4$·7H$_2$O, 1.0 yeast extract, 30 glucose, 2 ml microelements solution, 1 ml Fe^{3+} solution. The flasks were incubated at 37°C for 24 h and inoculated into the fermentation medium at 5%

(v/v). The fermentation medium containing (g/L):4.0 $(NH_4)_2SO_4$, 0.69 $K_2HPO_4 \cdot 3H_2O$, 0.25 KH_2PO_4, 0.2 $MgSO_4 \cdot 7H_2O$, 1.5 yeast extract, 1 ml microelements solution, 1 ml Fe^{3+} solution, 60 glucose[9].

2.2 Batch fermentation experiment

The batch fermentation process was performed in 500 ml flasks containing 150 ml fermentation medium with initial pH 7.0 at 37°C and at a shaking speed of 180 rpm for 84 h. Samples were harvested every 12 hours and maintained the pH by using sterile $CaCO_3$ powder, which depending on the experiment[10].

2.3 Analytical methods

Cell growth was monitored by a measurement of optical density of the broth (OD_{600nm}) with a spectrophotometer[11]. The cell dry weight was calculated from the optical density using calibration curve for the strain. The concentrations of 2,3-BD, ethanol, acetate, succinate, glucose, and lactate, obtained by batch cultures, were determined using a high-performance liquid chromatography (HPLC) System (Shimadzu Corp. Kyoto, Japan). Aminex HPX-87H column (Bio-Rad, Hercules, CA, USA) was used with 0.005 M H_2SO_4 as elution solvent at a flow rate of 0.8 ml/min, keeping column oven temperature at 65°C with RI detector[12].

3 RESULTS AND DISCUSSION

The species affiliation and typical metabolites of the strain *K.oxytoca* 12 were studied in our previous work. It was noticed that 2,3-BD was obtained in significant amounts from glucose even when the culture conditions were not favorable for its synthesis. That is why knocking out of lactate dehydrogenase gene was performed. Then ability of producing 2,3-BD was compared between *K.oxytoca* 12-1 and *K. oxytoca* 12.

3.1 The production of 2,3-BD in glucose batch fermentation

2,3-BD and other metabolites especially lactate were observed under the following conditions: aeration, 1 vvm; agitation speed, 180 rpm and initial glucose, 60 g/L. The fermentation was carried out at least two times. Growth curve of strains and pH value in the fermentation process were shown in Figure 1. The result of typical end-products were listed in Table 1.

Theoretically, the formation of lactate can be prevented by inactivating lactate dehydrogenase. The *K. oxytoca* 12-1 was constructed by disrupting *ldh*A gene that encodes lactate dehydrogenase, which showed markedly lower lactate production (1.34 g/L) compared to the original strain (3.74 g/L). Concomitantly, the finally 2,3-BD titer and yield increased from 14.81 to 20.59 g/L and 0.382 to 0.607 g/g, respectively. In order to evaluate and compare the performance

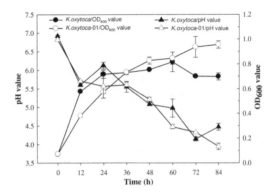

Figure 1. Batch fermentation of *K.oxytoca* 12 and *K.oxytoca* 12-1. Closed circle, cell growth of *K. oxytoca*; open circle, cell growth of *K. oxytoca* 12-1. closed Triangle-up, pH value of *K. oxytoca*. open Triangle-up, pH value of *K. oxytoca* 12-1. The values were the mean of two independent samples.

Table 1. Comparison of batch fermentation by *K. oxytoca* 12 and *K. oxytoca* 12-1.

	Strains	
	K. oxytoca 12	*K. oxytoca* 12-1
Glucose consumed (g/L)	38.75 ± 0.003	33.90 ± 0.02
2,3-BD titer (g/L)	14.81 ± 0.004	20.59 ± 0.012
Lactate titer (g/L)	3.74 ± 0.001	1.34 ± 0.004
Ethanol titer (g/L)	2.87 ± 0.002	5.19 ± 0.003
2,3-BD yield (g/g)	0.382 ± 0.003	0.607 ± 0.005
2,3-BD productivity (g/L·h)	0.176 ± 0.02	0.245 ± 0.13

of 2,3-BD fermentation between *K. oxytoca* 12 and *K.oxytoca* 12-1, the total glucose uptake concentration and 2,3-BD productivity were determined. And the total glucose uptake concentration of the mutant strain (33.90 g/L) was slightly lower than the wild-type strain (38.75 g/L). Maybe the residual glucose was advantageous to prolonged the stationary phase of the mutant strain, making the biomass of the mutant strain larger than the original strain. That's why the 2,3-BD productivity increased from 0.176 to 0.245 g/L·h.

The production of neutral metabolites such as 2,3-BD and ethanol is to prevent acidification resulting from the accumulation of acidic metabolites, such as acetate, lactate, and succinate from glucose in fermentation broth. The deletion of the *ldh*A gene resulted in much higher production of 2,3-BD as well as higher cell density at the expense of lactate. The reduced lactate in the *ldh*A-deficient *K. oxytoca* 12-1 seems to mitigate the inhibition of cell growth.

4 CONCLUSIONS

K.oxytoca 12 and *K.oxytoca* 12-1 have a comparative capability for the production of 2,3-BD through fermenting glucose. The experiment results showed that the ability in 2,3-BD production of genetic engineering strains was better than wild-type strain, simultaneously more suitable for further research of higher production of 2,3-BD.

ACKNOWLEDGEMENT

The research was supported by The National Natural Science Foundation of China (Grant No. 31270534), The National Natural Science Foundation of China (Grant No. 31270143), The National Natural Science Foundation of China (Grant No. 31470537), and the National Science Foundation for Distinguished Young Scholars of China (31300355).

REFERENCES

[1] E. Celińska. Biotechnological production of 2,3-butanediol—Current state and prospects. Biotechnology Advances, 27 (2009):715–725.

[2] Syu U M J. Biological production of 2,3-Butanediol [J]. Applied Microbiol and Biotechnol, 2001, 55(1): 10–18.

[3] Ji X J, Huang H, Ouyang P K. Microbial 2,3-butanediol production: a state-of-the-art review [J]. Biotechnology Advances, 2011, 29:351–364.

[4] Supaporn Suwannakham, Yan Huang, et al. Construction and Characterization of *ack* Knock-Out Mutants of *Propionibacterium acidipropionici* for Enhanced Propionic Acid Fermentation [J]. Biotechnology and Bioengineering, 2006, 94(2):383–395.

[5] Ji X J, Huang H, Du J, *et al*. Enhanced 2,3-butanediol production by *Klebsiella oxytoca* using a two-stage agitation speed control strategy [J]. Bioresource technology, 2009, 100(13): 3410–3414.

[6] Mas C de, Jansen N B, Tsao G T, Production of optically active 2,3-butanediol by *Bacillus polymyxa* [J]. Biotechnol Bioeng, 1988, 31:366–377.

[7] Duk-Ki Kim, Chelladurai Rathnasingh, et al. Metabolic engineering of a novel *Klebsiella oxytoca* strain for enhanced 2,3-butanediol production [J]. Journal of Bioscience and Bioengineering, 2013, 2:186–192.

[8] Ji X J, Huang H, Zhu J G, et al. Engineering *Klebsiella oxytoca* for efficient 2,3-butanecliol production through insertional inactivation of acetaldehyde dehydrogenase gene [J]. Appl Microbiol Biotechnol, 2010, 85:1751–1758.

[9] Jing-ping Ge. Comparison of capacity for 2,3-Butanediol production from corn cob hemicellulose hydrolysate fermentation by *Klebsiella oxytoca* HD79 and *Klebsiella pneumonia* [C]. Renewable energy and power technology II 672–674(2014):147–153.

[10] Soo-Jung Kim. Production of 2,3-butanediol by engineered *Saccharomyces cerevisiae*. Bioresource Technology. 146 (2013):274–281.

[11] Kaloyan Petrov, Penka Petrova. High production of 2,3-butanediol from glycerol by *Klebsiella pneumonia* G31 [J]. Appl Microbiol Biotechnol (2009): 84:659–665.

[12] Ke-Ke Cheng, Qing Liu, et al. Improved 2,3-butanediol production from corncob acid hydrolysate by fed-batch fermentation using *Klebsiella oxytoca* [J]. Process Biochemistry 45 (2010): 613–616.

Biomedical Engineering and Environmental Engineering – Chan (Ed.)
© 2015 Taylor & Francis Group, London, ISBN: 978-1-138-02805-0

Preparation of high specific surface area activated carbon from coconut shell with KOH activation by microwave heating

Shengzhou Zhang, Jian Wu, Hongying Xia, Libo Zhang & Jinhui Peng
Yunnan Provincial Key Laboratory of Intensification Metallurgy, Kunming University of Science and Technology, Kunming, Yunnan, China
National Local Joint Laboratory of Engineering Application of Microwave Energy and Equipment Technology, Kunming University of Science and Technology, Kunming, Yunnan, China
Faculty of Metallurgical and Energy Engineering, Kunming University of Science and Technology, Kunming, China

ABSTRACT: The high specific surface area activated carbon was prepared from coconut shell with KOH activation by microwave heating under a nitrogen atmosphere. The main effects on the adsorption properties of activated carbon such as microwave power, heating time, and KOH/C ratio were investigated. The optimal conditions were obtained as follows: microwave power is 700 W, heating time is 40 min, and KOH/C ratio is 4. Under the above conditions, the iodine number and the methylene blue adsorption value of the prepared activated carbon reached 2070 mg/g and 420 mg/g, respectively. The adsorption properties exceed the standard of activated carbon for an electric double-layer capacitor. The microwave heating time was shortened about 75% compared with that of the conventional process. Furthermore, the activated carbon was characterized by nitrogen adsorption isotherms at 77 K and the porous structures were analysed by BET and Non-local Density Function Theory (NLDFT) methods. The surface area and total pore volume are as high as 2199.36 m^2/g and 1.144 ml/g, respectively. Micropore volume accounted for 64.13% of the total pore volume. The results indicated that the high-surface-area activated carbon is predominantly microporous.

Keywords: High specific surface area, Activated carbon, Microwave heating, Coconut shell

1 INTRODUCTION

Coconut belongs to the palm coconut genus which is grown in more than 93 countries. South East Asia is regarded as the origin of coconut [1]. It is mainly grown in China in provinces such as Hainan, Yunnan, Guangxi and other places. An endocarp called coir is a major coconut byproduct. It is dark brown, hard and its weight is about 12–15% of a coconut. After the coconut is scraped out, the shell is usually discarded as waste [2]. Agricultural and industrial waste which can be used as potential material or replacement material in the construction industry is abundant in developing countries. This will have the double advantage of reducing the cost of construction material and also as a means of disposal of waste. It is also the most important source used to prepare activated carbon [3].

Activated carbons are porous materials with extremely good performance and they have been widely used in a variety of important fields such as chemistry, metallurgy, food processing, pharmaceutical, military, etc. [4, 5]. With the continuous development of industrial technology, activated carbons are showing wider and wider application prospects in many areas such as double electrochemical capacitors

material, storage of gas fuel, etc. [6, 7]. However, common activated carbon cannot satisfy their needs because they put forward higher requirements to the adsorption properties of activated carbon products.

Activated carbon can be prepared by different raw carbon resources like lignite, peat, coal, and biomass resources such as wood, bagasse, sawdust, etc. [8, 9]. Compared with other activated carbons, coconut shell activated carbon has a lot of advantages including high purity, high density, and a virtually dust-free nature [10]. Zinc chloride and phosphoric acid are used for the activation of no carbonized lignocellulosic materials. At the same time, metal compounds such as KOH are used for activation of coal precursors or chars. A gasification reaction results in the removal of carbon atoms and in the process, which results in porous activated carbon, simultaneously produces a wide range of pores [11].

To date, research on high specific surface area activated carbon has mainly used conventional heating methods [12]. Heating time is usually 40–90 min and heat insulation time is 60–240 min. Conventional heating is according to heat conduction, convection, and radiation theory and spreads heat from the outside to the inside. However, it has a lot of problems such

Table 1. Proximate analysis of char.

Moisture [%]	Volatile matter [%]	Fixed carbon [%]	Ash [%]
6.86	20.25	78.46	1.29

as long heating time, uneven heating, low thermal efficiency, and high energy consumption.

However, it has a lot of problems such as long heating time, uneven heating, low thermal efficiency, and high energy consumption. Recently, some research about the preparation of activated carbon by microwave heating has shown that microwave heating can overcome those disadvantages of conventional heating [13]. This paper describes high specific surface area activated carbon which is prepared from coconut shell with KOH activation by microwave heating using a research system that investigates the main factors affecting of activated carbon adsorption and discusses the pore structure of activated carbon, respectively.

2 EXPERIMENTAL

2.1 Materials

Coconut shell carbon is used for raw material. The proximate analysis is shown in Table 1. The reagents have analytical-grade potassium hydroxide (KOH) and analytical-grade hydrochloric acid (HCL).

2.2 Experimental

The carbonized material is crushed to less than 150 μm, and the broken material weighed and mixed with KOH according to a certain ratio. The carbonized material is heated in a microwave furnace under a nitrogen flow. The as-heated samples are first washed with deionized water then washed with HCl of a 1:1 ratio, and finally washed with deionized water to neutrality. The washed samples are dried at 120°C for more than 4 hours and the high specific surface area activated carbon is obtained. The frequency of microwave furnace is 2.45 GHz and the power is in the range of 0 to 700 W. The power can be set continuously.

2.3 Characterization of activated carbon

Iodine adsorption and methylene blue number determined by PR China (GB/T 12496.8-1999 and GB/T 12496.10-1999) for testing activated carbon. The BET surface area of the high specific surface area activated carbon is measured through nitrogen adsorption isotherm at 77 K (U.S. Contador company, Nova2). The total pore volume is determined by the nitrogen adsorption isotherm generated until a relative pressure of p/p_0 of 0.99 is reached. The pore size distribution

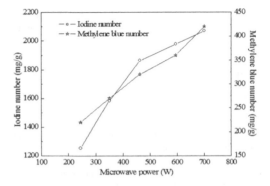

Figure 1. Effect of microwave power on adsorption properties of activated carbon.

is determined using the non-local density functional theory (NLDFT).

3 RESULTS AND DISCUSSION

3.1 Effect of microwave power

The effect of microwave power on the iodine number and methylene blue number of the activated carbon has been investigated under optimal conditions such as KOH/C ratio of 4 and microwave heating time of 40 min. The results are shown in Figure 1.

Figure 1 shows that the iodine number and methylene blue number of activated carbons increases with the increasing of microwave power because microwave power determines the temperature of the material. When power is in the low range, the material cannot get enough energy. It causes a low temperature so the adsorption reaction becomes slow. With the microwave power increasing, the material can get enough energy so that the activated reaction rate can be accelerated and the adsorption ability of the samples is also increased. The adsorption properties of the activated carbon are improved and the iodine number and methylene blue number of activated carbons are increased because the activated carbons have more abundant pores. When the power increases, the activation temperature exceeds the boiling point of potassium. Potassium vapour can enter the carbon layer. It is good for forming more pores and increasing adsorption properties of activated carbon.

3.2 Effect of microwave heating time

The effect of microwave power on the iodine number and methylene blue number of activated carbon has been investigated under optimal conditions such as KOH/C ratio of 4 and a microwave power of 700 W. The results are shown in Figure 2.

Figure 2 shows that the iodine number and methylene blue number of activated carbons increase with the increasing of microwave heating time. The iodine number and methylene blue number reach the optimal value when the heating time is 40 min and 50 min,

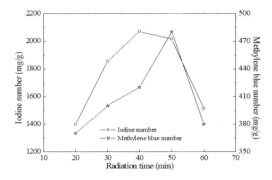

Figure 2. Effect of heating time on the adsorption properties of activated carbon.

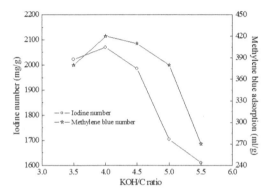

Figure 3. Effect of KOH/C ratio on adsorption properties of activated carbon.

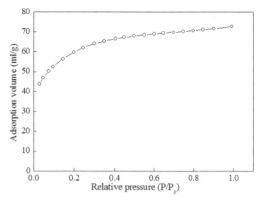

Figure 4. Nitrogen adsorption isotherm of activated carbon.

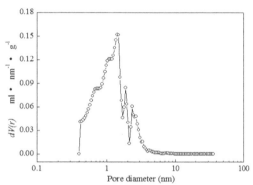

Figure 5. Pore size distribution of activated carbon.

respectively. After that, the iodine number and methylene blue number will decrease. This is because the heating time directly determines the extent of the activated reaction. When heating time becomes short, the extent of the activated reaction also becomes low. With the microwave heating time increasing, the degree of activation becomes deeper and the activated reaction proceeds more completely. The iodine number and methylene blue number are increased because activated carbon has more abundant pores. However, after the activated reaction has been finished completely, some pores will loss on ignition and cause the aperture to become wider or collapse if we continue to extend the heating time.

3.3 Effect of the KOH/C ratio

The effect of microwave power on the iodine number and methylene blue number of activated carbon has been investigated under optimal conditions such as a microwave heating time of 10 min and microwave power of 700 W. The results are shown in Figure 3.

Figure 3 indicates that the iodine number and methylene blue number of activated carbon increase until a KOH/C ratio of 4 is reached. When the KOH/C ratio reaches 4, the iodine number and methylene blue number of activated carbon has the best value. If the KOH/C ratio exceed 4, the iodine number and

methylene blue number of activated carbon decrease. Because less activated carbon joins in the activation of KOH when the KOH/C ratio is low, the pore is decreased and the adsorption properties of the activated carbon are low. With the KOH/C ratio increases, the degree of activation becomes deeper and a number of pores increase. The iodine number and methylene blue number increase because activated carbon has a lot of pores. When the KOH/C ratio is 4, the activated reaction is basically finished and the adsorption properties of activated carbon are at their best. Some pores will loss on ignition and cause the aperture to become wider or collapse if we continue to increase the KOH/C ratio.

3.4 Surface area and pore structure characterization

The nitrogen adsorption isotherm of activated carbon at optimal conditions is shown Figure 4. The surface area of the activated carbon is estimated to be 2199.36 m^2/g, while the total pore volume is 1.144 ml/g. Pore size distribution which is analysed by Non Local Density Function Theory (NLDFT) methods is shown Figure 5.

Figure 4 shows that the trend of the adsorption isotherm pertains to type I isotherm under the IUPAC

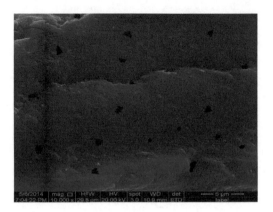

Figure 6.   SEM images of activated carbon.

classification of isotherms, bases on the progressive increase in the adsorption capacity beyond the relative pressure of 0.1 and saturates adsorption capacity is 72.22%. The adsorption capacity continues to increase with the increase of the relative pressure and adsorption capacity. The curve is an upward convex and it means that the adsorption translates from monolayer to multi-molecular. It becomes horizontal when the relative pressure is high. It shows that it is microporous activated carbon which has few mesopores and macropores.

Adsorbent pores are divided into three categories according to the IUPAC [14]. The pore size of the micropore is less than 2 nm. The pore size of a mesopore distributes in a range of 2–50 nm. The pore size of a macropore is more than 50 nm. Figure 5 shows that the pore size distribution of activated carbon is concentrated in a microporous range which is less than 2 nm. At the same time, there are few mesopores and macropores. Micropores are mainly concentrated in 1.41 nm–1.85 nm and mesopores are mainly concentrated in the range of 2–4 nm. It is majority near the 2.31 nm. Micropores accounted for 64.13%. This shows that it is microporous activated carbon which has few mesopores and macropores. It is same to the analysis of adsorption isotherm of activated carbon. Micropores have a great effect on the activated carbon surface area and adsorption properties. It is illustrated that high specific surface area activated carbon has a great application prospect.

3.5 *SEM analysis*

Figure 6 is an SEM image of the activated carbon prepared at the KOH/C ratio of 4, with an MW heating time of 40 min, and an MW power of 700 W. As can be seen from Figure 6, the activated carbon surface has abundant pores characterized by thick pore walls and circular pores. The pores on the surface could be macropores which leads to the branching micropores in the interiors of the activated carbon.

4 CONCLUSIONS

The high specific surface area activated carbon from coconut shell has good adsorption properties via microwave heating under KOH activation. The optimal conditions are found to be a power of 700 W, a heating time of 40 min, and a KOH/C ratio of 4, with an iodine number of 2020.53 mg/g and methylene blue number of 420 mg/g. The adsorption index of activated carbon product exceeds the standard (the iodine number is 1350 mg/g and number blue adsorption is 180 mg/g) of activated carbon for double electrochemical capacitors specified by PR China (LY/T 1617–2004). The heating time is shorter than the conventional heating more than 75% under optimal condition. The BET surface area is evaluated using a nitrogen adsorption isotherm for the optimal sample corresponding to 2199.36 m^2/g with the pore volume of 1.144 ml/g. Results show that the high specific surface area activated carbon microporous style.

REFERENCES

[1] K. Gunasekaran, R. Annadurai and P. S. Kumar: Constr. Build. Mater Vol. 208–215 (2012), p. 28.
[2] K. Gunasekaran, P. S. Kumar and M. Lakshmipathy: Constr. Build. Mater Vol. 92–98 (2011), p. 25.
[3] G. Afrane and O. S. Achaw: Bioresource. Technol Vol. 6678–6682 (2008), p. 99.
[4] R. Pietrzak and T. J. Bandosz: Carbon Vol. 2537–2546 (2007), p. 45.
[5] H. G. Park, T. W. Kim, M. Y. Chae and I. K. Yoo: Process. Biochem Vol. 1371–1377 (2007), p. 42.
[6] B. Kastening and M. Heins: Electro. Acta Vol. 2487–2498 (2005), p. 50.
[7] H. Jin, Y. S. Lee, I. Hong: Catal. Today Vol. 399–406 (2007), p. 120.
[8] W. H. Chen and B. J. Lin: Appl. Energ Vol. 551–559 (2013), p. 120.
[9] Z. H. Wang, Y. Chen, C. Zhou, R. Whiddon, Y. W. Zhang, J. H. Zhou and K. F. Cen: Int. J. Hydrogen. Energ Vol. 216–223 (2011), p. 36.
[10] M. K. B. Gratuito, T. Panyathanmaporn, R. A. Chumnanklang, N. Sirinuntawittaya, A. Dutta: Bioresource. Technol Vol. 4887–4895 (2008), p. 99.
[11] K. B. Yang, J. P. Peng, C. Srinivasakannan, L. B. Zhang, H. Y. Xia and X. H. Duan: Bioresource. Technol Vol. 6163–6169 (2010), p. 101.
[12] R. L. Tseng and S. K. Tseng: J. Hazard. Mater Vol. 671–680 (2006), p. 136.
[13] X. H. Duan, C. Srinivasakannan, W. W. Qu, X. Wang, J. H. Peng and L. B. Zhang: Chem. Eng. Process Vol. 53–62 (2012), p. 53.
[14] IUPAC: *Manual of Symbols and Terminology of Colloid Surface* (London, Butterworths 1982).

Author index

T - #0279 - 101024 - C0 - 246/174/19 [21] - CB - 9781138028050 - Gloss Lamination